Horváth / Möller (Hrsg.)
Intangibles in der Unternehmenssteuerung

Intangibles in der Unternehmenssteuerung

Strategien und Instrumente zur Wertsteigerung des immateriellen Kapitals

herausgegeben von

Péter Horváth und Klaus Möller

Verlag Franz Vahlen München

VERLAG
VAHLEN
MÜNCHEN
www.vahlen.de

ISBN 3 8006 3035 4

Wilhelmstraße 9, 80801 München
Satz: DTP-Vorlagen der Herausgeber
Druck und Bindung: Druckhaus Nomos
In den Lissen 12, 76547 Sinzheim
Gedruckt auf säurefreiem, alterungsbeständigem Papier
(hergestellt aus chlorfrei gebleichtem Zellstoff)

Vorwort

Was bestimmt den Erfolg bzw. Wert einer Unternehmung? Schon *Erich Gutenberg* stellte fest: Nicht die physischen Produktionsfaktoren allein sind ausschlaggebend, sondern ihre neuartige dispositive Kombination. Das klassische Rechnungswesen und das traditionelle Controlling tragen dem nicht Rechnung. Sie konzentrieren sich auf die Abbildung physisch sichtbarer Sachverhalte. Damit fehlen Informationen, die im Rahmen des Kombinationsprozesses und unter Einbezug nicht-physischer Faktoren entstehen. Sie sollten konsequenterweise definieren, was entscheidungsrelevante Informationen sind und wie diese zu messen sind. Inzwischen hat sich die Erkenntnis durchgesetzt, dass es dieses Faktoren in verschiedenen Ausprägungen gibt: Marken, Image, Know-how, Netzwerke, Patente etc. Es ist unstrittig, dass diese Parameter – allein oder im gegenseitigen Wechselspiel – den zukünftigen Erfolg bzw. Wert einer Unternehmung beeinflussen. Das Controlling steht nun vor der Aufgabe, sie einer systematischen Steuerung durch die Unternehmensführung zugänglich zu machen.

Seit den 1990er Jahren hat ausgehend von den angelsächsischen und skandinavischen Ländern eine intensive Auseinandersetzung mit den `nicht-physischen Produktionsfaktoren´ und ihrer internen Steuerung statt gefunden. Sie stellt die Begrifflichkeiten Intangible Assets, Intangibles, Immaterielles Vermögen oder Intellectual Capital in den Mittelpunkt der Analyse. Auch die externe Rechnungslegung hat sich dem Thema angenommen. So widmen sich US-GAAP und IAS/IFRS in separaten Richtlinien der Bilanzierung von immateriellen Vermögenswerten. Damit wird die Beschäftigung mit den Intangibles zu einer zentralen Frage der Unternehmenssteuerung. Die unreflektierte Anwendung traditioneller Erfassungs-, Bewertungs- und Steuerungskonzepte bietet allerdings keine tragfähige Antwort. Notwendig ist vielmehr eine tiefer gehende Auseinandersetzung mit der Thematik.

Vor diesem Hintergrund verwundert es umso mehr, dass sich in der deutschsprachigen Literatur nur wenige Publikationen zum Thema Intangibles finden. Der große Erfolg des Sonderheftes 3/4 2003 der `Zeitschrift Controlling´ (Controlling von Intangibles) zeigte das starke Interesse an dem Gegenstand und bestärkte uns darin, das Thema auch über den Leserkreis der Zeitschrift hinaus bekannt zu machen. Unser Ziel ist es daher, dem interessierten Leser einen Einstieg in die Thematik zu ermöglichen. Notwendig dazu ist ein breiter Überblick zu den verschiedenen Facetten der Intangibles-Diskussion. Gerade um hier den zum Teil deutlich weiter fortgeschrittene Erkenntnisstand in den angelsächsischen und skandinavischen Ländern einzubinden, sind entsprechende Beiträge unverzichtbar. Wir freuen uns sehr, dass wir hierzu die international renommiertesten Autoren gewinnen konnten.

Gemäß dem Anliegen einer Übersichtsdarstellung ist der Band in fünf Abschnitte gegliedert: Im Einführungskapitel setzen sich wesentliche Protagonisten der Intangibles-Diskussion mit der *Bedeutung* des Themas und Weiterentwicklungsmöglichkeiten auseinander. Der zweite Abschnitt ist der *Messung und Steuerung* gewidmet. Der dritte Abschnitt widmet sich *Rechnungslegung und Reporting*. Dabei findet sich neben Überblicksdarstellungen zum State of the art auch vertiefte Darstellungen von Teilaspekten. Im vierten Abschnitt werden *spezielle Intangibles* im Detail behandelt. Den Abschluss bildet ein fünftes Kapitel zu *Empirie und Praxisbeispielen*. Neben einzelnen Fallstudien findet sich auch eine empirische Querschnittsanalyse zum Reporting von Intangibles. Zusammenfassungen am Anfang der jeweiligen Kapitel erläutern deren Aufbau.

Den mitwirkenden Autoren gebührt unser besonderer Dank. Das vorliegende Werk konnte nur durch ihr großes Engagement entstehen. Außerdem danken wir Herrn *Alexander Wichmann* für die Unterstützung bei der redaktionellen Bearbeitung der Beiträge. Von Seiten des Verlags gebührt Herrn Dipl.-Vw. *Dieter Sobotka* für seine Kooperationsbereitschaft, Förderung und die gute Zusammenarbeit unser Dank.

In der Hoffnung, dass es uns mit der vorliegenden Auswahl an Aufsätzen gelungen ist, die Diskussion zu den Intangibles auch in Deutschland voranzubringen, wünschen wir dem Leser eine interessante Lektüre.

Stuttgart, im Dezember 2003

Péter Horváth

Klaus Möller

Inhaltsverzeichnis

3. Rechnungslegung und Reporting

4. Spezielle Intangibles

5. Empirie und Praxisbeispiele

1. Bedeutung

Buch- und Marktwert vieler Unternehmen klaffen erheblich auseinander. Im Bereich des internen Rechnungswesens finden verstärkt mehrdimensionale Steuerungskonzepte (wie die Balanced Scorecard) erhebliche Verbreitung. Ganz offensichtlich reichen die traditionellen – finanziell fokussierten – Steuerungs- und Berichtssysteme der Unternehmen in beiden Bereichen inzwischen nicht mehr aus, um angemessen zu kommunizieren und zu führen. In Zeiten einer globalisierten und weltweit vernetzten Wirtschaft haben materielle Vermögensgegenstände im Hinblick auf die Erreichung und Sicherung von Wettbewerbsvorteilen sowie zur Erhöhung des Unternehmenswertes immer mehr an Bedeutung verloren. In den Mittelpunkt der Betrachtung rücken zunehmend immaterielle Vermögenswerte, Intangible Assets bzw. Intellectual Capital. Die Ansätze zur Berücksichtigung von Intangibles in der Unternehmenssteuerung sind dabei vielfältig: Sie reichen von der Sichtweise einer zentralen Bedeutung der Organisationsstruktur über Intangibles als Grundlage einer neuen Generation von Performance Measurement Systemen bis zu Intangibles als Voraussetzung für Innovationen. Einigkeit herrscht jedoch in einem Punkt: Intangibles besitzen für die Unternehmenssteuerung eine zentrale Bedeutung. Offenbar reichen traditionelle Messkonzepte zur Bewertung und Steuerung nicht mehr aus!

Intangibles at a Crossroads

Baruch Lev [1]

---

[1] Prof. Dr. *Baruch Lev,* Chair of Accounting and Finance, Stern School of Business, New York University

The extensive work on intangible (intellectual) assets performed in the 1990s succeeded in creating wide awareness of the vast magnitude and substantial impact of intangibles, and the serious harms from the information deficiencies related to these assets. Prof. Baruch Lev – an internationally recognized expert on intangible assets characterizes this awareness-creation as Phase I of the „Intangibles Movement". This movement is currently at a crossroads, raising the question of what's next? Lev addresses this question, offering first a new perspective on the attributes of intangible assets, followed by a proposed shift of focus for the future work on intangibles.

1. Phase I: Documentation and Awareness Creation

The pioneers of the intangibles movement strove in the 1990s to alert managers, investors, and policymakers to the dramatic shift in the production functions and asset compositions of business enterprises the fast-increasing role of intangibles, becoming the primary drivers of corporate value and growth. This awareness creation effort was initially based on conceptual developments in the macroeconomic theory of growth („endogenous growth models"), and subsequently on the empirical process of documenting the magnitude and impact of intangibles. Research, for example, indicated that in the late 1990s the annual U.S. investment in intangible assets (R&D, business processes and software, brand enhancement, employee training, etc.) was roughly $1.0 trillion, almost equal to the total investment of the manufacturing sector in physical assets ($1.1 trillion). Furthermore, intangible capital currently constitutes between one-half and two-thirds of corporate market value, of both „old-" and „new-economy" enterprises (see *Lev*, 2001).

At the dawn of the 21st century, the prominence of intangibles as value and growth creators, at both the corporate and national economy levels, is widely accepted. Furthermore, there is general agreement that traditional (accounting-based) information systems fail to provide adequate numerical underpinning of intangibles and their economic impact. Research indicates that these „information failures" cause serious private and social harms, such as excessive cost of capital to intangibles-intensive enterprises, hindering their investment and growth; abnormally high volatility of stock prices, causing undue losses to investors and misallocation of resources in capital markets; systematic biases in managerial decisions; and excessive insider trading gains to corporate executives, eroding investors' confidence (see *Lev*, 2001).

While there is reasonable agreement about the causes and consequences of the intangibles-related information deficiencies, there is a heated controversy about the necessary remedies, ranging all the way from doing nothing (let free market forces work their way to improve the situation), to a significant overhaul of corporate accounting and financial reporting practices. In between, are various proposals for encouraging voluntary corporate disclosure of intangibles-related information.

Thus, the „intangibles movement" has largely succeeded in the first phase of its mission: Creating wide awareness and active discourse about the economic role of intangible assets and their consequences. This effort, however, appears now to face a crossroads.

What should follow the awareness-creation phase? Some hoped that by observing how firms manage intangibles and report internally about their values, we would be able to develop „best practice" cases for managers and design reporting modes for information disclosure about intangibles. Others believed that we could develop optimal valuation and disclosure practices for these assets by surveying the valuation practices of financial analysts and fund managers. My reading of the numerous surveys and questionnaire studies in this area leads me to conclude that, with but few exceptions, the persons being asked both managers and investors are at a very early stage of grappling with the management, valuation, and reporting of intangibles. To be sure, some companies report internally, and sometimes even externally, on certain aspects of intangible assets (e.g., employee and customer satisfaction), but such haphazard, nonstandardized reporting is of little value for resource allocation or investment decisions. It is therefore questionable whether significant advances in the work on intangibles can be based on such observational studies.

Another route of investigation involved empirical research on intangibles, which provided significant contributions to the understanding of the causes and consequences of intangible investments. But this effort too, in my opinion, reached the stage of diminishing marginal returns, due primarily to the paucity of raw material for research: public information on intangible assets. Research and development in the United States is the only intangible investment required to be publicly reported by companies, and while we learned a lot from R&D (and related, e.g., patent-based) research, R&D is just one component of intangibles. There are no comprehensive data on brands, information technologies (IT) investment, employee training, business processes, and other important intangibles to support empirical research on measurement and valuation. So, where do we go next?

And go we must, because the available knowledge about intangibles is insufficient to address fundamental concerns of managers and investors, as the following examples demonstrate.

2. The Limits of the Current Intangibles Knowledge Base

I will use two controversies to demonstrate the need for refocusing the investigative efforts in the intangibles area. The first concerns the proposed (approved in May 2002) merger between *Hewlett-Packard (HP)* and *Compaq*, pitting major shareholders against their CEOs. The CEOs, promoting the virtues of the merger, claim that the new organization will cut costs substantially, streamline products, and become a major player in the highly profitable tech services sector. Antagonists, led by the *Hewlett Packard* heirs, counter that practically all past big mergers between technology companies (e.g., *AT&T* and *NCR* in 1991) failed dismally, that *HP* and *Compaq* lack a cadre of business service professionals poised to develop a meaningful services business, and that the merger's raison détre is flawed, since PCs, a major product of both companies, are essentially commodities with razor thin profit margins, unlikely to create much value.

This controversy seems ripe for experts on intangibles to analyze and resolve. The two companies are intangibles-intensive; in fact most of their assets are intangible: valuable patents, widely recognized brands, and highly qualified employees. Alas, the two sides to the controversy hired investment bankers to make their case to shareholders, rather than experts on intangibles (e.g., intangibles valuation consultants). Did they hire the wrong people? I am not sure. I must confess that I am hard-pressed to see what the accumulated work on intangible (intellectual) capital could have contributed to the *HP/Compaq* merger and similar ones.

At issue are fundamental questions of corporate structure and performance: Why did *HP* and *Compaq*, with all their valuable intangibles and knowledge management systems lose market share consistently in recent years, and why have they seen their growth stagnate? Why did the innovation processes of these organizations erstwhile innovation leaders atrophy? Is a merger between two giants better suited to jump-start the companies than small, targeted acquisitions coupled with organic growth? Sadly, in my opinion, the concepts and tools developed to date in the intangibles field are not ready for „prime time“ in several respects: for resolving fundamental issues of corporate structure and performance; or for addressing questions of innovation going awry, loss of competitive edge, or how to rejuvenate intangibles' value.

The second case I consider, is, of course, *Enron*: a company that was hailed by „new economy“ gurus and other pundits as the epitomization of value creation by intangibles energy and broadband markets, patents and brands, and advanced knowledge management systems. All those intangibles, valued not long ago at $60 billion by investors, vanished into thin air. Abstracting from the thorny issues of alleged fraudulent reporting to investors, careless auditing, and insiders' self-dealing, one should still ask, „what happened to the intangibles?“ Were our tools for intangibles valuation or knowledge management capable of predicting the implosion of Enron? Are we able to predict now who will be the next „*Enron*“? Is the accumulated knowledge about intangibles sufficient to provide policymakers with guidelines to prevent future massive losses of intangibles?

This is what I mean by „intangibles efforts at a crossroads“ the difficulties addressing fundamental issues, such as those raised by the *HP/Compaq* and *Enron* cases, with the concepts, models, and tools developed in the intangibles (intellectual-capital) field. Why do we encounter these difficulties?

3. The Inertness and Commoditization of Intangibles

Intangibles are inert by themselves; they neither create value nor generate growth. In fact, without efficient support and enhancement systems, the value of intangibles dissipates much quicker than that of physical assets. Examples of inertness:

- Highly qualified scientists at *Merck*, *Pfizer*, or *Ely Lilly* (human capital intangibles) are unlikely to generate consistently winning products without innovative processes for drug research, such as the „scientific method“ which is based on the biochemical roots of the target diseases. Even exceptional scientists, using the traditional „random search“ methods for drug development will hit on winners only randomly.
- A large patent portfolio at *DuPont* or *Dow Chemicals* (intellectual property) is by itself of little value without a comprehensive decision support system that periodically inventories all patents, slates them by intended use (internal or collaborative development, licensing out, or abandonment), and systematically searches and analyzes the patent universe to determine whether the company's technology is state-of-the-art and competitive, and who infringes on the firm's intellectual property.
- A rich customer database (customer intangibles) at *Amazon* or *Circuit City* will not generate value without efficient, user-friendly distribution channels and a highly trained and motivated sales force.

Worse than just inert, intangibles are very susceptible to value dissipation (quick amortization); much more so than other assets. Patents that are not constantly defended against infringement will quickly lose value due to „invention around“ them. Highly trained employees will defect to competitors without adequate compensation systems and attractive work place conditions. Valuable brands may quickly deteriorate to mere „names“ when the firm loses competitive advantage (e.g., *Xerox*, *Yahoo!*, *Polaroid*). The absence of active markets for most intangibles strips them of value on a stand-alone basis (certain patents and trademark are exceptions).Evidence the billions of dollars of intangibles (R&D, customer capital, trained employees) lost at all the defunct dotcoms, telecom companies, or *AOL*, which announced in January 2002 a whopping write-off of $54 billion of mostly intangibles.

Intangibles are not only inert; they are also, by and large, commodities in the current economy, meaning that most business enterprises in a given sector have equal access to them. *Baxter*, *Johnson & Johnson*, along with the major biotech companies have similar access to the best and brightest of pharmaceutical researchers (human capital); every major retailer can acquire state-of-the-art supply chains and distribution channel technologies, capable of creating supplier and customer-related intangibles (e.g., mining customer information); most companies can license-in patents or acquire R&D capabilities via corporate acquisitions, and brands are frequently traded. The sad reality about commodities, tangible or intangible, is that they cannot create considerable value. Since competitors have equal access to such assets they, at best, return the cost of capital (zero value added).

4. Phase II: Focusing the Intangibles Efforts

The inertness and commoditization of most intangibles have important implications for the „intangibles movement. “ They imply that corporate value creation depends critically on the organizational infrastructure of the enterprise on the business processes and systems that transform „lifeless things,“ tangible and intangible, to bundles of assets generating cash flows and conferring competitive positions. Such organizational infrastructure, when operating effectively, is the major intangible of the firm. It is by definition noncommoditized (i.e., unique, idiosyncratic), since it has to fit the specific mission, culture, and environment of the enterprise. Thus, by its idiosyncratic nature, organizational infrastructure is the major intangible of the enterprise.

So what should follow Phase I of the intangibles work, which was primarily directed at documentation and awareness-creation? I propose a focus on organizational infrastructure, the intangible that counts most and about which we know least: the engine for creating value from other assets. Like breaking the genetic code, an understanding of the „enterprise code“ the organizational blueprints, processes, and recipes will enable us to address fundamental questions of concern to managers and investors, such as those raised above in relation to *HP/Compaq* and *Enron*. Success of such effort will elevate the work on intangibles to the highest managerial echelons (after all, organizational infrastructure is what management is all about); it will provide investors with an understanding and with analytical tools concerning the elusive concept of „quality of management”.

5. Conceptual Framework

My notion of organizational infrastructure derives from macroeconomic growth theory, particularly from the „total factor productivity" (TFP) concept. Economic growth theory deals with the economic development of nations, with special focus on the drivers of growth. The drivers of economic growth, measured for example, by Gross Domestic Product change, can be divided into: (a) increases in factor inputs, such as growth of the labor force and capital investments (property, plant and equipment), and (b) improvements in TFP – the productivity of the combined factor inputs. For example, the gross product of U.S. corporate businesses increased in 2000 (relative to 1999) by $443.2 billion, while consumption of physical capital increased by $57.9 billion and the compensation of employees by $289.1 billion. How did aggregate spending of $347.0 billion on capital and labor in 2000 increase gross product by $443.2 billion? Where did the $100 billion value added come from? The answer: Total factor productivity business processes, organizational designs and incentive systems, enabling the corporate sector to generate an output level substantially higher than invested inputs.

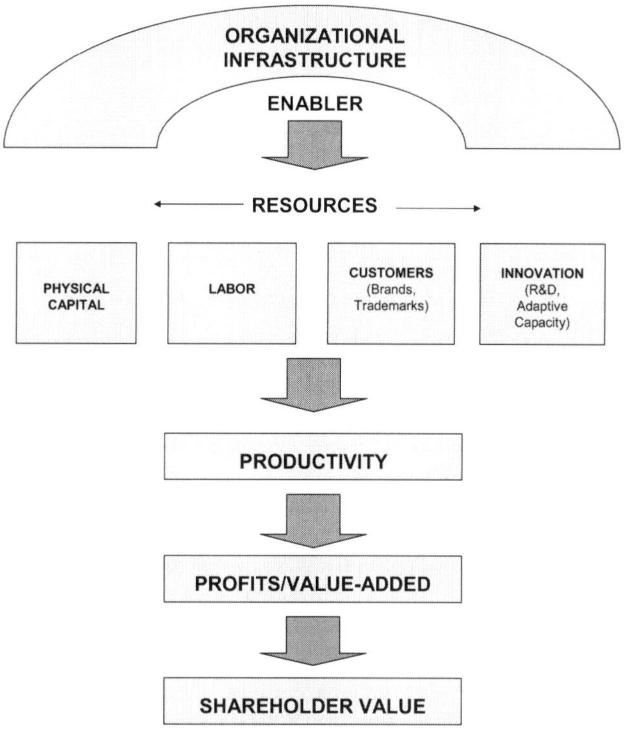

Fig. 1: The Value-Creation Chain

I define corporate organizational infrastructure, with required modifications, as the micro-economic analogue of total factor productivity (TFP). Organizational infrastructure that is, managerial processes, organizational blueprints, incentive and control (corporate governance) systems when operating successfully, enables management to generate excess product out of invested resources. This excess product relative to invested inputs reflects the productivity edge of the enterprise, the source of shareholder value and employee welfare. *figure 1* depicts the central role of organizational infrastructure in the business value-creation chain. It enables the enterprise to use its resources in a productive manner, which in turn enhances company productivity, profits, and shareholder value. This approach considers both tangible and intangible capital (R&D, customer-related intangibles), isolating organizational infrastructure from all other corporate resources, most of which, I argued, are inert and commoditized. This emphasizes the unique nature of organizational infrastructure. It's not an asset or capital, as many believe; rather it's an enabler of all other assets, tangible and intangible, performing their intended roles in the value creation chain (see *figure 1*).

6. Measurement

What do we know about that all important intangible organizational infrastructure (OI)? Admittedly, very little. The management literature and business media frequently tout new managerial processes and organizational designs, some of which fade away after the initial excitement. But systematic knowledge about OI is scarce. To gain quantifiable insights into the value and impact of OI, I developed with *Suresh Radhakrishnan* a methodology for valuing firm-specific OI, depicted in *figure 2* (see *Lev/Radhakrishnan* 2002).

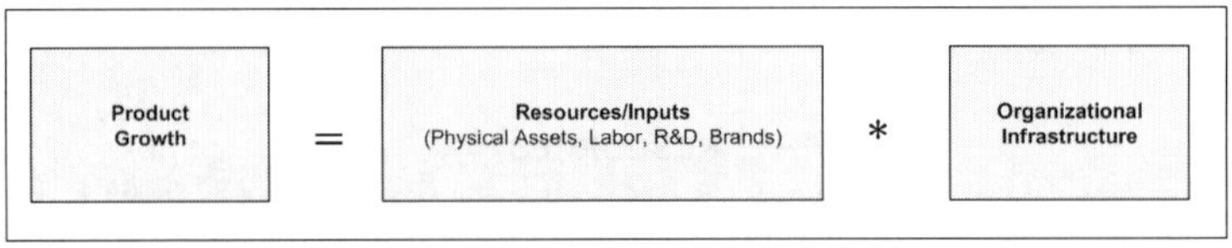

Fig. 2: The Valuation of Organizational Infrastructure

We estimate statistically for a sample of about 300 public companies the contribution to revenue growth of the four major resources: physical assets, labor, brands, and R&D the investment in each is derived from companies' financial reports. It is clear from the data that some companies are more productive than others, being able to generate higher revenue growth from a given level of resources. We attribute this firm-specific productivity to organizational infrastructure, and estimate it statistically by quantifying the contribution of organizational infrastructure to product growth. Our methodology thus defines OI in terms of the productivity of deploying the company's resources, and values OI from observed sales growth, given invested resource, as is done in national accounts with total factor productivity (TFP).

Figures 3 and *4* present for selected companies the average annual contribution of OI to product growth over 1994–97 (dark grey bar) against subsequent stock performance over 1997–99 (light grey bar). The three companies in *figure 3* are in the chemical industry. *Dow Chemicals* had a larger increase in OI than *DuPont* during 1994–97, and was accordingly rewarded by a larger shareholder value increase in 1997–99. *FMC Corp.* had a decrease in OI and a subsequent negative stock return. The companies in *figure 4* are in the retail sector. *Wal-Mart* and *Home Depot* with large increases in OI and commensurately large subsequent stock returns, while *K-Mart* suffered a significant loss in OI and a subsequent loss in shareholder value. Our measure of organizational infrastructure thus captures real growth potential, which is reflected in subsequent years by shareholder value creation.

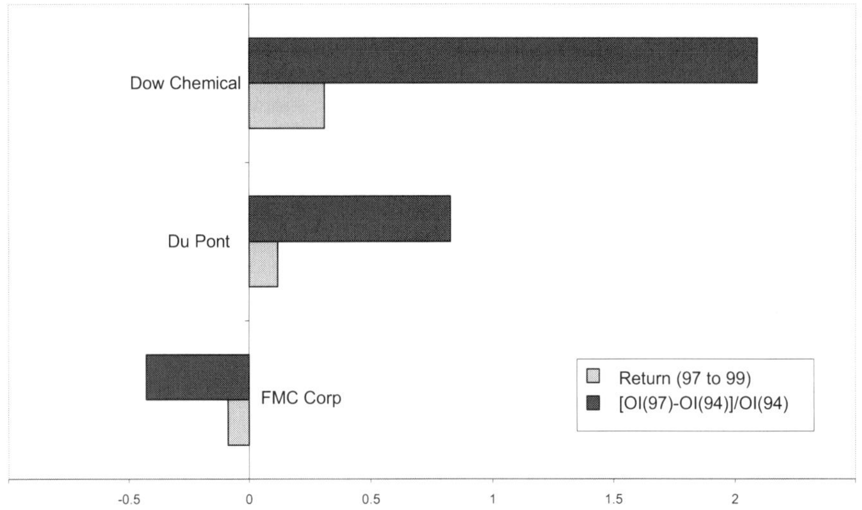

Fig. 3: Chemical Organizational Infrastructure and Shareholder Value

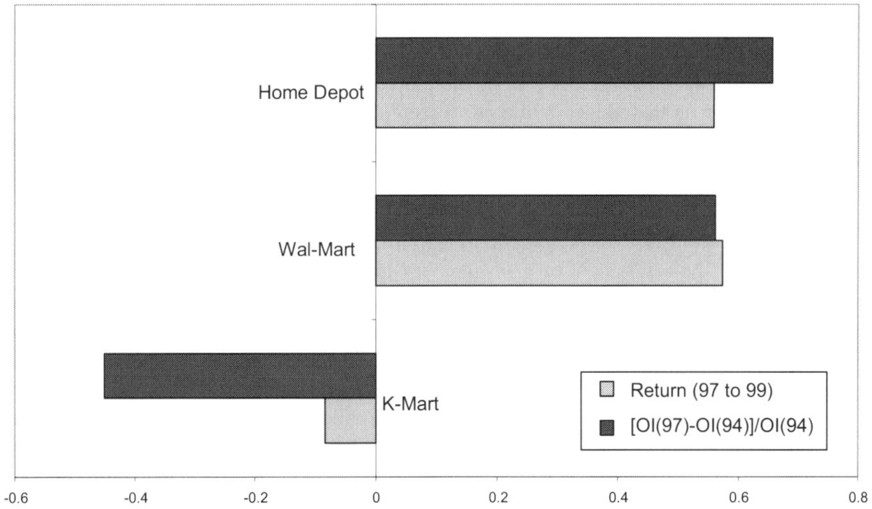

Fig. 4: Retail Organizational Infrastructure and Shareholder Value

7. Managerial and Investment Applications

The contribution of organizational infrastructure to enterprise productivity is essentially a reflection of managerial capabilities and their execution. Organization and execution are what management is all about. Despite oceans of ink, the quantification of managerial capabilities remains elusive. Vague terms like leadership, focus, or knowledge workers abound, but they don't significantly advance our ability to evaluate managers' capabilities and success. The OI concept discussed here does just that. Grossly simplifying, if both companies A and B increase capital investment from 100 to 110 units, while A's product grows 10% and B's 20%, the inescapable conclusion is that B's management was more capable and successful than A's. Thus, the extent of managerial capabilities and execution success can be measured by the OI approach of relating productivity growth to resource increases, and focusing on the residual-enabler.

The ability to estimate the value of companies' organizational infrastructure provides new valuation and control tools for managers and investors. Given space constraints, I will briefly describe just three examples.

Productivity of IT: The over-investment of the corporate sector in information technology during the 1990's and the almost complete cessation of IT growth in the last two years, emphasize the need for a rational assessment of the productivity (return on investment) of IT. Since IT primarily enables organizational infrastructure (e.g., the IT underlying supply chains or distribution channels), its productivity can be inferred from the relation between IT expenditures and the corresponding growth in the value of OI. For example, in our sample, *Home Depot* had a ratio of 1997–95 increase in OI to total IT investment of 0.51, while *Wal-Mart's* ratio is 0.16 only. Thus, *Home Depot's* return on IT in the mid-1990s appears to have been substantially higher than that of *Wal-Mart's*.

- **Management compensation:** The contribution of organizational infrastructure to the company's product growth is, as I argued above, a quantification of management's capabilities and execution success. It is, therefore, compelling to base part of managers' compensation on that value. While our research shows that there is a strong correlation between OI and stock performance, it's not a perfect correlation. OI is less susceptible to factors beyond the control of management (e.g., interest rate changes, tax changes) than stock prices, and therefore is a more suitable driver of mangers' compensation.
- **Investors' valuation of management capabilities and growth prospects:** Surveys indicate that the assessment of management capabilities is at the top of the wish list of financial analysts and institutional investors. Traditional performance measures, such as earnings or cash flows are subject to widely known limitations (e.g., prone to manipulations, biased by the expensing of intangibles, ignore the cost of equity capital, etc.), and are therefore noisy indicators of managerial capabilities and performance. The organizational infrastructure measure discussed above provides a sharp focus on managers' performance and the company's growth potential, by indicating abnormal productivity gains derived from given company resources. Our research indicates that when we form portfolios of companies with high and low OI values, the abnormal (risk-adjusted) returns on these portfolios in the two years subsequent to portfolio formation are: 3.0% vs. 1.5% and 2.5% vs. 1.3% in favor of the high OI companies. Investment in good management, as indicated by the OI measures, clearly pays.

8. Agenda for the Intangibles Movement

Productivity of scarce resources is the top priority of nations and business enterprises. Increased productivity of resources (capital, labor, R&D) is the major driver of nations' welfare (standard of living, employment), and of companies' profits and shareholder value. A focus on productivity at the national and business enterprise levels appears, therefore, appropriate for people concerned with intangibles, given that in modern economies and companies intangibles are the major drivers of productivity.

Current concerns, such as the valuation of specific intangibles (patents, brands), or the codification of tacit knowledge, while useful, are in my opinion of secondary relevance, subordinate to the overall productivity or efficiency enhancement goal. An intellectually-stimulating and highly useful agenda for the Intangibles Movement is accordingly to „peel off" the OI onion. Here are several concrete issues for study.

- What are the organizational structures of R&D activities most conducive to success? What specific blueprints and underlying processes of R&D yield the highest number of patents and innovations, and the largest share of revenues from recently introduced products?
- What are the systems used by companies to successfully enhance brand values? Why do some brands (e.g., *Coca Cola*, *Marlboro*) maintain their value, while others (*Polaroid*, *Xerox*), fall by the wayside?
- What is the impact of corporate networking alliances, partnerships and joint ventures on the productivity of companies? What is the optimal structure of such networking for achieving specified objectives (e.g., conducting research, sharing risk, selling products)?
- Develop measurement and reporting systems for evaluating the productivity of resources, particularly intangible ones: Assess the returns on investment in R&D, brands, IT, and human resources; and develop reporting modes for the productivity of these assets.
- What are the compensation systems and work practices that generate the highest productivity of human resources? How can employee productivity, particularly in service organizations be measured?
- How do successful enterprises develop their organizational structures and keep them unique? How to prevent competitors from quickly imitating such designs?
- The pace of knowledge accumulation: How are new knowledge and technologies adopted and diffused within the organization? Under what circumstances are old technologies abandoned and new ones adopted?
- What are the organizational impediments to knowledge creation and the adoption of new technologies and operating systems? How best to overcome the forces of the status quo?
- The collapse of Nasdaq demonstrated the vulnerability of intangibles, particularly OI. How can companies protect their intangibles? How can the risks inherent in intangibles be shared and minimized? What are successful measures of risk exposure?

Certain aspects of this research agenda are conducted in other areas economics, management, sociology and organizational behavior. It will be highly useful to assemble available research and emerging work under the umbrella of the Intangibles-Organizational Infrastructure undertaking.

The proposed Phase II of the Intangibles Movement the focus on organizational infrastructure and its role in enhancing the productivity of nations and business enterprises, and the welfare of citizens and shareholders offers a socially useful and intellectually stimulating agenda for people interested in intangible assets. Using *Nike's* slogan "Just Do It!".

9. References

Lev, B., Intangibles: Management, Measurement and Reporting, Washington 2001.

Lev, B., Intangibles at a Crossroads, in: Controlling, 15. Jg. (2003), H. 3/4, S.121-127.

Lev, B. and *Radhakrishnan, S.*, Structural Capital, http://www.haas.berkeley.edu/accounting/seminars/Lev%20Paper.doc

New Perspectives of Leadership for Value Creation

Leif Edvinsson[1] *and Mart Kivikas*[2]

---

[1] Prof. Dr. *Leif Edvinsson*, University of Lund, Sweden
[2] *Mart Kivikas*, WEKE GmbH, Germany

Knowledge has become the key source of value creation and wealth not only at an organisational but also on a national level. And enterprises, particularly those in the public sector, are in danger of undermining their future success by focusing exclusively on financial assets in their accounting procedures. It is time for a quest for a new approach for gaining back the sense of meaning making, based on lateral thinking, rather than just money making. It is a challenge not to be prisoners of vertical thinking, called bottom line syndrom. This might lead up to the longitude perspectives of organised intelligence, brain gym and the skill of quizzics -questioning skill with the basic question: Where is value being created in your company and enterprising? Inside or outside the companies boundaries? By out-sourcing or in-sourcing of strategic knowledgeable partners? Is there another value logistics emerging, calling for another knowledge and Intellectual Capital (IC) entrepreneurship? Can the corporate longitude perspective be another view and thought pattern to shape the value platforms for the New Wealth? How is the intelligence systems designed for leader-ship, around entreprenurship, culture and innovations for turning the Future into an Asset for the stakeholders?

1. The Knowledge Capital Economy

In the knowledge economy, the value of corporations, organisations and individuals is directly related to their knowledge and their intellectual capital. But broaden the perspective a little wider and you begin to understand the real possibilities. Think of entire nations and the public sector, not just traditional enterprises in the business sector. If intangibles and intellectual capital are important to private enterprise organisations, they are also important to the productivity and competitiveness of the public sector, and so to nations as a whole. So how can we seek to understand the dynamics of the intangibles at work on a national scale? Can the new corporate longitude be translated into a new perspective on national performance? Can the Scandinavian perspective of "naringsliv" as nourishment for life be a clue to the knowledge capital economy beyond the concept of business? Could there be a new eco system or DNA for the knowledge economics?

According to the „OECD report, Scoreboard 2001 – Towards a Knowledge-Based Economy", which is conducted every second year, the countries with knowledge intensive activities will be the winners in terms of future wealth creation. In this report, 30 member countries are scored according to intellectual capital investments such as R&D, education, patents, ICT (Information and Communication Technology) etc. The top-ten list of countries according to their potential future wealth looks like this:

1. Switzerland
2. Sweden
3. USA
4. Ireland
5. The Netherlands

6. Hungary
7. Canada
8. Belgium
9. UK
10. South Korea

Only knowledge will give us the opportunity to improve the wealth of nations. As such, we need intelligence systems to develop a new map of knowledge assets and intellectual capital; a map of regional intellectual capital, instead of the old agricultural and industrial plans so often found in regional planning offices. The key mapping dimension and intelligence dimensions should stem from the need to locate where wealth is created in a given framework, sometimes called enterprise. Could this process perhaps reveal a huge knowledge repository in the external sector, with a significant – but idle – potential for collective wealth creation? In other words it can be viewed as Wealth in waiting.

Adding value in the knowledge economy is inextricably linked to radical change in both societal assumptions and business models. A new political leadership agenda is evolving around the intellectual capital and Knowledge Management (KM) of nations, with the focus on:

- How to visualise the knowledge capital of nations.
- How to develop intelligence flows within and between knowledge capital clusters.
- How to cultivate efficiency and renewal of the knowledge capital of nations.
- How to capitalise on knowledge capital, by new innovative social systems, in terms of the collective wealth of nations.

In Norway several interesting initiatives have been applied to above issues. One is IC-Rating™ (see more under www.intellectualcapital.se) of *Larvik Kommune*. Another is an IC-Rating™ of the *Norwegian oil plateu*. Another is the prototyping of a Future Center to nourish society entrepreneurship. Based on the experiences we have made on Future Centers in Scandinavia, the initiative for this has been taken in Germany (www.wissenskapital.info/IC Akademie).

During the 1990's the research by Professor *Paul Romer* at Stanford University, California, has highlighted the exponential value of multiplying knowledge recipes as an exponential value curve, called the law of increasing marginal utility. In other words, the more connections, relationships and interactions in a network society or organisations, the higher potential value is emerging. This is evident in the case of development of software and knowledge assets. Furthermore the value is growing while used, in contrast to the more tangible hardware. A personal laptop, for example, is decreasing rapidly in value while used, a value process by accountants labelled amortising.

2. The Future as an Asset

The current market and economics development is at an inflexion point on its life cycle curve. As stated long ago by *Jay Forrester*, at MIT, United States, in his research of world dynamics, most industries, as well as most nations are going through a life cycle curve. On this curves there are turning points called inflexions points. Dr. *Debra Amidon* (www.entovation.com) is in her research and development work stating a number of phases in the process of knowledge innovation as well. *Figure 1* below shows a kind of gradual evolution of assets as knowledge innovation progresses from tangibles to intangibles: We move along a line that starts with product as an asset, then moves to the project, the company, the client, and finally the knowledge itself as an asset

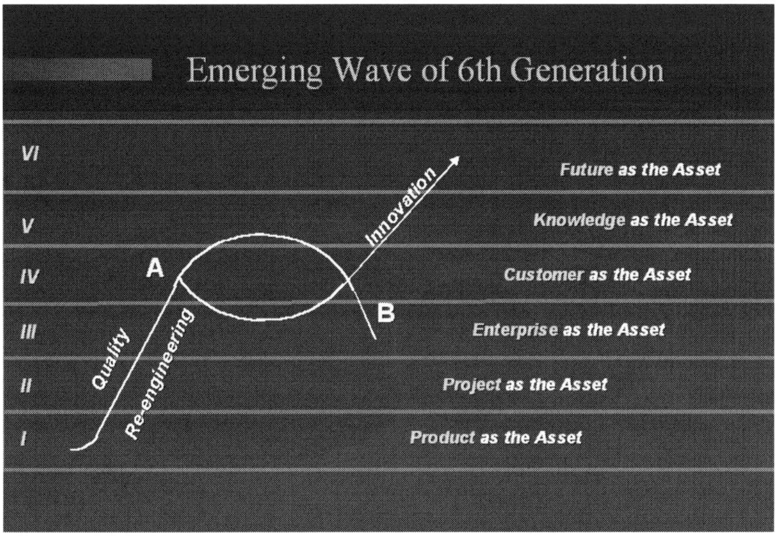

Fig. 1: Life cycle curve of assets

In this life cycle perspective we added on top of the curve the „Future" as an Asset, i.e. we acknowledge the potential value of innovation as an investment value. Right now, we assess present position - where we are today. This present position, can be assessed and appreciated, but not changed. That value is in other words a given asset, shaped by historical activities.

Now change the perspective. Look forward instead of backward and the creation of value is revealed to exist in the future, i.e. the time line and your management of the future "space" from the next few seconds to the eternity. However in most economical dimensions the „Future" is accounted for, in most balance sheets and profit & loss statements, as an expense and then turning up as a liability, not as an asset. Hence investing in the future as intangibles is eroding the equity. A paradox is emerging then - the more spent on the future, the more poverty or less equity in the balance sheet. And all futurizing is starting with intangible investment. Consequently we need both a better intelligence and a reshaped entrepreneurial approach for these new value spaces.

The most recent analogies on innovation is coming from life science and biology, as elaborated by among others one of the world's leading futurists *Stan Davis*, USA. Life science is highlighting the need for continuous mutation and cell evolutions. As a comparison look at the bacteria, which is mutating faster than the pharmaceutical industry is innovating. On a corporate level the financial service company in USA Capital One can be an illustration with about 2000 innovations per year as well as *IBM* with about 3400 filed patent applications per year. This calls for another type of organisational approach with very high intelligence agility.

Furthermore it looks at renewal/innovation and being in continuous move, is of essence for the organisational sustainability. If the enterprise stops or is frozen, there is no life. So the trust is in the belief of the emergent, or the time line towards the future. The value in the knowledge economy is in the time line.

From an enterprise perspective the shareholder value is in the capability and capacity for continuous innovations. For a nation it might be in the same innovation perspective beyond the traditional economical growth perspective. This will challenge the traditional theory of the firm, to become much more "unfirm". The new theory of the firm and intelligent enterprising is about continuous evolution, calling for not only incremental change but also radical change by longitude leadership.

3. The new Corporate Longitude

The logic behind the idea of corporate longitude is based on a story from the 18th century and presented more in detail in my recent book (see www.corporatelongitude.com). At that time, the British Navy could not navigate with any precision from east to west. Consequently, many 18th century ships lost their bearings, in the same way that today's analysts who only concentrate on financial capital do. The cases of *Enron* and *Worldcom* illustrate a recent example of a similar problem in the world of accounting. In the public sector it might be even worse, as the focus today is strongly on balancing the budget of financial capital. The result is that public sector organisations are being starved of the assets that will safeguard their future, into an anorectic behaviour. Extensive cutbacks are robbing them of the crucial nourishment offered by intangibles such as experts, knowledge recipes, R&D (research & development), learning as well as alliances and networks for social innovations.

The solution to the longitude problem was discovered not by Navy professionals or by leading university academics, but instead by a knowledge outsider: a watchmaker by the name of *John Harrison*, who discovered an innovative answer to the problem. In order to get a deeper understanding about the principles of wealth creation, it is essential to develop a similar lateral perspective. An approach whereby assets are simply recorded on a balance sheet is far too narrow. A lateral approach to accounting is necessary to include the intangibles and non-financial assets of knowledge creation, networks and relationships. The wealth of organisations, as well as the wealth of nations, lies in the space in which human capital and structural capital interact.

Currently, organisations, as well as societies as a whole, are like 18th century ships, charting their positions with only north/south navigational tools. Plotting a course solely based on traditional financial reference points leaves them blind to the opportunities on the lateral horizon. Lost on a turbulent sea of change and without a lateral navigational tool to guide them, they cannot navigate the uncharted challenges the management of intangibles is presenting, in particular to the public sector.

Most attention is still devoted to the financial map, in a world of corporations where the tangible assets account for often from 1 % to 25 % of a company's stock market value. For these assets the organization has a CFO and controllers, software systems called ERP (Enterprise Resource Planning) and the whole profession of auditors. But what do we have for the intangibles and intellectual capital? So far it is mainly CRM - customer relationship management systems? In 1991 *Skandia* appointed *Edvinsson* as the world's first director of intellectual capital to develop another logic for the renewal of *Skandia*. It is however still too rare as a function, for organizational value creation.

The financial reporting and its position on the global news agenda have become an obsession, which furthermore is becoming more and more illogical. What we now need is much more intelligence and know how about the intangibles and where value is created. For example, if you are investing in your competence upgrading is that a loss or something of value? In the old accounting practice it is regarded as a loss. It is the same way with research and development. Is that a loss or a potential future? This abuse of the management accounting is resulting in anorectic leadership behaviour, trying to starve the organisations into the future, especially by savings on the important nourishment of intangibles such as knowledge, R&D and learning.

Until sailors could measure longitude, they were frequently lost at sea. The same is the case for our value generating leadership of today. The difference between knowledge economics and tangible product based economics demands a similar shift in the way we view the corporation and track its value creating progress. We have to take a lateral perspective regarding alliances, networks, cultural context, knowledge recipes and other intangibles, usually outside the classical balance sheet perspective. We also have to shift to other deeper indicators such as those about time, rhythm, brain, relationships, knowledge recipes, innovations or in other words, tacit knowledge indicators.

Figure 2 shows an illustration of Corporate Longitude as a lateral dimension of intangibles inside and outside the firm´s vertical balance sheet. What is needed is an intelligence window for those new vaue-creating spaces.

The value is created in the interaction between people as human capital and the organizational structural capital such as R&D processes. Professor *Nonaka* is referring to this as knowledge creating dialectics or "Kenetics". Those dynamic spaces are also referred to as „Ba" by professor *Nonaka*, that literally is said to mean a space for appreciation in Japanese. In *Skandia* we labelled this prototyping space for shared intuition to Future Center (more about this in the case study „Future Culture and Earnings Capabilities"). The *Skandia* Future Center established under *Edvinssons* leadership in 1996 was focused on the value adding by experiential knowledge exploration. It became an arena to enter into the future and then return to the present with new insights (see also www.skandiafuturecenter.com).

Another such interesting space for renewal is shaped by the Danish government. In February 2002 the *Ministry of Economics* launched „Mind Lab", very much similar to the ideas prototyped at *Skandia* Future Center or the Japanese concept of "BA". "Mind Lab" is however a centre with the purpose to nourish knowledge management in the public sector (see www.mind-lab.org).

Now the internet and telecom evolution offers even more opportunities to invite brain power to come along virtually to such a space for imagination, where structural capital and human capital meet in different forms to shape the value added. Furthermore the experiences from prototyping within *Skandia* Future Center, as well as health research shows that attention to space and architecturial design to release brainpower potential is of utmost leadership importance.

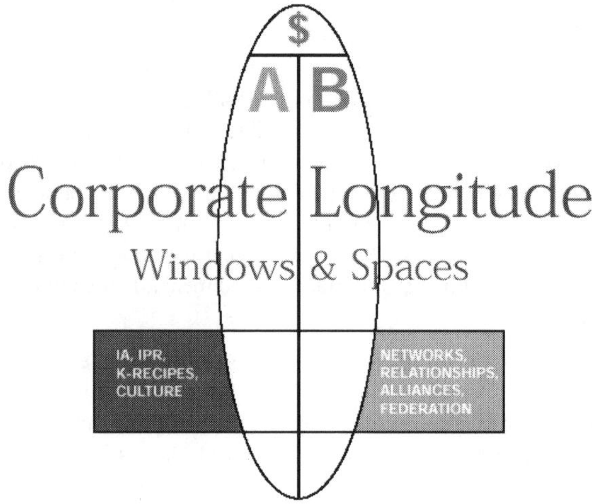

Fig. 2: The Corporate Longitude

4. Intellectual Capital Mapping and Measurement

The 500-year-old accounting system is focused on historical costs and transaction reporting. This backward looking metrics approach leads to growing inaccuracies in the value-creating map of today. Furthermore it results in misallocation of resources by investment institutions. It is like asking what is the cost of good or bad weather, instead of asking for the weather forecast.

What is needed is a longitudinal system to visualize, cultivate and capitalize on these value creation interactions? We have to go from cash flow measurement by cash registers, to a time flow based measurement by chronometers of innovation and networking speed. The broadening intellectual capital evolution outside the corporate world has already begun. Intellectual capital reporting of nations, regions and, indeed, the public sector, is also already happening.

In Denmark progress is now rapidly being made, and the country has long been at the forefront of examining the role of intangibles. *Jan Mouritsen*, professor at the *Copenhagen Business School*, has been working in the area for a number of years and has carried out several surveys and literature studies, a most recent of which was published in 2001. At the beginning of 1998, Denmark launched a project looking at intellectual accounting that aimed to help transform Denmark from an industry to a knowledge-based economy. Special Competence Council was organised, with *Lars Kolind* as chairman. This has produced a great deal of interesting work relating to Denmark's position in the new global knowledge competition, with the most recent report from October 2002. In 2000, the Danish government also published guidelines for intellectual capital statements, similar to those produced by *Skandia*. In December 2002 they published an upgraded version (see more on www.vtu.dk/icaccounts or download from www.danmark.dk/netboghandel)

In Norway several initiatives are also in progress. The local municipality of *Larvik* has been prototyping both annual reports as well as now IC-Rating™ for its activities. A Norwegian cabinet initiative is working on the intellectual capital aspects of the public sector. *The Norwegian Association of Financial Analyst* has in December 2002 launched guidelines for reporting on knowledge capital (see www.finansanalytiker.no). The *Nordic Industrial Fund*, based in Oslo, is working with several intiatives, among others two major reports; Nordeka and Frame.

In the USA the *Securities and Exchange Commission* (SEC) has appointed a task force to deal with the need for improved reporting. This work is also accompanied by the work of the *Federation of Accounting Standard Board* (FASB) to develop a white book on the subject, together with among others professor *Baruch Lev*. On the European scene the IAS rules are requesting all companies to have some kind of intangible reporting by 2005.

In the years 2001-2002, Dr *Nick Bontis*, McMaster University, Canada, conducted some outstanding research relating to intellectual capital of Nations. Based on *Edvinssons* prototyping work with the *Skandia* Navigator model, *Bontis* and his colleagues, sponsored by the United Nations Development Programme, conducted a study of ten Arab states. In the study, *Bontis* quantifies the state of intellectual capital for each nation and outlines an intellectual capital index, which can be used by each nation to rank themselves against their peers, and indeed to learn from the experiences of other countries.

By looking at the intellectual capital wealth of nations rather than standard measures of national competitiveness, we gain new insights into where a country's enablers for the wealth creating might lie. We can then more effectively nourish social innovation and society entrepreneurship.

Bontis' research suggests that the following represent the key areas for the political agenda to address:

- National agenda for renewal, research and development, ie, innovation capital;
- National agenda for education, ie, human capital;
- National agenda for foreign trade, ie, relationship capital;
- National agenda for industrial productivity, ie, process capital.

One of the most recent and very insightful reports on intellectual capital of Nations presented now in Jan 2003, is done in Croatia, by professor *Ante Pulic* and his team (see www.vaic-on.net), called „Efficiency in Croatian Economy". It is reporting a value creation efficiency index.

5. Me Assuring and Me-asurement

Why is measurement so important? Simply, because lack of intelligence and communicable information affects trust as well as the efficient supply and distribution of future resources. What is needed in a growing complexity is a clarifying supportive system to sensitise our minds to perceive the best options. In other words we need mapping systems to assure us on the angle of longitude navigation into the future.

What measurement systems can support this sea change in orientation? Certainly the contribution from the essential associated intangibles such as trust, brain efficiency, and healthy collaboration are well beyond the scope of conventional accounting systems. The value of relationships needs to be measured, as does the contribution from knowledge recipes.

Such knowledge creating dialectics, kenetics, will be shaped by the new potential interactive combinations, rather than old repititive best practice approach. In his research, *Edvinsson* calls it the IC-Multiplier, i.e multiplying the human capital potential with the surrounding structural capital – (SC/HC > 1). A good approximation for this is value adding per head. It shows how good the staff is in using the structural captial provided by the stakeholders. Consequently this will also be an interesting key indicator of the leadership power to release the brain power potential of the clusters of people inside and outside the firm.

Communicating these intangible interactions is crucial to attracting the right resources for wealth creation. Such creative modelling routines are by professor *Nonaka* called „KATA" and could as well stand for intellectual capital accounting. In other words me-assuring to create trustworthiness. A recent report from University Lund shows very high correlations between the IC-Multiplier and the stock market value.

As the new economical value is in the longitude, i.e. lateral dimensions instead of vertical dimensions, we have to develop more lateral, benchmarking, accounting of value creation potential of intangibles. We have to acknowledge such new intangible indicators and get the accountants to audit those, as well as annual reports to present transparency of such intellectual capital, to be able to navigate these new organizational value creations. One of the most refined recent such intellectual capital reports, following very much the experiences from *Edvinssons* prototyping intellectual capital reporting at *Skandia*, has during 2002 been presented by *Alexander Welzl*, then at *Seibersdorf Research Center*, and his intellectual capital pioneering colleagues in Austria. Furthermore during 2003 there has been implemented in Austria a law requiring all Universities and Colleges to publish a knowledge capital report annually, showing knowledge goals, knowledge processes as well as knowledge indicators. The very first prototype done among others by University of Kremz, Austria

IC-Rating™ is another important tool, complimentary to the *Standard & Poors'* of *Moody's* rating of financial capital in progress since 1997. IC-Rating™ is about benchmarking the perspectives of efficiency, renewal and risks on intellectual capital components, for future earnings potential (see more on www.intellectualcapital.se). It is now used by more than 200 organisations both in Europe and Japan. It seems to be especially interesting for public organisations such as schools and hospitals that do not have the public stock market as a grading reference point. This approach is now also being applied for regional rating of intellectual capital of cities and regions. The IC-Rating™ gives both a map for benchmark versus best in class, but also a platform for assessing the future earnings capabilities, thereby creating an intelligence trust for the future. As experienced in one company in Germany, the method was used to visualize a successful company culture and make it available for knowledge sharing to other entities within the group trough making an internal benchmarking possible.

In terms of measurement, the traditional reporting model mostly represents the tangible goods sector and has begun to encroach into the intangible economy sector, based on emerging requirements from among others IASB (*International Accounting Standards Board*) in London, and its rule IAS § 36 and § 38. However, it is not yet able to represent the full potential of the networked intangible business model of the 21st century based on intellectual capital. In large part it is restricted by the concept of ownership and confined to reporting on elements it can identify within a firm's legal boundaries. This is inadequate in the knowledge economy. What is needed is the lateral accounting perspective of the corporate longitude intangible items.

The new unit of analysis in the knowledge era will be innovations and sense making. It is about the roots meeting its context. It is about cultivating spaces for such meetings to happen. It is about the intangible indicators for kenetics and longitude leadership.

6. Longitudinal Leadership in the Knowledge Economy

The new corporate longitude is focusing on the lateral dimensions, as well as time to the future. This calls for another type of leadership role than traditional management. On the basis of an ongoing research project with the EU called *PRISM*, in which *Edvinsson* participates as advisor, some modelling has emerged about the value chain of the new theory of the firm. This modelling is highlighting the value creating areas of the enterprising as well as the key focus area for leadership (see *figure 3*).

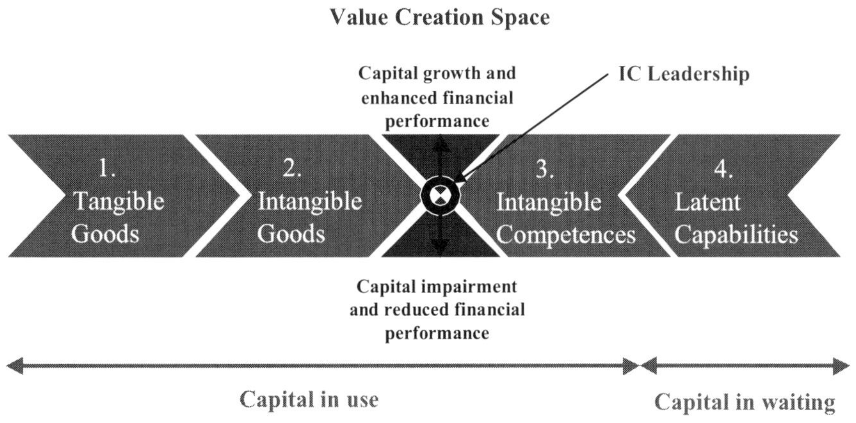

Fig. 3: Value areas for intellectual capital entrepreneurship

The four blocks are based on taxonomy from an earlier EU project. This project group was lead by *Clark Eustace* under the name HLEG (*European High Level Expert Group*) on intellectual capital. They are intended to represent the strategic assets of an enterprise or a nation. They are laid out horizontally to represent the fact that such assets or capital can be accessed from both within and without the traditional legal boundaries of the enterprise. The value adding genetic space is on the borderline where the internal intellectual capital meets the external intellectual capital.

To the left of the "value creation space" lie the tangible and intangibles assets over which ownership rights can - more or less - be elaborated. Tangible goods would include physical assets such as land and buildings, and plant, machinery and equipment. Intangible goods would include packaged and codified assets such as software, brands, trademarks, licenses, and legal intellectual property rights over scientific discoveries.

To the right of the "value creation space" lie the intangible competences and latent idle capabilities or in other words capital in waiting. Intangible competences embrace the organisational capital such as culture, networks and the human capital, which is effectively, leased for productive use from the individual knowledge workers. Latent capabilities are what investors, in particular venture capitalists, are interested in. The discovery and exploitation of this value shaping space is where top leadership truly differentiates itself from others.

At the core is the "value creation space" where intellectual capital leadership faces the challenge of leveraging these longitudinal resources and to create economic value adding. This is the dialectic space or kenetics for knowledge entrepreneurship. It might lead to growth of capital on the balance sheet as well as impairment of the balance sheet. In such a situation value destruction will occur. A critical question will emerge - what is the knowledge leadership of today doing to avoid erosion and leverage the idle intellectual capital in waiting and how do we know about this from the reporting?

This modelling is a starting point for further refinement. The view is very much based on the practice from how to shape new interactive spaces for knowledge creation and knowledge sharing for the mutual benefit of those involved. This is the space where 1+1 =11 or in other words that the value of knowledge is growing when shared. It is the space for the exponential value creation, as research by Professor *Paul Romer*. It is also a space where the traditional firm barriers are opening up for the value of networking and shaping alliances. In terms of Professor *Karl Erik Sveiby,* knowledge is capacity to act and consequently the storage of such capabilities is to a large extent to be found outside the focus of the firm. And in this model leadership will be the hot bed space, Ba, for synthesising the various dimensions of the SECI modelling from Professor *Nonaka*, on Socialisation, Externalisation, Combine and Internalisation, and get together to act for the Future. In further elaborated terms it can be viewed as intellectual capital in waiting now being externalised for society innovations.

An interesting emerging prototyping work is done by *Dave Snowdon* at *Cynefin Center for Organisational Complexity* at *IBM.* The focus is on the need for another approach than the linear management model. The leadership spaces in this *Cynefin* model are very much about sense making. They are based on a biological analogy as well as complexity and chaos research. The leadership is very much about a balance between abstract and concrete communication as well as balance between learning and teaching. Based on those dimensions it is emerging four different spaces for leadership practice as follows:

- Traditional leadership based on linear best practice.
- Expert leadership based on reductionistic best analysis.
- Shadow leadership based on best option prototyping and pattern recognition.
- Chaordic leadership based on probing best responsiveness in complexity.

This very early longitude leadership thinking, perhaps prototyped as in Scandinavia with a special longitude board, might assist us in getting out of the vacuum of lack of leadership in the present fog of economics. It might help us to see the need for much more focus on developing the organisational capital as multiplier for the human capital and its inherent potential for future earnings capabilities.

A new role of „society entrepreneurship" is needed for clarifying and shaping those organizational bridges and alliances for a greater wealth contribution from the brainpower potential of global citizens. This calls for a lot of entrepreneurial inititatives and prototyping for social innovations. Society entrepreneurship is an interesting concept that is gaining credence in countries such as England and Norway. It describes a role for society renewal that fits in the space between the business community and society; a space for collaborative prototyping based on the structural capital of each partner, but leveraged by individual human capital and innovative talent. Society entrepreneurs are often community entrepreneurs. Their output is social, the core asset might be social capital and the organisations they found are often non-traditional share holding enterprises. The challenge is to speed up the returns we are realising from the existing societal intellectual capital, or in other words to increase the existing intellectual capital of societies as corporations, or in other words increasing the turbo or productivity on existing knowledge investments for the new welfare state. A kind of intellectual capital in waiting is there to be tapped by challenging society entrepreneurs.

7. Longitude Board

"Leader-ship", contrary to management, is an interesting label, according to the myth related to the navigator in front of a Viking ship loading or measuring the depth of the water. The German word "Vorsitzender" or "Vorstand" has probably the same epistemological background. The task of the corporate board is to navigate the future of the company. The common practice of most board work of today is about fulfilment of legal requirements related to the maintenance of the firm. Therefore the metaphor of the longitude is interesting as has been pointed out earlier. Combined this is resulting in a prototyping work that we are exploring with Scandinavian enterprises today, i.e. a longitude board of directors.

The background was the quest for a supplementary work to the traditional board with an amplified focus on renewal and intelligence. This is a continuous process in order to be informed, alert and hedging what could be around the corner. In other words a capability to "jump" to the second life cycle curve, as illustrated above in *figure 1*. In Germany it could be called "intellektuelles Vermögen des Unternehmens".

For corporate leadership today it is essential to be able to look in all directions, ahead as well as backwards. This is the longitude perspective. And the key indicator is the angle for futurizing; showing the rhythm of the organisations renewal versus its context, i.e. speed of moving ahead into the future.

All mapping is based on a kind of coordinate system for navigation. The oldest is said to have emerged during the ancient days in Greece. The rotation of the globe is scored since the days of Babylon as 24 hours similar to 360 degrees, this gives that 1 degree is equivalent to 60 min.

Today we have meridians going from pole to pole, north - south. This refers to the meaning of midday line, or that the sun's position at midday is the same for all places along this line. GMT is the standard of today as zero meridians.

Longitude emerges from the Latin word longitude, meaning length (or degree). It has a distinction as a degree of length, or the angle between the meridian where you are at the moment and another selected meridian of the origin. It is measured both east and west. Maximum is 180-degree east or west. The measurement of longitude is tempered by time.

Latitude is the degree of width, or the angle between the equator and the parallel circle where you stand.

The zero degree parallel of latitude is given by nature, and the zero degree meridian of longitude shifts like the sands of time. And corporate life is a living process in the space between, as illustrated below (see *figure 4*).

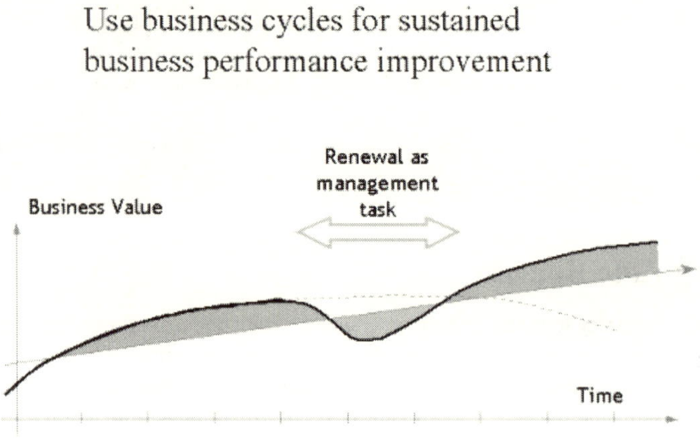

Fig. 4: The Corporate Longitude Board

But traditional boards are required by law to present the past. Often it resembles a budget meeting where the financial outcome is compared to forecast or budget. Consequently a number of interesting questions emerge. How much time in the board time is devoted to renewal and innovations? What is dominant in the board perspective, intelligence or ignorance?

The way to get around this is to initiate a supplementary intelligence force to the traditional board of directors - a temporary group of navigators with outsights, networks, relationships etc. Like the old success story of the city Ragusa in Croatia, sustaining for 500 years as one of the wealthiest places in the Mediterranean. It had a network of more than 150 intelligence reporters all around Europe. How many enterprises have such a systematized intelligence force for the board? How many countries, especially the new countries from Eastern Europe now entering the EU, actually take advantage of native knowledge workers outside its national existing borders? And here we do not refer to traditional strategy consultants or traditional advisory boards.

The role of the Longitude Board is to take the sometimes-challenging role of reporting, prototyping and recommending a tack, like a sailing ship, turning 90 degrees versus the forecasted financial cash flow route, expected by the stock analysts or government advisors. The purpose for this is to get up on a second life cycle curve of innovation as illustrated below. And this tack takes courage that might not be present in the board.

Furthermore the most essential subject for futurizing is not cash flow but time flow, or knowledge advantage based on intelligence. As long as the enterprise has this information or knowledge, it will become an advantage through a time monopoly that represents a strategical value.

Therefore one of the core distinctions of a longitude board is not to give advice, but to ask questions, raising the level of intelligence. This is also called "quizzics" - the art and science of questioning. Still another role than the traditional board that is more executive, i.e. focusing on decision taking. This might be contra productive as it is giving more attention of the board to present issues than to the futurizing dimensions. The board should be ahead of the execution and look for the opportunity spaces as well as threats. Then we cannot have a team at the top focusing on just staying on board, when we have to change course.

One-way to this is to have also a team of navigators for the longitude representing different age groups. We found in our experience in Skandia and then supported by research that there is a correlation between age and innovations, like illustrated below. The team should combine different age groups to get the intelligence, i.e. both seniors at 55 years of age and older as well as youngster up to 35 years of age.

8. Key Message – Longitude Value Creation

The wave of intellectual capital is increasing. It is evolving within universities, accounting standards groups, political and business communities. The message is that we need to understand and follow the wave of knowledge economics. The alternative is perishing by riding the life cycle curve of industrial economics down. It is a leadership liability not to address the potential or intellectual capital in waiting. A new type of Society entrepreneurship might be the key role to nourish this longitude value! Where do we find and develop the training camps for such kind of leader-ship? Whom would you like to nominate, inside or outside the firm, as one among the top ten intellectual capital entrepreneurs of today – for shaping the longitude value of tomorrow?

9. References

Amidon, D., The Innovation Superhighway, Woburn, Mass. 2003.

Bontis, N., National Intellectual Capital Index – IC Development in the Arab Region (UNDP 2002).

Clark E. et al., IC-report 2000 by HLEG, Brussels, 2000, (www.euintangibles.net).

Danish Ministry of Industry Guidelines for Knowledge Accounts (2001), www.videnskapsministeriet.dk/videnregnskaper or www.vtu.dk/icaccounts.

Davis, S., Lessons from the Future, Oxford 2001.

Edvinsson, L., Corporate Longitude, Stockholm 2002, (www.corporatelongitude.com).

Jay F., World Dynamics, Shelton 1971.

Mouritsen, J., IC and the Capable Firm, Copenhagen Business School 2001.

Nonaka, I., paper presented at Global Knowledge Forum, Tokyo, Oct. 24/25, 2002.

Romer, P., Increasing Returns and New Developments in the Theory of Growth, Cambridge, MA 1991.

Welzl, A., Intellectual Capital Report 2001, *ARC Seibersdorf Research Gmbh*, Austria 2001, (www.arcs.ac.at).

Some links for further reading

www.corporatelongitude.com
www.intellectualcapital.se
www.iccommunity.com
www.oecd.org/publications
www.entovation.com
www.kmcluster.com
www.isa.se
www.bontis.com
www.minez.nl
www.blev.stern.ny
www.ll-a.fr/intangibles/
www.vaic-on.net)
www.vtu.dk/icaccounts
www.monday.dk
www.larvikkommune.no
www.nordicinnovation.net
www.finansanalytiker.no
www.skandiafuturecenter.com
www.mind-lab.org

The Evolution of Measuring Intangible Assets – Past, Present and Future

Andy Neely [1], *Bernard Marr* [2], *Göran Roos* [3], *and Stephen Pike* [4]

[1] Prof. Dr. *Andy Neely*, Centre for Business Performance, Cranfield School of Management, UK

[2] Prof. *Bernard Marr*, Centre for Business Performance, Cranfield School of Management, UK

[3] Prof. *Göran Roos* Intellectual Capital Sevices, Ltd., London, UK

[4] Dr. *Stephen Pike*, Intellectual Capital Services, Ltd., London, UK

In this article we describe the evolution of approaches for measuring intangible assets. This evolution stated with first generation measurement systems, which were based on the assumption that financially biased measurement systems, should be supplemented with non-financial indicators, including intangibles. Whilst this was a valuable development, the problem with these first generation approaches was that they were static and failed to illustrate adequately the linkages between different areas of organizational performance. Second generation measurement systems addressed this issue by using strategy and/or success maps to take into account the dynamic nature of performance and the transformation processes linking objectives and resources. Third generation measurement systems will build on these developments and seek to link explicitly the non-financial and intangible dimensions of business performance to the generation of free cash flow.

1. First Generation: The Static View of Organisational Performance

The 1980s and early 1990s saw an explosion in the number of authors writing about and criticising the performance measurement systems used by organisations. Highly vocal and influential critics, such as *Robert Kaplan* and *Thomas Johnson* (1987), argued that traditional accounting methodologies were fundamentally flawed and that they had lost their relevance, because they were developed for an age when organisations generated value through labour, rather than the application technology or knowledge. Others identified inappropriate measurement methodologies and the short-term behaviours that they encouraged as a primary cause of the United States' economic decline *(Hayes/Abernathy,* 1980*)*. So powerful and influential were these critics that academics and practitioners began in earnest to seek new and better ways of measuring organisational performance. Some tried to improve methods of measuring financial performance by developing and applying concepts such as activity based costing, activity based management, economic profit, free cash flow analysis and shareholder value analysis. Others tried to supplement the traditional financial measures with non-financial (and often intangible) measures (*Starovic/Marr,* 2003). In doing so they ended up developing more rounded and comprehensive measurement frameworks, such as the Balanced Scorecard *(Kaplan/Norton,* 1992*)*, Skandia's Navigator (*Edvinsson/Malone*, 1997), and the Performance Prism (*Neely* et al., 2002*)*.

In effect, what these first generation approaches achieved was to supplement the traditional financial measures with non-financial measures and to provide a framework for prompting people to think through which non-financial measures they should include in their organisation's performance measurement system. The Balanced Scorecard approach suggests supplementing financial performance measures with indicators for customer satisfaction, internal processes, and innovation and learning. Interestingly the more recent frameworks that have been developed, such as the Performance Prism, have adopted a much broader perspective on stakeholders than the earlier ones, such as the Balanced Scorecard and the *Skandia* Navigator. The Performance Prism approach starts with asking who are the key stakeholders of the organizations and what do these key stakeholders want and require from the organisation? Here, stakeholders could include customers, shareholders, suppliers, alliance partners, and others. Whilst this broader perspective better reflects the current challenges facing organisations, it also brings with it an increasingly complex problem – namely how to link together and integrate all of the individual measures.

2. Second Generation: The Dynamic View of Organisational Value Creation

As performance measurement methodologies capable of assisting the decision maker, second generation approaches made a significant step forward as they started to address the dynamics of value creation by investigating transformations of intangible assets as well as the stocks of these resources. Indeed, the emphasis in second generation practices is on the transformations rather than the individual stock measures (*Pike/Roos*, 2001). Examples of second generation measurement approaches include strategy maps developed by *Kaplan* and *Norton* (2000), success and risk maps developed by *Andy Neely* and colleagues (2002), and the IC-Navigator model developed by *Göran Roos* and colleagues (*Roos* et al., 1997; *Chatzkel*, 2002).

Strategy maps are a natural extension to balanced scorecards and enable them to operate as second generation measurement frameworks. Although the strategy map follows the logic of the scorecard, it offers a different visualisation of the four scorecard perspectives. In this way it reflects the assumed causal relationships between the objectives on the scorecard (see *figure 1*). Although strategy maps are relatively easy to produce, they can be constraining if they are bound too closely to the four balanced scorecard perspectives. Most organisations today are more complex than the four perspectives included in the scorecard and executives are required to address the needs of stakeholders other than just customers and shareholders (*Freeman*, 1997).

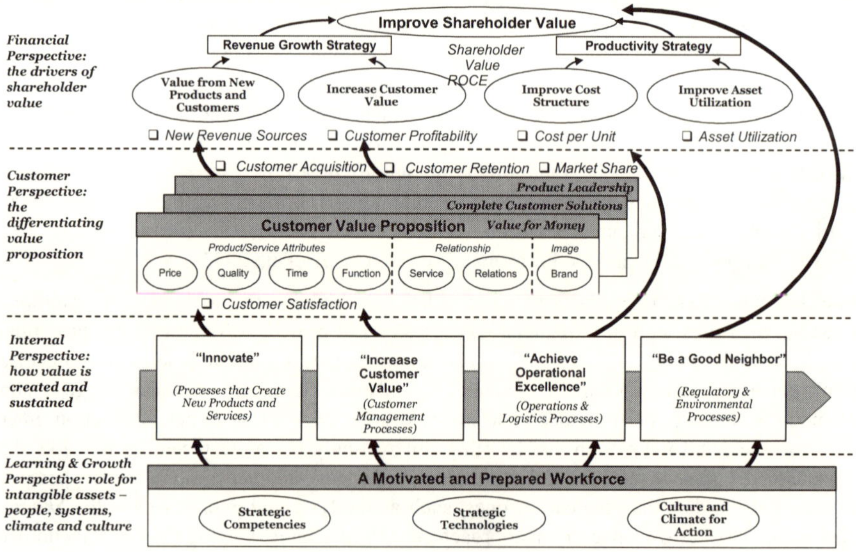

Fig. 1: Strategy Map (Kaplan/Norton, 2000)

To overcome the shortcomings in the balanced scorecard approach, the Performance Prism was developed. The Performance Prism deliberately takes a broader view of stakeholders and encourages organisations to address the following questions:

1. Who are our key stakeholders and what do they want and need?
2. What strategies do we have to put in place to satisfy these needs?
3. What process do we need to have in place to execute our strategy?
4. Which capabilities do we need to perform our processes?
5. What do we expect from our stakeholders in return?

Addressing these five questions allows organisations to build comprehensive success maps, sometimes for each major stakeholder (see *figure 2* for an example of a customer success map). A further refinement suggested by *Neely* et al. (2002) is the notion of failure or risk maps. These identify the potentially critical failure points in the organisation, that if unmonitored could lead to excess exposure to risk. The broader stance adopted by the Performance Prism and its reliance on success and failure maps provides a flexible structure that enables organisations to map everything that is important to them in their success and failure maps.

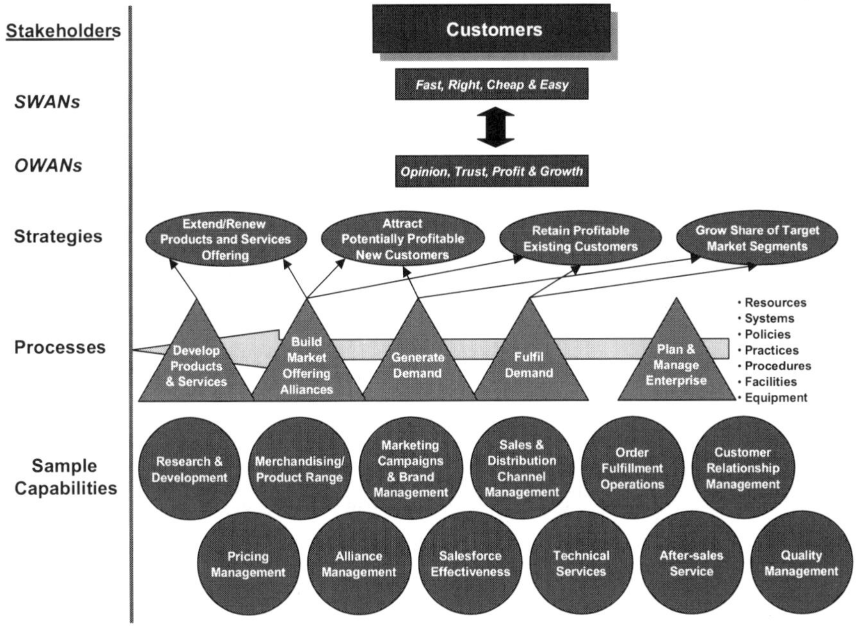

Fig. 2: Success Map (Neely et al., 2002)

Another methodology to unearth the most influential value creating pathways in organisations is the *IC-Navigator* model. As a conceptual map, the Navigator depicts the presence and importance of tangible and intangible resources and the transformations of these resources in accordance with achieving the organisation's strategic intent. In the Navigator (see *figure 3* for an example) the size of the circles represents the stocks of resources ranked according to their relative importance to the strategic objectives and the width of the arrows represents the importance of the transformations from one resource into others, again in accordance with the strategic objectives.

In contrast to first generation approaches, the Navigator has a stronger focus on long-term value creation potential. By focusing on the resource level, increased clarity is given to the processes through which an organisation's resources contribute to the development and deployment of strategic capabilities. With the Navigator, only proxy measures are taken of resources and transformations, hence all metrics are dimensionless, ordinal numbers (*Roos* et al., 1997). Evaluating performance through the Navigator approach brings some structure to the complexities and peculiarities involved with the measurement and management of intangibles. Subsequently, the strength of the Navigator lies in its ability to emphasize subjectivity, by simultaneously offering an objective means of looking at the critical resources, which the firm can leverage in order to define its strategic territory. The Navigator accounts for all the resources that contribute to value in an organisation however unique they may be and omits or discounts those that have little importance in value creation, no matter what emotional value is attributed to them. As a result, the analysis provided by the Navigator does not get bound to a set of connotations that may exist in first generation approaches.

According to *Pike* and *Roos* (2001), second generation performance measurement has `won its spurs' as far as a practical business management methodology is concerned but it has a fundamental weakness that needs to be recognized. The weakness is that whilst the existence of monetary resources is explicitly included in the mapping approaches, there is no ability to link the business-orientated methodology to real free cash flow, which is the current cornerstone of market valuation. In other words, second generation performance measurement makes no attempt to enter the `no- man's- land' between business orientated and financially orientated management and measurement methodologies. The fundamental shortcoming of second generation approaches provides the onset for the development of third generation performance measurement approaches.

3. Third Generation: Linking Value Creation with Free Cash Flow

The third generation of performance measurement requires organisations to seek greater clarity about the linkages between the non-financial and intangible dimensions of organisational performance and the cash flow consequences of these. Before such models can be developed it is essential that three fundamental criteria be satisfied (*Pike/Roos*, 2001).

1. Appropriateness and adequacy – the model must reflect reality.
2. Information adequacy – the right information must be provided.
3. Practicality and organisational alignment – the outcomes must be practical insights that will enable action.

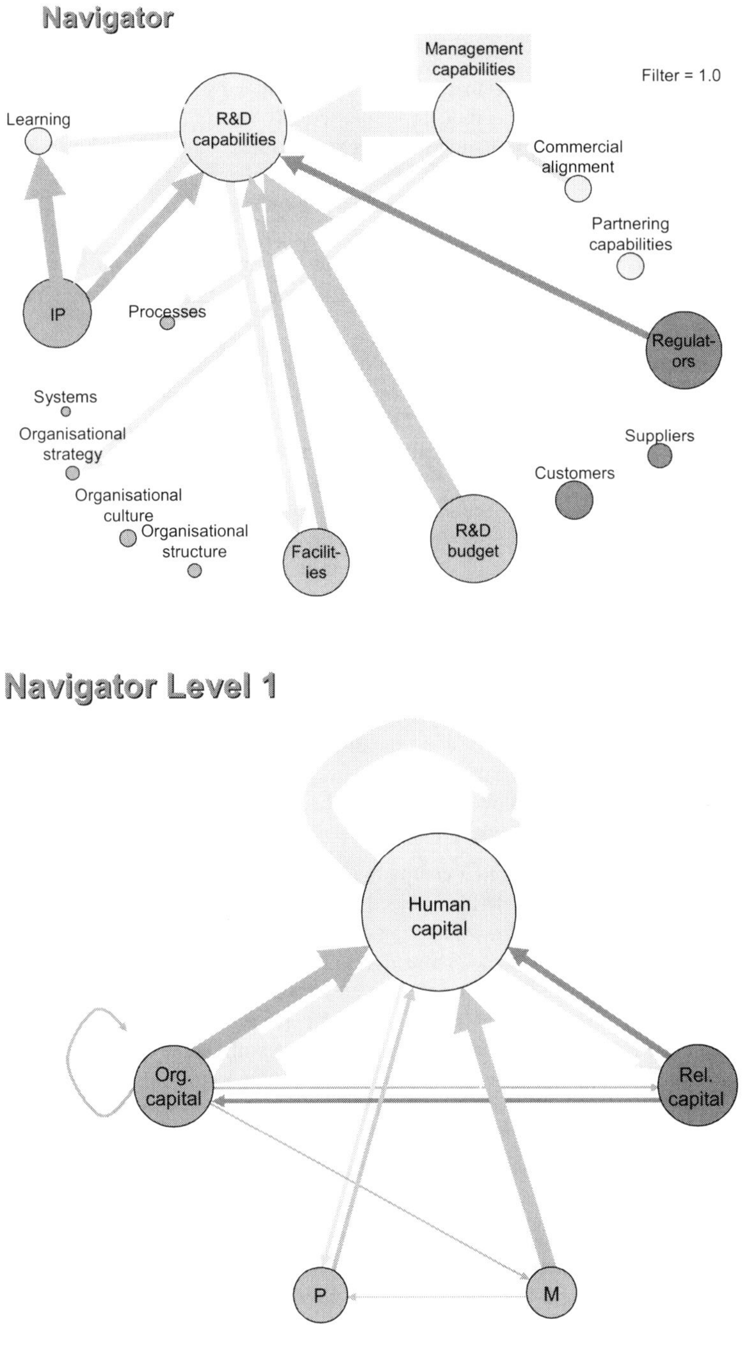

Fig. 3: Navigator Model

3.1 Appropriateness and Adequacy

In the evolution from purely financially biased to first and second generation measurement approaches, the models have gained appropriateness in how they reflect the realities of organisations. Second generation models are particularly valuable in that they allow visualisation of the linkage between intangible assets and business value. The challenge for third generation performance measurement approaches is to maintain the usefulness of the second generation approaches in addressing the key business areas but to do so in a way which extends the measurement to flows of cash. Most organisations today are subject to dynamic environments and subject to constant change. Subsequently, in order to keep any model relevant third generation frameworks must evolve with the change that takes place in organisations.

If measurement tools are to be used to support decision-making then they must give people the confidence that the models used reflect reality. If they do not then managers will revert to `rule-of-thumb' measures, outdated techniques, or simple `gut feel'. There is a clear trade-off between reflecting the complex reality in order to gain confidence in the model and being able to use the tool to manage the business and to take decisions. Somewhere in between the two extremes of reflecting all complexities of reality and reducing the model down to a manageable set of key performance areas lies the `pareto point' which allows for enough complexity in order to have confidence in the model but does not make it too impractical to manage. Taken as a whole, there is a decreasing return on investment in the development of complex models above the pareto point whereas models below the pareto point may be cheap and simple but are effectively useless. Above the `pareto point' there appears to be a point where model complexity matches the degree of inherent uncertainty that exists simply because the world is not structured. Models, which seek to support structured decisions beyond the quality and quantity of the input information, again represent poor investments and are difficult to produce. A solution to this balancing problem is to build a structured model and allow for the adding back of uncertainty to match the quality of the information available for decision-making.

In general, information internal to a company is well structured since managers are in control of the company and as a defined object, it can be measured. Of course there will be uncertainty such as the price of raw materials but uncertainty is different from risk. Information whose origin is external to the company often carries risk and this can be introduced as an external modifier to the decision aid. The advantage of such an approach is that it can introduce rigour to the main part of the internal performance measurement model. This means that there is little need for an exhaustive audit of both model and information. Those wishing to compare companies such as analysts and the investing community need only seek assurance that the (rigorous) model was used. There is evidence from the recently completed work of *CEST* (*CEST*, 2000) that amongst companies, audits of intellectual capital models used in companies to support strategic or operational decisions would be unwelcome. On the other hand, there was no objection to assurance.

One route to gain confidence in such models is to empirically test them. If organisations have sufficient measurement data for each of their performance items then they can start testing their assumptions. *Shell International* recently conducted empirical analysis to verify their Success Maps (*Marr* et al., 2002, 2003). However, to be able to do this organisations need adequate information which brings us to the next challenge.

3.2 Information Adequacy

Looking at corporate disclosure patterns, and with the exception of those organisations involved in delivering public services, the only information that seems to have traditionally been required by stakeholders to be reported in a standardised form has been financial information. Financial stability has long been a key performance indicator but along with that have been the quality, level and extent of the service the organisation provided. In the knowledge era, the concept of stakeholder and shareholder value extends far beyond simple financial performance measures. To communicate with stakeholders, all organisations are now required to deeper understand the attributes of value creation from the stakeholders' point of view. Internally in the company this means that the management's strategy must be more sophisticated and the levers that management pull to improve their performance are much more numerous and complex.

In terms of determining the extent of measurement and the information needed by decision makers, this is the defining stage. Decisions based on the necessary and sufficient levels of information are required. Less than this leads, after the `Pareto point', to unnecessarily unstructured decision-making. More than this leads to wasteful and unnecessary measurement. Key attributes that have to be measured, especially the non-financial attributes, should be broken down only as far as required. Two dangers emerge from over-measurement. The first is that the cost of data collection far outweighs the benefits of having it and that its collection also causes considerable irritation amongst those doing the measurement and those being measured. This is especially the case if the redundancy of the measurement information is obvious. The second danger is that management based on over-measurement leads to justified accusations of micro-management and the tendency to instil unwanted behaviours. People who want to improve performance will be dragged down to focus on many trivial elements in an over-elaborate measurement system. In doing this they lose sight of the bigger and more important picture.

One of the biggest challenges for performance measurement is to realise the difference between data and information. Far too often decision-makers are presented with raw performance data and are expected to analyse and translate the data into information straightaway. Scientists would never make a presentation to an audience without first analysing the data and understanding the messages it contains. Yet far too often performance data is presented without prior analysis. We give people figures on profitability by customer segment, on absenteeism levels, figures on productivity, but nobody in advance has been through the data and extracted the insights from it. *David Coles*, the Managing Director of *DHL* UK, described this as `numerical crosswords'. *DHL*'s board used to spend all of their time at performance reviews trying to join up the pieces of the numerical jigsaw that they were presented with. Individual directors would be looking at performance reports trying to draw spurious correlations between different events to offer explanations for unusual observations. More consideration must be given to presenting information instead of just raw data.

A further major challenge remains the struggle organisations face with measuring intellectual capital resources and transformations. Although second-generation approaches such as Success Maps or the Navigator are already adept at identifying what needs to be measured and elucidating measures, organisations continue to face difficulty in areas of metric quality, and the interpretation and understanding of what they are measuring.

Consideration must also be given to the users of information at lower levels in companies and the effects they have on information as it passes along to those who make strategic decisions. What happens is that as the levels are ascended within the company there is an inevitable blurring as data is interpreted and re-interpreted. Information handling is a continuum problem beginning with those who collect the raw data inside the company and ending with worldwide reporting and disclosure. Information is processed and used at all levels in organisations, any third generation measurement system must be take this into account.

The challenge for third generation approaches is to make information handling for decision-making as transparent as possible by concentrating on communicating information instead of raw data. Organisation must also integrate rigorous measurement of their intangible assets and fully integrate this information into the models. In this way the use of information will turn into a powerful resource itself adding to the competitive advantage of any organisation.

3.3 Practicality and Organisational Alignment

The condition of `practicality' requires that the third generation performance measurement models must be able to inform the operation of the organisation and deliver insights about the real levers of value creation. If performance measurement and management is to be of any real value to organisations then it must help management to translate their strategic intent into appropriate actions and deliver feedback information showing whether these actions are working or not. *The Boston Consulting Group* report has suggested that managers have only three effective levers in the creation of value namely, margin, asset productivity and investment (*BCG*, 1999). However others give more extensive lists adding innovation, attraction of talented people, alliances and customer satisfaction (see for example *Baum* et al., 2000).

In terms of decision-making, one of the major influences that affect managers is the external perception of actions they take, especially from a future cash flow perspective. This concentration on cash-flow, usually calculated using some kind of *NPV* techniques, within a specified and often short time period has tended to mask the importance of the intellectual capital transformation processes which create that cash flow. The flaws of *NPV* have been summarised by *Lewis* and *Lippitt* (1999) who suggest various modifications to overcome those shortcomings. However, the best solution might be to avoid NPV techniques by considering alternative futures as options and hence make this an opportunity to integrate real options theory or scenario planning (*Fink* et al., 2002). The most important impact is in the planning of projects to generate options and also to find ways of realising the value of the options. The challenge for third generation methodologies is to be flexible enough to incorporate these techniques within the model.

Practicality and organisational alignment also means that the measurement system is fully integrated with other organisational processes such as planning, budgeting, reporting, as well as performance reviews. If systems are not aligned and integrated it creates confusion and can jeopardise the entire implementation. Too often organisations put comprehensive performance measurement systems in place but when you attend meetings the only thing people talk about is budgets that are not aligned or liked to the performance measurement system. In the case of *DHL* UK they realised that their performance review meetings were not aligned with their performance measurement system. They decided to change the structure of their board meetings and defined specific questions that they wanted answers to. They now ask their performance analysts to come to the board meeting armed not with raw data or excuses, but instead with presentations which address questions of fundamental concern to the board – e.g. are we going to hit budget this year, how are our customers feeling, how are our employees feeling. The analyst's role at the board meeting is to present their answer to the question. The board's role is to probe the quality of the analysis and once they are comfortable with it decide what they are going to do to move performance in the desired direction. In changing the structure of their board meetings the board of *DHL* UK has been able to eliminate the defensive behaviours associated with performance reviews and encourage the creative dialogue associated with performance planning.

In adopting this new structure and format *DHL* recognised that they had to upgrade the skills of their performance analysts. But at least *DHL* had performance analysts. Many organisations do not and in these cases they need to appoint them. These performance analysts need not only to be able to manipulate performance data, but also interpret it and present it in a way that engages and provides insight to others. Research being undertaken in the *Centre for Business Performance* into this issue has resulted in the development of a concept called *The Performance Planning Value Chain*, which effectively encapsulates a systematic process for extracting insights from performance data. The analogy underpinning *The Performance Planning Value Chain* is that of a journalist. If you think about what a journalist does when presenting a story he or she is very careful to identify the „hook", or headline that will capture the reader's attention and then flush out the detail in the small print. Rarely do we do this with performance reports. Rarely do we ask our performance analysts to tell us the headline. In fact, rarely do we ask our performance analysts to really analyse data. Instead we expect them to spend all of their time pulling, collecting, and collating data.

This issue becomes even more important when the focus of measurement is shifted to systems not functions. The reality of organisations, as every executive knows, is that they consist of complex inter-dependencies. Marketing relies on Operations. Operations rely on Human Resources. Human Resources rely on Finance, etc. Yet when it comes to measurement we often ignore these inter-dependencies. Marketing looks at the marketing and customer satisfaction data. Human resources look at the people data. Operations look at the operational data, etc. It is as if we have functionalised measurement, just as we have functionalised everything else in organisations. Yet the functionalisation of measurement is a mistake. If there is a downturn in employee satisfaction, everyone assumes this will have an adverse impact on customer service. If the operation becomes too inefficient everyone knows this will impact the financial results. The reality of organisations is that the activities being undertaken in different parts of them interact and we have to recognise this interaction if we are to get the most from our measurement data.

The challenge for third generation approaches is to overcome these functional silos. We need to use our measurement data to understand the big picture, the big story of what is happening inside the organisation. And this requires collecting the right data, providing the right information and integrating the performance measurement model into organisational processes so performance analysts can understand the complexities of today's organisations but still gaining valuable insights that inform actions in order to add real value.

4. Conclusion: Towards the Third Generation

To date research in the field of performance measurement has mainly focused on first generation frameworks. Increasing attention is paid to second generation approaches, but to really overcome the measurement challenges approaches need to evolve towards third generation approaches and address the following issues:

1. Models must reflect the static and dynamic realities of organisations but at the same time not lose appropriateness as a managerial tool.
2. We must move from data to information and must provide rigorous information especially for the intangible value drivers in organisations.
3. The models must be practical and aligned with other organisational processes in order to allow actions to be taken.
4. And most fundamentally of all, we must seek increasingly robust ways of demonstrating the cash flow implications of the non-financial and intangible organisational value drivers.

Guiding a business towards strategic goals requires third generation performance measurement approaches that are soundly based and can reliably measure progress in order for organisations to gain real value from measuring business performance, be it tangible or intangible.

5. References

Baum, G., Ittner, C., Larcker, D., Low, J., Siedfield, T. and *Malone, M.*, Introducing the new value creation index, Forbes ASAP, 2000, pp. 140–143.

Boston Consulting Group, The value creators: A study of the world's top performers. http://www.bcg.com, 1999.

Carlucci, D., Marr, B. and *Schiuma*, G., The Knowledge Value Chain – How Knowledge Management Impacts Business Performance, in: International Journal of Innovation and Learning, Vol. 1 (2003), No. 3.

Chatzkel J., A Conversation with *Göran Roos*, in: Journal of Intellectual Capital, Vol. 3 (2002), No. 2, pp. 97–113.

Edvinsson, L. and *Malone, M. S.*, Intellectual Capital: The Proven Way to Establish Your Company's Real Value by Measuring Its Hidden Values, London 1997.

Fink, A., Siebe, A. and *Marr, B.*, The Future Scorecard: The Benefits of a Combination of Future Scenarios with Performance Measurement, in Proceedings: PMA 2002 – Research and Action, Boston 2002.

Freeman, R.E., A Stakeholder Theory of the Corporation, in: *Bowie, N. E.* and *Beauchamp, T. L.* (Eds.), Ethical Theory and Business, Englewood Cliffs 1997.

Hayes, R. H. and *Abernathy, W.J.*, Managing Our Way to Economic Decline, in: Harvard Business Review, July-August, 1980, No. 7/8, pp. 67–77.

Johnson, H. T. and *Kaplan, R. S.*, Relevance Lost – The Rise and Fall of Management Accounting, Boston 1987.

Kaplan, R. S. and *Norton, D. P.*, The Balanced Scorecard – Measures that drive performance, in: Harvard Business Review, Vol. 70 (1992), No. 1, pp. 71-79.

Kaplan, R. S. and *Norton, D. P.*, The Balanced Scorecard – Translating Strategy into Action, Boston 1996.

Kaplan, R. S. and *Norton, D. P.*, The Strategy Focused Organization: How Balanced Scorecard Companies Thrive in the New Business Environment, Boston 2000.

Lewis, E. and *Lippitt, J.*, Valuing intellectual assets, in: Journal of Legal Economics, Vol. 9 (1999), No. 1, pp. 31–48.

Marr, B. and *Schiuma, G.*, Measuring and Managing Intellectual Capital and Knowledge Assets in New Economy Organisations, in *Bourne, M.* (ed.), Handbook of Performance Measurement, London 2001

Marr, B. and *Schiuma, G.*, Business Performance Measurement – Past, Present, and Future, in: Management Decision, Vol. 41 (2003), No. 10.

Marr, B., Gupta, O., Roos, G. and *Pike, S.*, Intellectual Capital and Knowledge Management Effectiveness, in: Management Decision, Vol. 41 (2003), No. 10.

Marr, B., Gray, D. and *Neely, A.*, Why do firms measure Intellectual Capital, in: Journal of Intellectual Capital, Vol. 4 (2003), No. 4.

Marr, B. and *Neely, A.*, The Balanced Scorecard Software Report, Stanford, USA, 2002.

Marr, B., Neely, A. and *Thomas, G.*, Balanced Scorecard and Strategy Maps: How Intangibles Drive Corporate Performance at Shell International, Research and Action, Boston 2002.

Marr, B. and *Schiuma, G.*, Intangible Assets – Defining Key Performance Indicators for Organisational Knowledge Assets, in: Business Process Management Journal, Special Issue, Vol. 10 (2003), No. 4.

Neely. A., Bourne, M., Jarrar, Y., Kennerley, M., Marr, B., Schiuma, G., Walters,A. H., Sutcliff, M., Heyns, H., Reilly, S. and *Smythe, S.*, Delivering Value Through Strategic Planning and Budgeting, London 2001.

Neely, A., Adams, C. and *Kennerley, M.*, The Performance Prism: The Scorecard for Measuring and Managing Business Success, London 2002.

Neely, A., Marr, B., Roos, G., Pike, S. and *Gupta, O.*, Towards Third Generation Performance Measurement, in: Controlling, Vol. 15 (2003), No. 3/4, pp 129-135.

Pike, S., Rylander, A. and *Roos, G.*, Intellectual Capital Management and Disclosure, in: *Bontis, N.* and *Choo, C. W.* (Eds.), The Strategic Management of Intellectual Capital and Organizational Knowledge – A Selection of Readings, New York 2002.

Pike, S. and *Roos, G.*, Measuring and decision support in the knowledge society; The 4th World Congress on Intellectual Capital, Hamilton 2001.

Roos, J., Roos, G., Dragonetti, N. C. and *Edvinsson, L.*, Intellectual Capital: Navigating the new business landscape, London 1997.

SEC, Regulation FD, see http://www.sec.gov, 2000.

Starovic, D. and *Marr, B.*, Understanding Corporate Value: Managing and Reporting Intellectual Capital, CIMA, London, 2003.

Transparenzproblem Intangible Assets: Intellectual Capital Statements und der Neuentwurf eines Frameworks für Unternehmenssteuerung und externes Reporting

Jürgen H. Daum[1]

---

[1] *Jürgen H. Daum*, SAP AG Walldorf

Durch die zunehmende Dominanz der Intangible Assets werden Unternehmen zur „Black Box". Denn die Informationen des klassischen Rechnungswesens geben nur sehr partiell Auskunft über das Zukunftspotenzial und Risikoprofil eines Unternehmens und versagen immer mehr auch als alleinige Basis für die interne Steuerung, wenn die wahren Erfolgsfaktoren heutiger Unternehmen eher in ihrem immateriellen Wissens- und Beziehungskapital zu suchen sind als in ihrem Finanzkapital und den Sachanlagen. Vielversprechender erscheinen Ansätze in Dänemark, wo mittels Intellectual Capital Statements die Entwicklung der immateriellen Kompetenzplattform eines Unternehmens transparent gemacht werden soll. Das Konzept der Intellectual Capital Statements könnte Ausgangspunkt für die Lösung der heutigen internen und externen Transparenzprobleme in der Unternehmenswelt sein, die sich aus der zunehmenden Dominanz der Intangible Assets ergeben. Es ergänzt die Finanzperspektive, die das traditionelle Rechnungswesen bereitstellt, um eine weitere Perspektive, die ein Sicht auf die Wissensressourcen und auf die immaterielle Kompetenzplattform eines Unternehmens erlaubt, die den Unternehmenserfolg in der Zukunft begründen. Gelingt es beide Perspektiven, die Finanzperspektive und die Wissenskapitalperspektive mit der Steuerungsperspektive, also dem Unternehmenssteuerungssystem, zu einem umfassenden Framework zu verknüpfen, wird der Weg frei, um den Zusammenhang zwischen immateriellen Potenzialaufbau, Verwertung im Rahmen des operativen Wertschöpfungssystems und der Finanzperformance erkennen und optimieren zu können. In diesem Beitrag wird sowohl das dänische Konzept der Intellectual Capital Statements als auch der Entwurf eines umfassenden - das Intellectual Capital Statements Konzept integrierenden - Frameworks für Unternehmenssteuerung und externes Reporting vorgestellt, das Unternehmen sowohl für das Management als auch für Unternehmensexterne wieder transparent machen soll.

1. Die Problemstellung: Durch die heutige Dominanz der Intangible Assets werden Unternehmen zur „Black Box"

Immaterielle Vermögenswerte, auch Intangible Assets genannt, haben in den letzten Jahrzehnten deutlich an Bedeutung gewonnen. Zunehmend basieren effektive wirtschaftliche Wertschöpfungsprozesse auf Wissen, auf produktiven Beziehungen zu Geschäftspartnern, auf dem Bekanntheitsgrad des Firmennamens, der Produkte und Marken sowie vor allem auf Innovationskraft und neuen Technologien. Intangible Assets begründen damit heute die Wettbewerbsfähigkeit von Unternehmen und Volkswirtschaften: produktive Kopfarbeit schlägt produktive manuelle Arbeit. Bei Standortentscheidungen für ein Hochlohnland wie Deutschland spielen diese Faktoren inzwischen die entscheidende Rolle. So baut beispielsweise *General Electric* sein europäisches Forschungszentrum im teuren München, da dem Unternehmen und seinem Konzernchef *Jeffrey Immelt* Technologie und die Nähe zu den entsprechenden Universitäten wichtiger sind als billige Arbeitskräfte. *BMW* baut das Werk für den geplanten 1er *BWW* bei Leipzig und nicht in Tschechien. Die Facharbeiterlöhne sind in Tschechien zwar deutlich niedriger, aber es mangelt dort an qualifizierten Managern und

Spezialisten, etwa für die Qualitätssicherung, die für BMW wichtiger sind, als billige Facharbeiter (vgl. *Lamparter, 2003*).

Denn Mehrwert, den man finanzwirtschaftlich als eine Rendite definieren kann, die über den Kapitalkosten liegt, lässt sich heute nur noch über laufende Innovation und den Aufbau guter Beziehungen zu Kunden, Partnern und wichtigen Meinungsbildnern schaffen – also durch Investionen in Intangible Assets, vor allem in Wissens- und Beziehungskapital. Sie erwirtschaften heute auch in traditionellen Branchen, wie z.B. in der chemischen Industrie, Renditen, die weit über den Kapitalkosten und denen von Investitionen in Sachanlagen liegen (vgl. *Aboody/Lev*, 2001, S. 18-21). So ist es auch nicht verwunderlich, dass Unternehmen aller Branchen heute mehr in Intangible Assets investieren als je zuvor – mit weiter steigender Tendenz. In den OECD-Ländern sind Investitionen in die Produktion von Wissen, wie zum Beispiel in Forschung und Entwicklung, bereits 1999 mit dem Volumen der Investionen in Sachanlagen gleichgezogen (vgl. *OECD*, 1999, S. 2). Seit dem wird in diesen Volkswirtschaften mehr in Intangibles investiert als in Sachanlagen. Auch Unternehmen traditioneller Branchen verfügen deshalb heute oft über signifikante Intangible Assets und übertreffen dabei teilweise noch die typischen wissensintensiven Branchen wie die Softwareindustrie (vgl. *Gu/Lev*, 2001, S. 12), ohne dass dies allerdings aus ihren Bilanzen ersichtlich wird.

Unsere Management-, Rechnungslegungs-, und Controllinginstrumente haben jedoch mit dieser Entwicklung nicht Schritt gehalten. Sie stammen aus einem anderen Wirtschaftszeitalter und orientieren sich immer noch an den industriellen Wertschöpfungssystemen der Vergangenheit und deren Produktivfaktoren: an Kapital und manuelle Arbeit. Sie liefern deshalb eine viel zu begrenzte Sicht und klammern die wichtigsten Produktivfaktoren unserer zunehmend wissens- und dienstleistungsorientierten Unternehmen und Volkswirtschaften aus – eben die Intangbile Assets: nichtmonetäre, nichtmaterielle, »ungreifbare« Produktivfaktoren und die ihnen innewohnenden Produktivkräfte und Risiken. Dies zeigt auch die wachsende Lücke zwischen dem Marktwert eines Unternehmens und dem, was im traditionellen Rechnungswesen ausgewiesen wird: in den letzten Jahrzehnte hat sich der Anteil der nicht in der Bilanz ausgewiesenen Intangible Assets am Marktwert von Unternehmen kontinuierlich vergrößert. Für S&P500 Unternehmen etwa ist dieser in den letzten zwanzig Jahren im Durchschnitt von einem Anteil am Marktwert von 38% (1982) auf über 80% (2002) angewachsen.

Wenn aber immer mehr Werttreiber durch die klassischen Management- und Bewertungsinstrumente, also durch Bilanz und Gewinn- und Verlustrechnung, nicht mehr erfasst werden, ergibt sich als Konsequenz, dass Unternehmen allein auf der Basis dieser Informationen nicht mehr steuerbar oder extern zu beurteilen sind. Unternehmen werden immer undurchschaubarer, ihre Produktiv- und Risikofaktoren und die wirtschaftliche Leistungsfähigkeit ihrer Geschäftssysteme werden immer schwieriger fassbar – sie werden zur „Black Box“. Für Manager, Investoren und Ökonomen ist diese Situation alarmierend. Da ein großer Teil der Aktivitäten, mit denen sie Mehrwert für ihre Aktionäre und Stakeholder schaffen, im Vergleich mit der Situation im Industriezeitaltern nicht mehr systematisch erfasst wird, bleiben diese Aktivitäten und vor allem ihre Ergebnisse sowohl intern als auch extern intransparent. Dieses Informationsdefizit führt zu einer erhöhten Volatilität der Aktienkurse, was sowohl Verluste bei Investoren als auch Fehlallokation von Ressourcen auf den Kapitalmärkten und damit makroökonomische Ineffizienzen verursacht. Als direkte Folge sind Intangibles-intensive Unternehmen mit hohen Kapitalkosten konfrontiert, was ihre Investitionstätigkeit und damit ihr Wachstum behindert. Intern erhöht sich durch das Fehlen verlässlicher Informationen über den größten Teil der Unternehmens-Aktiva das Risiko von Fehlentscheidungen des Managements und die Wahrscheinlichkeit, dass Optionen für Wachstums und Produktivitätsverbesserungen nicht erkannt und genutzt werden. Wir laufen so Gefahr, dass Unternehmen nicht in der Lage sind, ihr volles wirtschaftliches Potenzial zu entfalten, um Mehrwert und Wohlstand zu schaffen.

Eine intensivere Auseinandersetzung mit der Behandlung dieser immateriellen Wertreiber in Rechnungswesen, Controlling und Management als auch in der externen Berichterstattung ist im Ausland, vor allem in den angelsächsischen und skandinavischen Ländern bereits seit Anfang der 1990er Jahre im Gang. Lange hatte dies keine konkreten Auswirkungen auf die Management- und Rechnungslegungspraxis in Unternehmen. Gerade im Bereich der Rechnungslegung ist hier aber in jüngster Zeit einiges in Bewegung geraten – eine Entwicklung, die nun das Thema Intangible Assets im Rahmen diverser Novellen internationaler Rechnungslegungsstandards (US GAAP bzw. IAS/IFRS) in den Aufmerksamkeitsfokus einer breiteren Gruppe an wichtigen und mächtigen internen und externen Unternehmensstakeholdern rückt, die u.a. die Finanzchefs, Wirtschaftsprüfer, Finanzanalysten und das übrige Top-Management der in Europa oder den USA börsennotierten Unternehmen umfasst.

Auslöser für diese Entwicklung waren die aus der Übernahmewelle der 1980er und 1990er Jahre resultierenden hohen Goodwill-Beträge in den Bilanzen der Unternehmen. Denn im Rahmen der Kaufpreisfindung kam es de facto zu einer Aufdeckung und Bewertung der seit Anfang/Mitte der 1980er Jahre stark anwachsenden Werte der Intangible Assets der übernommenen Unternehmen. Die Differenz zum Buchwert, der Goodwill, addierte sich mit der in der Bilanz des übernehmenden Unternehmens aus früheren Übernahmen bereits vorhandenen Goodwillposition zu teilweise enormen Beträgen auf, die an sich nichts mehr aussagen. Ein immer größerer nicht unbeträchtlicher Teil der Aktivseite der Bilanz blieb damit völlig intransparent und entzog sich jeglicher Beurteilung durch Unternehmensexterne, wie etwa Finanzanalysten (und wahrscheinlich auch des Managements).

Um die im Goodwill aggregierten Intangibles wieder transparent zu machen, wurden 2001 die entsprechenden neuen US GAAP Regeln erlassen. Gemäß SFAS 141 ist bei einem Unternehmenserwerb die „Pooling of Interest"-Methode nun verboten - der bezahlte Goodwill muss also in jedem Fall aufgedeckt werden. Auch darf der in der Bilanz des Übernehmers ausgewiesenen Goodwill künftig nicht mehr planmäßig abgeschrieben werden, sondern muss in einzeln identifizierbare Intangible Assets zerlegt und diese auf die „Cash"- generierenden Unternehmenseinheiten heruntergebrochen werden. Auf dieser relativ detaillierten Ebene müssen die ausgewiesenen Werte dann mindestens einmal jährlich einem so genannten *Impairment-Test*, also einer Fair-Market-Bewertung, unterzogen werden (SFAS 142). Liegt ein Impairment vor, liegt also der Marktwert unter dem Buchwert, muss die Differenz ergebniswirksam abgeschrieben werden. Die genannten US GAAP-Regeln sollen im Rahmen des so genannten Business Combinations Project in ähnlicher Weise in den Neuentwurf der IAS/IFRS (International Financial Reporting Standards) übernommen werden, die ab 2005 für den Konzernabschluss börsennotierter Aktiengesellschaften in Europa gelten werden.

Mit den neuen internationalen Rechnungslegungsstandards erhofft man sich eine höhere Transparenz bei den Intangibles – sowohl intern, für das Management, also auch extern für Finanzanalysten und Investoren. Auf jeden Fall werden die betroffenen Unternehmen künftig höhere Schwankungen beim Ergebnis als auch beim ausgewiesenen Eigenkapital ausweisen. Ob dadurch allerdings die Transparenz tatsächlich erhöht wird, kann bezweifelt werden, solange die Unternehmen nicht verpflichtet werden, umfassende Informationen zur Werthaltigkeit der entsprechenden Intangibles strukturiert und standardisiert im Rahmen eines Zusatzreportings, zusätzlich zum rein finanzorientierten Jahresabschluss zu veröffentlichen. Eine so erweiterte externe Veröffentlichungspflicht würde nicht zu letzt auch eine ausreichende Motivation schaffen, um Initiativen in den Unternehmen zu fördern, die die Entwicklung von entsprechenden internen Steuerungsinstrumenten, die Intangible Assets mit ein beziehend, voran treiben, um damit auch die Steuerungsfähigkeit des Managements und die gesamte Corporate Governance eines Unternehmens zu verbessern.

Zwar werden bereits heute bei Unternehmen, die nach US GAAP berichten müssen, die Geschäftspläne, die oft die einzige Basis für die für den Impairmenttest erforderliche Bewertung bilden, in die Prüfung durch den Wirtschaftsprüfer miteinbezogen. Diese Planungen und die darauf aufbauende, meist auf Basis der Discounted Cashflow Methode erstellten, Bewertungen sind in den meisten Fällen jedoch zu grob, um gesicherte Aussagen über die Werthaltigkeit zuzulassen bzw. die Willkürlichkeit bei der Bewertung in ausreichendem Maß zu verhindern. Ohne die Existenz verfeinerter Planungs- und Steuerungsinstrumente, die die ausgewiesenen Intangibles in den Zusammenhang mit anderen internen und externen Einfluss- und Produktivfaktoren stellen, sind die Planzahlen bzw. die Bewertungen weder von Unternehmensexternen in ausreichendem Maß interpretier- und beurteilbar, noch vom Management in Entscheidungsprozessen nutzbar. Denn gerade für Intangible Assets gilt, dass sie keinen Wert per se generieren, sondern ihre Werthaltigkeit immer nur im Zusammenhang mit anderen Produktivfaktoren entfalten. bzw. systematisch nur im Rahmen des individuellen Zusammenhangs im Wertschöpfungssystems des Unternehmens durch Management-Initiativen beeinflussbar und Management-Entscheidungen zugänglich sind.

Um zu einer zufriedenstellenden Lösung des Rechnungslegungs-, Transparenz- aber auch Managementproblems Intangible Assets zu kommen, muss deshalb das traditionelle Finanzreporting erweitert und um ein Zusatzreporting ergänzt werden, das die Intangibles des Unternehmens im Zusammenhang mit dem gesamten Unternehmenswertschöpfungssystem darstellt. Darüber hinaus muss das Unternehmenssteuerungssystem um Instrumente erweitert werden, dass Intangible Assets Management-Entscheidungen und -Initiativen zugänglich macht.

Die Grundzüge für einen solchen Neuentwurf von Unternehmenssteuerungssystem und externem Reporting sollen in diesem Beitrag ausgehend vom Beispiel des dänischen Konzepts der Intellectual Capital Statements dargestellt und diskutiert werden. Zuvor sind jedoch einige grundsätzliche Überlegungen erforderlich. Denn ein Neuentwurf von Unternehmenssteuerungssystem und externem Reporting, der Intangible Assets greifbar macht, muss zunächst mit dem Neuentwurf des Unternehmensmodells starten, das zumindest implizit jedem betriebswirtschaftlichen „Repräsentationssystem", wie sie Rechnungswesen, Controlling und Unternehmensreporting darstellen, zugrunde liegt.

2. Die Aufgabe: Umfassender Neuentwurf von Rechnungswesen, Unternehmenssteuerungssystem und externem Reporting

Die Metapher der „Black Box" wurde im Zusammenhang mit der wirtschaftlichen internen und externen Beurteilung von Unternehmen bereits erwähnt. Diese gilt prinzipiell: Außenstehenden als auch dem Management selbst erscheint ein Unternehmen aus ökonomischer Sicht zunächst einmal als „Black Box", denn man kann ohne geeignete Hilfsmittel und Instrumente die Wirtschaftlichkeit und Wirtschaftskraft eines Unternehmens durch die bloße

Beobachtung der operativen Unternehmensaktivitäten nicht beurteilen. Um derartige Instrumente entwerfen und richtig einsetzen zu können, bedarf es jedoch eines impliziten oder expliziten Modells, das von den operativen Aktivitäten eines Unternehmens abstrahiert und diese in eine betriebswirtschaftliche Sicht projiziert, die die dem System „Unternehmen" innewohnende wirtschaftliche Logik - die „Business Economics" - transparent macht. Nur so kann Licht in die „Black Box" Unternehmen gebracht werden, so dass das Management bei der Steuerung des wirtschaftlichen Ergebnisses weiß, wo anzusetzen ist, und Investoren fundierte Investitionsentscheidungen treffen können. Betrachten wir einmal den letzten Fall, den eines Investors:

Einem Investor etwa stellt sich eine Aktiengesellschaft als eine Einheit dar, deren Zweck das Erzielen einer möglichst hohen Rendite auf das eingesetzte Kapital ist, die sich für ihn in möglichst hohen Dividendenzahlungen und Wertzuwächsen der Aktie realisieren soll. Bevor er in die entsprechende Aktie investiert, wird er deshalb abzuschätzen versuchen, über welche Wirtschaftskraft, also über welche Fähigkeiten ein konkretes Unternehmen im Vergleich mit anderen verfügt, die heute getätigten Investitionen in einen positiven und möglichst hohen „Return" in Form von Dividendenzahlungen und Aktienkurssteigerungen in der Zukunft zu transformieren. Dabei entsteht für ihn ein Transparenz- und Informationsproblem: er kann nämlich in die „Black Box" Unternehmen nicht hineinsehen (vgl. *Abb. 1*).

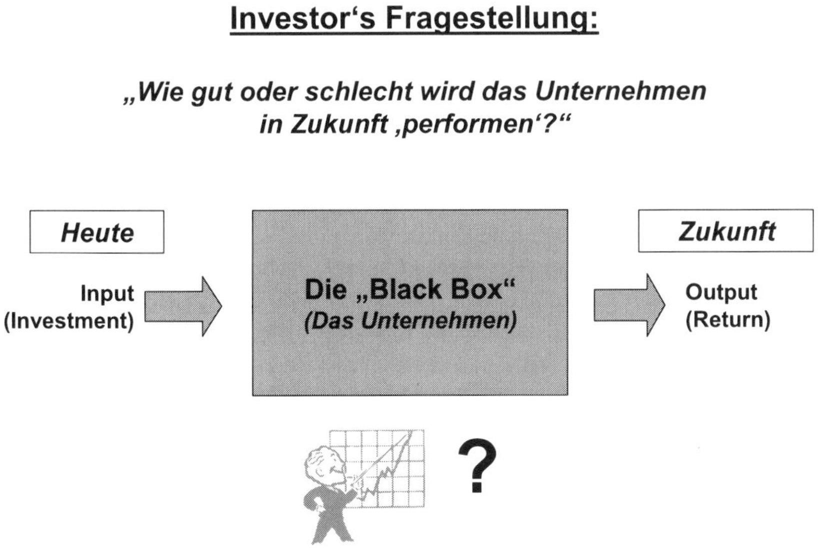

Abb. 1: Das Unternehmen als „Black Box" aus Sicht eines potenziellen Investors

Die durch das traditionelle Rechnungswesen zu gewinnenden Informationen waren in der Vergangenheit ausreichend, um dieses Problem zu lösen. Die Gewinn- und Verlustrechnung gab Auskunft über die Kosteneffizienz mittels des Ausweises von Herstellkosten, Marge und Gemeinkosten, wobei die Kosten als die wesentlichen Treiber operativer Produktivität im fertigenden Industrieunternehmen angesehen wurden. Diese Art der periodischen Erfolgsrechnung ließ auch aufgrund der, verglichen mit der heutigen Situation, relativ stabilen Marktverhältnissen bzw. den nur langsamen Veränderungen, einigermaßen gesicherte

Schlüsse auf die Zukunft zu. Kombiniert mit der Bilanz erhielt man zusätzlich Auskunft über die Effektivität der eingesetzten Sachanlagen und des Finanzkapitals. All dies erlaubte es einem Investor, mit relativ großer Sicherheit Rückschlüsse auf die künftige Performance des Unternehmens zu ziehen.

Ganz anders die Situation heute: Das aktuelle Ergebnis sagt immer weniger über die zukünftige Situation des Unternehmens aus, die Ermittlung eines korrekten Periodenergebnisses wird immer schwieriger, wie die zunehmend von Unternehmen ausgewiesenen „Pro Form Earnings" und die damit verbundene Diskussion unter Rechnungslegungsexperten zeigt. Die Bilanz belegt im Durchschnitt nur noch einen Bruchteil des tatsächlichen Unternehmenswertes, wie die, trotz der daniederliegenden Börsen, immer noch großen Market-to-Book-Ratios belegen.

2.1 Die Transformation der industriellen Wirtschaft zur Wissensökonomie

Der Grund für das Versagen der bislang üblichen Rechnungslegung liegt darin, dass das dem traditionellen Rechnungslegungsansatz zugrunde liegende betriebswirtschaftliche Modell, nämlich das des Industrieunternehmens mit dem Fokus auf effizienter Massenfertigung, nicht mehr trägt. Spätestens seit den 1980er Jahren befinden sich die entwickelten Volkswirtschaften in einem sich immer mehr beschleunigenden umfassenden Transformationsprozess, der die Wertschöpfungssysteme ihrer Unternehmen fundamental verändert. Die Entwicklung geht weg vom Unternehmensmodell der massenfertigenden, effizienten und produktorientierten „Maschine", die das industrielle Zeitalter geprägt hat und hin zu einem kunden- und serviceorientierten Unternehmensmodell, das von Flexibilität und hoher Anpassungsfähigkeit auf externe Einflüsse und Bedarfsveränderungen geprägt ist und in dem die Beziehungen zu den Stakeholdern den Charakter von „Assets" annehmen.

Diese Entwicklung ist aber durchaus nichts grundsätzlich Neues. Jede wirtschaftliche Epoche schaffte ihre eigenen Wertschöpfungsstrukturen in Form von Unternehmensmodellen, die jeweils eigener Organisations- und Steuerungsprinzipien und -instrumenten bedarf:

Ende des 15. Jahrhunderts etwa erfand der sich ausbreitende internationale Handel die doppelte Buchführung, um die entscheidenden wertschaffenden Transaktionen, nämlich Einkäufe (möglichst preisgünstig) und Verkäufe (möglichst teuer) systematisch und nachvollziehbar in „Soll" und „Haben" zu erfassen. Die neue Buchführungstechnik, die sich im 16. Jahrhundert dann rasch verbreitete, schuf erstmals die Grundlage für die Steuerung von Handelsimperien, wie desjenigen von *Jakob Fugger*, der über ganz Europa verteilt Handelsniederlassungen, die so genannten Faktoreien, betrieb. Über deren Geschäftstätigkeit war *Fugger* in der Zentrale in Augsburg immer aufs beste informiert, da die Faktoreien alle paar Wochen Auszüge ihrer Buchhaltung, die um Berichte über das lokale Marktumfeld ergänzt wurden, nach Augsburg senden mussten.

In der zweiten Hälfte des 19. und Anfang des 20. Jahrhunderts brachte die Industrialisierung das Industrieunternehmen hervor und mit ihm das Wertschöpfungsmodell der industriellen Massenfertigung sowie seine spezifischen Rechnungswesen- und Managementtechniken wie etwa die industrielle Kosten- und Leistungsrechnung und das General-Management-Konzepts, dessen Grundlagen *Taylor* Ende des 19. /Anfang des 20. Jahrhunderts mit seinen Arbeitsstudien geschaffen und das dann bei Unternehmen wie *Dupont* und *General Motors* in Richtung einer umfassenden „Financial Control" weiter entwickelt wurde.

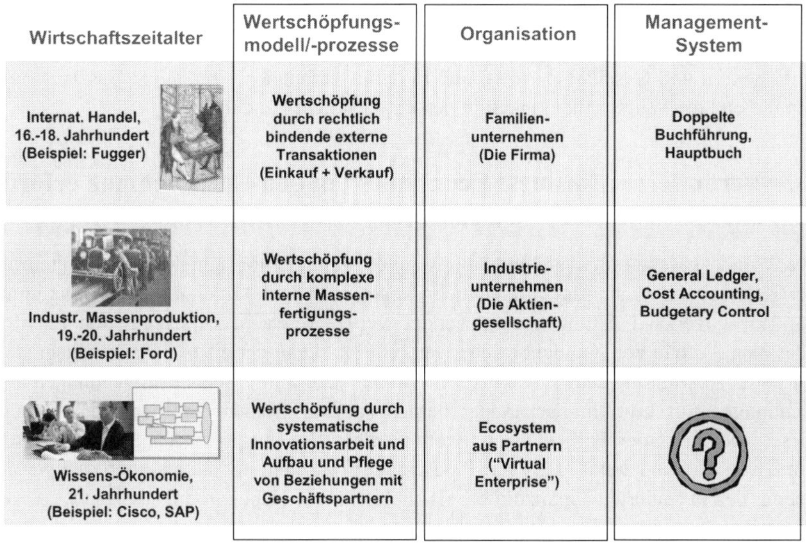

Abb. 2: Die Evolution von unternehmerischen Wertschöpfungsmodellen, Organisationsformen und Managementwerkzeugen von Wirtschaftsepoche zu Wirtschaftsepoche

Der Beginn jeder neuen Epoche war mit großen wirtschaftlichen und sozialen Umwälzungen verbunden. Auch heute befinden wir uns wieder mitten in einer ähnlichen großen Transformation unserer Wirtschaft und Gesellschaft. Die Verkäufermärkte, die durch einen Mangel an Produkten und Servicequalität geprägt waren, haben sich zu Käufermärkten gewandelt, deren wesentliche Eigenschaften der intensive Wettbewerb der Anbieter um die Kunden, das Überangebot an Waren und Dienstleistungen und die Notwendigkeit zur Differenzierung des Angebots sind. Innovation wird damit zum entscheidenden Wettbewerbsfaktor und Intellektuelles Kapital zur kritischen Ressource. Als Konsequenz hat sich über die letzten Jahrzehnte die Wertschöpfungsbasis von Unternehmen aber auch von Volkswirtschaften kontinuierlich weg von den traditionellen Produktivfaktoren Finanzkapital und manueller Arbeit, hin zu einer Dominanz der immateriellen Produktivfaktoren entwickelt (vg. *Abb. 2*).

Auch in traditionellen Branchen der Fertigungsindustrie hat sich der Anteil an Wissenskapital und damit an Intangible Assets signifikant erhöht. Sie fließen in die Entwicklung und den Designprozess neuer Produkte ein und bilden die Grundlage für moderne, wettbewerbsfähige Fertigungs- und Supply-Chain-Prozesse. Gleiches gilt für die Service- und Kundenbeziehungsmanagement-Prozesse, die oft entscheidend zum Unternehmensergebnis beitragen – und die gerade auch für Fertigungsunternehmen wichtiger werden.

Gleichzeitig sind eine Reihe neuer Branchen entstanden bzw. sind zu höherer volkswirtschaftlicher Bedeutung gelangt, weil ein immer größerer Anteil des Bruttosozialproduktes von ihnen erwirtschaftet wird. Dies betrifft Branchen, die fast ausschließlich auf Wissen und Intellektuellem Kapital basieren, wie etwa die Software-, Medien- oder Pharmabranche. In der Konsequenz bedeutet dies, dass im Gegensatz zur Blütezeit des Industrialismus, die entwickelten Volkswirtschaften heute zum größten Teil ihren Reichtum und ihren Wohlstand dem Wissenskapital zu verdanken haben, das sich in den Köpfen ihrer Bevölkerung bzw. in ihren Unternehmen, Forschungs- und Bildungseinrichtungen angesammelt hat und weniger dem vorhandenen Sachkapital, etwa Fabrikanlagen. Vor allem Investitionen in Innovation und in den Aufbau von einzigartigen und dauerhaften Beziehungen zu Kunden, Mitarbeitern und Geschäftspartnern und in damit verbundene Intangible Assets dominieren zunehmend die Wirtschaftstätigkeit in den entwickelten Ländern.

2.2 Veränderte „Business Economics" in den Unternehmen erfordern ein neues betriebswirtschaftliches Unternehmensmodell

Wachsende Investitionen in Innovations- oder Veränderungsfähigkeiten bedeuten aber in der Konsequenz auch, dass immer mehr Ressourcen und Mittel für die Vorbereitung der kundenbezogenen Leistungen aufgewendet werden (etwa in Forschung und Entwicklung oder zum Aufbau von Kundenbeziehungen, von Marken oder eines Bekanntheitsgrades), als für deren Ausführung selbst. Als Ergebnis stellen sich immer mehr Aufwendungen als Aufwendungen mit Investitionscharakter heraus und verfälschen die Erfolgsrechnung, wenn diese, wie meist der Fall, nicht aktiviert werden. Die entsprechenden neuen „Assets" in Form von Humankapital, der F&E-Pipeline, von Marken etc. stellen aber weniger ein Ergebnis der investierten Finanzmittel, als vielmehr ein Nebenprodukt von gut gemanagten operativen Aktivitäten dar. Das Operationsmodell heutiger Unternehmen hat sich damit, im Gegensatz zum industriellen Modell, das eine klare Trennung zwischen kurzfristig orientiertem Tagesgeschäft und langfristiger (stabiler) Assetbasis aufweist, zu einem komplexen Operationsmodell gewandelt, wo die Grenzen zwischen Potenzialaufbau und Verwertung verschwimmen..

Das im Vergleich zum industriellen Modell komplexere Wertschöpfungssystem erhöht auch die „Trade-Offs", die gegenseitigen Abhängigkeiten und Wechselbeziehungen sowie auch die Spannungen und „kreativen Konflikte" im Geschäftssystem, und führt zu erhöhtem Abstimmungsbedarf. Denn Unternehmen müssen heute sehr unterschiedliche Dinge gleichzeitig tun, die oft miteinander in Zielkonflikte geraten: die richtigen Produkte entwickeln (langfristig orientiert), gute Beziehungen zu Kunden, Mitarbeitern und Geschäftspartnern aufbauen (mittelfristig), und effizient und profitabel operieren (kurzfristig) (vgl. *Abb. 3*).

Gleichzeitig erhält die Außenorientierung mehr Gewicht und damit kommt natürlich auch mehr Dynamik und laufende Veränderung ins System: Da sich die Unternehmensumwelt ständig wandelt, wie auch die Intentionen und Ziele der einzubeziehenden Stakeholder, kann die wertschaffende Konstellation, die dem Operationsmodell des Unternehmens zugrunde liegt, nicht stabil bleiben. Diese muss stattdessen immer wieder neu hergestellt werden. Der Geschäftszweck des Unternehmens, das diesem zugrunde liegende Wertesystem, welches mit den Stakeholdern geteilt wird und die damit gewollten Potenziale als auch die Unternehmensstrategie, die die Umsetzung dieser Werte und Potenziale in die reale Wirtschaftswelt immer wieder in neue dem sich verändernden Umfeld angepasste Rezepte fasst, kommt dann als Teil eines modernen Unternehmensmodells entscheidende Bedeutung zu.

War das Unternehmensmodell und die Unternehmensorganisation des Industrialismus durch das Bild der Maschine charakterisiert, ist heute eine Metapher aus der Biologie passender: der Organismus. Ein biologischer Organismus ist selbstadaptiv, d.h. er passt sich evolutionär den sich verändernden Umweltbedingungen an. Ähnlich müssen sich heute die Unternehmen flexibel an Marktveränderungen und neue Kundenbedürfnisse anpassen, indem sie mit Produkt-, Service- und Prozessinnovationen reagieren, indem die vorhandenen Fähigkeiten und Aktivitäten möglichst effizient und dynamisch auf aktuelle und konkrete Kundenbedürfnisse hin immer wieder neu ausgerichtet werden. Dies sollte quasi automatisch und evolutionär erfolgen. Damit ist es nicht mehr die interne Effizienz eines Unternehmens, die fast ausschließlich über den wirtschaftlichen Erfolg entscheidet, sondern vielmehr dessen externe Effektivität, mit der es die Erwartungen seiner Stakeholder auch in einem sich dynamisch verändernden Umfeld erfüllen kann, der entscheidende Erfolgsfaktor.

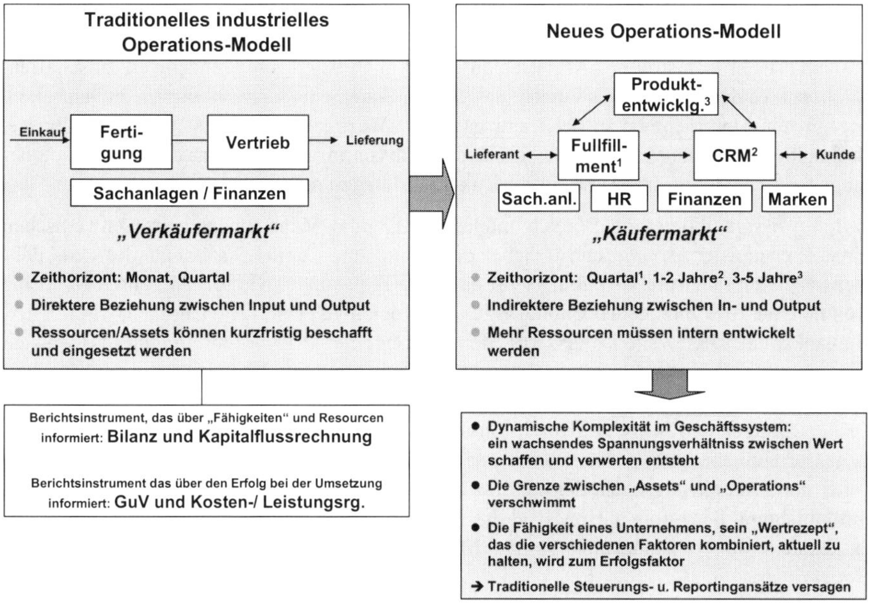

Abb. 3: Der Wandel des operativen Unternehmenswertschöpfungssystems

Wenn jedoch die dauernde Wandlungs- und Innovationsfähigkeit zum entscheidenden Wettbewerbsfaktor wird, kippt damit auch das bislang gültige strategische Paradigma für gute Unternehmensführung. Denn strategische Differenzierung kann dann nicht mehr vorrangig durch einzigartige Produkte oder Technologien entstehen, die ja durch den immer schnelleren allgemeinen Innovationsdruck auch schneller obsolet werden. Stattdessen gewinnt die allgemeine Wissens- und Kompetenzplattform über die ein Unternehmen verfügt, an Bedeutung, die es diesem ermöglicht, immer wieder und quasi automatisch mit Veränderung und Innovation auf Umfeldveränderung in antizipativer Weise zu reagieren. Diese Wissens- und Kompetenzplattform wird dann zum eigentlichen Wettbewerbs- und Differenzierungsfaktor und erfordert die Aufmerksamkeit durch das Management und von Investoren.

Genau diese Überlegungen stehen hinter dem in Skandinavien entstandenen Konzept des Intellectual Capital Statements. Es handelt sich dabei um einen Reporting- und Steuerungsansatz, der einem solchen dynamischen Unternehmensmodell besser gerecht werden möchte als das traditionelle Rechnungswesen und rein finanzorientierte Ansätze. Intellectual Capital Statements etablieren, zusätzlich zur finanzorientierten Sicht, eine weitere wirtschaftliche Gesamtperspektive auf ein Unternehmen, die dieses hinsichtlich seiner intellektuellen Ressourcen, seines Wissenskapital, also seiner Potenziale für die Anpassung an eine noch ungewisse Zukunft beurteilbar machen soll.

3. Das dänische Konzept des Intellectual Capital Statement (ICS): Ziele, Philosophie, Elemente eines ICS und praktische Anwendung

Ziel eines Intellectual Capital Statements ist es, den Mechanismus darzustellen, der die Beziehungen zwischen Mitarbeitern, externen Geschäftspartnern und dem Strukturkapital des Unternehmens, also seinen Geschäftsprozessen, organisatorischen Verfahren, informationstechnischen Infrastruktur, Arbeitsabläufen etc. mobilisiert, um Kundenwert zu schaffen. Damit soll es dem Leser eines Intellectual Capital Statements möglich werden zu beurteilen, wie gut ein Unternehmen in der Lage ist, sein „Wertrezept" an neue Marktverhältnisse, Technologien etc. anzupassen und so seine Wertbasis im Rahmen eines solchen Veränderungs- und Anpassungsprozesses in die Zukunft „rüber zu retten".

Leif Edvinsson, Pionier im Bereich Intellectual Capital Management beim schwedischen Finanzdienstleister *Skandia* und Urheber der Intellectual Capital Statements des Unternehmens (vgl. zum Beispiel *Skandia*, 1998) hat diese Philosophie wie folgt beschrieben: „Eine Komponente [des Intellectual Capital eines Unternehmens] sind die Menschen, Mitarbeiter, Humankapital. Die zweite ist das, was Menschen in einer Organisation umgibt, das was ich 'Structural Capital' nenne – all die Intangible Assets, die übrig bleiben, wenn die Mitarbeiter nach Hause gehen. Dazu zähle ich: interne Prozesse und Strukturen, Datenbanken, Kundenbeziehungen und so weiter. Durch 'Structural Capital' setzen sie eine Organisation in die Lage, ihr Humankapital produktiver zu machen. Es geht dabei nicht darum, dass die Mitarbeiter härter arbeiten. Es geht darum, dass sie „smarter" sind bei ihrer Arbeit, indem sie das Strukturkapital der Firma nutzen. Und das ist es, was den wirklichen Wert einer Organisation darstellt: nicht Finanzkapital, nicht Humankapital, sondern ihr Strukturkapital" (vgl. *Daum/Edvinsson*, 2001, S. 4–6).

Wertschöpfung wird hier als ein Effekt des Verbindens von Human- mit Strukturkapital verstanden. Dies erfordert eine besondere Art von Management, da intellektuelle Ressourcen nicht per Befehl und Gehorsam ihrer produktiven Verwendungen zugeführt werden können, sondern dazu motiviert werden müssen. Dies verändert das Anforderungsprofil und das Aktionsmodell des Managements. Wenn dieses die Unternehmensstrategie und das eigene Geschäftsmodell kontinuierlich an Marktveränderungen anzupassen hat, ohne das Unternehmens-Ecosystem und die darin agierenden Stakeholder-Akteure vollständig kontrollieren zu können, erfordert dies andere Techniken und Fähigkeiten von Managern im Umgang mit Mitarbeitern, Geschäftspartnern, Kunden, externen Aktivistengruppen, der Öffentlichkeit und letztlich auch mit Investoren, als diejenigen, die auf Basis des industriellen Unternehmensmodells entstanden sind. Der wesentliche Grundgedanke hinter dem Konzept des Intellectual Capital ist, die Wurzeln des Wertes eines Unternehmens, dessen verborgene dynamischen Faktoren, die dem sichtbaren Unternehmen mit seinen Gebäuden und Produkten zugrunde liegen, unter den neuen Bedingungen transparent und damit „manageable" zu machen (vgl. *Edvinsson/Malone*, 1997, S. 10–11).

3.1 Der skandinavische versus dem amerikanischen Ansatz

Dieser Ansatz, mit seinem Fokus auf das Zukunftspotenzial und auf das auf dem Wissenskapital basierende grundlegende „Produktionsmodell" eines Unternehmens, wird von den Vertretern des Intellectual Capital-Konzeptes als „typisch skandinavisch" bezeichnet (vgl. *Mouritsen* et al., 2001a, S. 401). Im Gegensatz dazu sieht man als den „typischen amerikanischen" Ansatz denjenigen an, der lediglich auf die Außenseite der „Black Box" fokussiert, indem man versucht, management-relevante Erkenntnisse durch eine Korrelationsanalyse des Inputs, in Form von Aufwendungen bzw. von investiertem Kapital, mit dem Output, in Form des finanziellen Ergebnisses, zu gewinnen.

Als Vertreter des amerikanischen Ansatzes wird hier das Konzept des Wertmanagement auf Basis des Economic Profit oder EVA™ genannt (vgl. *Mouritsen*, 1998). Während das EVA-Konzept auf die bereits vorhandenen Unternehmensaktivitäten fokussiert und versucht, deren zukünftige Erträge in eine einzige Zahl in Form eines Market Value Added (MVA) zu fassen, versucht, das Intellectual Capital-Konzept, die Entwicklung der Kompetenzen-Basis des Unternehmens für das Erwirtschaften zukünftiger Erträge in allen dafür relevanten Dimensionen darzustellen. EVA versucht die Zukunft auf Basis von bereits vorhandenen Kosten- und Erlösstrukturen abzuschätzen, dass Intellectual Capital-Konzept hingegen versucht, Rechenschaft über die Zukunftsaussichten eines Unternehmens auf Basis seiner Kompetenzen und organisatorischen Möglichkeiten abzulegen.

Das Intellectual Capital-Konzept ist so weniger mit dem Wachstumspotenzial befasst, das sich aus bestehenden und bereits vorhandenen Märkten, Technologien und Produkten ergibt, sondern mit der Mobilisierung von internen Basis-Fähigkeiten und Kompetenzen, die ein langfristiges Wachstum und Überleben des Unternehmens ermöglichen. Strukturkapital, also die Fähigkeiten und Kompetenzen des Unternehmens als Organisation, stellen aus dieser Sicht die wesentliche Basis für Wettbewerbfähigkeit dar, die über einen langen Zeitraum aufgebaut wird. Die wesentliche zugrunde liegende Annahme ist, dass Wachstum und Wertschöpfung nicht durch bestimmte Produkte und Märkte entsteht, sondern durch breites organisatorisches Know-how, das von einzigartigem Charakter für das jeweilige Unternehmen ist, und das es diesem erlaubt, sich laufend an veränderte Umfeldbedingungen anzupassen.

Die Kritik am EVA-Konzept seitens der Vertreter des Intellectual Capital-Konzeptes konzentriert sich vor allem darauf, dass es unter der EVA-Brille nicht um die Firma als Firma, sondern ausschließlich um deren Cashflows geht. Die Firma als Firma, in Form ihrer Produkte, Mitarbeiter, Innovationen, Geschäftsbeziehungen etc., also all der Dinge, die zusammen erst die Sphäre des Unternehmens als Organisation begründen, ist nicht Gegenstand der Betrachtung. Aus einer reinen EVA-Sicht heraus kann deshalb auch keine Unternehmensstrategie begründet werden (vgl. *Mouritsen*, 1998, S. 469).

Intellectual Capital Statements sollen stattdessen eine breitere Sicht auf ein Unternehmen gestatten und vor allem dessen Zukunftspotenzial beleuchten, um so eine nachhaltige Entwicklung zu fördern. Vor allem in Dänemark hat sich die Praxis, dass Unternehmen Intellectual Capital Statements erstellen und diese auch veröffentlichen, inzwischen – nicht zuletzt durch eine Reform der Berichtspflichten von Unternehmen – weit verbreitet. Im Nachfolgenden soll der dänische Ansatz entsprechend dem von der dänischen Regierung herausgegebenen „Guidelines for Intellectual Capital Statements" beschrieben werden (vgl. *Danish Agency for Trade and Industry*, 2000 und *Danish Ministry of Science, Technology and Innovation*, 2003a und 2003b).

3.2 Intellectual Capital Statements in Dänemark

Ähnlich wie in den übrigen skandinavischen Ländern wird von der dänischen Regierung Wissen als eine kritische Ressource für das Sichern der wirtschaftlichen Leistungsfähigkeit des Landes und für das Schaffen zukünftigen Wohlstandes angesehen. *Helge Sander*, Minister für Wissenschaft, Technologie und Innovation schreibt dazu: „Unsere Fähigkeit Wissensressourcen zu managen und zu verwalten bestimmt, ob wir das bedeutende Potenzial, das in unseren Unternehmen und öffentlichen Organisationen liegt, freisetzen können" (vgl. *Danish Ministry of Science, Technology and Innovation*, Intellectual Capital Statements – The New Guideline, Kopenhagen 2003a, S. 3).

Deshalb werden von der dänischen Regierung bereits seit einigen Jahren Forschungsprojekte zur Entwicklung von Konzepten gefördert, die Unternehmen und anderen Organisationen dabei helfen, Wissen systematischer und umfassender zu managen und zur wirtschaftlichen Wertschöpfung zu nutzen. Als Ergebnis eines solchen Forschungsprojektes, geleitet von *Jan Mouritsen*, Professor an der *Copenhagen Business School*, wurden im November 2000 die ersten „Guidelines" für Intellectual Capital Statements veröffentlicht (vgl. *Danish Agency for Trade and Industry*, 2000). Inzwischen haben etwa 100 Unternehmen auf Basis dieser Guidelines Intellectual Capital Statements erstellt. Die Erfahrungen dieser Unternehmen wurden im Rahmen eines weiteren Forschungsprojektes, das wiederum von *Jan Mouritsen* geleitet wurde, untersucht. Die Ergebnisse sind in die neue Version der „Guidelines" eingeflossen, die im März 2003 veröffentlicht wurden und die die erste Version verfeinern und vor allem um konkrete Tipps und Anweisungen für die Erstellung von Intellectual Capital Statements ergänzen (vgl. *Danish Ministry of Science, Technology and Innovation*, 2003a). Seit Anfang 2002 gilt zusätzlich in Dänemark eine neue Berichtspflicht, die Unternehmen mit „signifikantem" Wissenskapital dazu verpflichtet, mit dem Jahresabschluss ein Intellectual Capital Statement zu publizieren. Wie mir *Jan Mouritsen* mitteilte, hat dies die Publikation von Intellectual Capital Statements in Dänemark weiter gefördert, was sich auch daran zeigt, dass sich die Wirtschaftsprüfer in Dänemark inzwischen darauf eingestellt haben, Intellectual Capital Statements in ihre Jahresabschlussprüfung mit einzubeziehen.

Ein Intellectual Capital Statement wird in Dänemark als ein integraler Bestandteil des Wissensmanagements eines Unternehmens angesehen. Es wird dabei als ein Produkt einer zweiten, weiterentwickelten Wissensmanagementwelle betrachtet - im Vergleich zu dem, was gemeinhin unter Wissensmanagement verstanden wird: es geht nicht nur um Wissen, sondern um Wissen über Wissen, um eine systematische Sicht auf die Wissensressourcen eines Unternehmens zum Zweck der Management-Intervention. Ein Intellectual Capital Statement wird als eine „Management-Technologie" verstanden, die Wissen darüber produziert, wie Wissen in einem Unternehmen geschaffen, weiterentwickelt und angewendet wird. Es stellt zusammenfassend die Bemühungen eines Unternehmens dar, seine Wissensressourcen zu entwickeln und anzuwenden. Dazu schafft das Intellectual Capital Statement, ähnlich wie die Financial Statements, durch die Abstraktion von den laufenden operativen Aktivitäten im Unternehmen Distanz zu diesen und ermöglicht es so, indem diese quasi „eingefroren" werden und in eine andere Perspektive transportiert werden, dass diese aus einer Wissensmanagement-Perspektive unabhängig von den Anforderungen und Dringlichkeiten des Tagesgeschäfts analysiert und beurteilt werden können. Durch die Repräsentation der Zusammensetzung der Wissensressourcen und deren Entwicklung über die Berichtsperioden stimuliert ein Intellectual Capital Statement Fragen hinsichtlich deren Qualität und der Erfordernis zu deren Weiterentwicklung und gibt dafür die notwendigen Anregungen und Impulse für das Management. Außenstehenden soll ein Intellectual Capital Statement Einblick in die Wissensressourcen und deren geplanten Weiterentwicklung geben (vgl. *Mouritsen*, 2003).

Dazu muss ein Intellectual Capital Statement die Wissensmanagement-Strategie des Unternehmens identifizieren und beschreiben, indem die Ziele hinsichtlich der Zusammensetzung, Anwendung und Entwicklung der Wissensressourcen und die davon abgeleiteten Initiativen als auch deren Ergebnisse dargestellt werden. Ein Intellectual Capital Statement ist sowohl ein internes strategisches Wertmanagement-Werkzeug, als auch ein Kommunikationsinstrument, um Mitarbeiter, Kunden und anderen Stakeholder für diesen Prozess zu motivieren und zu engagieren.

Mit Hilfe eines Intellectual Capital Statements lassen sich die Initiativen und Ergebnisse zur Verbesserung der Wissensressourcen eines Unternehmens vom Management überwachen und steuern. Es gibt den Unternehmen quasi einen Rahmen vor, der sie dabei unterstützt, Wissensmanagement in Einklang mit der Unternehmensstrategie im Unternehmen aufzusetzen und zu organisieren. Sollte ein Unternehmen bereits eine Wissensmanagement-Strategie verfolgen, wird der Vorteil der Anwendung von Intellectual Capital Statements vor allem darin gesehen, dass sie helfen Wissensmanagement systematisch zu betreiben, dass es Wissensmanagement in die Unternehmensstrategie einbindet, und dass es ein Management-Werkzeug für die Überwachung und Steuerung der entsprechenden Initiativen bereit stellt.

Ein Intellectual Capital Statement soll zudem den Stakeholdern zeigen, ob das Unternehmen sich bei der Entwicklung und beim Management seiner Wissensressourcen verbessert hat, indem es die unsichtbaren „Assets" in Form des Wissenskapitals, also das Zukunftspotenzial des Unternehmens für diese sichtbar gemacht wird. Intellectual Capital Statements werden damit auch als Marketinginstrument genutzt, um die Attraktivität des Unternehmens für wichtige Stakeholder wie Mitarbeiter, Kunden und Investoren zu erhöhen.

3.3 Elemente eines dänischen Intellectual Capital Statements

Theoretische Grundlage des dänischen Konzeptes ist eine Wissensmanagementphilosophie, die auf der Annahme basiert, dass Wissen an sich nicht beschrieben, beurteilt und gemanagt werden kann. Es muss dazu zunächst in so genannte Wissensressourcen „übersetzt" werden, auf die man zeigen und sagen kann: das ist Wissen. Erst diese Wissensressourcen können entwickelt, beurteilt und auf neue Weise zur Unterstützung neuer „Wertrezepte" kombiniert - also aus einer ökonomischen Sicht gemanagt werden. Wissensressourcen werden in den dänischen Guidelines in vier Kategorien einteilt: Mitarbeiter, Kunden, Prozesse und Technologien:

- Die Ressource „Mitarbeiter" umfasst dabei die Fähigkeiten und Kompetenzen der Mitarbeiter, ihre Erfahrung, die Mitarbeiterstruktur, also die Kombination verschiedener Mitarbeitertypen und Ausbildungsstände, die Motivation der Mitarbeiter, ihr Commitment, ihr Wille zur Anpassung und zum Wandel.
- Die Ressource „Kunden" umfasst den Kundenmix, die Beziehungen zu Kunden und Anwendern, ihr Zufriedenheitsgrad und ihre Loyalität, inwieweit sie das Unternehmen weiterempfehlen, der Grad des Einblicks in die Bedürfnisse der Anwender und Kunden, der Grad der Zusammenarbeit mit Kunden und Anwendern in der Produktentwicklung.
- Die Ressource „Prozesse" repräsentiert das Wissen, das den stabilen Prozessen und Abläufen des Unternehmens zugrunde liegt. Dabei kann es sich um die Innovationsprozesse, Qualitätssicherungsprozesse, Management- und Controllingprozesse und um die Informationsbearbeitungs-Mechanismen handeln.
- Aufgabe der Ressource „Technologien" ist es, die übrigen Wissensressourcen mit geeigneten Technologien zu unterstützen. Der Fokus liegt hier üblicherweise auf der IT-Infrastruktur des Unternehmens (Software und Hardware), auf der IT-Intensität, den IT-Kompetenzen und den IT-Wirkungsgrad.

Diese Wissensressourcen werden im Rahmen dieser Philosophie in den Kontext des unternehmensspezifischen „Wissensmanagement-Szenarios“ gestellt, das beschreibt, wie das Wissenskapital des Unternehmens mit der speziellen Art von Organisationsidentität und dem spezifischen Managementansatz der Firma in Beziehung steht und welche Rolle die Ressource „Wissen“ dabei spielt, dass das Unternehmen in einer unsicheren wissensbasierten Zukunft im Wettbewerb bestehen und eine zu laufenden Wandel fähige Firma sein kann (vgl. *Abb 4*).

Abb. 4: Die Idee eines „Intellectual Accounts“ als Basis für das Modell der Intellectual Capital Statements (vgl. Mouritsen J. et al., 2001b, S. 17)

Ausgehend von dieser Grundlage soll ein Intellectual Capital Statement darstellen,

- wozu das Unternehmen seine Wissensressourcen weiterentwickeln möchte, d.h. auf Basis welcher Wissensmanagement-Strategie („Knowledge Narrative“),
- wo dabei die Aktionsbereiche liegen, also welche Management-Herausforderungen die Strategie mit sich bringt,
- wie deren Umsetzung erfolgen soll, also mit welchen Initiativen das Unternehmen die genannten vier Ressourcen auf Basis der Strategie weiter entwickeln will, und schließlich
- wie erfolgreich es bei der Umsetzung ist, also welches der Status der relevanten Performance-Kennzahlen ist, die den Fortschritt der Initiativen abbilden

Nach den neuesten Guidelines vom März 2003 besteht ein Intellectual Capital Statement deshalb aus den vier Elementen Knowledge Narrative, das die Wissensmanagement-Strategie beschreibt, der Beschreibung der Management-Herausforderungen, der sich daraus ergebenden Initiativen zur Verbesserung des Wissensmanagements bzw. zur Entwicklung der Wissensressourcen und schließlich den Kennzahlen für die Erfolgsmessung.

Knowledge Narrative

Das Knowledge Narrative beschreibt, wie ein Unternehmen durch seine Produkte oder Dienstleistungen Wert in Form eines „Gebrauchswert" (use value) für seine Kunden schafft bzw. schaffen möchte. Außerdem beschreibt es, welche Wissensressourcen zum Schaffen dieses Gebrauchswertes benötigt werden und wie die Firma ihre Ressourcen so organisiert, dass sie dies erreicht. Es stellt somit dar, wie Wissen zu Verbesserungen, also zu Mehrwert, für die Anwender der Produkte bzw. Dienstleistungen des Unternehmens führen soll. Dabei werden typischerweise Begriffe wie „weil", „deshalb" und „um zu" verwendet. Ein Knowledge Narrative setzt sich zusammen aus einer Beschreibung

- der Mission des Unternehmens mit einem speziellen Fokus auf den Anwender (Kunden) und die Produkte des Unternehmens
- des Gebrauchswertes der Produkte und Dienstleistungen des Unternehmens aus Sicht der Anwender,
- der Wissensressourcen, die benötigt werden, um die Bedürfnisse der Anwender und Kunden zu befriedigen,
- wie Gebrauchswert und Wissensressourcen zueinander in Beziehung stehen, wie also die grundsätzlichen „Produktionsbedingungen" des Unternehmens aus einer Wissensperspektive aussehen, wie also Wissen im Unternehmen „arbeitet".

Management-Herausforderungen

Die Management-Herausforderungen ergeben sich, indem man versucht, die im Knowledge Narrative formulierten Absichten zu implementieren und sich fragt, welche bereits vorhandenen Wissensressourcen dazu gestärkt bzw. welche neu erworben werden müssen. Die im Intellectual Capital Statement beschriebenen Management-Herausforderungen haben eine längerfristige Gültigkeit und ändern sich typischerweise nicht jedes Jahr. Sie sind eng mit dem Knowledge Narrative und mit den individuellen Wissensressourcen im Unternehmen verknüpft. Die Beschreibung und Strukturierung der Management-Herausforderungen soll dem Unternehmen dabei helfen, die darauf bezogenen Initiativen zu organisieren und deren Zusammenhang zur Wissensmanagement-Strategie herzustellen. Sie soll aufzeigen und darstellen, wie die Initiativen im Zusammenhang wirken. Management-Herausforderungen beziehen sich typischerweise auf die folgenden Bereiche:

- Einstellung, Bindung und Entwicklung von Mitarbeitern und Kompetenzen;
- Entwicklung der Geschäftsprozesse;
- Teilen und Verbreiten von Wissen und Kompetenzen im Unternehmen;
- Sichtbarkeit im Markt;
- Aufbau von Partnerschaften mit Kunden;
- Aufbau von Wissen über die Bedürfnisse der Anwender;
- Entwicklung des Zugangs zu externen Wissensressourcen.

Die Transformation des Knowledge Narrative in die Management-Herausforderungen stellt dabei einen kreativen Akt dar, bei dem die strategischen Wissensmanagement-Programme des Unternehmens definiert werden, um Wert („Gebrauchswert") zu schaffen.

Wurden in der Version der Guidelines von 2000 die Initiativen (Maßnahmen) noch als Bestandteil der Management-Herausforderungen dargestellt, sind diese im Neuentwurf der Guidelines von 2003 als eigener Bereich in den Intellectual Capital Statements etabliert.

Initiativen

Als Initiativen werden die Maßnahmen bezeichnet, die sich als notwendig erweisen, um auf die Management-Herausforderungen zu reagieren. Die in den Intellectual Capital Statements dargestellten Initiativen sind darauf ausgerichtet, Wissensressourcen neu zu kombinieren, zu entwickeln und zu beschaffen und diesen Prozess sowie die erzielten Effekte zu überwachen. Initiativen sind im Gegensatz zu den Management-Herausforderungen, die längerfristig gültige strategische Wissensmanagement-Programme des Unternehmens darstellen, eher kurzfristig orientiert. Sie sollen Ergebnisse innerhalb eines Jahres zeigen, auch wenn sich bestimmte Initiativen von Jahr zu Jahr wiederholen sollten. Initiativen beschreiben die konkreten Aktionen des Managements in Bezug auf die Wissensressourcen des Unternehmens. Für jede Initiative gibt es einen Verantwortlichen. Beispiele für Initiativen sind:

- Einstellung von mehr F+E Beratern oder Entwicklern;
- Fortbildungsprogramme in bestimmten Prozessbereichen;
- Einführung eines Qualitätsicherungssytems;
- Einführung finanzieller und/oder symbolischer Incentives für das Teilen von Wissen im Unternehmen;
- Aufbau bestimmter Anwender-Arbeitskreise;
- ...

Kennzahlen

Hier wird ein Set von Kennzahlen dargestellt, die sich auf die Initiativen beziehen und das es dem Leser möglich machen soll, festzustellen, wie weit eine Initiative vorangekommen ist und welche Effekte erzielt wurden. Durch die Kennzahlen soll eine Initiative „sichtbar" werden, indem sie messbar gemacht wird. Sie wird damit gewissermaßen formalisiert und institutionalisiert. In den Guidelines werden die drei wichtigsten Funktionen der Kennzahlen folgendermaßen dargestellt:

- Definition: Sie dienen der Definition (Spezifizierung) der Management-Herausforderungen und Initiativen.
- Beurteilung: Sie ermöglichen die Beurteilung ob Initiativen bezüglich der Management-Herausforderungen gestartet wurden und welchen Effekt sie erbringen.
- Reporting: Sie stellen ein wichtiges Verbindungsglied zum Design des externen Intellectual Capital Statements dar.

Alle vier Elemente eines Intellectual Capital Statements repräsentieren zusammen eine Analyse des Wissensmanagements einer Firma. Sie hängen eng miteinander zusammen und jedes Element erschließt sich nur im Zusammenhang mit den übrigen Elementen. Die Kennzahlen alleine sind schwierig zu interpretieren und das Knowledge Narrative erweckt leicht den Eindruck einer „freien Prosa", wenn es nicht durch die Indikatoren, Initiativen und Management-Herausforderungen erläutert wird. Alle vier Elemente wirken zusammen. Die Kennzahlen zeigen, ob Initiativen gestartet wurden und wie sie wirken. Die Initiativen führen die in den Management-Herausforderungen identifizierten Probleme durch konkrete

Aktivitäten einer möglichen Lösung zu. Die Management-Herausforderungen wiederum zeigen auf, wo angesetzt werden muss um die Wissensressourcen des Unternehmens zu entwickeln. Und das Knowledge Narrative fasst zusammen, kommuniziert und reorientiert, wo das Unternehmen welche Kompetenzen benötigt, um Wert für die Anwender ihrer Produkte zu schaffen. Die vier Elemente wirken also in einem Wissensmanagement-Modell zusammen, in dem die dargestellten Beziehungen zwischen ihnen sichtbar werden (vgl. *Abb 5*).

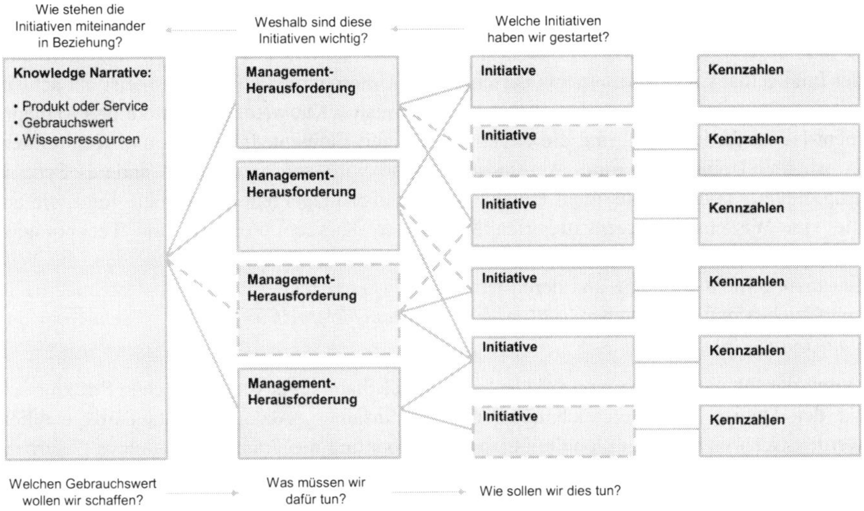

Abb 5: Das Wissensmanagement-Modell der dänischen Intellectual Capital Statements (vgl. Danish Ministry of Science, Technology and Innovation, 2003a, S. 13 u. 20)

3.4 Intellectual Capital Statements in der Praxis - Fallbeispiele

Vergleicht man die Intellectual Capital Statements dänischer Firmen aus der ersten Welle miteinander, also bevor bzw. kurz nach dem die erste Version der Intellectual Capital Statements aus dem Jahr 2000 vorlag, stellt man noch große Unterschiede zwischen diesen fest, auch wenn sie einem ähnlichen gemeinsames Muster folgen (vgl. *Mouritsen* et al., 2001b). Dies hat sich in der zweiten Welle seit dem Jahr 2000 verändert. Die Statements sind sich, was die Struktur betrifft, von Firma zu Firma ähnlicher geworden. Die Struktur der Intellectual Capital Statements hat sich seitdem offenbar stabilisiert und folgt dem Muster die in den Guidelines vom März 2003 beschrieben werden.

In allen Fällen stellt das Intellectual Capital Statement im Rahmen des Knowledge Narrative die Wissensressourcen dieser Firmen dar und wie diese interagieren, um das Unternehmen in die Lage zu versetzen, bestimmte Leistungen für externe Anwender, die Kunden, zu erbringen. Es identifiziert die damit verbundenen Management-Herausforderungen und stellt die Maßnahmen des Managements dar, um die Wissensressourcen des Unternehmens weiterzuentwickeln und auf die zu lösenden Kundenprobleme hin „zu konditionieren". Mittels Kennzahlen soll sowohl für die vorhandenen Ressourcen, die eingeleiteten Maßnahmen und über deren Ergebnis „Rechnung gelegt" werden. Der Sinnzusammenhang der Intellectual Capital Indicators wird durch das Knowledge Narrative, der Beschreibung der Management-Herausforderungen und der Initiativen hergestellt, die ja auch erst die Relevanz der Kennzahlen begründen, indem der Geschäftszweck der Firma und deren strategischen Ziele dargestellt werden. Dies erfolgt durch Beantwortung von Fragen wie: Wofür sind wir da? Wofür strengen wir uns an? Welche Effekte sollen für unsere Kunden erzielt werden? Die Kennzahlen zeigen, dass es das Management mit den dargelegten Absichten und eingeleiteten Maßnahmen ernst meint und dass es sich daran messen lässt.

Da Intellectual Capital Statements aber in hohem Masse von den sehr individuellen Knowledge Narratives, also von der Wissensstrategie einer Firma und ihrer Management-Herausforderungen geprägt sind, stellt dies den externen Leser vor ein Problem: die Inhalte der Intellectual Capital Statements verschiedener Unternehmen lassen sich meist nur schwer miteinander vergleichen, da die gestaltenden Elemente, Knowledge Narrative und Management-Herausforderungen, und die davon abhängigen Element, Initiativen und Kennzahlen, ja prinzipiell einzigartig sind. Aus diesem Grund wird zur Analyse ein anderes Schema empfohlen, das der bereits oben erwähnten Wissensmanagement-Philosophie folgt, die auf die vier Wissensressourcenkategorien Mitarbeiter, Kunden, Prozesse und Technologien abhebt. Innerhalb dieser Struktur soll es einem Leser leichter gemacht werden, die Wissensmanagement-Strategie und deren Umsetzung verschiedener Firmen im Vergleich nachzuvollziehen und zu verstehen. (vgl. *Abb. 6* und *Danish Ministry of Science, Technology and Innovation, 2003b*).

Durch die für die Analyse von Intellectual Capital Statements vorgeschlagene Struktur soll für den Unternehmensvergleich quasi ein gemeinsamer „Accounting-Standard" etabliert werden. Während das traditionelle Financial Accounting die Transaktionen eines Unternehmens in den Bereichen Erlöse, Kosten, Anlagen und Verbindlichkeiten aufzeichnet, diese also in eine eher gegenwartsorientierte wirtschaftlich-finanzielle Perspektive abstrahiert, um Aussagen und Entscheidungen hinsichtlich der finanziellen Solidität, der Liquidität und der Profitabilität eines Unternehmens zu ermöglichen, soll das Intellectual Capital Statement hier die Aktivitäten des Unternehmens in den Bereichen Mitarbeiter, Kunden, Prozesse und Technologie transparent und steuerbar machen, indem die operativen Aktivitäten in eine zukunftsorienterte Wissensmanagement Perspektive abstrahiert wird. Dazu wird der Stand der Wissensressourcen des Unternehmens, die Aktivitäten zur Verbesserung bzw. Weiterentwicklung dieser Ressourcen, sowie deren „Produktivität", also die erzielten Effekte durch Kennzahlen dargestellt.

Im Folgenden werden zwei Fallbeispiele von Intellectual Capital Statements dänischer Firmen beschrieben. Beim ersten, dem von *Systematic*, handelt es sich um ein Fallbeispiel aus der ersten dänischen Intellectual Capital Statement Welle, die zu der ersten Version der „Guidelines" geführt hat. Beim zweiten, dem Fallbeispiel von *Maxon Telecom A/S*, um ein Fallbeispiel aus der zweiten Welle, die zur letzten Version der „Guidelines" vom März 2003 geführt hat.

Fallbeispiel *Systematic*

Eines der 17 dänischen Unternehmen, das im Rahmen des ersten Forschungsprojektes von *Jan Mouritsen* untersucht und beschrieben wurde, ist das dänische Softwarehaus *Systematic Software Engineering Ltd.*, das seit 1999 externe Intellectual Capital Reports, wie diese von *Systematic* genannt werden, veröffentlicht. Der Nutzen des Intellectual Capital Reports wird bei *Systematic* darin gesehen, dass die Aufmerksamkeit von Managern und Mitarbeitern auf die Entwicklung von wichtigen Kompetenzen, auf das Verbinden dieser Kompetenzen mit Prozessen und auf die Notwendigkeit der Einführung von so genannten Wissensagenten und Mentoren gelenkt und damit die Wissensmanagement-Initiative des Unternehmen entscheidend unterstützt werden. Der wesentliche externe Effekt wird darin gesehen, dass *Systematic* sich einen Namen in der für *Systematic* wichtigen Forschungs-Community, bei der Presse und den lokalen Behörden gemacht hat. Die neue Reporting- und Kommunikationspraxis hat auch dabei geholfen, führende IT-Spezialisten als Mitarbeiter zu gewinnen und ein Netzwerk an Beziehungen mit anderen Unternehmen aufzubauen, die ebenfalls am Thema „Kompetenzentwicklung" interessiert sind.

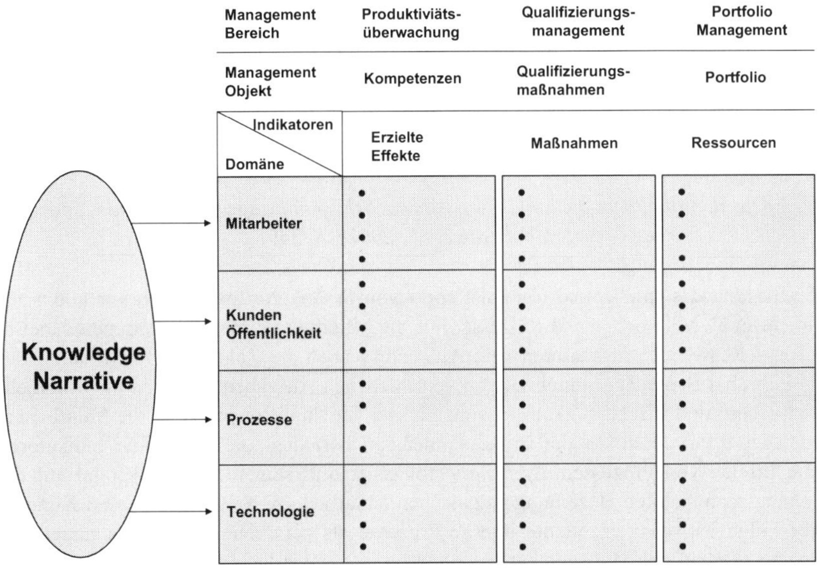

Abb. 6: Klassifikationsschema für die Analyse von Intellectual Capital Statements (vgl. Mouritsen et al., 2001b, S. 20)

Der Intellectual Capital-Report von *Systematic* gliedert sich in drei Teile. Zunächst werden die „Grundlagen" der Firma, bestehend aus Mission, Vision, Zielen und Werten, beschrieben. Dann werden die Aktivitäten des Unternehmens zum Kompetenzaufbau in den Bereichen Menschen, Prozesse und Infrastruktur präsentiert, um schließlich im letzten Teil die erzielten Ergebnisse in den Bereichen Kunden, Mitarbeiter, Innovation und externe Beziehungen und letztlich auch das finanzielle Ergebnisse darzustellen. Das in *Abb. 7* dargestellte Schema kann als Modell der wissensbasierten „Economic Engine" des Unternehmens angesehen werden, die durch den Intellectual Capital-Report transparent gemacht werden soll (vgl. *Mouritsen* et al., 2001c, S. 370).

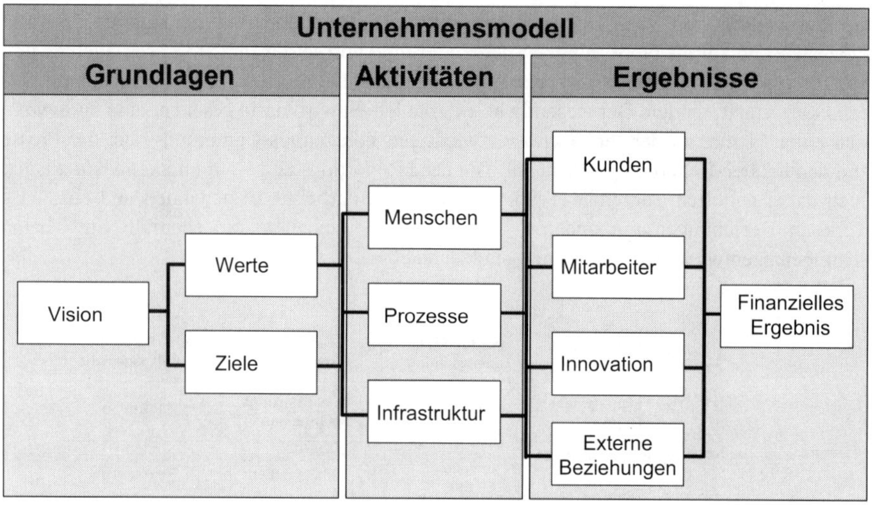

Abb 7: Das den Intellectual Capital-Reports von Systematic zugrunde liegende Modell (vgl. Mouritsen et al., 2001c, S. 369)

Die Kennzahlen des Intellectual Capital-Report von *Systematic* fokussieren vor allem auf die Dimensionen Mitarbeiter und Prozesse. Für die Mitarbeiterdimension kommen dabei in erster Linie Ressourcen-Indikatoren (die Aufschluss über die Zahl und Qualifikation der Mitarbeiter, über deren Zuordnung zu den verschiedenen Aufgabenbereichen im Unternehmen etc. geben) und Effekt-Indikatoren zum Einsatz, die zum Beispiel über die Mitarbeiterzufriedenheit informieren. In der Prozessdimension berichten die gewählten Indikatoren über die Effekte von Prozessen. *Systematic* fokussiert mit seinem Wissenskapital auf die Entwicklung von stabilen Beziehungen zwischen Mitarbeitern, Technologien und Kunden, was zur Folge hat, dass exzellente interne Prozesse als zentraler Erfolgfaktor angesehen werden. Als wesentliche Management-Herausforderungen werden dargestellt:

- die Einstellung der besten Softwareentwickler im Markt und diese an das Unternehmen zu binden;
- die laufende Weiterentwicklung ihrer beruflichen und persönlichen Kompetenzen;
- der Aufbau partnerschaftlicher Beziehungen zu den Kunden;
- die laufende Verbesserung der Entwicklungsprozesse, um Software im Rahmen eines integrierten Prozesses effizient und effektiv zu entwickeln (vgl. *Abb. 8*).

Alle vier Management-Herausforderungen stehen miteinander in Beziehung: Wenn es keine neuen Mitarbeiter gäbe, wäre der Fokus auf die Software-Entwicklungsprozesse nicht so wichtig, da eingearbeitete Mitarbeiter diese beherrschen, es keine „Störung" durch neue Mitarbeiter gäbe und der Ausbau der Prozesse und deren Anpassung an eine wachsende Mitarbeiterzahl nicht erforderlich wäre. Auch wenn die Kundenbasis immer die gleiche bleiben sollte, würde es keine Notwendigkeit geben, neue partnerschaftliche Kundenbeziehungen zu entwickeln. Andererseits erfordert die Veränderung von Prozessen und deren Ausbau auch eine entsprechende Entwicklung der Mitarbeiter. Die Management-Herausforderungen spiegeln so die Strategie der Firma wieder, zu wachsen und Potenziale für dieses zukünftige Wachstum aufzubauen. Dabei hat sich der Fokus des Managements von *Systematic* über den Zeitablauf gewandelt, je nach dem Entwicklungsfortschritt des Unternehmens und den sich daraus ergebenden Anforderungen. Im Intellectual Capital Report 2002 wurde beispielsweise die Verbesserung des Software-Entwicklungsprozesses als Hauptthema für dieses Jahr benannt. *Systematic* lässt seine Intellectual Capital-Reports auch von Wirtschaftsprüfern testieren, um die Ernsthaftigkeit des Unterfangens zu dokumentieren (vgl. *Systematic*, 2002).

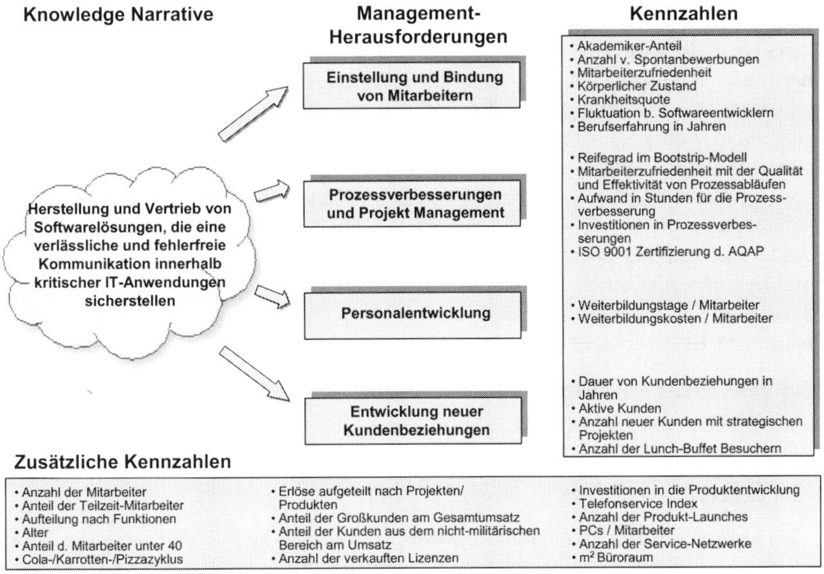

Abb. 8: Das Intellectual Capital Statement von Systematic (vgl. Mouritsen et al., 2001c, S. 379)

Fallbeispiel *Maxon Telecom A/S*

Maxon Telecom A/S in Dänemark (im Folgenden *Maxon* genannt) operiert seit mehr als einem Jahrzehnt als Forschungs- und Entwicklungsabteilung für die Muttergesellschaft *Sewon/Maxon Telecom Co. Ltd.*, das zweitgrößte IT-Konglomerat Koreas. *Maxon* entwickelt mobile Kommunikationslösungen, die von der Muttergesellschaft im Rahmen der von ihr produzierten Mobiltelefone vermarktet werden. Dabei erhält es aus Korea die Basis-Produktspezifikationen und entwickelt dann, als Sparringspartner und im laufenden Dialog mit der Mutter, die technischen Spezifikation und das technische Design des Endproduktes. Abgedeckt wird dabei sowohl der mechanische Bereich (Hardware) als auch der Softwarebereich. Zusätzlich zur Rolle als ausgelagerte F&E-Abteilung der Mutter übernimmt *Maxon* auch Aufträge von dritten. Das Unternehmen bezeichnet sich dabei selbst als „Entwicklungshaus für mobile Innovationen“ bzw. als ein „Mobile Technologies Think Tank".

Um als „Think Tank“ erfolgreich zu sein und es zu bleiben, muss *Maxon* immer auf dem Laufenden über vorhandene und zukünftige Technologien im Bereich Mobile Communication sein. Hochqualifizierte Mitarbeiter, die über dieses Wissen verfügen und darüber hinaus in der Lage sind, mit neuen Technologien umzugehen, die die Anwenderbedürfnisse verstehen, und neue Technologien in funktionsfähige Produkte transformieren können, welche Gebrauchsmehrwert schaffen, stellen deshalb eine besonders wichtige Wissensressource für das Unternehmen dar. Deren Fähigkeiten müssen laufend ausgebaut und an neue Technologien bzw. Marktanforderungen angepasst werden. Darüber hinaus muss sichergestellt werden, dass die Mitarbeiter von ihren Führungskräften motiviert und gefördert werden. Dazu sind entsprechende Maßnahmen und Initiativen aufzusetzen.

Aber nicht nur motivierte Mitarbeiter und das Wissen über zukunftsfähige bzw. zukünftige Technologien und darüber, wie diese in für den Anwender wertschaffende Produkte umgesetzt werden können, stellen eine wichtige wettbewerbsentscheidende Ressource für das Unternehmen dar. Auch die Fähigkeit Neuentwicklungen schnell auf den Markt bringen zu können, also möglichst schnell in marktfähige Produkte umzusetzen, spielt als kritischer Erfolgsfaktor für das Unternehmen eine entscheidende Rolle. Durch die hohe Innovationsrate und Innovationsgeschwindigkeit in der Branche sinkt der Gebrauchswert einer neuen Technologie oder Produktgeneration für die Anwender, je länger es dauert diese auf den Markt zu bringen. Time-to-Market hat also direkt einen Einfluss auf den Mehrwert, den das Unternehmen in der Lage ist zu produzieren. Deshalb stellt die Fähigkeit, Entwicklungs-Projekte auch parallel pünktlich, mit der ausreichenden Qualität und unter erreichen der Kostenziele abschließen zu können, selbst eine wichtige Wissensressource für das Unternehmen dar. Diese Fähigkeit zu sichern bzw. kontinuierlich zu verbessern ist eine weitere Management-Herausforderung bei *Maxon*.

Somit wurden folgende Management-Herausforderungen identifiziert:

- Wissen über und Fähigkeiten bezüglich neuer Technologien systematisch aufbauen;
- Produktentwicklung (Fähigkeit Produkte mit Gebrauchsmehrwert für die Anwender zu entwickeln);
- Verbesserung und kontinuierlicher Ausbau der persönlichen Fähigkeiten der Mitarbeiter;
- Sicherstellung, dass Produkte pünktlich, also zeitnah auf dem Markt kommen (Time-to-Market).

Zu allen Management-Herausforderungen hat das Unternehmen entsprechende Initiativen aufgesetzt und dafür jeweils Kennzahlen definiert. Ziel der Initiativen ist es, die Management-Herausforderungen durch entsprechende Maßnahmen zu adressieren. Im Bereich Produktentwicklungsfähigkeiten etwa, soll durch den Aufbau eines Netzwerkes mit bzw. durch enge Beziehungen zu Anwendern die Fähigkeit verbessert werden, marktfähige Produkte mit Gebrauchsmehrwert zu entwickeln. Die Projektmanagement-Kompetenzen im Unternehmen sollen durch Fortbildungsmaßnahmen und eine passende Organisation, durch die Implementierung entsprechender Prozesse in der Produktentwicklung und durch den Einsatz passender Werkzeuge verbessert werden. Die entsprechenden Kennzahlen sollen den Status bzw. Fortschritt der jeweiligen Initiative aufzeigen (vgl. *Abb. 9*).

Knowledge Narrative	Management-Herausforderungen	Initiativen	Kennzahlen
• Produkt / Service: Maxon Telecom entwickelt und designt Mobiltelefone auf der Basis führender Technologie • Gebrauchstwert: Kompetenter Sparringspartner (für die Mutter oder Dritte) um "Communication, any-time, anywhere" möglich zu machen • Wissensressourcen: Mitarbeiterspezialistenwissen u. –fähigkeiten, Einblick in die Anwender- und Kundenbedürfnisse, Einblick in existierende und zukünftige Technologien und die Fähigkeit, Projekte zu managen	• Produktentwicklung	• Evaluierung der Anwendererwartung und -zufriedenheit	• Anzahl von durchgeführten Zufriedenheitsstudien (u. Marktbefragungen) • Zufriedenheit der Kunden mit.der Qualität • Anzahl "bestellter" Projekte / Jahr
	• Verbesserung u. kontinuierlicher Ausbau der persönlichen Fähigkeiten der Mitarbeiter	• Einführung von Mitarbeiter-Performance -Reviews • Erstellen von Mitarbeiter-Kompetenzentwicklungsplänen • Einführung von Tutor-Schemes • Einführung von Manager-Trainings • Einführung von CASE-Trainings • Einführung von Führungskräfte-Coaching	• Abwesenheitsquote • Erfüllungsgrad der Fortbildungsanforderungen der MUS-Ergebnisse • Mitarbeiterzufriedenheit mit den Kurs- und Trainingsinitiativen • Anzahl von nach Plan absolvierten Performance-Reviews • Assessments d. Sozialkompetenz durch Kollegen • Fluktuationsrate • Anzahl der Mitarbeiter m.Kompetenzen-Entw.plan • Anzahl der Mitarbeiter in Jobrotation und im Auslandseinsatz • Anzahl Mitarbeiter, von denen wir glauben, dass sie sich bei Maxon weiter entwickeln können • Anzahl Mitarbeiter die aussagen, dass ihr direkter Vorgesetzter sie ausreichend motivieren kann • Anzahl neuer Mitarbeiter im Verhältniss zu der Anzahl Tutor-Schemes
	• Sicherstellung, dass Produkte pünktlich, also zeitnah auf dem Markt kommen (Time-to-market)	• Starten von Microsoft-Project Trainings • Einführung einer Projektorganisation • Implementierung eines Teambuilding-Prozesses	• Anzahl pünktlich abgeschlossener Projekte • Anzahl im Budget abgeschlossener Projekte • Anzahl intern besetzter Junior-Projektmanager-Posten • Anzahl Mitarbeiter mit Projektmanager-Qualifikat. • Zufriedenheit der Mitarbeiter mit der Entscheidungs- und Handlungsgeschwindigkeit • Anzahl Projektgruppen mit weniger als 16 MA • Anzahl Projektgruppen ohne eigenen Projektraum
	• Wissen über und Fähigkeiten bezüglich neuer Technologien systematisch aufzubauen	• Mitarbeiter i. neuen Technologien ausbilden • Roadmaps einführen • Teilnahme an Konferenzen • Teil d. Netzwerks v. Operator/Entw.h. sein	• Teilnahme an d. CEBIT und Cannes • Anzahl v. Koordinationsmeeting pro Jahr • Anzahl v. Abteilungsleitern / Technologie-Scouts • Anzahl von Entwicklern in externen Netzwerken

Abb. 9: Das Intellectual Capital Statement von Maxon (vgl. Danish Ministry of Science, Technology and Innovation, 2002a, S. 15)

Das Schema des Intelectual Capital Statement von *Maxon* folgt damit dem in der neuen Version der Guidelines dargestellten Reporting-Schema. Der Vergleich mit dem hier dargestellten Schema des Intellectual Capital Statements von *Systematic*, das noch aus der ersten Welle stammt, zeigt, dass die nun eingefügt Spalte mit den Initiativen für mehr Klarheit sorgt, da sowohl die ergriffenen Maßnahmen des Managements zur Verbesserung der in den Management-Herausforderungen benannten kritischen Wissensressourcen des Unternehmens transparent werden und gleichzeitig so auch die Interpretation der Kennzahlen erleichtert wird.

Die neuen Guidelines, die unter Mitarbeit von einer ganzen Reihe von Unternehmen unterschiedlichster Branchen (z.B. aus den Bereichen Fertigung, Banken, Versicherung, Versorgung, Management-Beratung, Öffentliche Verwaltung, Bildung, IT u.a.) und anderer Organisationen (z.B. des dänischen Gewerkschaftsverbands, des dänischen Verbands eingetragener Wirtschaftsprüfer, des dänischen Manager Verbands, der Copenhagen Stock Exchange, der dänischen Handelskammer, der dänische Industrieverband u.a.) entstanden sind (vgl. *Danish Ministry of Science, Technology and Innovation, 2003a, S. 4)* stellen so einen wichtigen Schritt nach vorne im Bemühen um die Herstellung von Transparenz und um die Verbesserung der Steuerungsfähigkeit wissensintensiver Unternehmen dar, wie man sie heute in so gut wie allen Branchen antrifft.

4. Entwurf eines Frameworks für die Neugestaltung von Unternehmenssteuerung und externem Reporting und die Rolle von Intellectual Capital Statements

Das Intellectual Capital-Konzept unterscheidet sich wesentlich von den klassischen rein finanzorientierten Berichts- und Steuerungsansätzen. Während das Financial Accounting und darauf basierende Wertmanagement-Konzepte, wie etwa das EVA-Konzept, nur eine Momentaufnahme des finanziellen Status des Unternehmens wiedergeben können, versucht das Intellectual Capital-Konzept die dahinter liegenden Faktoren, die als (Wissens)Potenziale oder Kompetenzen zu zukünftigen finanziellen Ergebnissen führen, zu identifizieren und darüber „Rechnung zu legen". Dahinter steht die Einsicht, dass Unternehmen sich heute immer wieder und vor allem häufiger an veränderte Umweltbedingungen mit ihrem gesamten internen „Produktionssystem" und ihren Außenbeziehungen anpassen müssen, um überlebensfähig und erfolgreich zu sein und es vor allem zu bleiben. Diese Anpassungsfähigkeit, auf der Basis einer breiter angelegten Kompetenzplattform, wird als mindestens genauso wichtig angesehen, wie die Fähigkeit eines Unternehmens, aktuell finanzielle Performance zu produzieren. Denn letztere kann schnell verloren gehen, wenn sich das Geschäftsumfeld des Unternehmens ändert. Es ist deshalb mindestens genauso wichtig zu verstehen, durch welche Maßnahmen und Verfahren und mit welcher relativen Verbesserung oder Verschlechterung das Unternehmen Wert für Kunden in der Zukunft generieren kann bzw. wie gut es über Basisfähigkeiten verfügt, diese Kompetenzen immer wieder in neue unerwartete Marktkonstellationen „hinüber zu retten", als zu verstehen, was das Ergebnis einer konkreten Periode ist und wie viel zu einem bestimmten Zeitpunkt an finanziellem Ergebnis produziert wurde.

Mit dem Intellectual Capital Statement soll deshalb die Aufmerksamkeit des Managements, der Investoren und anderer Stakeholder auf die immateriellen Ressourcen eines Unternehmens, auf sein Wissenskapital gelenkt werden, das in der heutigen globalen und dynamischen Wissensökonomie für das Überleben und den Erfolg von Firmen wichtiger geworden ist als Finanzkapital und Sachanlagen. Financial Accounting wird durch Intellectual Capital Statements aber keineswegs ersetzt. Die „Financial Results" spielen auch aus der Intellectual Capital Perspektive als letztendliche Prüfsteine des Erfolgs eines Unternehmens eine wichtige Rolle. Investoren und Anteilseigner wollen ja schließlich auch nachgewiesen bekommen, dass das Unternehmen nicht nur für Kunden Wert schafft, sondern auch aus der ökonomischen Perspektive erfolgreich und in der Lage ist, Gewinne und Free Cashflows mit seinen Kompetenzen zu erwirtschaften, um neue Projekte und die eigene Zukunft finanzieren und natürlich den Investoren eine gute Verzinsung ihres Kapitals bieten zu können.

Intellectual Capital Statements leisten so einen wichtigen Beitrag beim Lösen des Transparenzproblems Intangible Assets. Sie beschreiben einen Ansatz, der eine neue Perspektive für die externe Beurteilung und für die interne managementorientierte Steuerung mit Fokus auf die Wissensressourcen eines Unternehmens eröffnet. Diese neue Perspektive tritt dabei neben die finanzielle Perspektive und ergänzt diese, welche durch die traditionellen Rechnungswesen, Controlling- und Rechnungslegungsinstrumente etabliert wird und die vor allem die Finanzressourcen des Unternehmens und dessen Effektivität und Effizienz in der Verwertung dieser Ressourcen im Hinblick auf das Schaffen von Mehrwert transparent machen soll. Aber die Ergänzung der traditionellen Management- und Reportinginstrumente um Intellectual Capital Statements wird nicht genügen, um das Transparenz- und Steuerungsproblem Intangible Assets in den Griff zu bekommen.

Intellectual Capital Statements stellen zwar im Vergleich zu den traditionellen, auf die Ressource Finanzen fokussieren Rechnungswesen, Steuerungs- und Reportinginstrumente, eine neue wichtige ergänzende Perspektive bzw. Sicht auf das Unternehmen für Management und Unternehmensexterne zur Verfügung, die auf die heute so wichtigen Wissensressourcen und ZukunftsPotenziale eines Unternehmens fokussiert. Aber Intellectual Capital Statements fokussieren eben nur auf diese. Dem Management und vor allem natürlich auch Unternehmens-Outsidern wie Finanzanalysten und Investoren fällt es so in der Regel schwer, zu verstehen, wie sich die im Rahmen eines Intellectual Capital Statements dargestellten Wissensmanagement-Strategie und die daraus abgeleiteten Initiativen im konkreten Tagesgeschäft des Unternehmens auswirken, um dann tatsächlich am Ende zu einer Verbesserung der Finanzperformance zu führen (oder auch nicht).

Diese Beziehung herzustellen ist umso wichtiger, da gerade Wissensressourcen, wie auch Intangible Assets generell, immer nur ökonomisch wertschöpfend im Zusammenhang mit anderen Produktivfaktoren sein können. Sie müssen deshalb im Kontext des gesamten Wertschöpfungssystems des Unternehmens betrachtet werden. Andererseits gibt es aber keine direkte Kausalbeziehung zwischen einer Maßnahme zur Verbesserung der Wissensressourcen und der Verbesserung der finanziellen Performance, denn im Bereich Wissensmanagement haben es Unternehmen mit einer ganzen Reihe von „weichen" Einflussfaktoren zu tun. Wissensressourcen, die ja zu einem großen Teil an Menschen gebunden sind, benötigen z.B. eine bestimmte „Atmosphäre", eine entsprechende Unternehmenskultur etc. für ihren produktiven Einsatz bzw. zu ihrer effizienten Verwendung, was sich nur zum Teil eindeutig deskriptiv (beschreibende historische Analyse) und preskriptisch (handlungsorientierte planende Analyse) darstellen und in Kausalketten abbilden lässt. Dies herauszustreichen ist der Verdienst der dänischen Entwicklung im Bereich Intellectual Capital Statements und vor allem von *Jan Mouritsen*, der in seinen Aufsätzen und Vorträgen auf dieses Phänomen immer wieder hinweist. Gerade deshalb wird ja auch betont, dass Wissensmanagement in einer eigenen, von den Bedingungen der Finanzsphäre abgekoppelten Perspektive bearbeitet werden sollte. Die Evaluierung und das Management von Wissensressourcen erfordert eine Sicht, die von der Perspektive der operativen Steuerung eines Unternehmens genauso abstrahiert wie die Finanzperspektive, aber in der Lage ist, (zunächst) unbeeinflusst von den Restriktionen der Finanzperspektive, ihrer eigenen Logik zu folgen. Diese Argumentation ist richtig. Trotzdem muss es dem externen Interessierten und dem Management möglich gemacht werden, Wissensmanagement im Kontext der operativen Aktivitäten eines Unternehmens und deren Steuerung zu sehen, um so verstehen zu können, wie sich die im Rahmen des Wissensmanagements geschaffenen Potenziale in wirtschaftliche und finanzielle Performance transformieren können. Dazu muss zwischen die Wissensmanagement-Perpektive und die Finanzperspektive, quasi als „Übersetzer", die Steuerungsperspektive treten, die die operativen Aktivitäten und Managementprozesse des Tagesgeschäfts im Unternehmen abbilden und steuerbar machen.

Alle drei Perspektiven, die Sicht auf die Wissensressourcen, die (eher operative) Steuerungssicht, und die Finanzsicht stehen zunächst einmal lose nebeneinander, bilden aber dennoch zusammen die gesamte „Produktionsfunktion" eines Unternehmens für die Generierung von Mehrwert, vom Wissensmanagement, über die operativen Aktivitäten und Managementprozesse, bis zur Finanzperformance ab. Jede ist dabei durch ihre jeweils eigene Logik bestimmt. Die Verbindung zwischen den Sichten ist dann intern wie auch extern durch entsprechende Kommunikationsprozesse (Management- bzw. Stakeholder-Dialoge) sicherzustellen. Denn die Fakten allein, die in jeder Perspektive berichtet werden, reichen für die umfassende Interpretation für eine Interventionsabsicht (Management) oder für eine zukunftsgerichtete vergleichende Beurteilung (Investoren) nicht aus. Erst durch Kommunikationsprozesse werden verschiedene Standpunkte von betroffenen Personen und Parteien so zusammengeführt, dass eine entsprechende Interpretation der Fakten und eine „Theoriebildung“ möglich werden, die ein einigermaßen gesichertes Handeln oder Beurteilung zulässt. Um die dafür erforderlichen Fakten strukturiert und in einer einigermaßen vergleichbaren Form bereit zu stellen, bedarf es eines Frameworks für die Re-Präsentation von relevanten Management-Informationen und für das externe Reporting, das alle drei Perspektive umfasst und es – im Rahmen internen und externen der Kommunikationsprozesse – möglich macht, die Verbindungen zwischen den jeweils prinzipiell eigenständigen Perspektiven interpretativ herzustellen (vgl. *Abb. 10*).

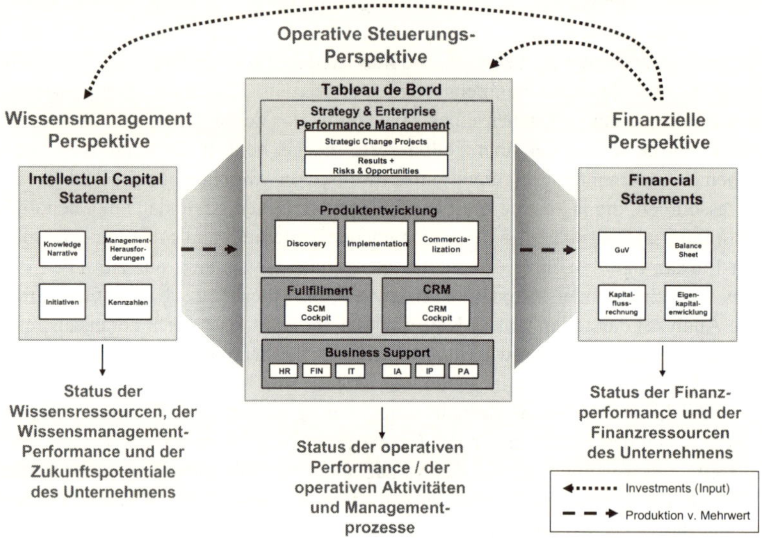

Abb. 10: Framework für internes und externes Reporting und die Unternehmenssteuerung

Dieses Framework soll die Sicht auf die Performance eines Unternehmens sowohl mit einem Blick auf den immateriellen Potenzialaufbau (Wissensmanagement-Perspektive), wie im Konzept der Intellectual Capital Statements dargestellt, als auch mit der Perspektive auf das Operations-Modell des Unternehmens (Steuerungssicht) kombinieren, über das quasi als Transmissionsriemen die Potenziale in konkrete operative Ergebnisse und schließlich in Erlöse, Deckungsbeiträge und Gewinne (Finanzperspektive) umgesetzt werden. Das neue Modell muss also, neben der Sicht auf die Finanzeffizienz, die nur auf die Ränder der „Black Box“, nämlich auf finanziellen Input und Output fokussiert, auch den Vorstoß in das Innere der „Black Box“ über die beiden übrigen Perspektiven erlauben. Dabei muss die gesamte Wertschöpfungsproduktionsfunktion des Unternehmens transparent werden: von den strategischen WissensPotenzialen und deren Weiterentwicklung, über die Maßnahmen zur Strategieumsetzung im Rahmen der strategischen Change Management Projekten, den Management-Prozesse im Bereich Corporate Performance Management und allgemeiner Ressourcensteuerung, den operativen Verwertungsaktivitäten, bis zu Aktivitäten in der Finanzsphäre. Um dies möglich zu machen, muss neben der Entwicklung von Instrumenten für ein umfassendes strategisches Wissensmanagement, was der Fokus des Intellectual

Capital Statement Konzepts ist, ein umfassender Neuentwurf des Unternehmenssteuerungs- und Performance Management-Systems treten, das den Bedingungen der Wissensökonomie mit der Dominanz von Intangible Assets gereicht wird (vgl. *Daum*, 2002a). Darüber hinaus bedarf es auch einer Neugestaltung des Financial Accounting und Reporting, die über die Eingangs erwähnten Reformen der internationalen Rechnungslegungsstandards hinaus geht.

4.1 Von der traditionellen Kosten-, Leistungs- und Ergebnisrechnung zu einem umfassenden neuen Performance-Management- und Steuerungsansatz

Im Bereich Performance Management- und Steuerungssystem geht es um die Perspektive „wirtschaftliche Performance operativer Aktivitäten". Während es aus der Wissensmanagement-Perspektive darum geht, die breite Kompetenzplattform, also Wissensressourcen und -potenziale des Unternehmens transparent und „manageable" zu machen, um dessen Zukunftsfähigkeit zu sichern, ist die Zielsetzung des Steuerungssystems, durch operative Aktivitäten wirtschaftliche Performance zu generieren bzw. diese aus einer solchen Sicht zu optimieren.

Das Performance Management- und Steuerungssystem im Wesentlichen nur auf die traditionelle industrielle Kosten-, Leistungs- und Ergebnisrechnung zu stützen reicht aufgrund der Eingangs beschriebenen Veränderungen der Wertschöpfungsmodelle der Unternehmen nicht mehr aus. Die Kosten-, Leistungs- und Ergebnisrechnung als Instrument für die interne Steuerung und Optimierung von Ressourcenverbräuchen, also der Effizienz, sollte deshalb zu einem das gesamte Spektrum des jeweiligen Wertschöpfungssystems eines Unternehmens umfassenden Unternehmenssteuerungssystems ausgebaut werden, das auf dessen Effektivität im Hinblick auf Markt- und Stakeholder-Erwartungen abhebt. Dass dies viele Unternehmen bereits seit längerem erkannt haben, zeigt das Aufkommen neuer Performance-Management Konzepte in den letzten Jahren.

Konzepte wie beispielsweise die Balanced Scorecard, Portfoliomanagementtechniken im F&E-Bereich von Pharmaunternehmen, Customer Lifetime Value Management als strategischer Rahmen für die Steuerung und Optimierung operativen Aktivitäten im Bereich Kundenbeziehungsmanagement (CRM) oder jüngst das Dynamic Performance Management – oft im Rahmen des Beyond Budgeting Modells genannt, sind dafür ein Beleg. Was hier allerdings bislang fehlte, war ein systematischer umfassender Ansatz für das Performance Management, der diese neuen Konzepte und Techniken in Beziehung zu den neuen Operations-Modellen der Unternehmen setzt.

Wird dann ein solches Steuerungsmodell im Rahmen des hier vorgeschlagenen Frameworks in den Zusammenhang mit der auf den Potenzialaufbau fokussierenden Wissensmanagement-Perspektive und der finanziellen Perspektive gestellt, wird erst deutlich, wie ein Unternehmen konkret seine Wissensressourcen nicht nur an sich aufbaut (Wissensmanagement-Perspektive), sondern wie dies im Rahmen der operativen Aktivitäten und Managementprozesse geschieht. Erst so kann transparent werden, wie Wissensmanagementstrategie und -initiativen dort, im operativen Tagesgeschäft, durch eine entsprechende Ausgestaltung der Aktivitäten gefördert (oder eben behindert) wird, und vor allem auch, wie die geschaffenen Potenziale im Rahmen der operativen Tätigkeit und Steuerung so verwertet werden, dass sowohl die kurz- als auch die langfristige Finanzperformance optimiert wird.

Um ein geeignetes Performance Management- und Steuerungssystem entwerfen zu können, muss aber zunächst einmal besser verstanden werden, wie Unternehmen heute mit ihren

Wertschöpfungssystemen Mehrwert schaffen. Im Zentrum solcher Analysen steht hier meist der (Produkt)Innovationsprozess, der in vielen Unternehmen heute zumindest für eine gewisse Zeit Wettbewerbsvorteile schafft und so für einen Großteil des in der so genannten Competitive Advantage Period geschaffenen Mehrwerts verantwortlich ist. *Lev* hat deshalb zurecht darauf hingewiesen, dass dieser Prozess mehr Aufmerksamkeit seitens des Managements aber auch seitens der Investoren verdient und deshalb durch ein kennzahlbasiertes Zusatzreporting, das er „Value Chain Blueprint" nennt, transparent gemacht werden sollte (*Lev*, 2001, S. 110-118). Der Innovationsprozess muss allerdings auch immer in Zusammenhang mit den anderen beiden großen operativen Aktivitätsbereichen in Unternehmen betrachtet werden, die ja erst die Verwertungsmöglichkeit von Produktinnovationen im Hinblick auf realisierbare Produktverkäufe und damit Erlöse begründet: das Supply Chain Management, oder allgemein, das Fullfillment, dass die neuentwickelten Produkte in ihre physische Existenz und zum Kunden bringt und das Customer Relationship Management, das durch das Schaffen von Kundenbeziehungen oder Markenreputation erst die Gelegenheit für ausreichende, die Entwicklungskosten deckende Verkäufe schafft. Alle drei Aktivitäts- bzw. Prozessbereiche müssen im Rahmen eines „Value Networks" zusammenwirken, um tatsächlich Mehrwert für Kunden, Investoren und andere Unternehmensstakeholder zu schaffen.

Zusätzlich werden, quasi in einer Sub-Perspektive des Steuerungs- bzw. Performance Managementsystems („Business Support"), Instrumente und Prozesse dargestellt, die für die Steuerung und das Management von Ressourcen (Finanzen, IT-Infrastruktur, Personal) und von Assets, also von einzeln identifizierbaren Produktivvermögen im weiteren Sinne (Intangible Assets im Sinne der Rechungslegungsstandards, Intellectual Property und Sachanlagen) zuständig sind. Aufgabe der entsprechenden Managementprozesse ist die Beschaffung, der systematische Aufbau bzw. Weiterentwicklung dieser Ressourcen oder Assets durch entsprechende Maßnahmen als auch deren optimale Verwertung im Rahmen der operativen Aktivitäten. Beides reicht in das Operations Management hinein, wo diese Faktoren ja verwertet und auch teilweise (bei Intangible Assets und Intellectual Property) erst geschaffen werden.

Darüber hinaus bedarf es einer Gesamtsteuerung („Enterprise Management"), um die Trade-Offs im System laufend zu optimieren. Dies umfasst das kontinuierliche Strategiemanagement, das das gesamte „Wertschöpfungsrezept" des Unternehmens steuert und über entsprechende strategische Programme immer wieder an neue Gegebenheiten anpasst (inkl. der Wissensmanagement-Initiativen) und das Corporate Performance-Management, das die Gesamtperformance des Unternehmens über alle ökonomischen Subsysteme und Wertschöpfungsaktivitäten hinweg laufend optimiert.

Damit sind die wesentlichen Elemente eines umfassenden, neuen Performance-Management bzw. Steuerungs-Systems beschrieben, die beim Schaffen und der Optimierung von wirtschaftlicher Performance zusammenwirken müssen (vgl. *Abb. 11*). Die wichtigsten Bausteine für den Betrieb eines solchen Performance-Management-Systems sind einmal ein umfassendes, aber trotzdem möglichst konzentriertes Mess- und Kennzahlsystem in Form eines so genannten Tableau de Bord und Strategie-, Performance-Management und operativer Management-Prozesse, die ineinander greifen (vgl. *Daum*, 2002b, S. 255–372).

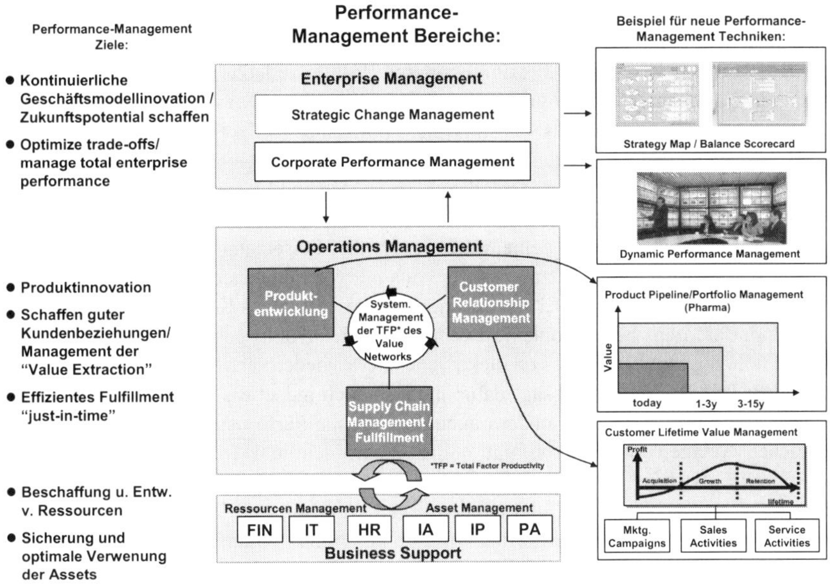

Abb. 11: Holistisches Performance Management Modell

4.2 Vom Financial Accounting zum Business Accounting und zu einer neuen Messgröße für die Gesamtperformance von Unternehmen

Außerdem sind auch in der Finanzperspektive, repräsentiert durch das traditionelle externe Rechnungswesen, Erweiterungen und Verbesserungen notwendig, die nur teilweise durch die Eingangs erwähnten Reformen der internationalen Rechnungswesenstandards adressiert werden. Im wesentlichen geht es hier einmal um die Erweiterung des Financial Accounting zu einem umfassenderen Business Accounting und zum Zweiten um die Entwicklung einer neuen Messgröße für den wirtschaftlichen Gesamterfolg von Unternehmen - jenseits der klassischen Messgrößen wie Gewinn oder auch Economic Value Added, die nicht mehr in der Lage sind, die heutigen komplexeren Business Economics eines Unternehmens in einer Kennzahl zu reflektieren.

Das traditionelle Financial Accounting muss zu einem erweiterten Business Accounting ausgebaut werden, welches das Business Modell des jeweiligen Unternehmens wieder spiegelt. So sollen etwa in der Gewinn- und Verlustrechnung, neben den Produktherstellkosten, auch die Kosten anderer wichtiger Wertschöpfungsaktivitäten im Wertschöpfungsmodell des Unternehmens separat ausgewiesen werden, wie die der Produktentwicklung, des Kompetenzaufbau von Humankapital, des Ausbaus der Vermarktungspartnernetzwerkes etc. Aufwendungen mit Investitionscharakter könnten, wie von *Lev* vorgeschlagen (vgl. *Lev*, 2001, S. 122–127), in der Bilanz aktiviert werden, sobald entsprechende gesicherte Erkenntnisse über deren Nützlichkeit und Erfolg vorliegen. Sie sollten in Raten abgeschrieben werden, wenn die entsprechenden Erlöse fließen. Dies würde die Gewinn- und Verlustrechnung in ihrer Aussagekraft erhöhen, da diese nicht mehr durch Aufwendungen mit Investitionscharakter verzerrt wird. Zudem kann durch den Ausweis aller Investitionen in der Bilanz der „Return on Investment" der wichtigsten Investitionsmaßnahmen ermittelt werden.

Zusätzlich wird eine neue Gesamtmessgröße für die wirtschaftliche Performance von Unternehmen benötigt, denn die traditionellen wie operatives Ergebnis, EBIT, ROI oder auch neuere Konzepte wie Economic Profit / Economic Value Added, versagen hier angesichts der heutigen Dominanz von Intangible Assets. Eine solche neue Messgröße muss darstellen, wie effizient ein Unternehmen als Gesamtsystem mit seinen Teilkomponenten, den operativen und Management-Prozessen, seiner Organisationsstruktur, seiner Unternehmenskultur und seiner Wissens- und Kompetenzplattform so im Verbund zusammen wirken kann, so dass Inputfaktoren möglichst optimal in wirtschaftlichen Mehrwert heute und in der Zukunft transformiert werden können. Als eine Metapher für die Zielsetzung einer solchen Messgröße könnte der Begriff des „Wirkungsgrades" aus der Verfahrenstechnik dienen. Der Wirkungsgrad bezeichnet dabei die Effizienz einer Anlage, z.B. eines Kohlekraftwerks, mit der aus den Inputfaktoren, beim Kohlekraftwerk die in der Kohle gebundene Energie, einen möglichst hohen Output in Form von elektrischer bzw. wieder verwendbarer Wärmeenergie zu erzeugen. Der Wirkungsgrad sagt dabei nicht einfach nur etwas über die aktuelle „Performance" des Kraftwerks aus, sondern auch über dessen Performance, die auch zukünftig bei gleicher Anlagenkonfiguration mit hoher Wahrscheinlichkeit erreicht werden kann. Genauso sollte es diese neue Gesamtperformance-Messgröße im betriebswirtschaftlichen Bereich möglich machen, eine Unternehmensorganisation, mit all ihren wirtschaftlich verwertbaren Potenzialen und Fähigkeiten, hinsichtlich ihrer zukünftigen wahrscheinlichen wirtschaftlichen Gesamt-Performance beurteilen zu können. Eine mögliche Lösung stellt hier das Konzept der Enterprise Total Factor Productivity dar (vgl. *Lev/Daum*, 2003 sowie *Daum*, 2003a und 2003b). Allerdings steht hier die betriebswirtschaftliche Forschung noch ganz am Anfang. Denn im Gegensatz zur Verfahrenstechnik, die sich auf bereits gut erforschte Grundlagen der Physik und Chemie stützen kann, ist die Forschung im Bereich der organisatorischen Effizienz, gerade moderner wissensbasierter Unternehmen, noch wenig vorangeschritten. Auch greift der Vergleich mit der Technik hier nur bedingt, da es sich bei Unternehmensorganisationen um sozio-technische Gebilde, in denen neben technischen, vor allem organisationspsychologische und soziale Mechanismen wirken.

4.3 Erweiterte externe Berichterstattung und Kommunikation: Vom Financial Statement Reporting zum Business Reporting

Darüber hinaus muss das traditionelle externe Reporting von Unternehmen, das bislang weitgehend nur auf den Financial Statements basiert, zu einem umfassenderen Business Reporting ausgebaut werden. Dieses soll Investoren und anderen externen Stakeholdern Einblick in die tatsächlich wertschöpfenden Aktivitäten heutiger Unternehmen gewähren und es diesen erlauben, die wahren Potenziale und Risiken des Unternehmens, als auch dessen Fähigkeit, diese in nachhaltige Ergebnisse umzusetzen, einschätzen zu können.

In den USA hatte sich in den letzten Jahren die allgemeine Meinung herauskristallisiert, dass Unternehmen zusätzlich zu den Financial Statements mittels eines so genannten „Supplemental Reporting" über ihre Geschäftsstrategie, ihr Geschäftsmodell als auch über darauf bezogenen Operational und Intangible Key Performance Indicators berichten sollten (vgl. *SEC*, 2001, und *FASB*, 2001). Damit eine derartige Praxis eine größere Verbreitung finden kann, sind darüber hinaus branchenspezifische Standards für derartige Kennzahlen und Berichtsschemata erforderlich (vgl. *DiPiazza/Eccles*, 2002, 56 ff.). Die dänischen Intellectual Capital Statements gehen prinzipiell in die gleiche Richtung wie die amerikanischen Ansätze, nehmen aber, was oft übersehen wird, eine ganz andere Perspektive ein. Während die Amerikanischen Inititiativen sehr nah am Performance-Management und an der Finanzperformance sind, versucht man mit dem dänischen Ansatz, wie oben beschrieben, bewusst dazu etwas Distanz zu nehmen und das Unternehmen aus einer Wissensmanagement-Perspektive, die eher auf das langfristig wichtige Potenzial und die entsprechenden Kompetenzen abhebt, darzustellen.

All diese Initiativen versuchen in der Tat für die externe Beurteilung von Unternehmen relevante Sachverhalte transparent zu machen. Was jedoch bislang fehlt, ist eine Systematik, die die verschiedenen Perspektiven und Sichten in einen Gesamtzusammenhang stellt. Als Lösung dafür soll das oben beschriebene Framework vorgeschlagen werden, das als Rahmen nicht nur für die interne Steuerung, sondern auch für das externe Berichtswesen und für die eine zielgruppenspezifische externe Kommunikation dienen kann. Die wichtigsten Element sind hier (vgl. *Abb. 12*):

- Der Bericht über die Entwicklung der Wissenspotenzial und der breiten Kompetenzplattform des Unternehmens (Wissensmanagement-Perspektive) – z.B. in Form der in Dänemark praktizierten Intellectual Capital Statements.
- Der Bericht über den Status sowohl der strategischen Initiativen, der Gesamtperformance und der operativen Aktivitäten und Business Support Prozesse.
- Der Bericht über die finanzielle Gesamtperformance, bestehend aus Financial Statements und einem Branchenbenchmark der wirtschaftlichen Gesamteffizienz - z.B. auf Basis des Enterprise Total Factor Productivity Konzepts.
- Hintergrundinformationen zur „Theory of Business", zur Strategy und zum Geschäftsmodell, zu den allgemeinen Management-Herausforderungen und zu Risiken und Chancen ergänzt, um Fallbeispiele aus der täglichen Management-Praxis.

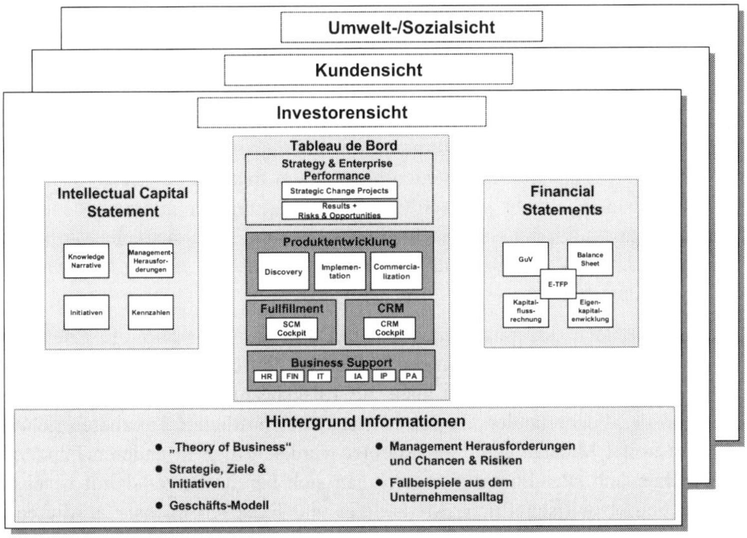

Abb. 12: Framework für einen neuen Ansatz im Bereich externes Berichtswesen und Unternehmenskommunikation

Wichtige Stakeholder-Gruppen müssen außerdem in einen laufenden aktiven Dialog einbezogen werden, um diese dauerhaft für das Unternehmen zu engagieren: Stakeholder Relationship-Management wird zu einer wichtigen laufenden Managementaufgabe und muss in die internen Managementprozesse integriert werden (vgl. *Daum*, 2000).

5. Fazit

Durch die zunehmende Dominanz der Intangible Assets werden Unternehmen zur „Black Box“: die traditionellen Steuerungs- und Reportinginstrumente versagen. Diese Dominanz der Intangibles ist Ausdruck des Wandels vom industriellen Unternehmenswertschöpfungsmodell zu den komplexeren wissensbasierten Wertschöpfungsmodellen heutiger Unternehmen, in denen laufende Innovation und Beziehungsaufbau und –pflege zu Geschäftspartnern zu den Mehrwert schaffenden Aktivitäten wurden. Dieser Wandel hat die internen Business Economics der Unternehmen so nachhaltig verändert, dass die traditionellen Management-, Rechnungswesen- und Reportinginstrumente nicht mehr „greifen“. Das Ergebnis ist ein internes Steuerungs- und externes Transparenzproblem. Das in Dänemark entwickelte und praktizierte Konzept der Intellectual Capital Statements adressiert diese Probleme und soll dafür eine Lösung bereitstellen – was teilweise gelingt.

Intellectual Capital Statements sind Unternehmensberichte, die zusätzlich zur traditionellen Rechnungslegung veröffentlicht werden, um Investoren, Finanzanalysten und andere Stakeholder über die nicht in der Bilanz sichtbaren Potenzialaufbau (oder -abbau) eines Unternehmens zu informieren. Dies soll eine bessere Einschätzung der zukünftigen finanziellen Performance eines Unternehmens, aber auch die gezielte Entwicklung, Optimierung und Steuerung dieser Kompetenzen und Wissensressourcen möglich machen. Damit leisten Intellectual Capital Statements einen wichtigen Beitrag beim Lösen des Transparenzproblems Intangible Assets. Sie beschreiben einen Ansatz, der eine neue Perspektive für die externe Beurteilung und auch für die interne managementorientierte Steuerung der Wissensressourcen eines Unternehmens eröffnet. Diese neue Perspektive tritt dabei neben die finanzielle Perspektive und ergänzt diese. Aber die Ergänzung der traditionellen finanzorientierten Management- und Reportinginstrumente um Intellectual Capital Statements wird nicht genügen, um das Transparenz- und Steuerungsproblem Intangible Assets vollständig in den Griff zu bekommen. Vielmehr ist auch eine Neugestaltung der internen Steuerungsinstrumente, die auf der traditionellen Kosten-, Leistungs- und Ergebnisrechnung basieren, und des externen Rechnungs- und Berichtswesens erforderlich, das bislang weitgehend auf den Financial Statements aufbaut.

Die Praxis der Intellectual Capital Statements in Dänemark setzt zwar ein Zeichen, dass das Thema Wissensmanagement und Wissenskapital als betriebswirtschaftliches Thema heute ernst genommen wird. Darauf deutet auch die Tatsache hin, dass zunehmend, wie z.B. in Schweden und den Niederlanden, an den betriebswirtschaftlichen Fakultäten Lehrstühle für „Intellectual Capital Management“ eingerichtet werden. Auch in anderen Ländern als den angelsächsischen und skandinavischen, wo man sich bereits länger damit beschäftigt hat, arbeiten nun betriebswirtschaftliche Methodiker und Experten intensiv an diesem Thema. Dazu gehören die Niederlande, Österreich, Italien, Spanien, Frankreich, Australien und in jüngster Zeit auch Deutschland. Was allerdings bislang noch fehlt, ist die Einordnung von Wissens- oder Intellectual Capital Management in eine betriebswirtschaftliche Gesamtsystematik, die dieses Thema in einen konkreteren Zusammenhang mit der (operativen) Unternehmenssteuerung und der Finanzsphäre bringt.

Dies ist gerade beim Thema Wissenskapital bzw. Intangible Assets so wichtig, da sie ihre ökonomische Wirkung nur im Zusammenhang mit anderen Produktivfaktoren entfalten können. Diesen Zusammenhang transparent zu machen, ist eine wichtige Aufgabe der Betriebswirtschaft der nächsten Jahre. Der Herstellung einer Gesamtsicht auf die gesamte „Produktionsfunktion" eines Unternehmens für die Generierung von Mehrwert Unternehmens inklusive aller Produktivfaktoren und Geschäfts- und Managementprozesse, kommt dabei entscheidende Bedeutung zu. Dazu bedarf es neben geeigneter Instrumente für die Darstellung des Unternehmens aus der Finanz-Perspektive und aus einer Wissensmanagement-Perspektive auch eines Steuerungsansatzes, welcher die wichtigsten Wertschöpfungsaktivitäten in ihrer heute typischen Ausprägung in Unternehmen abdeckt. Die wichtigsten Bausteine eines entsprechenden Performance-Management-Systems sind einmal ein umfassendes, aber trotzdem möglichst konzentriertes Mess- und Kennzahlsystem in Form eines so genannten Tableau de Bord, und Strategiemanagement, Corporate Performance-Management sowie operativer Management-Prozesse, die ineinander greifen und die Trade-Offs im Geschäftssystem beherrschbar machen. Wird die Steuerungsperspektive im Rahmen des in diesem Beitrag vorgeschlagenen Frameworks in den Zusammenhang mit der auf den Potenzialaufbau fokussierenden Wissensmanagement-Perspektive und der finanziellen Perspektive gestellt, wird deutlich, wie ein Unternehmen konkret seine Wissensressourcen nicht nur an sich aufbaut, sondern wie dies im Kontext der operativen Aktivitäten und Managementprozessen geschieht, wie also die dargestellte Wissensmanagement-Strategie und die daraus abgeleiteten Initiativen sich im konkreten Tagesgeschäft des Unternehmens auswirken, um dann tatsächlich am Ende zu einer Verbesserung der Finanzperformance zu führen (oder auch nicht). Erst durch eine solche Darstellung der gesamten „Produktionsfunktion" eines Unternehmens für die Generierung von Mehrwert, vom Wissensmanagement, über die operativen Aktivitäten und Managementprozesse bis zur Finanzperformance, wird es möglich, einem externen Stakeholder, aber auch dem Management einen Vorstoß in die Black Box „Unternehmen" möglich und diese transparent zu machen. Wird dies um ein erweitertes Konzept für die Gesamtperformance-Messung, wie etwa das Enterprise Total Factor Produktivity Konzept, ergänzt, steht dann auch eine einzelne Messgröße für den schnellen externen Unternehmensvergleich zur Verfügung, der in einer Welt, in der Intangible Assets dominieren, besser geeignet ist, die ökonomische Effizienz, den „Gesamtwirkungsgrad" eines Unternehmens bei der Verwertung von Ressourcen, zu beurteilen, als dies traditionelle Messgrößen wie Gewinn, EBIT, ROI oder auch EVA leisten können.

Die Notwendigkeit eines durch ein solches Framework geschaffenen neuen Systematik der Betriebswirtschaft als Grundlage für die Entwicklung geeigneter Performance Management/Steuerungs- und Reportingmodelle wächst mit jedem Aufkommen eines neuen Konzeptes, das ein Einzelproblem im Management oder der Rechnungslegung heutiger Unternehmen adressiert. Die „Inflation" derartiger Konzepte ist Ausdruck des Versagens unserer traditionellen Management-, Controlling-, Rechnungswesen- und Reportingkonzepte vor der Herausforderung Intangible Assets. Ohne eine solche neue betriebswirtschaftliche Systematik und Theorie droht ein noch verschärfteres Management- und Governanceproblem für die Unternehmen. Denn diese sind dann gezwungen, ohne die Unterstützung einer soliden und auf breiter Basis entwickelten und erprobtem Systematik, wie sie etwa die traditionelle Betriebswirtschaftslehre mit dem traditionellen Rechnungswesen bereitstellt, eigene Wege per Versuch und Irrtum zu finden. Dabei ist nicht nur das Risiko groß, Fehltritte zu tun, sondern durch das Fehlen allgemeiner Standards wird die Vergleichbarkeit zwischen Unternehmen erschwert und die externe Transparenz wird nicht verbessert, sondern sinkt.

Das hier vorgestellte Framework, als systematischer Rahmen, sowohl für die Unternehmenssteuerung, als auch für das externe Reporting, versteht sich jedoch nur als erster Entwurf und erhebt noch keinen Anspruch auf Vollständigkeit. Es soll damit die notwendige Diskussion zwischen den entsprechenden Fachleuten in den Unternehmen, Managern, der betriebswirtschaftlichen Wissenschaft, Managementberatern, Wirtschaftsprüfern und anderen relevanten Stakeholdern eröffnet werden. Es versteht sich als Appell, über die Lösung von Einzelfragen und Problemen im Zusammenhang mit Intangible Assets in der Rechnungslegung, im Controlling und im Unternehmensberichtswesen hinaus am Gesamtbild und an einer umfassenden betriebswirtschaftlichen Systematik zu arbeiten, die in der Lage ist, heutige Unternehmen wirklichkeitsnäher abzubilden und somit auch einen wichtigen Beitrag zur Lösung der aktuell diskutierten Corporate Governance Problematik zu leisten, deren Ursache zu wesentlichen Teilen auch im Transparenzproblem Intangible Assets zu finden ist.

6. Literatur

Aboody, D. and *Lev. B.*, R&D Productivity in the Chemical Industry, New York 2001, (die Studie ist erhältlich über Baruch Lev's Website www.baruch-lev.com).

Danish Agency for Trade and Industry, A Guideline for Intellectual Capital Statements, Kopenhagen 2000 (http://www.efs.dk/download/pdf/videnUK.pdf).

Danish Ministry of Science, Technology and Innovation, Intellectual Capital Statements – The New Guideline, Kopenhagen 2003a (http://www.videnskabsministeriet.dk/fsk/publ/2003/guideline_uk/guideline_uk.pdf).

Danish Ministry of Science, Technology and Innovation, Analysing Intellectual Capital Statements, Kopenhagen 2003b (http://www.videnskabsministeriet.dk/fsk/publ/2003/analyse_uk/analyse_uk.pdf).

Daum, J. H., SEM-SRM Stakeholder Relationship Management, in: *Küting, K.* und *Weber,C.-P.* (Hrsg.), Wertorientierte Konzernführung. Kapitalmarktorientierte Rechnungslegung und integrierte Unternehmenssteuerung, Stuttgart 2000, S. 467-516.

Daum. J. H., Werttreiber Intangible Assets: Brauchen wir ein neues Rechnungswesen und Controlling? – Ein Ansatz für ein verbessertes Managementsystem, in: Controlling, 14. Jg. (2002a), H. 1, S. 15-24.

Daum. J. H., Intangible Assets oder die Kunst, Mehrwert zu schaffen, Bonn 2002b.

Daum, J. H., Intangible Assets und die Optimierung der Enterprise Total Factor Productivity, in: Controlling&Management, 47. Jg. (2003a), H. 2, S. 129-135.

Daum. J. H., Intellectual Capital Statements, in: Controlling, 15. Jg. (2003b), H. 1, S. 143-153.

Daum, J. H. und *Edvinsson L.*, Intellectual Capital: Aktivposten Wissenskapital – Interview mit *Leif Edvinsson*, in: Controlling & Finance, o. Jg. (2001), H. 10, S.4f.

DiPiazza, S. A. Jr. and *Eccles, R. G.*, Building Public Trust: The Future of Corporate Reporting, New York 2002.

Edvinsson, L. and *Malone M. S.*, Intellectual Capital, New York 1997.

FASB Business Reporting Research Project, Improving Business Reporting: Insights into Enhancing Voluntary Disclosures, 2001 (http://www.fasb.org/brrp/BRRP2.pdf).

Gu, F. and *Lev. B.*, Intangible Assets: Measurement, Drivers, Usefulness, New York 2001 (Die Studie ist erhältlich über Baruch Lev's Website www.baruch-lev.com).

Hand, J. and *Lev, B.*, Intangible Assets: Values, Measures, and Risks, Oxford, U.K., 2003.

Lamparter, D. H., Wer bleibt am Standort D?, in: DIE ZEIT, Nr. 38, 11. September 2003, S. 23-24.

Lev. B. und *Daum, J. H.*, Intangible Assets: Neue Ansätze für Unternehmenssteuerung und Berichtswesen, in: *Horváth, P.* und *Gleich, R.* (Hrsg.), Neugestaltung der Unternehmensplanung, Stuttgart, 2003, S. 33-50.

Lev. B., Intangibles: Management, Measurement, and Reporting, Washington, D.C., 2001.

Mouritsen, J., Driving Growth: Economic Value Added Versus Intellectual Capital, in: Management Accounting Research, Vol. 9 (1998), H. 12, S. 461–482.

Mouritsen, J., Measuring and Intervening: Indicating and Crafting Intellectual Capital Effects, Working Paper und Vortrag beim PMA Intellectual Capital Symposium "Research and Theory in Intellectual Capital", 1.-2. Oktober 2003, Cranfield School of Management, UK, S.4.

Mouritsen, J., Larsen, H. T. and *Bukh, P. N.*, Valuing the Future: Intellectual Capital Supplements at Skandia, in: Accounting, Auditing and Accountability Journal, Vol. 14 (2001a), No. 14, S. 399–422.

Mouritsen, J., Larsen, H. T. and *Bukh, P. N.*, Intellectual Capital and the 'Capable Firm': Narrating, Visualising and Numbering for Managing Knowledge, in: Accounting, Organizations and Society, Vol. 26 (2001b), No. 7/8.

Mouritsen, J., Larsen, H. T. and *Bukh, P. N.*, Reading Intellectual Capital Statements: Describing and Prescribing Knowledge Management Strategies, in: Journal of Intellectual Capital, Vol. 2 (2001c), No. 4, S. 359–383.

OECD (Organisation for Economic Cooperation and Development) Directorate Science, Technology & Industry, Science, Technology and Industry Scoreboard 1999: Benchmarking Knowledge-based Economies", Paris 1999 (http://www.oecd.org//dsti/sti/stat-ana/prod/scorebd_summ.htm).

SEC Taskforce, Strengthening Financial Markets: Do Investors have the Information they need?, New York 2001 (www.fei.org/finrep/files/SEC-Taskforce-Final-6–6-2k1.pdf).

Skandia, Human Capital in Transformation: Intellectual Capital Prototype Report, Stockholm 1998 (http://www.skandia.com/en/includes/documentlinks/annualreport1998/ supplements1998.zip).

Systematic, Intellectual Capital Report 2002, (www.systematic.dk).

Nachhaltige Wertsteigerung mit immateriellem Vermögen

Hans-Gerd Servatius [1]

1. Dritte Welle der Führung
2. Immaterielles Vermögen als komplexes adaptives System
3. Prozessmodell der strategischen Führung
4. Entwicklung von neuen Geschäftskonzepten
5. Nutzung von Potenzialen als Werttreiber
6. Verbesserung des Corporate Reporting und nachhaltige Wertsteigerung
7. Fazit
8. Literatur

---

[1] Prof. Dr. *Hans-Gerd Servatius*, Horváth AG, Düsseldorf

Die Bedeutung des immateriellen Vermögens für die Unternehmensführung hat in den letzten Jahren stark zugenommen. Wissenschaft und Praxis sind daher auf der Suche nach neuen Konzepten zum Management und Reporting von Intangibles. Wir betrachten immaterielles Vermögen als komplexes adaptives System, das im Rahmen leistungsfähiger Geschäftskonzepte das Potenzial für eine nachhaltige Wertsteigerung bildet. Die Aufgabe des Managements gleicht dabei einer Navigation in dynamischen Wertlandschaften. Auf dieser Grundlage skizzieren wir ein Prozessmodell der strategischen Führung. Anschließend beschreiben wir die Teilprozesse Geschäftskonzept-Innovation, Nutzung von Potenzialen als Werttreiber und Verbesserung des Corporate Reporting.

1. Dritte Welle der Führung

Die betriebswirtschaftliche Führungslehre wurde in den letzten Jahrzehnten durch drei große Wellen geprägt. In den 80er-Jahren begann der Siegeszug des strategischen Managements mit seinen verschiedenen Schulen und konzeptionellen Ansätzen. Die 90er-Jahre waren das Jahrzehnt der Informationstechnik. Zunächst stand das Reengineering der operativen Geschäftsprozesse im Mittelpunkt des Interesses. Danach begann der E-Business Hype, der mit dem Zusammenbruch der New Economy endete. In der Schlussphase dieser zweiten Welle konzentrieren sich etablierte Unternehmen auf ein Collaborative Business, das auf die Verbesserung der kooperativen Prozesse mit Kunden und Wertschöpfungspartnern abzielt.

Gegenwärtig zeichnet sich eine dritte Welle ab, die durch eine zunehmende Bedeutung des immateriellen Vermögens gekennzeichnet ist (vgl. *Lev*, 2001; *Daum*, 2002; *Servatius*, 2003). Die Unternehmen erkennen, dass die so genannten „Intangibles" einen wichtigen und oft vernachlässigten Werttreiber darstellen. Angesichts der Krise der Aktienmärkte erhoffen sie sich von einer bewussteren Gestaltung des immateriellen Vermögens einen entscheidenden Beitrag zur Leistungserhöhung und nachhaltigen Wertsteigerung (vgl. *Abb. 1*).

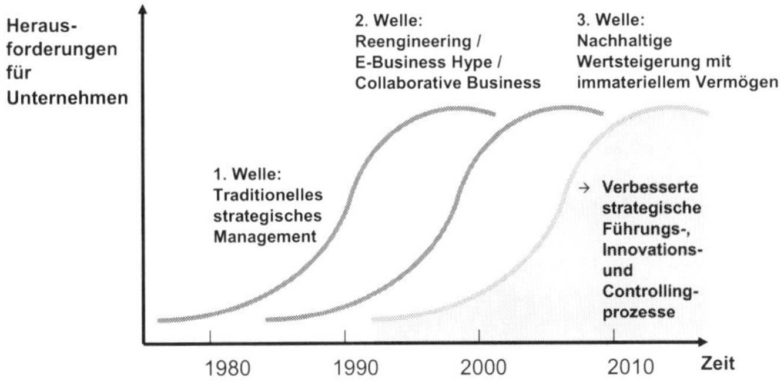

Abb. 1: Drei Wellen der Führung

In den letzten Jahren hat es zahlreiche Versuche gegeben, das immaterielle Vermögen zu strukturieren. In Balanced Scorecards finden sich Intangibles in der Kunden-, der Prozess- und vor allem in der Potenzial-Perspektive. Eine verbreitete Gliederung des so genannten intellektuellen Kapitals unterscheidet zwischen Humankapital, Kundenkapital, Prozesskapital sowie Erneuerungs- und Entwicklungskapital (vgl. *Edvinsson/Malone*, 1997).

Bei diesen frühen Ansätzen steht die Messung des immateriellen Vermögens im Vordergrund (vgl. *Stewart*, 2001, S. 247 ff.). Es stellt sich aber die Frage, ob die Erfassung von über hundert Indikatoren, wie z.B. beim schwedischen Versicherungsunternehmen *Skandia*, oder deren Zusammenfassung zu Indizes wirklich den angestrebten wirtschaftlichen Erfolg bringen.

Wir vertreten in diesem Beitrag die These, dass sich der Wert des immateriellen Vermögens aus seiner Rolle im Geschäftskonzept eines Unternehmens ergibt (vgl. *Palass/Servatius*, 2001). Dabei ist zu berücksichtigen, dass die einzelnen Vermögensbestandteile ein komplexes adaptives System bilden. Die Herausforderung für das Corporate Reporting liegt darin, den Kapitalmärkten zu vermitteln, wie das immaterielle Vermögen im Rahmen von neuen Geschäftskonzepten zur Wertsteigerung beiträgt. Dies erfordert eine Weiterentwicklung der strategischen Führungs-, Innovations- und Controllingprozesse innerhalb der skizzierten dritten Welle.

2. Immaterielles Vermögen als komplexes adaptives System

Die Ökonomie des immateriellen Vermögens ist durch spezifische Kosten-Nutzen-Relationen gekennzeichnet, die als Wertreiber bzw. Wertreduzierer wirken (vgl. *Lev*, 2001, S. 21 ff.). Wichtige Werttreiber sind

- die häufig unbegrenzte Multiplizierbarkeit („Scalability") eines einmal entwickelten immateriellen Vermögensbestandteils und
- Netzwerkeffekte aufgrund des mit der Anzahl der Kunden steigenden Nutzens.

Eine Wertreduktion kann sich ergeben aus:

- den eingeschränkten Ausschlussmöglichkeiten eines Einsatzes durch Wettbewerber,
- den inhärenten Risiken z.B. von F&E-Projekten und
- der begrenzten Handelbarkeit von Intangibles.

Diese spezifischen Werttreiber und Wertreduzierer sind beim Management und Reporting des immateriellen Vermögens zu beachten. So erfordert z.B. die Behandlung einer Ausgabe als Investition die volle Kontrolle über das Investitionsobjekt, was bei vielen Intangibles nicht der Fall ist.

Was gehört nun alles zum immateriellen Vermögen? Eine relativ weite Interpretation des Begriffs unterscheidet zwischen den folgenden Elementen (vgl. *Low/Cohen Kalafut*, 2002):

- Führung
- Reputation
- Marken (Brands)
- Strategieumsetzung

- Organisation und Kultur
- Technologiepotenziale
- Innovationspotenziale
- Kundenpotenziale
- Kommunikation und Transparenz
- Anpassungsfähigkeit
- Humanpotenziale
- Wissenspotenziale sowie
- Netzwerkpotenziale.

Eine Untersuchung des Entscheidungsverhaltens von professionellen Anlegern kommt zu dem Ergebnis, dass deren Empfehlungen etwa zu einem Drittel auf den Informationen des Unternehmens über sein immaterielles Vermögen basieren (vgl. *Low/Cohen Kalafut*, 2002, S. 41). Dies sollte allen Führungskräften zu denken geben, die glauben, ihre Aktien-Performance hinge allein von den finanziellen Ergebnissen ab.

Es reicht jedoch nicht aus, Informationen über das immaterielle Vermögen als bloßen Anhang zum traditionellen Reporting zu betrachten. Entscheidend sollte die Empfangsperspektive sein. Es ist daher von großer Bedeutung, das Entscheidungsverhalten der wichtigsten Anleger zu kennen. Diese orientieren sich in der Regel an unterschiedlichen Zeithorizonten und Informationskategorien, die neben der finanziellen Perspektive im Mittelpunkt ihres Interesses stehen. Hierzu zählen z.B. Informationen über erfolgreiche Innovationsaktivitäten, neue Strategien oder die Besetzung von Führungspositionen (vgl. *Coyne/Witter*, 2003).

Ebenfalls relevant für das Entscheidungsverhalten der Anleger sind Defizite bei wichtigen Elementen des immateriellen Vermögens. Eine entscheidungsschwache Führung, demotivierte Mitarbeiter oder eine beschädigte Reputation können als Leistungspathologien wirken, die die Entwicklung des Unternehmens entscheidend beeinträchtigen. Die Intangibles haben also sowohl den Charakter von Potenzialen als auch von Barrieren, die es zu überwinden gilt.

Die große Herausforderung beim Management der Intangibles besteht in den vielfältigen Wechselwirkungen zwischen den einzelnen Elementen. Das immaterielle Vermögen bildet ein komplexes adaptives System. Ein solches nicht-lineares, dynamisches System ist aus einer Vielzahl miteinander interagierender Komponenten zusammengesetzt, die sich in einem permanenten Prozess der Ko-Evolution in und mit ihrer Umwelt entwickeln (vgl. *Kelly*, 1995; *Servatius*, 2001, S. 56 ff.). Wesentliche Impulse zum Verständnis komplexer adaptiver Systeme gingen in der Vergangenheit vom 1984 gegründeten *Santa Fe Institute* aus. Die dritte Welle der Führung baut auf diesen Erkenntnissen auf. Von besonderer Bedeutung im Rahmen von wissensbasierten Geschäftskonzepten sind zunehmende Grenzerträge, die über positive Rückkopplungen hohe Wertsteigerungspotenziale ermöglichen.

Bei der Beantwortung der Frage, wie immaterielle Vermögensbestandteile einen Beitrag zur nachhaltigen Wertsteigerung leisten, hilft eine Beschäftigung mit den zugrunde liegenden Lenkungsprinzipien. Kennzeichnend für komplexe adaptive Systeme ist eine „Lenkung von unten herauf", bei der das System sich selbst unter Kontrolle hält, indem die Aufgabenzuweisung auf den unteren Ebenen beginnt und durch übergeordnete Ebenen unterstützt wird. Eine solche Bottom up Control ähnelt dem Subsidiaritätsprinzip politischer Institutionen (vgl. *Kelly*, 1995, S. 36 ff.). Ein entsprechendes Beispiel aus der Unternehmenspraxis ist das

VISA-Netzwerk, das mehr als 360 Millionen Kreditkarteninhaber betreut. Dieses Netzwerk ist schichtenartig aufgebaut, und die einzelnen Mitglieder-Organisationen besitzen alle Rechte und Pflichten, die sinnvollerweise nicht von der nächsthöheren Ebene wahrgenommen werden sollten.

Bei der Entwicklung neuer Geschäftskonzepte führt die Anwendung von Bottom up Control-Prinzipien zu einem intensiven Austausch zwischen dezentral agierenden Innovatoren und ihren Förderern auf den Führungsebenen. Ein solches Gegenstrom-Verfahren sichert einerseits den Nachschub an neuen Ideen und gewährleistet andererseits die Bewahrung stabiler Ordnungsmuster. Das komplexe adaptive System der Intangibles bildet so das Potenzial, aus dem über Bottom up Control-Mechanismen erfolgreiche innovative Geschäftskonzepte entstehen (vgl. *Abb. 2*).

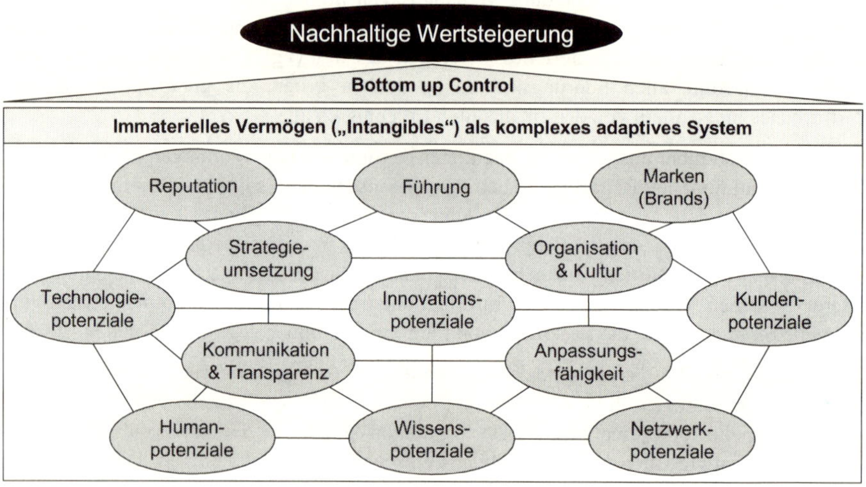

Abb. 2: Intangibles als Potenziale

An eine Phase der Geschäftskonzept-Innovation mit neuen Nutzenkategorien für den Kunden, anderen Vertriebswegen oder veränderten Wertschöpfungsstrukturen schließt sich dabei häufig eine Phase der Geschäftskonzept-Optimierung an, in der die operativen Prozesse verbessert werden (vgl. *Servatius*, 2002a).

3. Prozessmodell der strategischen Führung

In der Komplexitätstheorie werden die Erfolgsaussichten von Strategien in Form von Fitness-Landschaften dargestellt. Diese Landschaften sind ständig in Bewegung und bilden ein dynamisches System. Übertragen auf den Erfolg von Geschäftskonzepten sprechen wir von Wertlandschaften. Bergspitzen repräsentieren erfolgreiche Geschäftskonzepte, Täler Misserfolge. Innerhalb dieser Gebirgswelt entspricht das immaterielle Vermögen der Wissens- und Kompetenzbasis des Bergsteigers auf dem Weg zum Gipfel. Für den Erfolg seiner Strategie oder verschiedener Strategie-Optionen kommt es auf die richtige Mischung aus Anpassung an das Umfeld und aktiver Auswahl der Route an („Balance of Adapting and Shaping") (vgl. *Oliver/Roos*, 2000). Wir sprechen daher von adaptiven und gestaltenden Strategien. Diese Modellvorstellung bildet die Grundlage für unser evolutionäres Strategieverständnis einer Navigation in dynamischen Wert-Landschaften. Dabei sind die Intangibles als integraler Bestandteil des Geschäftskonzepts wichtige Determinanten seines Wertes (vgl. *Abb. 3*).

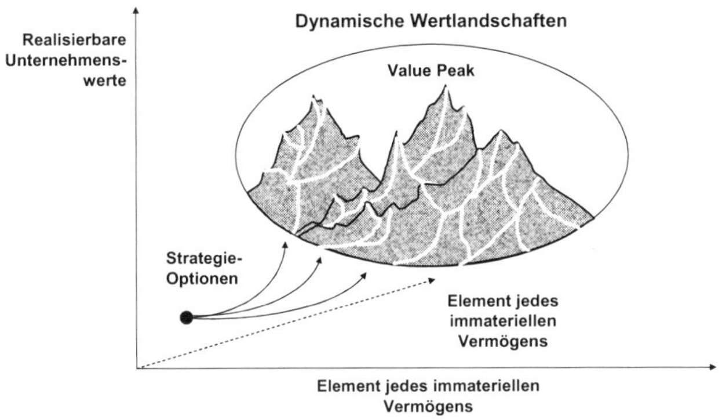

Abb. 3: Dynamische Wertlandschaften

Die nachhaltige Wertsteigerung auf der Basis von immateriellem Vermögen erfordert häufig eine Neugestaltung der strategischen Führungsprozesse (vgl. *Kaplan/Norton, 2001; Servatius*, 2002 b). Den Ausgangspunkt hierzu bildet in der Regel die Formulierung oder Überarbeitung eines wertorientierten Unternehmensleitbildes. Hieran schließt sich die Entwicklung von Strategien und neuen Geschäftskonzepten an. Die oft beklagte Umsetzungslücke haben viele Unternehmen durch eine Operationalisierung der Strategien geschlossen, indem sie z.B. Balanced Scorecards als Performance Management-Konzept einsetzen. Auf dieser Grundlage erfolgen dann die strategische Programmplanung und Budgetierung, bei der zurzeit neue Ansätze, wie das Beyond Budgeting, diskutiert werden (vgl. *Hope/Fraser*, 2003).

Vielfach vernachlässigt wurde in der Vergangenheit der Aufbau von Innovations-, Kunden-, Human- und Netzwerkpotenzialen sowie weiteren Elementen des immateriellen Vermögens. Daher stellt sich die Aufgabe, diese Intangibles besser in den strategischen Führungsprozess zu integrieren und als Werttreiber zu nutzen (vgl. *Abb. 4*).

Der emotional intelligenten Personalführung kommt dabei eine Resonanzfunktion zu, die Potenziale verstärkt, Zielvereinbarungen zu Commitments macht und erfolgreichen Wandel ermöglicht. So haben der Führungsstil und die Kompetenz der Führungskräfte einen entscheidenden Einfluss z.B. auf das Innovationsklima und die Kundenbeziehungen eines Unternehmens (vgl. *Goleman/Boyatzis/Mc Kee*, 2002). Eine Leistungssteigerung der Organisation und Kostensenkungen erfordern die Fähigkeit zum Abbau von Widerständen gegen Veränderungen. In der Vergangenheit sind leider viele der so genannten Change Management-Programme an Kompetenzdefiziten der Führung gescheitert.

Aus der Synchronisation dieser lose gekoppelten Teilprozesse der strategischen Führung resultieren die Möglichkeiten zu einer nachhaltigen Wertsteigerung. Das Unternehmen muss den Kapitalmärkten, Investoren und Analysten durch geeignete Formen des Corporate Reporting vermitteln, dass es diese Möglichkeiten auch ausschöpft.

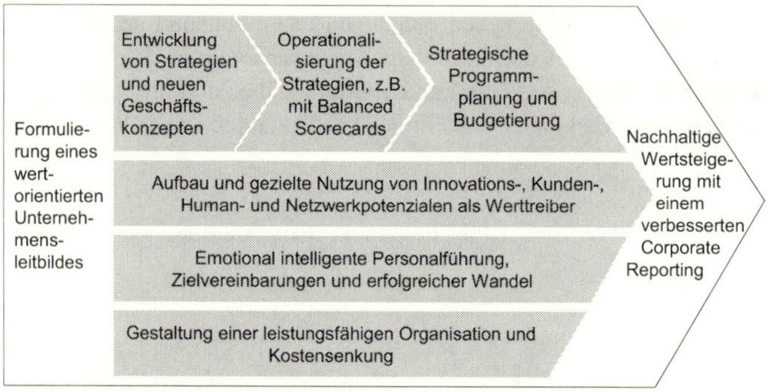

Abb. 4: Prozessmodell der strategischen Führung

Natürlich ist dieses Prozessmodell der strategischen Führung situativ anzupassen. Es bildet einen konzeptionellen Rahmen für die Beantwortung der Frage, welche Teilprozesse den größten Handlungsbedarf für Verbesserungen ausweisen. Dies ist häufig bei der Nutzung von Potenzialen im Rahmen von neuen Geschäftskonzepten der Fall, deren Erfolgsaussichten an die Kapitalmärkte zu kommunizieren sind. Im Folgenden gehen wir daher auf diese Teilprozesse näher ein.

4. Entwicklung von neuen Geschäftskonzepten

Das Geschäftskonzept eines Unternehmens lässt sich anhand der Perspektiven seiner Balanced Scorecard beschreiben: der Finanz-Perspektive, der Kunden-Perspektive, der Prozess-Perspektive und der Potenzial-Perspektive. Bei der Entwicklung von neuen Geschäftskonzepten hat sich ein Vorgehen in drei Phasen bewährt. In der ersten Phase wird ein umfassendes Verständnis der Ausgangssituation erarbeitet. Hier kommt es darauf an, möglichst frühzeitig Innovations-Chancen und -Risiken zu erkennen. In der zweiten Phase entwickelt ein Team in einem kreativen Prozess das eigentliche Innovationskonzept. Dieses wird dann in Pilotanwendungen getestet und detailliert. Eine solche Erprobung erfolgt zunehmend in Innovationslaboren, die Feldtests ergänzen. Zwischen den Phasen gibt es in der Regel zahlreiche Rückkopplungen (vgl. *Abb. 5*).

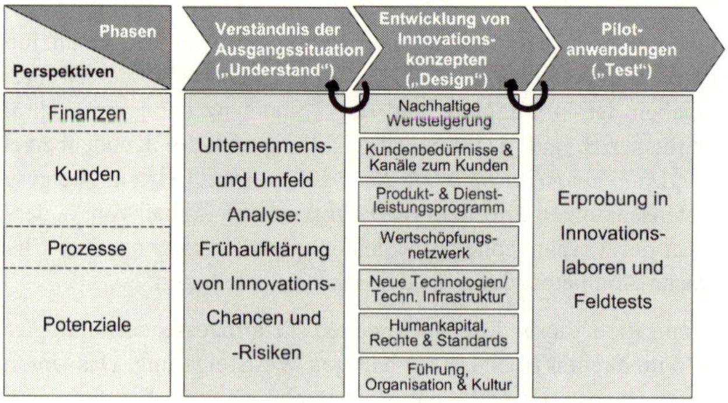

Abb. 5: Geschäftskonzept-Innovation

So erhoffen sich viele produzierende Unternehmen Wachstumsimpulse von einer Ergänzung ihres Leistungsprogramms um Dienstleistungen. Dies ist in der Regel mit einer Erneuerung des Geschäftskonzepts verbunden. Den Ausgangspunkt bildet meist das Erkennen bislang unbefriedigter Kundenbedürfnisse. Neben den Produkten suchen viele Kunden nach individuell zugeschnittenen Problemlösungspaketen, z.B. in den Bereichen Anwendungstechnik und Prozessoptimierung. Teilweise steht sogar ein Outsourcing ganzer Bereiche zur Diskussion. Das Angebot derartiger Dienstleistungen erfordert jedoch eine Erweiterung der traditionellen Vertriebskanäle.

Stärker problemlösungsorientierte Geschäftskonzepte führen in der Regel auch zu einer Neugestaltung des Wertschöpfungsnetzwerks. Es stellt sich die Frage, welche Kompetenzen im eigenen Hause vorhanden sind und mit welchen Partnern eine Zusammenarbeit angestrebt werden soll. Neuen Informationstechnologien kommt dabei eine „Enablerfunktion" zu. So erfordert z.B. das Angebot von Servicepaketen ein leistungsfähiges Kundenbeziehungsmanagement. Neben die traditionellen technischen Anlagen tritt daher eine Software-Infrastruktur, deren Entwicklung vom Rechnungswesen häufig nicht als Investition behandelt wird. Über den Erfolg in wissensintensiven Geschäften können darüber hinaus Ausgaben für Humankapital, den Erwerb von Rechten oder die Etablierung von Standards entscheiden, die in einigen Branchen den Großteil des Kapitalbedarfs ausmachen. Der Führung kommt die Aufgabe zu, eine Organisation und einen kulturellen Rahmen zu schaffen, in denen aus diesen Potenzialfaktoren innovative Geschäftskonzepte entstehen.

Die traditionellen Methoden des Wertmanagements erfassen derartige „Potenzialinvestitionen" jedoch nicht als wachstumsrelevante Werthebel. Aufgrund des Fokus auf Investitionen in physisches Vermögen kommt es in wissensintensiven Geschäften zu einer systematischen Fehlallokation von Ressourcen. Neue Ansätze einer nachhaltigen Wertsteigerung müssen daher dem Potenzialcharakter des immateriellen Vermögens besser Rechnung tragen.

5. Nutzung von Potenzialen als Werttreiber

In vielen Unternehmen ist es erforderlich, die Werttreiberfunktion von Intangibles transparenter zu machen. Genauso wichtig ist es, neben diesen Potenzialen auch Barrieren zu erkennen, die als Leistungspathologien wirken können. Wir sprechen daher von immateriellen „Aktiva" und immateriellen „Passiva". Zu den „Passiva" zählen z.B. Organisations- und Kulturelemente, die eine nachhaltige Wertsteigerung behindern können (vgl. *Slywotzky/Wise/Weber*, 2003, S. 51 ff. und S. 173 ff.). In einem ersten Schritt wird daher eine Bilanz der immateriellen „Aktiva" und „Passiva" erstellt. Dies erfordert eine ganzheitliche Sicht des Geschäftskonzepts, der Organisation und des Führungsverhaltens. Eine „Bilanzanalyse" zielt darauf ab, die Wechselwirkungen zwischen den verschiedenen Elementen besser zu verstehen (vgl. *Abb. 6*).

Aus dieser Analyse leitet man das strategische Zielsystem und die Ursache-Wirkungsbeziehungen zwischen den Zielen ab. Hierzu bedient man sich so genannter Strategy Maps, um die Vernetzung zwischen einzelnen Zielen zu veranschaulichen. Aufgrund ihres Potenzial-Charakters wirken die Intangibles indirekt über die Kunden- und Prozess-Perspektive auf den Unternehmenswert. Bei der Ableitung geeigneter Messgrößen sind diese Kausalitäten zu berücksichtigen. Eine solche Analyse kann im Rahmen der Erarbeitung von Balanced Scorecards erfolgen (vgl. *Kaplan/Norton*, 2001, S. 69 ff.). Auch bei vorhandenen Scorecards bedarf das immaterielle Vermögen aber häufig einer vertieften Behandlung.

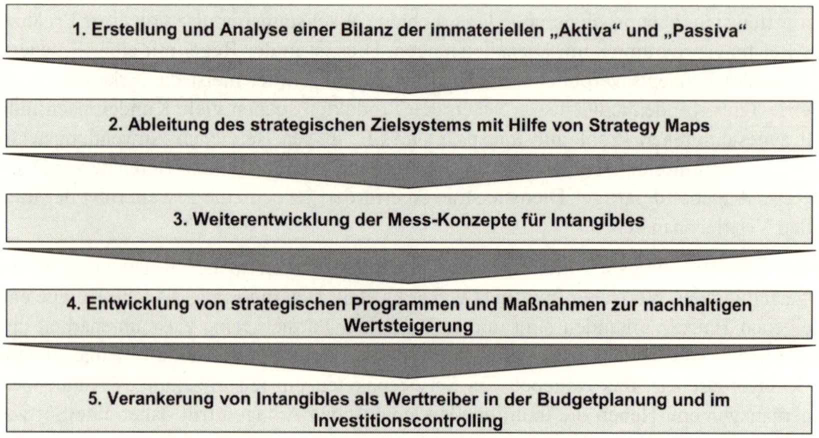

Abb. 6: Intangibles als Werttreiber

Dies gilt vor allem im Hinblick auf eine Weiterentwicklung der Mess-Konzepte für immaterielles Vermögen. Unternehmen, die in der Vergangenheit z.B. im Rahmen ihrer Ansätze zum Wissensmanagement an Mess-Konzepten gearbeitet haben, stehen vor der Aufgabe, die zum Teil sehr große Anzahl an Messgrößen zu reduzieren und stärker mit dem Geschäftskonzept zu verknüpfen. Insgesamt besteht hier allerdings noch Handlungsbedarf für die anwendungsorientierte Forschung.

Danach kann das Unternehmen dann gezielt strategische Programme und Maßnahmen zur nachhaltigen Wertsteigerung entwickeln. Hierzu zählen z.B.

- Ansätze zur Verbesserung der Innovationsprozesse oder
- Aktivitäten zur Ausschöpfung von Kundenpotenzialen.

Allen diesen Maßnahmen ist gemeinsam, dass ihr ökonomischer Nutzen häufig schwer zu quantifizieren ist.

Das Unternehmen muss daher die Frage beantworten, ob und wie welche Intangibles als Werttreiber in der Budgetplanung und im Investitionscontrolling verankert werden sollen. So existieren z.B. im Rahmen des Economic Value Added-Konzepts bei der Ermittlung des eingesetzten Geschäftsvermögens eine Vielzahl von Anpassungsmöglichkeiten. Diese betreffen unter anderem den Aufwand für Forschung und Entwicklung, der im Unterschied zu den gängigen Rechnungslegungsvorschriften als Investition behandelt werden kann. Angesichts der Bedeutung des Werttreibers Wachstum erscheint es aber notwendig, auch andere Bereiche des immateriellen Vermögens in das Wertmanagement zu integrieren. Dies könnte zu einer Erweiterung des Investitionsbegriffs führen.

Das unternehmensinterne Management der Intangibles bildet die Basis für eine verbesserte externe Kommunikation. Dieses Thema ist zurzeit Gegenstand einer Vielzahl von Initiativen.

6. Verbesserung des Corporate Reporting und nachhaltige Wertsteigerung

Unter Führungskräften besteht weitgehend Einigkeit darüber, was die Treiber einer nachhaltigen Wertsteigerung sind. In empirischen Untersuchungen werden genannt (vgl. *Di Piaza/Eccles*, 2002, S. 84 ff.):

- Produkt- und Servicequalität,
- Kundenzufriedenheit und -loyalität,
- operative Effizienz,
- finanzielle Ergebnisse,
- Produkt- und Service-Innovationen sowie
- Mitarbeiterzufriedenheit und -fluktuation.

Die traditionelle Unternehmensberichterstattung (Corporate Reporting) vernachlässigt aber weitgehend die immateriellen Werttreiber und konzentriert sich auf die finanziellen Ergebnisse. Diese Reporting-Lücke trägt zur Unsicherheit über die „wahre Leistungsfähigkeit" des Unternehmens bei und erschwert es Analysten, Unternehmen fair zu bewerten.

Es gibt daher eine Reihe von Ansätzen, immaterielles Vermögen stärker in das Corporate Reporting zu integrieren. Hierzu zählen z.B. die Überlegungen zu einer neuen Goodwill-Bilanzierung nach US-GAAP auf Grundlage des Statement of Financial Accounting Standard (SFAS) Nr. 142. Dieses sieht einen Impairment-Test vor, dessen Kernbestandteil eine Werthaltigkeitsprüfung des derivativen Goodwills ist. Die Prüfung bildet die Grundlage für die Ermittlung eines außerplanmäßigen Wertberichtigungsbedarfs (vgl. *Lorson/Heiden*, 2002).

Daneben gibt es Anregungen aus Wissenschaft und Praxis, wie ein erweitertes Corporate Reporting aussehen sollte. So sieht z.B. das Value Reporting™ von *PriceWaterhouseCoopers* eine Gliederung des Berichtswesens in die folgenden vier Säulen vor (vgl. *Eccles/Herz/Keegan/Phillips*, 2001):

- Marktübersicht,
- Strategie,
- wertorientiertes Management (finanzielle Performance, Finanzlage, Risikomanagement und segmentbezogene Leistungsdaten) sowie
- Wertebasis (Innovation, Marken, Kunden, Lieferkette, Human Resources und Reputation).

Es fehlen allerdings Aussagen zur Frage, wie das immaterielle Vermögen im Rahmen des jeweiligen Geschäftskonzepts als Werttreiber wirkt.

Baruch Lev von der New Yorker *Stern School of Business* orientiert sich stärker am Innovationsprozess und gliedert seine Value Chain Scorecard in die Phasen (vgl. *Lev*, 2001, S. 110 ff.):

- Forschung und Lernen,
- Implementierung und
- Vermarktung.

Für diese Phasen definiert er neun Felder mit entsprechenden Informations-Kategorien (vgl. *Abb. 7*).

Forschung und Lernen	Implementierung	Vermarktung
1. Interne Erneuerungsfähigkeit - Forschung - Training und Personalentwicklung - Organisationskapital, Prozesse	**4. Intellektuelles Eigentum** - Patente, eingetragene Marken und Copyrights - Lizenzvereinbarungen - Codifiziertes Know-how	**7. Kundenseite** - Marketing-Allianzen - Wert von Markennamen - Kundenwert/Kundenabwanderung - Online-Verkäufe
2. Erworbene Fähigkeiten - Zugekaufte Technologien - Ausnutzung von "Spillovers" - Investitionstätigkeit	**5. Technologische Machbarkeit** - Klinische Tests, behördliche Genehmigungen - Beta Tests, Prototypen - "First Mover"	**8. Performance** - Umsatz, Ergebnisse und Marktanteil - Innovationserlöse - Lizenzerlöse - "Wissenserträge"
3. Networking - F&E Allianzen und Joint Ventures - Lieferanten- und Kunden-integration - Communities of Practice	**6. Internetaktivitäten** - Threshold traffic - Online-Einkäufe - Wichtige Internet-Allianzen	**9. Wachstumsperspektiven** - Produktpipeline und Markteinführungstermine - Vorauss. Effizienzsteigerungen - Geplante Initiativen - Erwarteter Breakeven, Cash-Burn-Rate

Abb. 7: Value Chain Scorecard (Quelle: Lev, 2001)

Diese Ansätze liefern wertvolle Impulse zur Weiterentwicklung des Corporate Reporting. Sie ersetzen aber unseres Erachtens nicht die unternehmensspezifische Beantwortung der Frage, wie die verschiedenen immateriellen Vermögensbestandteile wirken. Im Rahmen der Geschäftskonzept-Kommunikation muss das Unternehmen seine Leistungsfähigkeit und die kausalen Zusammenhänge einer nachhaltigen Wertsteigerung deutlich machen. Ein solcher Ansatz ist längerfristig orientiert und betont die Entwicklung von Potenzialen. Er ist möglicherweise Erfolg versprechender als die einseitige, finanzwirtschaftliche Orientierung der traditionellen Shareholder Value-Konzepte.

7. Fazit

Unternehmen, die die Kapitalmärkte von ihrer Performance überzeugen wollen, sollten nicht auf die Verabschiedung allgemeingültiger Standards zum Management und Reporting von Intangibles warten. Das Thema bietet genügend Raum für Pioniervorteile und eine wettbewerbliche Differenzierung. Gefragt sind kreative Ansätze und eine offene Kommunikation. Die neuen Konzepte werden ihre Wirkung im Dialog mit den relevanten Akteuren des Corporate Reporting-Netzwerks entfalten. Eine aktive Rolle sollten dabei die Aufsichtsgremien übernehmen. Wenn diese in ihren Unternehmen die Zusammenhänge zwischen immateriellem Vermögen, neuen Geschäftskonzepten und Wertsteigerung kritisch hinterfragen, werden sie zu echten Sparringspartnern für den Vorstand. Der begrüßenswerte Corporate Governance-Kodex wird so mit Leben gefüllt. Aber auch für die Wissenschaft ist noch viel zu tun. In weiten Bereichen ist das immaterielle Vermögen noch eine Terra incognita. Neue Impulse auf dieser Entdeckungsreise werden von der Theorie komplexer adaptiver Systeme ausgehen.

8. Literatur

Coyne, K.P. und *Witter, J. W.*, Lernen Sie Ihre Aktionäre kennen, in: Harvard Businessmanager, o. Jg. (2003), H. 2, S. 35-47.

Daum, J. H. , Intangible Assets oder die Kunst, Mehrwert zu schaffen, Bonn 2002.

Di Piazza, D. A. and *Eccles, R. G.*, Building Public Trust. The Future of Corporate Reporting, New York 2002.

Eccles, R. G., *Herz, R. H.*, *Keegan, E. M.* and *Phillips, D. M. H.*, The Value Reporting™ Revolution. Moving Beyond the Earnings Game, New York 2001.

Edvinsson, L. and *Malone, M. S.*, Intellectual Capital. Realizing Your Company's True Value by Finding its Hidden Brainpower, New York 1997.

Goleman, D., *Boyatzis, R.* und *Mc Kee, A.*, Emotionale Führung, München 2002.

Hope, J. and *Fraser, R.*, Beyond Budgeting - How Managers Can Break Free from the Annual Performance Trap, Boston 2003.

Kaplan, R. S. and *Norton, D. P.*, Die strategiefokussierte Organisation. Führen mit der Balanced Scorecard, Stuttgart 2001.

Kelly, K., Out of Control. The Biology of Machines, Social Systems, and the Economic World, Reading 1995.

Lev, B., Intangibles. Management, Measurement, and Reporting, Washington 2001.

Lorson, P. C. and *Heiden, M.*, Intellectual Capital Statement and Goodwill-Impairment, in: *Seicht, G.* (Hrsg.), Jahrbuch für Controlling und Rechnungswesen, Wien 2002, S. 369–403.

Low, J. and *Cohen Kalafut, P.*, Invisible Advantage. How Intangibles are Driving Business Performance, Cambridge 2002.

Oliver, D. and *Roos, J.*, Striking a Balance. Complexity and Knowledge Landscapes, London 2000.

Palass, B. und *Servatius, H. G.*, WissensWert. Mit Knowledge Management erfolgreich im E-Business, Stuttgart 2001.

Servatius, H. G., Jenseits der New Economy. Transformation von Unternehmen mit E-Business Solutions, in: *Scheer, A. W.* (Hrsg.), Die E-Transformation beginnt. Lessons Learned, Branchenperspektiven, Hybrid Economy, M-Business, Heidelberg 2001, S. 51–90.

Servatius, H. G., Geschäftskonzept-Optimierung in der Netzwerk-Ökonomie, in: Controlling, 14. Jg. (2002a), H. 8/9, S. 437–445.

Servatius, H. G., Integration des Performance Measurement in die Führungsprozesse, in: *Horváth, P.* (Hrsg.), Performance Controlling. Strategie, Leistung und Anreizsystem effektiv verbinden, Stuttgart 2002b, S. 179–204.

Servatius, H. G., Immaterielles Vermögen, innovative Geschäftskonzepte und nachhaltige Wertsteigerung, in: Controlling, 15. Jg. (2003), H. 3/4, S. 155–161.

Slywotzky, A., *Wise, R.* and *Weber, K.*, How to Grow When Markets Don't, New York 2003.

Stewart, T. A., The Wealth of Knowledge. Intellectual Capital and the Twenty-First Century Organiszation, London 2001.

2. Messung und Steuerung

Die zielgerichtete Steuerung von Intangibles setzt deren Messung und Bewertung voraus. Die Wichtigkeit eines solchen kontinuierlichen Performance Measurement-Prozesses basiert auf drei Gründen zur Messung von Intangibles: (a) Verbesserung von Strategieformulierung, -bewertung und -umsetzung; (b) Verhaltensbeeinflussung; (c) externe Berichterstattung und Vergleiche. Dazu können verschiedene Meßmethoden herangezogen werden. Neben monetarisierenden Methoden stellen nichtmonetäre Mess- und Bewertungsmethoden einen wichtigen Bestandteil des Instrumentariums dar. Bisher existiert noch kein allgemeingültiger Ansatz zur Messung und Bewertung von Intangibles, vielmehr ist abhängig von Anwendungskontext und -ziel ein geeignetes Verfahren bzw. ein Verfahrensmix auszuwählen. Das Controlling steht hier vor der Herausforderung, dem Management auf der Basis geeigneter Bewertungs- und Auswahlansätze nachvollziehbare Empfehlungen zur Entscheidungsunterstützung zu geben. Auch die daraus abgeleiteten Steuerungskonzepte weisen eine erhebliche Spannweite auf. Sie reichen von qualitativ orientierten mehrdimensionalen Verfahren auf Basis eines Balanced Scorecard-Ansatzes bis zu quantitativen wertorientierten Verfahren zur Berechnung eines monetären Übergewinns im Bereich des Mitarbeiter-, Kunden- oder Lieferantenkapitals.

The Three Reasons why Organizations Measure Their Intellectual Capital

Bernard Marr [1] *and Dina Gray* [2]

---

[1] Prof. *Bernard Marr*, Centre for Business Performance, Cranfield School of Management, UK

[2] *Dina Gray,* Centre for Business Performance, Cranfield School of Management, UK

In this chapter we discuss the reasons why organizations measure their intellectual capital. It is generally believed, within the current literature, that an academic and practitioner focus on intellectual capital (IC) is important and that the measurement of a company's intangibles provides real business benefits. However, it is essential for practitioners and researchers in the field of IC to be able to justify these newly formed assumptions through rigorous empirical testing. This chapter reports on the results of a systematic investigation into the theoretical underpinnings of why firms measure their IC and existing empirical evidence that helps to prove that the measurement of IC is really worthwhile. The chapter defines the concept of IC and then identifies strategy, influencing behavior, and external validation as the key reasons for measuring IC. The chapter then critically reviews the state of research for each of the identified drivers. The major contribution is the classification of reasons for measuring IC and the finding that the majority of research within the IC measurement field is at the theory building stage, and that very little of the proposed measurement theory has yet been fully tested. With this chapter we outline possible avenues scholars might pursue in order to further the development of the IC measurement field.

1. Introduction

Intellectual capital (IC) is widely and increasingly recognized as a strategically important resource for organizations. This recognition has led to a rapidly growing body of research and published literature on measuring and reporting IC (e.g. *Cañibano* et al., 2000; *Guthrie* et al., 2001). Research into the general topic of IC began in the 1990s and was mainly concerned with raising awareness about the existence and value of intangible assets within organizations and about developing classification models for IC (*Hall*, 1989; *Itami* 1991; *Roos* et al., 1997; *Stewart*, 1997; *Brooking*, 1996). Following this initial work, the strategy and economics field began to carry out research into intellectual assets and formulated the concept of the knowledge-based organization (*Winter*, 1987; *Nonaka*, 1991; *Teece*, 1998, 2000b; *Spender/Grant*, 1996). Specific focus on the measurement of IC has been concerned with the creation of frameworks, indices and guidelines to support the initial concepts (*Sveiby*, 1997; *Mouritsen* et al., 2001; *Bontis* et al., 1999; *DATI*, 2000; *Lev*, 2001; *MERITUM*, 2002). More recently, the measurement and reporting of IC has been picked up by accounting researchers, even though none of the above guidelines were developed in accordance with accounting principles or standards. The literature in the area of IC measurement and the number of measurement frameworks is continuously growing as researchers attempt to develop metrics that inform strategy formulation and implementation, improve disclosure, benchmark performance, and predict future business performance.

Articles that have focused on the measurement of IC differ in their reasons as to why organizations should measure their IC. Much of the discussion is based on no coherent rationale or theory which often leads to confusion as to what drives organizations to measure their IC. Therefore, the aim of this paper is to shed some light onto this complex field, by segregating and explaining the different rationales for the measurement of IC. In order to achieve this aim we conducted a systematic literature review (*Tranfield* et al., 2002), which enabled us to identify a set of reasons and motives that drive the measurement of IC. These reasons include those that help develop, monitor or manage a business from an internal perspective, as well as externally driven reasons which focus on the ability to communicate measures and results to external stakeholders. Notwithstanding the confusion about the different reasons, researchers and practitioners alike continue to develop tools without questioning the premise that "measurement is assumed worthwhile" (*Guthrie*, 2001). Many research articles assume, but fail to provide evidence, that positive benefits accrue from measuring and reporting IC. In section 4 of this chapter we will review the theoretical underpinnings and the different rationales for measuring IC and critically review existing empirical evidence that purports to test the rationales put forward for each of the IC measurement drivers. The systematic identification of the various IC measurement drivers and the subsequent critical review of the empirical evidence have enabled us to put forward a detailed research agenda for the field of IC measurement.

Based on a systematic and critical review of the research to date in the field of IC measurement, the three contributions of this paper are (i) a taxonomy of drivers as to why organizations measure their IC, (ii) improved understanding of whether the extant research lends any credence to the thesis that measuring IC delivers business benefits and (iii) and finally recommendations as to a future research agenda.

2. Defining Concepts

The concept and definition of IC can be confusing. Fuelling this confusion is the fact that concepts such as intangible assets, knowledge assets, capabilities, competencies, intellectual property, and intellectual capital are often used interchangeably. This brings us to a significant problem many people face when they first delve into this area – the definitions are not very clear and neither are the boundaries. We believe that this is, to a large extent, due to the multidimensional nature of the topic. Some early definitions refer to IC as the difference between the book value of the company and the amount of money someone is prepared to pay for it (e.g. *Brooking*, 1997). Accountants, on the other hand, may refer to IC as 'non-financial fixed assets that do not have physical substance but are identifiable and controlled by the entity through custody and legal rights' as defined by the Accounting Standards Board in FRS 10, their main standard for reporting intangibles and goodwill. Such a stringent definition excludes many commonly accepted categories of IC like relationships, knowledge and skills of employees as they cannot be controlled by the firm in an 'accounting' sense (*Starovic/Marr*, 2003). HR professionals or academics might refer to IC as skills, knowledge, and attitude of employees whereas marketing managers might argue that intangibles such as brand recognition and customer satisfaction are at the heart of IC, and the IT manager might view key IC as being software applications and network capabilities. This language issue is further complicated by the fact that few organizations seem to use the phrase 'intellectual capital' and rather refer to its constituent parts such as people skills, knowledge, innovation capability, or relationships.

Many organizations involved in measuring intellectual capital develop their own idiosyncratic definitions of IC. *Skandia*, for example, defines IC as 'the possession of knowledge, applied experience, organizational technology, customer relationships and professional skills that provide *Skandia* with a competitive edge in the market'.

For the purpose of this chapter we follow the definition of *Marr* and *Schiuma* (2001) who define IC as a group of knowledge assets that are attributed to an organization and most significantly contribute to an improved competitive position of this organization by adding value to defined key stakeholders. Intangible assets, on the other hand, we define as only those that financial accounting standards would recognize as assets and allow on balance sheets. And finally intellectual property is a subset of intangible assets such as patents, trademarks and copyrights that can be included in traditional financial statements.

We therefore define IC broader than intangible assets and intellectual property, but emphasize the point that it has to add value to the organization. To make the concept of IC more comprehensible it is often split into different categories. A common and probably most commonly used distinction is between human, relational and structural capital (*MERITUM*, 2002):

- **Human capital** is defined as the knowledge that employees take with them when they leave the firm. It includes the knowledge, skills, experiences and abilities of people. Some of this knowledge is unique to the individual, some may be generic. Examples are innovation capacity, creativity, know-how and previous experience, teamwork capacity, employee flexibility, tolerance for ambiguity, motivation, satisfaction, learning capacity, loyalty, formal training and education.
- **Relational capital** is defined as all resources linked to the external relationships of the firm, with customers, suppliers or R&D partners. It comprises that part of Human and Structural Capital involved with the company's relations with stakeholders (investors, creditors, customers, suppliers, etc), plus the perceptions that they hold about the company. Examples of this category are image, customer loyalty, customer satisfaction, links with suppliers, commercial power, negotiating capacity with financial entities, environmental activities, etc.
- **Structural capital** is defined as the knowledge that stays within the firm at the end of the working day. It comprises the organizational routines, procedures, systems, cultures, databases, etc. Examples are organizational flexibility, a documentation service, the e-xistence of a knowledge center, the general use of information technologies, organizational learning capacity, etc. some of them may be legally protected and become Intellectual Property Rights, legally owned by the firm under separate title.

3. Research Approach

In order to fully map the prior research in the field of IC measurement a systematic literature review was undertaken (*Tranfield* et al., 2002). Traditional 'narrative' reviews often lack rigor, and in many cases are not undertaken as genuine pieces of investigatory science. *Tranfield* et al. (2002) recommend the specific principles of the systematic review methodology that are used in medical science in order to counteract bias and produce transparent, high-quality and relevant literature reviews in management research. Conducting a systematic review means adopting a replicable, scientific and transparent process, in other words a detailed process that minimizes bias through exhaustive literature searches of published and unpublished studies and by providing an audit trail of the reviewers decisions, procedures and conclusions (*Cook* et al., 1997)

The way in which the systematic review was undertaken, ensured that it was both methodical and replicable. In order to assess the relevance and size of the literature and to state clearly the focus of the research study, the scope of the literature review process was delimited by factors of disciplinary perspective, keywords and the quality of the research sources.

For replicability, the full research criteria are given in Appendix A to this chapter. The results of the initial search on papers using the specified keywords resulted in 707 papers being stored and their abstracts initially assessed. The initial assessment criteria used were studies specifically related to business, theoretical and empirical studies, quantitative and qualitative studies and only studies published in academic or high quality business journals. Following the initial assessment approximately 10% of the articles and books were deemed to be of a suitable quality to be fully read and analyzed further. The papers were studied and coded in regard to IC measurement drivers. The results of the coding formed the sub-headings of the various sections of this paper. This paper is intended to give an overview of the drivers of measuring IC and therefore not all the papers researched have necessarily been referenced in this paper.

4. The Three Reasons for Measuring Intellectual Capital

As already discussed the first aim of this paper is to identify the reasons why organizations are seeking to measure IC. Through the systematic literature review we were able to identify three main categories of reasons why firms measure their IC. These were (1) strategy, (2) Influencing behavior and (3) external validation (see *figure 1*). In the following sections we will explore the theoretical benefits that are proposed in the literature and contrast those with the empirical evidence found in our review.

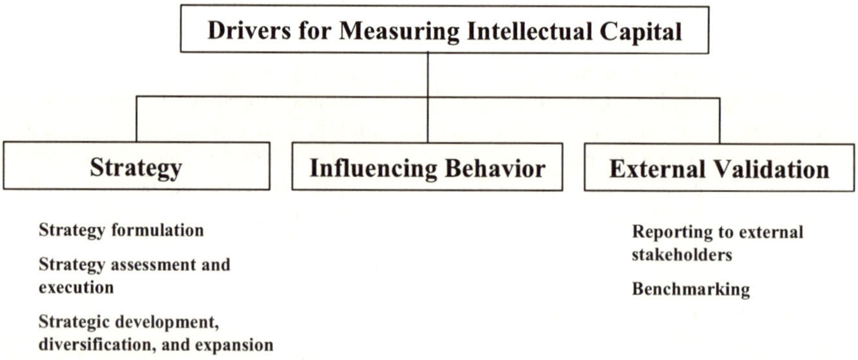

Fig. 1: Drivers for measuring IC

4.1 Strategy

The strategic importance of intellectual capital can be divided into three sub-categories, as each is sufficiently different. The first concerns the formulation of organizational strategies, the second is concerned with the execution and assessment of the strategy, and the third addresses the strategic expansion and diversification decisions.

Strategy formulation

Measuring IC can be used to help formulate business strategy. *Kenneth Andrews* defines corporate strategy as "the pattern of decision making in a company that determines and reveals its objectives, purposes, or goals, produces the principal policies and plans for achieving those goals, it defines the range of business the company is to pursue, the kind of economic and human organization it is or intends to be, and the nature of the economic and non-economic contribution it intends to make to its shareholders, employees, customers, and communities" (*Foss*, 1997, p. 52). IC of organizations is of strategic importance (*Grant*, 1991; *Stewart*, 2001; *Andriessen/Tissen*, 2001; *Teece*, 2000). Organizations realize that when formulating a corporate strategy it is not enough to just identify the competitive forces, opportunities, and threats of the industry as suggested by *Michael Porter* (1979), in addition organizations have to identify their corporate competence and resources in order to evaluate opportunities (*Andrews*, 1971). Different firms develop different distinctive competence (*Selznick*, 1957) and the question they have to ask themselves is: does the organization have the right competence to pursue certain opportunities? *Kenneth Andrews* brings the strategic importance of competences to a head when he states that "opportunism without competence is a path to fairyland" (*Andrews*, 1971).

This view of competence-based competition was first framed by *Edith Penrose* (1959) and then later enhanced by *Birger Wernerfelt* (1984) and *Richard P. Rumelt* (1984) who are seen as developers of the modern resource based view of the firm (*Foss*, 1997). Resource based theorists see firms as heterogeneous entities characterized by their unique resource base (*Nelson/Winter*, 1982, *Barney*, 1991) and this resource base consists increasingly of IC (*Stewart*, 1997; *Ross* et al. 1997; *Lev*, 2001; *Sveiby*, 2001, 1997; *Marr/Schiuma*, 2001).

This means that the IC of a firm should be one of the central considerations in formulating strategy and one of the primary constants upon which a firm can establish its identity and frame its strategy, as well as one of the primary sources of the firm's profitability (*Grant*, 1991). Therefore, firms need to strategically identify and develop their IC in order to gain a competitive advantage and to increase their performance (*Petergraf*, 1993; *Prahalad/Hamel*, 1990; *Teece* et al., 1997). The key to a resource-based approach to strategy formulation is to understand the relationships between IC, competitive advantage, and profitability (*Grant*, 1991).

Most research in this field seems exploratory in nature and not empirical. The field seems more concerned with an epistemological discussion (*Sveiby*, 2001) and a focus on theory building rather than theory testing. There is little empirical evidence of strategy formulations based on IC. An exception is the case of telecommunications software company *APiON* which documents how the company developed and implemented a growth strategy that allowed it to realize a dramatic increase in shareholder value through proactively focusing on harnessing its IC resources (*Peppard/Rylander*, 2001). Ten years ago *Richard Hall* (1992) conducted a survey of 95 UK chief executives to determine the relative contribution each intangible resource makes to the success of the business. Across all sectors the perception was that company reputation, product reputation and employee know-how were the most important intangible contributors for overall strategic success. *Hall* (1993) confirms his survey findings with research results from six case studies. He concludes that intangible resources should play a major role in the strategic management process. *Marr* and *Schiuma* (2001) have documented how *Lycos* and *Great Universal Stores* identified their knowledge-based assets to influence their strategy formulation process. *Collins* and *Montgomery* (1995) identified companies such as *Disney*, *Newell*, *Cooper*, and *Sharp* as ones that demonstrate the power of resource-based strategies and the return they generate, although neither of these companies may have set out explicitly to formulate their strategy around their key resources.

Strategy assessment and execution

The second driver for measuring IC is to develop key performance indicators to help evaluate the execution of strategy (*Neely* et al., 1996; *Kaplan/Norton*, 1996; *Meyer/Gupta*, 1994; *Bassi/Van Buren*, 1999). Experiences from the pioneers of reporting on IC (*Edvinsson/Malone*, 1997) indicate that information on IC has little value for users unless it is linked to the strategy of the firm. Any performance measurement system should be used to assess and challenge the assumptions underpinning the current strategic direction (*Neely*, 1998). Verifying or rejecting strategic assumptions will potentially impact the resource allocation in organizations. Therefore, the development of a set of performance measures should be guided by strategy (*Meyer/Gupta*, 1994; *Neely* et al., 1996). *Kaplan* and *Norton* (1996) elucidate that in order to execute a strategy, measurement has to be continuous and include single-loop as well as double-loop learning. However, many organizations seem to struggle with strategy execution and with the application of strategic single loop and double loop learning (*Argysis/Schon*, 1978).

Many measurement systems assume causal relationships or are based on a business hypothesis which enables organizations to test or challenge any underlying hypothesis (see for example *Kaplan/Norton*, 2000a). *Kaplan* and *Norton* (2000b) have shown many examples of such causal systems in the form of strategy maps. Strategy maps or success maps (*Neely* et al., 2002) manifest the strategic assumption and tell the story of how an organization transforms its IC into strategic objectives such as shareholder returns or market leadership. However, correlations between measures have rarely been empirically proven and some critics even doubt the assumption of causalities in widely implemented measurement systems such as the Balanced Scorecard (*Norreklit*, 2000).

Despite some doubt about causal relationships there is evidence of successful correlation analyses. *Ittner* and *Larcker* (1998), using customer and business unit data, find support for claims that customer satisfaction measures are leading indicators of customer purchase behavior (retention, revenue, and revenue growth), growth in the number of customers, and accounting performance (business unit revenues, profit margins, and return on sales). *Ittner* and *Larcker's* findings are supported by empirical evidence collected by *Helmi* (1998), and *Yeung* and *Ennew* (2001) who provide evidence for a positive impact of satisfaction on business performance using a range of different financial measures.

Furthermore, *Rucci* et al. (1998) developed a business model for *Sears* that tracked success from management behavior through employee attitudes to customer satisfaction and financial performance. The so-called employee-customer-profit chain verifies *Sears'* strategy to be a compelling place to work shop and invest. Under each of those headings they developed objectives and measures which allowed them to challenge and test their hypotheses. *Neely* and *Al Najjar* (2002) also empirically prove a positive link between employee satisfaction, customer satisfaction, and financial performance. *Marr* et al. (2002) demonstrate how *Shell International* verified the impact of intangible assets such as employee satisfaction, organizational culture, environmental and social responsibility, on their corporate strategy and financial performance. *Stivers* et al. (1998) find in their research that although 63% of CEOs think measurement of intangibles is important only 10% were using the results to evaluate strategic direction.

Strategic development, diversification and expansion

This driver for measuring rent-yielding IC resources builds on the rational of the first driver for the purpose of strategy formulation, but comes into play when firms seek to better exploit these resources when they plan to develop, diversify, or expand in forms of mergers and acquisitions (*Teece*, 1980; *Montgomery/Wernerfelt*, 1988). Where firms lack the critical resources (intangible and tangible), they often seek to access these from other organizations through a variety of inter-organizational links (*Markides/Williamson*, 1994). These include strategic alliances, joint ventures and mergers and acquisitions. *Lev* (2001) suggests that network economies and synergies associated with R&D and other intangibles are fundamental issues in corporate acquisitions, diversification and alliances. The takeover prices paid for targets in many of these deals, especially those in knowledge intensive industries, often include very large payments for goodwill and IC. Thus many of the deals seem to have been driven by the need for the acquiring firms to access IC.

Value creation in these acquisitions therefore depends critically upon the acquirer's ability to leverage the intangible assets that it has acquired. In acquisitions, the acquirer needs to combine its own intangible assets with those of the acquired firm (*Sullivan/Sullivan*, 2000). This requires an understanding of the nature and sources of intangible assets of the two firms, how they complement one another and how they can be leveraged to strengthen the acquirer competitive position (*Montgomery/Wernerfeld*, 1988). For organizations to effectively leverage intangible assets in acquisitions requires the ability to identify and measure its own IC as well as the IC of the target firm. Since many of the intangible assets are deeply embedded in organizational teams, routines and cultures (*Nonaka*, 1990) and many of these assets are highly dynamic in nature (*Grant*, 1996a; *Teece* et al., 1997; *Nonaka*, 1994) this process poses a formidable challenge. Recent reports of massive post-acquisition goodwill write-offs and restructuring charges amounting to billions of dollars by *Vodaphone*, *Vivendi*, *Marconi*, *AOL-Time Warner* and many others emphasize the challenge of this process.

Corporate acquisitions represent an exchange of both tangible assets and intellectual assets requiring these assets to be measured and valued (*Deng/Lev*, 1998). However, valuing intangible assets is often characterized by poor audits at the acquisition stage and poor post-acquisition strategy planning (*James* et al. 1998). Without correct assessment, measurement and valuation of IC, the acquirers may overvalue it thus causing value destruction for the acquiring firm's shareholders and other stakeholders (*Sullivan/Sullivan*, 2000). Post-acquisition integration is fraught with complexity and high risk of failure and acquisitions may fail to create new IC through innovation (*Hitt* et al., 2001). These failures have perhaps contributed to long run value destruction in corporate acquisitions (*Agrawal/Jaffe*, 2000). On the other hand acquirers might overlook potential acquisition targets because they do not understand the value of IC the potential target company might possess. Through the fact that organizations are not properly measuring their IC it is possible that millions of dollars in revenue are sitting unrecognized in companies (*Klaila/Hall*, 2000).

According to the literature, there is little empirical testing of theories in the area of strategy development, diversification and expansion. Theory development and anecdotal evidence seems to be predominant starting with concepts put forward by *Penrose* (1959) and further developed by *Teece* (1980), *Montgomery* and *Wernerfelt* (1988), and *Markides* and *Williamson* (1994). . Empirical evidence can be found in *Gupta* and *Roos* (2001) who use a case study approach to demonstrate how the measurement of intellectual capital can aid organization's mergers and acquisitions strategy. Furthermore, using data from the 1970s and early 1980s, *Morck* and *Yeung* (2003) produce empirical evidence, which demonstrates that diversification adds value in the presence of intangibles related do R&D or advertising, but destroy value in their absence. *Das* et al. (2003) report empirical evidence that on average strategic alliances do create value due to the creation of intellectual capital.

4.2 Influencing Behavior

A fair amount of research has been conducted to investigate the impact of performance measures on the behavior of managers and employees. Although a number of companies have started to link financial and operating efficiency to regular performance reviews, it has been shown that few rely on these measures to drive organizational change. *Lingle/Schiemann* (1996). The importance of measures in the area of employee performance and behavior has been shown by *Lingle* and *Schiemann* to be the biggest single measurement area that separates successful from less successful firms. However, metrics can have unintended and unanticipated consequences. Managers and employees who are judged on their success against performance measures will adapt their behavior and output accordingly. The firm eventually becomes exactly what it set out to measure. (*Hauser/Katz*, 1998).

An important question raised in the literature is whether the same measures or scorecard used to develop strategic priorities and monitor strategic actions should be used to evaluate managerial performance (*Ittner/Larker*, 1998). Analytical research by *Gjesdal* (1981), *Paul* (1992) and *Feltham* and *Xie* (1994) shows that an information system that is useful for valuing the firm need not be useful in assessing a manager's performance. Consequently, just because an economic value measure or a non-financial indicator predicts stock returns or accounting performance better than an alternative measure does not necessarily imply that the same measure or weight should be used to evaluate and reward managers (*Ittner/Larcker*, 1998).

Most organizations have realized that relying purely on financial measurement can encourage short-term thinking (*Johnson/Kaplan*, 1987; *Kaplan/Norton*, 1992), especially if those financial measures are linked to compensation systems (*Bushman* et al., 1995). Financial measures have been criticized for being too historical and "backward-looking", for encouraging dysfunctional behaviors, and for giving inadequate consideration to the development of intangible assets such as employee capabilities and customer satisfaction (*Ittner/Larcker*, 1998). Furthermore, agency models (*Feltham/Xie*, 1994; *Hauser* et al., 1994; *Hemmer*, 1996) suggest that financial measures in compensation plans alone are unlikely to be the most efficient means to motivate employees. Therefore, it has been suggested that financial performance measures should be supplemented or replaced by non-financial measures, which are more informative of employees' actions and can improve contracting (*Ittner/Larcker*, 2002).

Evidence of the incremental use of non-financial measures in managers' compensation plans has been noted during the past decade. For example: (1) *Chrysler Corporation* paid bonuses to its 200 top executives based on the attainment of vehicle quality and customer satisfaction targets in addition to measures of profitability (*Lavin,* 1994). (2) *Ford Motor Company* has an executive compensation plan, similar to the plan used by *Chrysler*, which includes non-financial customer satisfaction and operational measures (*New York Times*, 1998). (3) *Ittner* et al. (1997) report that 36 percent of the companies surveyed in their study used non-financial measures in executive compensation.

A principal justification for the use of non-financial performance measures in compensation schemes is that these measures are leading indicators of financial performance (*Banker* et al., 2000). A second justification for the use of non-financial performance measures in compensation plans is the higher level of information these measures provide about managerial effort and actions desired by the firm. *Ittner* and *Larcker* (1997) studied this phenomenon in a sample of 317 firms by analyzing the relationship between each firm's strategy and the performance measures included in its executive incentive plan. Specifically, the authors focused on the link between each firm's strategy and the relative weight assigned to non-financial measures. *Ittner* and *Larcker* found that firms pursuing an innovation-oriented strategy tended to place relatively greater weight on non-financial performance in their incentive plans. Similarly, firms following a quality oriented strategy placed relatively more weight on non-financial metrics. This results support the notion that firms where "intangible" assets are more valuable, such as innovation or quality oriented firms, tend to place greater emphasis on non-financial performance measures.

It seems generally accepted that measures affect managerial behavior and actions which in turn drive the strategy implementation (*Neely* et al., 1996). This implies that performance measurement system should evaluate the journey towards achieving their strategic goals (*Bassi/Van Buren*, 1999). *Jonathan Low* demonstrated that 70% of CEOs admit that there is a big gap between what gets measured and rewarded and what actually drives performance (*Chatzkel*, 2001).

4.3 External Validation

External validation refers to two major themes. (1) The first theme is the reporting to external stakeholders. One of the reasons for that is that market capitalization reflects the markets view of two things. First, it reflects the market's understanding of the value of the firm's fixed assets and secondly the market's intuition of the stock of intangibles and a company's ability to leverage value from those intangibles (*Sullivan/Sullivan*, 2000). In addition, organizations are increasingly legally required to report on their IC to various authorities. (2) A second theme is the external validation of own targets and goals through benchmarking with other organizations.

Reporting to external Stakeholders

Firms are sometimes legally required to communicate measures to external stakeholders, these requirements include for example financial accounts and regulator demands (*Neely*, 1998). Although it seems unlikely that all intellectual capital will soon be reported in financial statements as accountancy bodies, financial analysts and financial academics are still defining and arguing over IC metrics and no generally accepted accounting principles have been agreed upon thus far, the pressure on companies to account for and disclose the value of their IC is growing. This pressure will undoubtedly affect the way in which companies operate and measure their IC. Although the disclosure of intangibles assets is yet to be made mandatory a number of companies are attempting to make public more information about their intangible value drivers (*DATI*, 2000; *MERITUM*, 2002; *DMSTI*, 2003; *Williams* 2001). Failing to adequately communicate intellectual capital can have implications such as (1) smaller shareholders are disadvantaged, as they usually have no access information on intangibles often shared in private meetings with larger investors (*Holland*, 2001), (2) insider trading if managers would exploit internally produced information on intangibles unknown to other investors (*Aboody/Lev*, 2000), (3) volatility and the danger of incorrect valuations of firms, which leads to investors and banks placing a higher risk level to organizations, which in turn (4) increases the cost of capital (*Lev*, 2001, *Leadbeater*, 2000) Further, the communication gap between companies and the stock market seems to be larger in knowledge based industries than in consolidated industries (*Eccles/Mavrinac,* 1995).

The appropriate level of capital is essential for all organizations and with the increasing proportion of intangible assets to tangible assets organizations have to prove how their intangibles will translate into organizational performance. *Charles Leadbeater* (2000) has discussed how knowledge based companies have a higher cost of capital than companies with higher levels of tangible assets. The reason is that it is difficult to obtain security against intangible assets. In today's volatile market place it is becoming evident that those companies who were able to raise initial finance are now finding it more difficult given lower than expected results.

A recent survey of the UK market reported in the *Financial Times* of May 27th 1998 showed that most financial directors believe their company's shares have been undervalued in the past year. More than a third of companies said their share price had differed from what they perceived to be 'fair value' for more than six months and 84% said the price had been too low or too high at one point. Of these only a few thought they were overvalued. Most analysts included in the study said they firmly believed an open communication strategy would lead to a higher and more stable stock price (*Rylander* et al., 2000). In a survey by *Eccles* and *Mavrinac* (1995) 65% of the managers of high-tech companies thought their firm's shares were slightly to significantly undervalued, whilst the corresponding figure for consolidated industries was 20%. Therefore one of the drivers for a company to measure its IC is to ensure a fair and stable share price and therefore a more favorable cost of capital.

Studies in both the US and the UK have shown that analysts value information about intangibles (*Mavrinac/Seisfeld*, 1997; *Coleman/Eccles*, 1997). In addition a number of empirical studies have also demonstrated that companies who are able to make meaningful disclosures about their long-term prospects achieve more satisfactory market valuations (*Narayanan* et al., 2000; *Gu/Lev*, 2001). Consequently a growing number of companies are beginning to report their IC indicators in the annual report (see e.g. *Mouritsen* et al., 2001), although it has been shown that companies typically try to manage the information disclosed to investors in order that they do not disclose information which will erode their competitive advantage (*Williams*, 2001; *Narananan* et al. 2000). There are various empirical studies showing the impact of intangible assets on financial performance and stock price. The best documented and most widely researched is the area of R&D. *Aboody* and *Lev* (2001) show that one dollar invested in chemical R&D increases, on average, current and future operating income by two dollars. Econometric studies relating R&D intensities to corporate market value of book-to-market ratios yield consistently positive and statistically positive association estimates (e.g. *Hirschley/Weygandt*, 1985; *Bublitz/Ettredge*, 1989).

Based on *Deng* et al. (1999) and *Hirschey* (1998), *Lev* (2001) reports that three patent-related measures are predictive of subsequent stock returns and market-to-book values of public companies: (1) number of patents granted to the firm in a given year, (2) intensity of citations to a firm's patent portfolio by subsequent patents, (3) number of citations of a firm's patents to scientific papers. *Brynjolfsson* and *Yang* (1999) showed in a regression analysis that intangible assets such as R&D expenditure and investments in computers have a positive impact on market value of 1000 large firms.

Seethamraju (2000) finds positive and statistically significant investor reactions to acquisition announcements of companies that acquired trademarks from other companies. This indicates that investors expect added value from these trademarks that are above the acquisition value. Further, *Lev* (2001) shows that quantitative measures of customer related intangibles can be defined and estimated.

Whereas the areas of R&D and organizational capital have been empirically researched, other areas of intangibles have received substantially less research attention (*Lev*, 2001). *Bassi* et al. (1999) examined financial reports of forty large public organizations, and found no relevant, quantitative information about human resources, although these firms frequently reported that employees are their most valuable assets. Even though there is growing pressure on companies to disclose the role of intangibles within their organizations a *Brookings Institution* Project (2001) looking at the role of intangibles in the economy concluded that although markets need improved information disclosure managers have no incentive to improve the information about their IC. Nondisclosure of most expenditures for intangibles is a major impediment to the advancement of knowledge about intangibles in particular and corporate performance in general (*Lev*, 2001: 54).

Benchmarking

Benchmarking is generally recognized as a tool that enables a company to understand its current performance levels and set future targets (*Camp*, 1989). Target setting seems to be important especially in the area of IC, as there is little available indication of 'good performance'. Many firms already do benchmarking in areas such as finance and productivity, and in fact an industry has been created in providing benchmark statistics. Measurement of IC may well allow organizations to do the same for the less tangible parts of their performance.

Organizations can only gain a sustainable competitive advantage from – what it knows, how it uses what it knows, and how fast it can know something new (*Viedma*, 2001). However, according to *Viedma* very few know how to manage IC in an efficient way and benchmarking techniques can facilitate the process of learning from the best competitors.

Drew (1997) believes that benchmarking is one of the most popular tools for strategic management by demonstrating that all of the winners of the Baldridge award score highly on the use of benchmarking. Although the paper does not discuss how many of those aspects benchmarked are intellectual capital it may be an indication of positive effects from measuring IC. In order to help companies improve their IC management, and realize benefit from their intellectual measurement *Viedma* (2000) of the *Polytechnic University of Catalonia* has created an IC Benchmarking System as a tool to compare various aspects of IC against those who are considered best in class. However, *Marr* (2003) argues that due to the idiosyncratic nature of IC and the fact that IC is often interwoven with other tangible and intangible assets of the firm benchmarking IC can become a very difficult and often unsuccessful process.

5. State of the Art and Implications for the Future

In this section we discuss in turn the literature findings for each driver for measuring IC and focus on the state of validating theoretical concepts using empirical evidence. This allows us to analyze the position of research for each of the drivers, enables us to suggest some future research directions, and allows us to draw up some implications for practice.

5.1 Strategy

The state of the art in strategy as a driver to measure IC will be discussed below in the previously defined subsections 'strategy formulation', 'strategy assessment and execution', and 'strategy development, diversification and expansion'.

Strategy formulation

There is scope and opportunity to conduct empirical research into the usage of IC measurement and its influence on the strategy formulation process. Besides anecdotal evidence and a limited number of case studies there is little empirical evidence in the field of measuring IC to drive strategy formulation and deliver competitive advantage. Most research in strategic management is based on large sample, quantitatively-operationalized research designs (*Vankataraman/Grant*, 1986). *Rouse* and *Daellenbach* (1999) suggest changing the dominant research methodologies in the field of strategic management towards a higher level of intrusion, involvement or participation in an organization. Their argument is that intangible assets, which give organizations a competitive advantage, are often organizational in origin, tacit, highly inimitable, socially complex, probably synergistic, embedded in process, and often driven by culture (*Rouse/Daellenbach*, 1999: 489). Organizations seem to intuitively take their intellectual capital into account during their strategy formulations, however, well documented evidence is scarce and few rigorous empirical research data is available. This also means that large sample, cross-sectional studies are unlikely to disentangle the variety of effects associated with time, industry, environment, strategy, and the resource/capability of interest, whereas it is exactly this context of relevance that will provide empirical data needed to justify perceptions that the strategy formation based on intangible resources delivers competitive advantage (*Beyer*, 1997; *Miller* et al., 1997; *Rouse/Daellenbach*, 2002). We encourage more organizations to explicitly document and share their strategy formulation and suggest that future research studies provide deep context information following the suggestions put forward in *Rouse* and *Daellenbach* (2002). In addition, larger empirical studies could be used to identify the process of how firms measure their strategic intellectual capital to influence strategy formulation.

Strategy assessment and execution

Although there is some empirical evidence showing how IC drives performance and how its measurement can be used to challenge, verify, or reject strategic assumptions, this research area offers a myriad of opportunities for scholars to make contributions. Large-scale econometric studies based on publicly available data prove the value relevance of intangibles on a macro level (e.g. *Aboody/Lev*, 1998; *Hand*, 2003). However, there seems to be a lack of studies addressing the micro level which again could disentangle the context specific issues. One of the major problems here seems to be the availability of and access to relevant data. Few organizations have all the data available to run reliable econometric studies, and if they have the data it is rarely in a format that can be used in empirical studies. Furthermore, many organizations, especially those in competitive markets, might be reluctant to allow the publication of such data as it reveals their strategic direction to their competitors. Without better designed case study evidence and empirical research findings in this area it might be difficult for organizations to justify ongoing measurement of IC in individual firms. This area of research would benefit from more longitudinal and rich case studies, which define context and go beyond the classical interview based case research, in order to understand the process of IC measurement for strategic management purposes. In addition, more rigorous statistical correlation tests of measurement systems as seen in *Neely* and *Al Najjar* (2002) and *Marr* et al. (2002) might also inform this area of research and are strongly encouraged. This will enable organizations to confidently manage with measures and execute a 'tested' strategy.

Strategic development, diversification and expansion

Given the great number of mergers and acquisitions in which a significant proportion of the take-over price was paid for IC, this area offers great opportunities for further research into the process of how organizations assess their own and the IC of the target firm in a merger and acquisition context. Limited access to the IC of potential target firms in a merger or acquisition context makes the process difficult for the acquiring firm as well as for researchers. Better disclosure of IC would make this process easier, which links this field into the work on corporate reporting of IC conducted by *Mouritsen*, *Larsen* and *Bukh* (2001) as well as other researchers involved in the EU Meritum project on managing and reporting intangibles. There is a clear need for a better and a more transparent valuation process which might have an impact on company reporting and information exchange in merger and acquisition situations. The area of strategic development would benefit from further micro-level case studies of how organizations identify matching and complementary intangibles and how organizations measure the post merger integration success.

5.2 Influencing Behavior

The primary reasons suggested for the use of intellectual capital performance measures in compensation systems are that these measures are better indicators of future business performance than accounting measures, and they are valuable in providing information for the evaluation and motivation of managerial performance. However, the existing studies that support these reasons suffer from several limitations that should be taken into account. For example, *Ittner* and *Larcker* (1997, 2002) use independent variables in their structural model that are endogenous choice variables that can create inconsistencies. As argued by the authors, this limitation could be overcome by using simultaneous equations models (including the choice of exogenous instrumental variables for identification). Another example is the *Banker* et al. (2000) study, which is based on the hotel industry but generalization to other industries is not feasible. *Anderson* et al's (1994, 1997) and *Ittner* and *Larcker's* (1998) studies relied on short time series (the number of longitudinal observations is less than or equal to four), which might not be sufficient to fully confirm their findings. As a result, we believe that further research is needed to overcome the limitations of previous studies and to enhance our understanding of: (1) the complex relationship between IC measures and performance of individuals; (2) the applicability of IC performance measures that are useful for valuing the firm for assessing a manager's performance; and (3) the ideal structure of incentive plans and (4) how managers react to the use of less tangible measures in the evaluation of their performance.

5.3 External Validation

The state of the art in external validation as a driver to measure IC will be discussed below in the previously defined subsections 'reporting to external stakeholders', and 'benchmarking'.

Reporting to external Stakeholders

Overall the benefits of externally disclosing information on IC are not clear at this point. Thus far, the disclosure of information on IC is mainly voluntary. Reasons vary from general corporate governance to disclosing IC information to investors in order to improve stock valuation and cost of capital. There are potentially a number of downsides to such disclosure. The most common arguments against voluntary disclosure, from a managerial perspective, are fear of giving away sensitive information to competitors and the extra costs associated with the collection and disclosure of the information (*Rylander* et al., 2000). *Carroll* and *Tansey* (2000) warn that firms need to be careful about disclosure as internal measures are not yet tried and tested and they therefore run the risk of exposing the company to external criticism. Therefore companies need to plan carefully and must try to balance the risk of disclosure against its potential gains. Even though this area of research has had significant attention in terms of empirical research, there are many questions left unanswered. More empirical work focusing on the positive and negative effects of disclosing information on IC is strongly encouraged. There are intangible assets that have had less empirical research attention in the area of disclosure such as knowledge, skills, relationships, culture, practices and routines would contribute to this field of enquiry. Researchers are strongly encouraged to include those 'difficult to measure' intangibles in future studies. A major factor for analysts seems to be credibility of the information disclosed (*Neely* et al., 2001). The question is then raised as to how analysts and investors see the disclosure of IC in terms of credibility and economic relevance (*Narayanan*, 2000). Especially as disclosing information is frequently used by managers to manage investors' perceptions (*Lev* 2001). More research into how analysts see and use disclosed information on IC would contribute to this debate. An effective, strategically managed disclosure process can help analysts to understand the long term view of the company, which should help with a fairer share price valuation, and hence lower cost of capital. In order to measure this assertion and enable companies to realize the benefit of their IC measurement costs, more research is encouraged to investigate whether the disclosure really does provide benefits such as better cost of capital.

Benchmarking

Within the IC field there is much discussion on the difficulty of collecting IC measures, and even more difficulty in creating a standardized measurement system in order for companies to benchmark themselves against one another. Although the benefits of benchmarking may be realized, the costs and difficulty of the process may be prohibitive to companies. Even more problematic is the belief that firms can only gain sustained competitive advantage from increasing those assets that are inimitable, not substitutable, tacit in nature, and synergistic (*Wernerfelt*, 1994; *Barney*, 1991; *Rumelt* et al., 1994; *Teece* et al., 1997). If this is the case then it may be questionable as to the benefits accrued from benchmarking intangible assets against other companies. *Drew* (1997) argues that companies would not want to benchmark their core competencies with competitors because they could give away their competitive advantage. Therefore within the IC field companies may want to consider benchmarking their measures against their customers, partners or suppliers rather than their competitors. For these reasons research that shows the positive impact of benchmarking IC would contribute significantly to this debate. Furthermore, research on the role of benchmarking for the purpose of target setting in the context of measures for IC would also shed more light on this issue. For now there is only confusing evidence as to why organizations should benchmark their IC.

6. Conclusions

The discussion above showed the stage of research in each of the different areas that drive the measurement of IC. *Figure 2* summarizes the research agendas for each of the drivers identified. Areas such as stock price valuations show empirical research results which verify some of the theories put forward, other areas such as strategy development or benchmarking have had little empirical research attention. In summary there is much stage 1 (building theories / raising awareness) research and little stage 2 (theory testing). Baruch Lev states that to advance knowledge in the area of intangibles, "theoretical principles should be subjected to empirical examination and observation" *Lev* (2001, p. 51). The above discussion has revealed many research avenues which scholars might consider pursuing to take the field of IC measurement further.

Besides the obvious gaps in empirical research there are some fundamental questions raised regarding the research methodology (*Rouse/Daellenbach*, 1999; 2002). Empirical evidence should not just be provided by classical large sample, cross-sectional research projects but be complemented by rich, longitudinal case studies that will allow us to understand the specific context which seems to be critical for the analysis of IC. This kind of research will then allow us to answer questions of how the different drivers of IC measurement impact the way measurement is viewed and implemented in organizations and how it is perceived by externals stakeholders.

To date, we can be assured that awareness has been achieved and academics as well as practitioners agree that IC plays an important role in today's organizations. In this paper we have taken a step back and have looked at the existing empirical evidence that validates the theoretical drivers of IC measurement put forward in the literature. The authors feel that the field of IC could run into danger of losing credibility if researchers fail to produce more research that tests the theories put forward, rather then further adding to the large body of literature of theoretical discussions and theory building. Without more rigorous research we will not be able to move beyond the stage of only assuming that measurement of intangibles is worthwhile (*Guthrie*, 2001; *Lev*, 2001).

By separating out the various IC measurement drivers and then reviewing existing theoretical and empirical evidence supporting such drivers, this chapter has established a starting point from where we would like to encourage scholars and practitioners to pick up the themes identified in the research agenda in order to further the field of IC measurement.

Driver	Research Agenda
Strategy	
Strategy formulation	• Rich and context specific case studies of how intangibles inform strategy formulation. • Empirical studies following the *Rouse & Daellenbach* method. • Empirical studies to identify a process to measure resources to inform strategy formulation.
Strategy assessment and execution	• Econometric studies of relationships in measurement systems, in particular the impact of intangibles on strategic goals. • Empirical research into organizational processes using measures of intangibles to manage strategically. • Empirical investigations into contexts in which these processes work best.
Strategic Development, Diversification and Expansion	• Empirical research into how intangibles drive mergers and acquisitions. • Empirical studies of processes organizations use to identify intangibles of the target firm. • Rich case studies of how organizations identify matching intangibles. • Rich case studies of how organizations measure the post-merger integration success of intangibles.
Influencing Behavior	
	• Empirical research into the potential impact of linking intangible performance drivers to compensation systems. • Case studies discussing the applicability of IC performance measures for valuing individual's performance. • Empirical studies which identify the ideal structure of incentive plans and how managers react to the use of IC measures in the evaluation of their performance.
External Validation	
Reporting to external stakeholders	• More empirical research into how IC disclosure influences stock valuations (Especially for those intangibles that had little research attention so far). • Empirical research into the process analysts use to gather information on IC. • Empirical research on the link between credibility of IC information communicated and its potential impact. • Empirical research into how information on IC influences capital raising. • Rich case studies of how the usage of information on IC influenced capital availability.
Benchmarking	• Empirical research into the applicability of benchmarking of IC • Rich case studies of organizations using benchmarking of IC

Fig. 2: Research agenda for measuring IC

7. Appendix

This Appendix explains the criteria followed in the systematic literature review as well as further detailed information about the search results.

In order to assess the relevance and size of the literature and to state clearly the focus of the research study, the scope of the literature review process was delimited by the following factors:

- research sources,
- keywords/subject areas,
- journal impact factor score.

In order to summarize, integrate and collate the findings of the different studies on the research question three separate analyses were undertaken.

In the first stage of the analysis the specified categories of journals were subjected to a detailed key word search. In the second phase the relevant abstracts were read and scored with respect to the journal from which the paper emanated. For the third and final phase, the papers selected in the second phase were fully reviewed and a table was drawn up of what is known and established already, what were the core contributions and what were the key emerging themes and research questions. The results of this third phase are those discussed within the main body of this paper.

Phase 1 – Keyword search

Focusing the disciplinary perspective on the following factors delimited the search for the range of journals found within the following disciplines.

- Intellectual capital
- Measurement
- Business

For the initial phase of the research the search covered as wide an area as possible and covered the following possible publications:

- Published journals
- Unpublished studies
- Conference proceedings

For identification of relevant studies the search used the following database sources:

Electronic Service Databases:

- ProQuest
- Emerald
- Infotrac
- Ingenta
- Centre for Business Performance reference database

The following search terms were then used to search the abstract of papers within the research sources.

- Intellectual capital
- Human capital
- Structural capital
- Knowledge capital
- Resource based perspective
- Resource based theory
- Business performance
- Firm performance
- Competitive advantage
- Stock returns
- Shareholder returns
- Valuation
- Value
- Intangible
- Knowledge management
- Knowledge economy
- Benchmarking
- Metrics
- Measurement
- Indicators
- Dynamic capabilities
- Competences
- Resources
- Assets
- Patents
- R&D

The numerical results of the systematic literature search against the keywords defined in the scope of the research are shown in *figure 3*. The figures at the top of each of the "relevant" columns specify the total number of relevant papers for that category, the rows below then specify how these were split across the key words. Some papers covered more than one keyword and hence the columns do not necessarily sum to the total. Each of the cells represents a paper that contained both the word specified in the first column of the table and the word specified in the first row of the table.

	Intellectual capital		**Human capital**		**Structural capital**	
	Total	**Relevant**	**Total**	**Relevant**	**Total**	**Relevant**
Intellectual capital	378	37				
Human capital	25	8	320	31		
Structural capital	8	2	8	5	9	5
Knowledge capital	3	1	1	0	0	0
Resource based perspective	11	2	19	5	0	0
Resource based theory	11	0	23	2	0	0
Business performance	4	1	4	1	1	1
Firm performance	1	0	4	1	0	0
Competitive advantage	31	5	33	8	0	0
Stock returns	1	1	0	0	0	0
Shareholder returns	0	0	3	2	0	0

Valuation	27	7	30	6	0	0
Value	50	3	156	18	2	1
Intangible	79	17	21	10	5	3
Knowledge management	119	13	25	5	3	1
Knowledge economy	12	0	8	2	1	1
Benchmarking	3	1	1	0	0	0
Metrics	8	2	4	3	0	0
Measurement	30	9	26	9	1	1
Indicators	8	3	18	4	1	0
Dynamic capabilities	1	0	1	0	0	0
Competences	15	3	50	13	0	0
Resources	62	8	50	8	1	1
Assets	115	14	50	4	5	2
Patents	21	5	4	1	1	1
R&D	16	1	15	2	0	0

Fig. 3: Results of the initial systematic literature search

Phase 2 – Quality selection

Those abstracts that were of papers already published were selected firstly by identifying the quality of the journal with respect to the 2001 Journal Impact Factors list and FT ranked journals. A paper was only included if it emanated from a journal with an impact factor of 0.6 or greater. Any journals that were specifically focused on intellectual capital were included even if they did not appear in either of these lists.

Phase 3 – Categorization

Of the 73 articles and books deemed relevant to this research *figure 4* shows how the literature splits into the categories specified in this paper. The research revealed that 50% of the literature reviewed was classified under background material to the field and only 11 of the 73 articles and books were actual empirical studies.

Research category	Relevant papers
Background	31
Theoretical discussion	4
Literature reviews	5
Definition and classification	7
Frameworks	6

Indicators	3
Measurement	6
Disclosure and accounting	9
Statements	2
Empirical	11

Fig. 4: Research categories

Parts of this paper are based on *Marr* et al. (2003), "Why do firms measure their intellectual capital", published in the Journal of Intellectual Capital. The authors would like to thank the two anonymous reviewers for their valuable comments as well as reviewers and participants of the E*Know Net: Transparent Enterprise Conference in Madrid, 2002. Furthermore, we would like to thank *Monica Franco* for the valuable input on the compensation part of this paper.

8. References

Aboody, D. and *Lev, B.*, The Value Relevance of Intangibles: The Case of Software Capitalization, Journal of Accounting Research (supplement), Vol. 36 (1998), No. 2, pp.161-191.

Aboody, D. and *Lev, B.*, Information Asymmetry, R&D, and insider Gains, in: Journal of Finance, Vol. 55 (2000), No. 6, pp. 2747-2766.

Anderson, E.W., *Fornell, C.* and *Lehmann, D. R.*, Customer Satisfaction, Market Share, and Profitability, in: Journal of Marketing, Vol. 58 (1994), No. 7, pp. 53-66.

Anderson, E. W., *Fornell, C.* and *Rust, R.*, Customer Satisfaction, Productivity and Profitability: Differences between Goods and Services, in: Marketing Science, Vol. 16 (1997), No. 2, pp. 129-145.

Andrews, K. R., The Concept of Corporate Strategy, Dow Jones-Irwin, Homewood, IL 1971.

Andriessen, D. and *Tissen, R.*, Weightless Wealth: Find your real value in a future of intangible assets, London 2000.

Anonymous, Ford Is Changing Bonus Program, in: New York Times, Mar 26: C7 (N), 1998.

Argyris, C. and *Schon, D. A.*, Organizational Learning: A Theory of Action Perspective, Reading 1978.

Banker, R. D., *Potter, G.* and *Srinivasan, D.*, An Empirical Investigation of an Incentive Plan that includes Non-financial Performance Measures, in: The Accounting Review, Vol. 75 (2000), No. 1, pp. 65-92.

Barney, J. B., Firm Resources and Sustained Competitive Advantage, in: Journal of Management Vol. 17 (1990), No. 1, pp. 99-120.

Bassi, L. J. and *Van Buren, M. E.*, Valuing Investments in Intellectual Capital, in: International Journal of Technology Management, Vol. 18 (1999), No. 2, pp. 414-432.

Bontis, N., *Dragonetti, N. C.*, *Jacobsen, K.* and Roos, G., The Knowledge Toolbox: A Review of the Tools Available to Measure and Manage Intangible Resources, in: European Management Journal, Vol. 17 (1999), No. 3, pp. 391-402.

Brooking, A., Intellectual capital: Core Assets for the Third Millennium Enterprise, London 1996.

Brooking, A., The Management of Intellectual Capital, in: Long Range Planning, Vol. 30 (1997), No. 3, pp. 364-365.

Beyer, J., Research Utilization: Bridging the Cultural Gap Between Communities, in: Journal of Management Inquiry, Vol. 7 (1997), No.1, pp. 17-22.

Brynjolfsson, E. and *Yang, S.*, The Intangible Cost and Benefits of Computer Investments: Evidence From the Financial Markets, MIT Working Paper, 1999.

Bublitz, B. and *Ettredge, M.*, The Information on Discretionary Outlays: Advertising and Research and Development, in: Accounting Review, Vol. 64 (1989), No. 1, pp.108-124.

Bushman, R. M., *Indjejikian, R. J.* and *Smith, A.*, Aggregate Performance Measures in Business Unit Manager Compensation: the Role of Intrafirm Interdependencies, in: Journal of Accounting Research, Vol. 33 (1995), Supplement, pp. 101-128.

Cañibano, L. M., *García-Ayuso C.* and *Sánchez, P.*, Accounting for Intangibles: A Literature Review, in: Journal of Accounting Literature, Vol. 19 (2000), No. 1, pp. 102-130.

Carroll, R. F. and *Tansey, R. R.*, Intellectual Capital in the New Internet Economy - Its Meaning, Measurement and Management for Enhancing Quality, in: Journal of Intellectual Capital, Vol. 1 (2000), No.4, pp. 296-311.

Chatzkel, J., A Conversation with *Jonathan Low*, in: Journal of Intellectual Capital, Vol. 2 (2001), No. 2, pp. 136-147.

Coleman, I. and *Eccles, R.*, Pursuing Value: Reporting Gaps in the United Kingdom, PricewaterhouseCoopers, London, 1997.

Cook, D. J., *Mulrow, R. B.* and *Jaynes, B. B.*, Systematic Reviews: Synthesis of Best Evidence for Clinical Decisions, in: Annuals of Internal Medicine, Vol. 126 (1997), No. 4, pp. 367-380.

Das, S., *Sen, P. K.* and *Sengupta, S.*, Strategic Alliances: a Valuable Way to Manage Intellectual Capital, in: Journal of Intellectual Capital, Vol. 4 (2003), No. 1, pp.10-19.

(DATI) Danish Agency of Trade and Industry, A Guideline for Intellectual Capital Statements - A Key to Knowledge Management, *Ministry of Trade and Industry*, Copenhagen, 2000.

Deng, Z. and *Lev, B.*, Flash-then-Flush: The Valuation of Acquired R&D in Process, Working paper, Stern School of Business, New York University 1998.

Deng, Z., *Lev, B.* and *Narin, F.*, Science and Technology as Predictors of Stock Performance, in: Financial Analyst Journal, Vol. 55 (1999), No. 3, pp. 20-32.

Eccles, R.G. and *Mavrinac, S.*, Improving the Corporate Disclosure Process, in: Sloan Management Review, Vol. 36 (1995), No. 4, pp. 11-25.

Edvinsson, L. and *Malone, M. S.*, Intellectual capital: The Proven Way to Establish Your Company's Real Value By Measuring Its Hidden Values, London 1997.

Feltham, G. and *Xie, J.*, Performance Measure Congruity and Diversity in Multi-Task Principal Agent Relations, in: The Accounting Review, Vol. 69, Jul (1994), No. 5, pp. 429-453.

Foss, N., Resources Firms and Strategies: A Reader in the Resource-Based Perspective, Oxford 1997, pp. 3-18.

Gjesdal, F., Accounting for Stewardship, in: Journal of Accounting Research, Vol. 19, (1981), No. 3, pp. 208-231.

Grant, R. M., The Resource-Based Theory of Competitive Advantage: Implications for Strategy Formulation, in: California Management Review, Vol. 33 (1991), No. 1, pp. 14-35.

Grant, R.M., Prospering in Dynamically-Competitive Environments: Organizational Capability as Knowledge Integration, in: Organization Science, Vol. 7 (1996a), No. 4, pp. 375-387.

Grant, R.M., Toward a Knowledge-Based Theory of the Firm, in: Strategic Management Journal Vol. 17 (1996b), Special Issue, pp. 109-122.

Gu Feng and *Lev, B.* (2002), Intangible Assets: Measurement, http://pages.stern.nyu.edu/~blev/.

Gupta, O. and *Roos, G.*, Mergers and Acquisitions Through an Intellectual Capital Perspective, in: Journal of Intellectual Capital, Vol. 2 (2001), No. 3, pp. 297-309.

Guthrie, J., *Petty, R.* and *Johanson U.*, Sunrise in the knowledge economy: Managing, measuring and reporting intellectual capital, in: Accounting, Auditing and Accountability Journal, Vol.14 (2001), No.4, pp. 365-384.

Guthrie, J., The Management, Measurement and Reporting of Intellectual Capital, in: Journal of Intellectual Capital, Vol. 2 (2001), No.1, pp. 27-41.

Hall, R., The Management of Intellectual Assets: A New Corporate Perspective, in: Journal of General Management, Vol. 15 (1989), No.1, pp.53-68.

Hall, R., The Strategic Analysis of Intangible Resources, in: Strategic Management Journal, Vol. 13 (1992), No. 2, pp. 135-144.

Hall, R., A Framework Linking Intangible Resources and Capabilities to Sustainable Competitive Advantage, in: Strategic Management Journal, Vol. 14 (1993), No. 8, pp. 607-618.

Hand, J., The Increasing Returns-to-scale of Intangibles, In *Hand, J.* and *Lev, B.* (Ed.), Intangible Assets: Values, Measures, and Risks, Oxford, 2003, pp. 303-331.

Hauser, J., *Simester, D.* and *Wernerfelt, B.*, Customer Satisfaction Incentives, in: Marketing Science, Vol. 13 (1994), No. 4, pp. 327-350.

Helmi, M. A., Measuring the Effect of Customer satisfaction on Profitability: A challenging role for Management Accountants, in: The National Public Accountant, Vol. 43 (1998), No.10, pp. 8-11.

Hemmer, T., On the Design and Choice of Modern Management Accounting Measures, in: Journal of Management Accounting Research, Vol. 8, (1996), No. 3, pp. 87-116.

Hirschey, M., How Much is a Tulip Worth? In: Financial Analyst Journal, Vol. 54 (1998), No. 4.

Hirschley, M. and *Weygandt, J.*, Amortization Policy of Advertising and Research and Development Expenditure, in: Journal of Accounting Research, Vol. 23 (1985), No. 3, pp. 226-235.

Holland, J., Financial Institutions, Intangibles and Corporate Governance, in: Accounting, Auditing & Accountability Journal, Vol. 14 (2001), No. 4, pp. 479-529.

Itami, H., Mobilizing Invisible Assets, Cambridge MA 1991.

Ittner, C. D., *Larcker, D. F.* and *Rajan, M. V.*, The Choice of Performance Measures in Annual Bonus Contracts, in: The Accounting Review, Vol. 72 (1997), No. 2, pp. 231-255.

Ittner, C. D. and *Larcker, D. F.*, Are Nonfinancial Measures Leading Indicators of Financial Performance? An Analysis of Customer Satisfaction, in: Journal of Accounting Research, Vol. 36 (Supplement) (1998).

Ittner, C. D. and *Larcker, D. F.*, Determinants of Performance Measure Choices in Worker Incentive Plans, in: Journal of Labor Economics, Vol. 20 (2002), No. 2, pp. 58-90.

Johnson, T. H. and *Kaplan, R. S.*, Relevance Lost: The Rise and the Fall of Management Accounting, Boston 1987.

Klaila, D. and *Hall, L.*, Using Intellectual Assets as a Success Strategy, in: Journal of Intellectual Capital, Vo. 1 (2000), No. 1, pp. 47-53.

Kaplan, R. S. and *Norton, D. P.*, The Balanced Scorecard - Measures that Drive Performance, in: Harvard Business Review, Vol. 70 (1992), No. 1, pp. 71-79.

Kaplan, R.S. and *Norton, D.P.*, The Balanced Scorecard - Translating Strategy into Action. Boston, MA 1996.

Kaplan, R.S and *Norton, D.P.*, Having Trouble with Your Strategy? Then Map It, in: Harvard Business Review, Vol. 78 (2000a), No. 9/10, pp. 167-176.

Kaplan, R.S. and *Norton, D.P.*, The Strategy Focused Organization: How Balanced Scorecard Companies thrive in the New Business Environment, Boston 2000b.

Lavin, D., *Chrysler* Aides to Get Bonuses Equal to Salaries, in: Wall Street Journal, Feb 22, A3 (1994).

Leadbeater, C., New Measures for the New Economy, ICAEW, London 2000.

Lev, B., Intangibles: Management, Measurement, and Reporting, Washington 2001.

Marr, B., *Gray, D.* and *Neely, A.*, Why do Firms Measure their Intellectual Capital?, in: Journal of Intellectual Capital, Vol. 4, (2003), No 4.

Marr, B., *Schiuma, G.* and *Neely, A.*, Assessing Strategic Knowledge Assets in eBusiness, in: Int. J. Business Performance Management, Vol. 4 (2002), No. 2-4, pp. 279- 295.

Marr, B., *Neely, A.* and *Thomas, G.*, Balanced Scorecard and Strategy Maps: How Intangibles Drive Corporate Performance at *Shell International*, in: Proceedings: PMA 2002 - Research and Action, Boston MA 2002.

Marr, B. and *Schiuma, G.*, Measuring and Managing Intellectual Capital and Knowledge Assets in New Economy Organizations, In: *Bourne, M.* (Ed.), Handbook of Performance Measurement, London 2001.

Marr, B., Measuring and Benchmarking Intellectual Capital, working paper, CBP, Cranfield School of Management, 2003.

Markides, C. C. and *Williamson, P J.*, Related Diversification, Core Competences and Corporate Performance, in: Strategic Management Journal, Vol. 15 (1994), Special Issue, pp. 149-165.

Mavrinac, S. and *Siesfeld, T.*, Measures That Matter. An Exploratory Investigation of Investors' Information Needs and Value Priorities, *Ernst & Young LLP*, Boston 1997.

MERITUM, Guidelines for Managing and Reporting on Intangibles, Madrid 2002.

Meyer, M.W. and *Gupta, V.*, The Performance Paradox, in: Research in Organizational Behavior, Vol. 16 (1994), No. 4, pp. 309-369.

Miller, D., *Greenwood, R.* and *Hinings, B.*, Creative Chaos Versus Munificent Momentum: The Schism Between Normative and Academic Views of Organizational Change, in: Journal of Management Inquiry, Vol. 6, (1997), No.1, pp. 71-78.

Montgomery, C. A. and *Wernerfelt, B.*, Diversification, Ricardian Rents, and Tobin's q, in: Journal of Economics, Vol. 19 (1988), No. 4, pp. 623-632.

Morck, R. and *Yeung, B.*, Why Firms Diversify: Internalization vs. Agency Behavior", in: *Hand J.* and *Lev, B.*, Intangible Assets: Values, Measures, and Risk, Oxford 2003, pp. 269-302.

Mouritsen, J., *Larsen, H. T.* and *Bukh, P. N.*, Intellectual Capital and the 'Capable Firm': Narrating, Visualising and Numbering for Managing Knowledge, in: Accounting, Organizations and Society, Vol. 26 (2001), No. 7, pp. 735-762.

Narayanan, V. K., *Pinches, G, E.*, *Kelm, K. M.* and *Lander D. M.*, The Influence of Voluntarily Disclosed qualitative Information, in: Strategic Management Journal, Vol. 21 (2000), No. 7, pp. 707-722.

Neely, A., Measuring Business Performance: Why, What and How, London 1998.

Neely, A., *Adams, C.* and *Kennerley, M.*, The Performance Prism: The Scorecard for Measuring and Managing Business Success, London 2002.

Neely, A. and *Al Najjar, M.*, Linking Financial Performance to Employee and Customer Satisfaction, in: *Neely, A.*, Business Performance Measurement: Theory and Practice, Cambridge 2002.

Neely, A., *Mills, J.*, *Gregory, M.* and *Richards, H.*, Performance Measurement System Design: Should Process Based Approaches be adopted? In: International Journal of Production Economics, Vol. 46-47 (1996), pp. 423-431.

Neely, A. et al., Driving Value through Strategic Planning and Budgeting, in: Research Report Cranfield School of Management / Accenture, 2002.

Nelson, R. R. and *Winter, S.G.*, An Evolutionary Theory of Economic Change, Cambridge, MA 1982.

Nonaka, I., Redundant, Overlapping Organizations: a Japanese Approach to managing the Innovation Process, in: California Management Review, Vol. 32 (1990), No. 1, pp. 27-38.

Nonaka, I., The Knowledge-Creating Company, in: Harvard Business Review, Vol. 69 (1991), No. 1, pp. 96-104.

Nonaka, I., A Dynamic Theory of Organizational Knowledge Creation, in: Organization Science Vol. 5 (1994), No. 1, pp. 14-37.

Norreklit, H., The Balance on the Balanced Scorecard - A critical analysis of some of its assumptions, in: Management Accounting Research, Vol. 11 (2000), No. 1, pp. 65-88.

Paul, J., On the Efficiency of Stock-Based Compensation, in: Review of Financial Studies, Vol. 5 (1992), No. 3, pp. 471-502.

Penrose, E. T., The Theory of the Growth of the Firm, New York 1959.

Peppard, J. and *Rylander, A.*, Using an IC Perspective to Design and Implement a Growth Strategy: the Case of *APiON*, in: European Management Journal, Vol. 19 (2001), No. 5, pp. 510-525.

Petergraf, M. A., The Cornerstones of Competitive Advantage: A Resource-Based View, in: Strategic Management Journal, Vol. 14 (1993), No. 2, pp. 179-188.

Petty, R. and *Guthrie, J.*, Intellectual Capital Literature Review Measurement, Reporting and Management, in: Journal of Intellectual Capital, Vol. 1 (2000), No. 2, pp. 155-176.

Porter, M. E., How Competitive Forces Shape Strategy, in: Harvard Business Review, Vol. 57 (1979), No. 3/4.

Prahalad, C. K. and *Hamel, G.*, The Core Competence of the Corporation, in: Harvard Business Review, Vol. 68 (1990), No. 5/6, pp. 79-91.

Roos, J., *Roos, G.*, *Dragonetti, N. C.* and *Edvinsson, L.*, Intellectual Capital: Navigating the new business landscape, London 1997.

Rouse, M. J. and *Daellenbach, U. S.*, Rethinking research methods for the resource-based perspective: isolating sources of sustainable competitive advantage, in: Strategic Management Journal, Vol. 20 (1999), No. 5, pp. 487-494.

Rouse, M. J. and *Daellenbach, U. S.*, More Thinking on Research Methods for the Resource-based Perspective, in: Strategic Management Journal, Vol. 23 (2002), No. 10, pp. 963-967.

Rucci, A. J., *Kirn, S.P.* and *Quinn, R. T.*, The Employee-Customer Profit Chain at Sears, in: Harvard Business Review, Vol. 76 (1998), No. 1/2, pp. 82-98.

Rumelt, R. P., Towards a Strategic Theory of the firm. In: *Lamp, R.B.* (Ed.), Competitive Strategic Management, New Jersey 1984.

Rylander, A., *Jacobsen, K.* and *Roos, G.*, Towards Improved Information Disclosure on Intellectual Capital, in: International Journal of Technology Management, Vol. 20 (2000), No. 5-8, pp. 715-741.

Seethamraju, C., The value relevance of trademarks, in: New York University 2000.

Selznick, P., Leadership in Administration, San Francisco 1957.

Spender, J. C. and *Grant, R. M.*, Knowledge and the Firm: Overview, in: Strategic Management Journal, Vol. 17 (1996), Winter Issue, pp. 5-9.

Starovic, D. and *Marr, B.*, Understanding Corporate Value – Managing and Reporting Intellectual Capital, London 2003.

Stewart, T. A., Intellectual Capital: The New Wealth of Organizations, New York 1997.

Stewart, T. A., The Wealth of Knowledge – Intellectual Capital and the Twenty-First Century Organization, London 2001.

Stivers, B. P., Covin, T. J., Hall, N. G. and *Smalt S. W.*, How nonfinancial performance measures are used, in: Management Accounting, Vol. 79 (1998), No. 2, pp. 44-49.

Sullivan, P. H. Jr. and *Sullivan, P. H. S.*, Valuing intangible companies - An intellectual capital approach, in: Journal of Intellectual capital, Vol. 1 (2000), No. 4, pp.328-340.

Sveiby, K. E., The New Organizational Wealth: Managing and Measuring Knowledge-based Assets, San Francisco 1997.

Sveiby, K. E., A Knowledge-Based Theory of the Firm to Guide in Strategy Formulation, in: Journal of Intellectual Capital, Vol. 2 (2001), No. 4.

Teece, D. J., Economies of Scope and the Scope of the Enterprise, in: Journal of Economic Behavior and Organization, Vol. 1 (1980), No. 3, pp. 223-233.

Teece, D. J., Capturing Value from Knowledge Assets: The New Economy, Markets for Know-How, And Intangible Assets, in: California Management Review, Vol. 40 (1998), pp. 55-79.

Teece, D. J., Managing Intellectual Capital: Organizational, Strategic, and Policy Dimensions, Oxford 2000a.

Teece, D. J., Strategies for Managing Knowledge Assets: The Role of Firm Structure and Industrial Context, in: Long Range Planning, Vol. 33 (2000b), No. 1, pp. 35-54.

Teece, D. J., *Pisano, G.* and *Shuen, A.*, Dynamic Capabilities and Strategic Management, in: Strategic Management Journal, Vol. 18 (1997), No.7, pp.509-533.

Tranfield, D., *Denyer, D.* and *Smart, P.*, Undertaking Systematic Review: Developing an Evidence-Based Approach for Management Research, Denver, Colorado, Academy of Management Conference 2002.

Venkatraman, N. and *Grant, J.*, Construct Measurement in Organizational Strategy Research: A Critique and Proposal, in: Academy of Management Review, Vol. 11 (1986), No.1, pp. 71-87.

Viedma, J. M., ICBS – Intellectual Capital Benchmarking System, in: Journal of Intellectual Capital, Vol. 2 (2001), No 2, pp. 148-164.

W.A., Human Capital on Strategy and Performance in Professional Service Firms: A Resource-Based Perspective, in: Academy of Management Journal Vol. 44 (2001), No. 1, pp. 13-28.

Wernerfelt, B., A Resource Based View of the Firm, in: Strategic Management Journal, Vol. 5 (1984), No. 2, pp. 171-180.

Williams, M. S., Is Intellectual capital Performance and Disclosure Practices related? In: Journal of Intellectual Capital, Vol. 2 (2001), No. 3, pp.192-203.

Winter, S. G., Knowledge and competence as a strategic asset, in: *Teece, D. J.*, The competitive challenge, Cambridge 1987.

Intellectual Capital Management, Measurement and Disclosure

Göran Roos [1], *Stephen Pike* [2] *and Lisa Fernström* [3]

---

[1] Prof. *Göran Roos*, Intellectual Capital Sevices, Ltd., London, UK

[2] *Lisa Fernström*, Intellectual Capital Sevices, Ltd., London, UK

[3] Dr. *Stephen Pike*, Intellectual Capital Sevices, Ltd., London, UK

In this chapter, the authors give their view on the intellectual capital (IC) approach in terms of management, measurement and disclosure of IC. The chapter looks at the reasons why methodologies for the measurement of intangible assets of the last decade have failed to meet the expectations of managers and regulators. It is therefore most important that a systematic approach to measurement is taken by discussing measurement theory and the requirements on an IC measurement model. Based on the results of this analysis an examination of existing models is carried out. This leads to methodological and practical requirements on measurement models as well as recommendations for disclosure.

1. Introduction

Over the past decade, there has been a rapidly growing realisation of the importance of intangible assets and intellectual capital as a whole in the operation of organisations. With this realisation has come the need to manage companies in a new way and to measure their performance in a new way. A sharp illustration of the differences in the managerial attitudes of the industrial and post-industrial ages (*Sveiby*, 1997) has been in the appreciation that people can no longer be considered to be costs on the profit and loss statement but are, in fact, assets to be invested in, developed and deployed carefully. Many authors have written at length about the failure of accounting systems to assist managers with the management of their companies. The failure of accounting as a useful management aid spawned a variety of different approaches to company management. In the early 1990s, these involved analysis of company activities using the concept of competencies (*Kaplan/Norton*, 1996). Later in the decade (*Stewart*, 1997), the investigation of intangible assets and intellectual capital came to prominence with a very broad range of explanations and implementations. In the later part of the last decade and into the first few years of this one, there have been a number of high-profile financial collapses. In addition, the world has seen the rise and fall of the "dotcom" companies and the inability of the markets or venture capitalists to estimate their worth with safety. Regulatory bodies have been pressing and moving towards the mandatory disclosure of certain elements of intangible resources and the recording of elements of goodwill in mergers and acquisition (FAS 141, 142) (FASB, 2001).

2. Intellectual Capital Approach

The starting point of the IC reasoning is the resource-based view of the firm stating that any given firm or organisation is a system made up of a set of resources that are deployed in a given structure (e.g. *Penrose*, 1959; *Wernerfelt*, 1984; *Rumelt*, 1987; *Peteraf*, 1993; *Barney*, 1991 and 2001). Intellectual capital is the science and art dealing with how to extract maximum value from the resources you have and the way you have chosen to deploy them. Five resource categories can be used as a framework facilitating the identification of all the different resources. These categories are human, organisational and relational resources on the intangible side, and monetary and physical resources on the tangible side (*Roos/Roos*, 1997).

Human resources include e.g. the knowledge, competence, intellectual agility, relationship ability and attitude of the employees. These resources are by definition not owned by the firm. By contrast, organisational resources are owned by the firm and comprise e.g. all the structures, systems and processes that the company uses to support their operations, but also items such as brands, image, culture, prototypes, documented information and intellectual property. Then there are the relational resources, which embrace e.g. all external relationships, such as customers, suppliers, media, strategic partners and other types of alliances. These resources are not owned by the company and are, by definition controlled (at least partly) by the other party. As for the tangible resources there is a distinction between monetary and physical resources, which are the assets owned by the firm, shown in the traditional accounting.

The IC perspective not only provides a new and more holistic way of viewing the company and its resources, but it also offers a new language in which to talk about these resources that have always been there to a more or less extent, but that have not been articulated in internal management nor in disclosure to stakeholders. There is still not a generally adopted terminology around intellectual capital resources, although the IC community is starting to move in the direction of consensus. The holistic classification framework in *figure 1* sets out the different types of tangible and intangible assets that financial assets, or traditional accounting assets, and IC assets comprise respectively.

	Assets				
	Traditional Accounting Assets		**Intellectual Capital Assets**		
	Monetary	**Physical**	**Relational**	**Organizational**	**Human**
Tangible Assets	■ Cash ■ Investments ■ Receivables/Debtors ■ Payables/Creditors	■ Property ■ Plant ■ Equipment ■ Inventory - ➢ Finished Goods ➢ WIP ➢ Parts/Raw Materials	■ Customer Contracts ■ Formal Alliances (JVs, Supply Agreements)	■ Systems ■ Formalized Processes ■ Codified Knowledge ■ Patents ■ Brands ■ Mastheads	■ Acknowledged Skill Sets ■ Experience ■ Employee Loyalty
Intangible Assets	■ Credit Rating/s ■ Undrawn Facilities ■ Borrowing Capacity (relative to like companies – based on character) ■ Borrowing Covenant Slack ■ Receivables and Accruals Certainty ■ Quality of Earnings ■ Balance Sheet Strength	■ Plant Flexibility ■ Plant Modernity ■ Infrastructure Surrounding Plant/s ■ Stranded Assets? ■ Tradability" of Facilities? ■ Access Rights ■ Plant Regard ("Can Do"; "Will Do") ■ Inventory Good, Obsolete, Redundant)	■ Customer Loyalty ➢ Behavioral ➢ Attitudinal ■ Quality of Supply Contracts ■ Right to Tender"; Right to Compete"; "Right to Design" ■ Strength of Stakeholder Support (including Opinion Leaders) ■ Networks ■ Regulatory Imposts	■ Structural Appropriateness ■ Informal Processes ■ Organizational Reputation ■ Brand Meaning (Strength; Stature) ■ Productivity of R&D Process ■ Quality of Corporate Governance ■ Know How" ■ Tacit Knowledge	■ Top Management Quality ■ Top Management Experience ■ Ability to Execute on Strategy ■ Capabilities ■ Problem Solving Ability ■ Employee Loyalty - ➢ Behavioral ➢ Attitudinal ■ Personnel Reputation

Fig. 1: The holistic classification framework (Burgman/Roos, 2003)

2.1 Value Creation

Having a resource and the language to express the presence of a resource is however not enough to create value. In order to do that, the resources have to be deployed effectively and efficiently. By using the IC lens, the value creation path in the organisation can be determined. The example of *Company A* and *Company B* will illustrate the point. Both companies are consulting firms. *Company A* is made up of very bright people with a minimal set of physical resources (such as computers) and some working capital to keep the business going. They need numerous relationships with customers and other people, and they need to have an enormous amount of competence in themselves as individuals. As individuals, they build strong relationship with clients. There is no structure and there are no processes in this company. Every time they do something they tend to reinvent the wheel. They make money from two transformations: 1) by converting their time to cash, which is based on their competence, meaning that they can charge considerable amounts for their time; 2) since they have very strong personal relationships, they can charge even more than they otherwise would. That is, the relationship's contribution to the charging level is the second transformation.

Company B has a different approach and provides packaged solutions. They have some average skilled people who they can leverage through excellent structures and thereby produce high quality deliverables. Their processes, systems, brands and intellectual property allow them to use average people to provide good solutions. For *Company B*, relationships between themselves and client organisations are based on the *Company B* brand rather than any individual. They charge a certain amount for man-hours, although they cannot charge at very high rates for average people. Instead, they make their money by supplying a packaged solution. Hence, there are three contributing transformations to cash flow: 1) a small amount from man-hour sales; 2) significant amounts from sales of standardized solutions; and 3) some amount from the organisational relationship (brand) that they have with the client organisation.

Looking at these two firms from a financial perspective, they may appear exactly the same. They can have the same turnover, margin and balance sheet. Even if extending to asking clients about how satisfied they are, the companies may appear very similar, with the same customer satisfaction and retention ratio levels. However, when examining the IC of the two firms, looking at the type of competencies, processes, structures, systems, brands, IP, etc. they have and how they are used, the difference between them will be visible. In conclusion, the only way to discover the difference between these two firms, and analyse their effectiveness and efficiency by which these approaches actually create value, is by taking the intellectual capital perspective.

If illustrating the value creation of these firms graphically, the deployment of resources can be shown in a navigator. In the navigator, the key resources are represented by circles, whose size reflects the importance of the resource. The transformation of one type of resource to another is represented by the connecting arrows, whose thickness represents the importance of the transformation. In the two diagrams in *figure 2*, the difference between *Company A* and *Company B* is immediately obvious. The reliance of *Company A* is reflected in the size of the human resources (attitude, competence and intellectual agility) and the resulting transformations. In contrast, *Company B* relies more on written down processes, part of organisational capital.

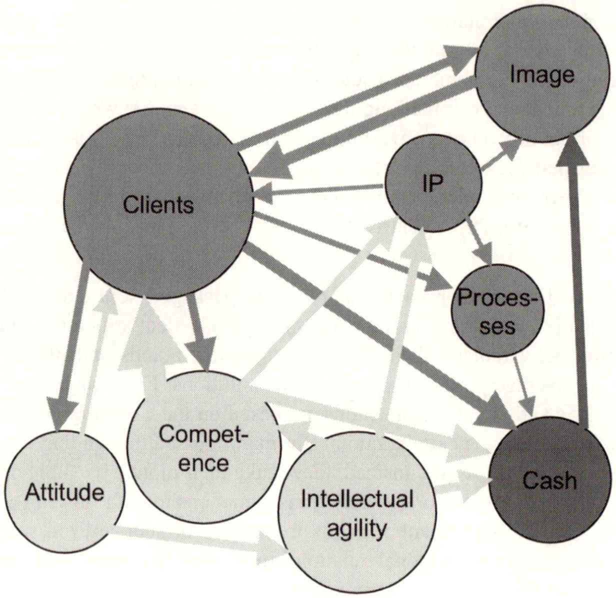

Company A

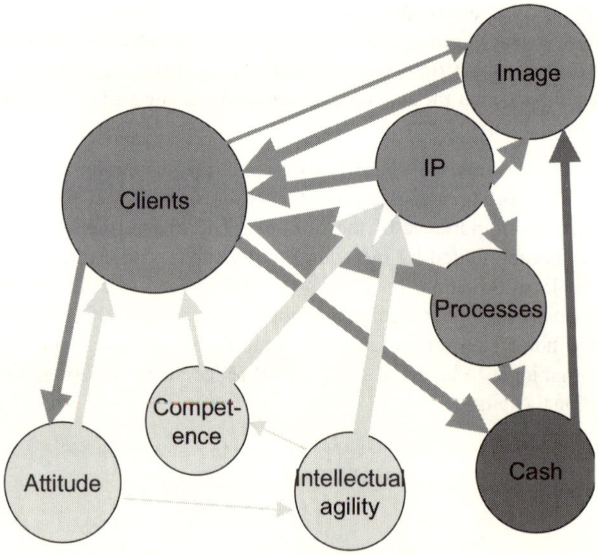

Company B

Fig. 2: The value creation paths of two different advisory firms

The above example with the two advisory firms shows how different seemingly similar companies within an industry can operate. This is because they have chosen different value creation logics for their activities. The value creation logics comprise the value chain, the value shop and the value network (see *figure 3*). In *Porter*'s conventional value chain (1985), the output alone is what brings value to the customer. This type of value creation relies on standardised processes and repetition (economies of learning) and mass production (economies of scale). In a value shop, the main focus is solving a problem for a client. In contrast to the value chain, the value in a value shop resides not only in the solution itself – the output – but also in the individuals who came up with the solution and the way they reached it. In a value network, the basis for value creation lies in connecting people who wish to be interdependent where the actual enabling when connecting customers to each other creates the value.

To illustrate this, a bank may serve as a good example. A bank can be a value chain when it converts inputs to outputs, in which case the focus will be on efficiency. It will look upon itself as a factory that processes loans, applications, or something similar. A bank operating as a value shop views itself as a financial service operator or a financial advisor that also has the ability to provide the money. It would identify the problems of a client, address them, and select a solution together with the client and help to implement it. It would have stringent quality controls. As part of its offering, it would possibly supply clients with some cash as a loan or accept cash for investment. The third option for the bank is the value network, in which case the bank would act solely as a conduit between people who do not have money and people who do. The bank then arranges the flow of cash between them.

For each of the activities in the different logics, there is a requirement that the activity is done well. Doing them well is the ability to transform different types of intellectual resources effectively and efficiently. For the consulting case above, it boils down to transforming the knowledge that resides in the individual into good, useful solutions for the client. Having the knowledge within the organisation is not enough. There is a crucial distinction between having a resource - be it competences, processes, systems, intellectual property, relationships, machines, cash - and putting it to value-creating use, which is the ability to transform one resource into another.

2.2 Value Drivers

In order to make progress in understanding how to turn the IC view of the company to practical advantages, the company "engine" will need to be coupled to its value drivers. Why are drivers of value important? They are drivers of perceived value in the minds of customers and are important because they impact two drivers of cash: revenue and margin. Revenue is driven by revenue drivers such as the number of client relationships, customer lifetime cycle, amount of each purchase, and purchase frequency. Those are all impacted by the perceived value of the customers. Margin is driven by the relative performers of perceived value and delivered costs and, therefore, perceived value is one of the two things to drive the margin of the organisation. Therefore, in order to manage margin and revenue well, an excellent understanding of the cost drivers and the perceived value drivers is needed. Most organisations have rigorous control of the cost drivers; hence that information is usually readily available in a company.

Based on several years research on executive MBA programs executed by the authors there are four identified categories of value drivers. The first is price, which are all the resources

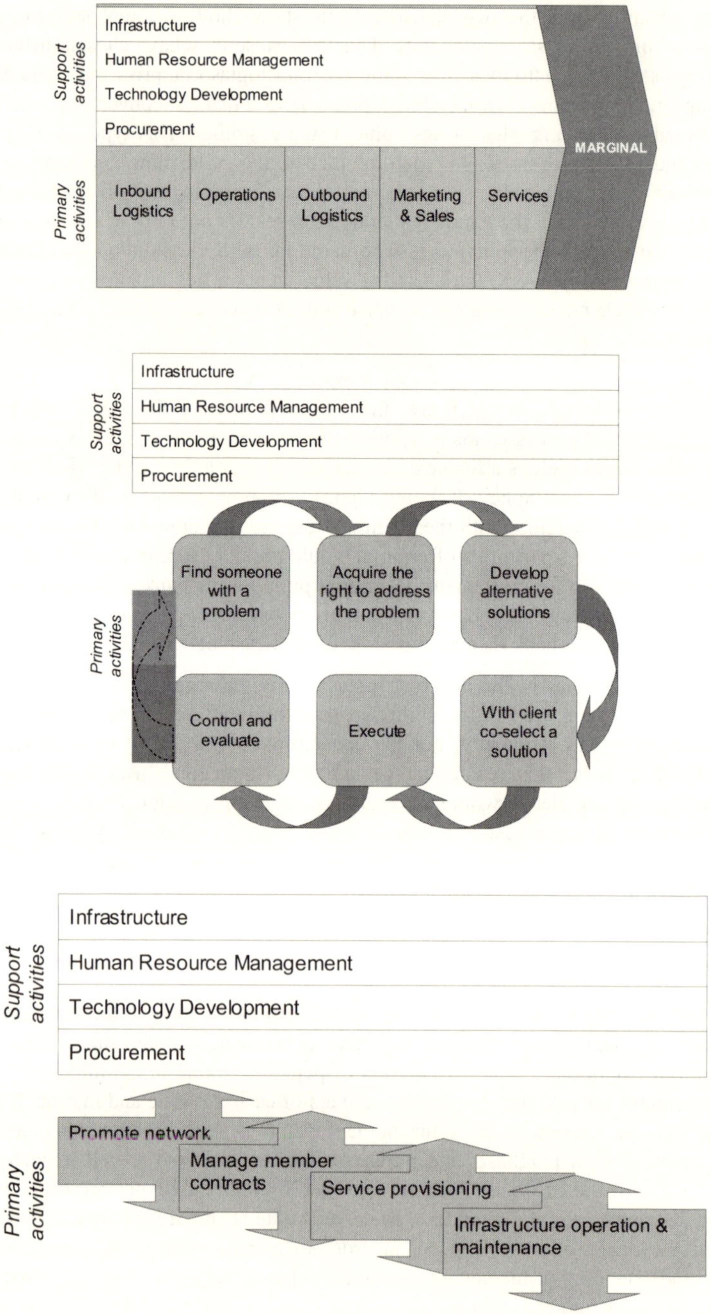

Fig. 3: The Value Creation Logics, the Chain, the Shop and the Network (based on Stabell and Fjeldstad, 1998)

that the organisation has to give up in return for the service or product that it purchases. This can be actual money, some other resource, time or something else. The second is functionality, which is everything objectively measurable surrounding the delivery of a product or service, such as on-time delivery, specification match (colour, features and functionality). The third is distinctness, which is every thing that generates mind share as a precursor to market share (brands, reputation and references). The fourth and final value driver category is mental proximity, which constitutes the basis for and lowers the barriers for building person-to-person trust, such as shared values and beliefs. The drivers of the situational relative importance of these drivers have also been identified in the same research and is summarised in *figure 4*.

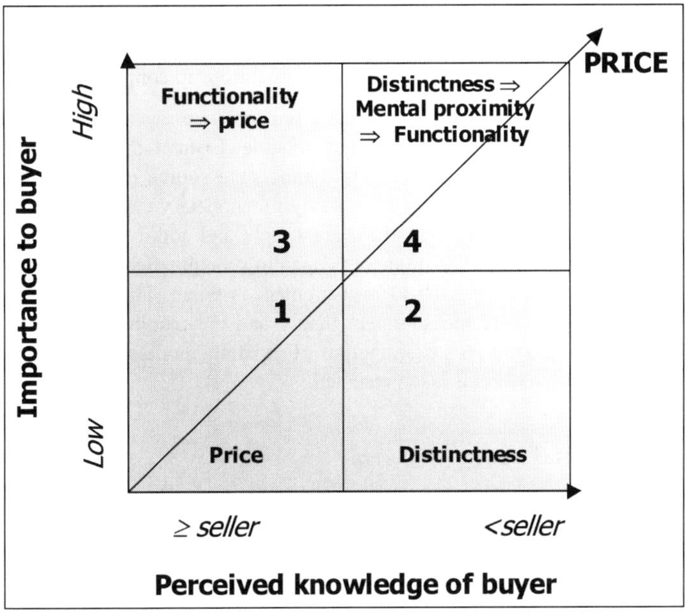

Fig. 4: Situational importance of different perceived value driver categories

Those are the value drivers. By knowing them and the cost drivers, and knowing what to do to improve on them, an organisation can move from where it is to where it strategically desires to be. This move would have an impacting consequence on margin, revenue, working capital, cash, taxes, and acquisitions, and investments, thereby directly impacting the market value of the company in a predictable and manageable way.

3. IC Measurement

As stated in the introduction to this chapter, the business community is today lacking powerful and standardised tools for measuring the performance of a company, that is, of all its resources and their transformations. The change in the key characteristics of companies in the knowledge era leads to one of the principal distinctions in the approaches to measurement and management. The most obvious approach is to try to retain as much of the rigour of conventional accounting by adjusting its traditional instruments. Where practitioners believe this to be impossible, they have resorted to measuring new things but retaining the forms of conventional accounting. The alternative to this is to abandon traditional accounting completely and base measurement and management on the attributes of the value generating processes of individual companies. In both methodologies, the aim is to allow managers to manage effectively internally, increasing the performance and competitive position of the company and also to communicate the performance, position and potential of the company externally to shareholders and the investing community at large. The issue is whether either of these approaches can meet the need in an auditable, useful and secure way. Internally, this means a measurement regime that gives managers the levers necessary to guide the business while not instilling bad behaviours through measuring the wrong things nor imposing a heavy burden of measurement of people who have better things to do. Externally it means providing stakeholders and potential investors with the often long-term information they need but without disclosing sensitive strategic intentions to competitors.

If managers are to make judgements based on a performance measurement model, the information it generates has to be auditable and reliable. Even if the interpretation of that information may differ from one manager to another, the source must be unimpeachable. Traditional accounting generally offers that quality for monetary capital and, to a considerable extent, physical capital since quantities are simple and additive. Intangible assets are not additive in nature and their measurement is open to semantic debate. It is therefore most important that a systematic approach to measurement is taken. The issue is developed in more detail here below, where measurement theory and the requirements on an IC measurement model are discussed. An examination of existing models and why they are not providing a satisfactory solution is also presented.

3.1 Measurement Theory

Serious thinking about measurement can be traced back to *Eudoxus* in ancient Greece but the ideas of the modern theory of measurement date from the 19th century work of *Helmholtz* (1887) and others. However, its formalisation is a surprisingly recent event. The primary motivation for the formalisation of measurement theory was the need to understand, using measurement in the physical sciences as a paradigm, what it means to measure things in the social sciences, things like preference, value, loudness, etc. The catalyst for the formalisation of measurement theory is generally accepted to be the psychologist *S.S. Stevens* with later interest from the field of quantum physics. However, it was not until the 1970s that measurement was fully axiomatised with the publications of *Scott* and *Suppes* (1958), and *Suppes* and *Zinnes* (1963).

The theory of measurement was developed to place issues of measurement on a more mathematical foundation and specifies the general framework in which measures should be defined. In pragmatic applications of measurement theory, a multi-stage process is generally involved in which the representation of the object to be measured and the measurements system are kept separate. Thus, in the first stage, an empirical relation system is specified to define the relations among the attributes of the studied entity, in this case, the company. Secondly, a numerical relation system is defined, to provide values for the measures of the attributes and relations among these values. Measures are then the link between the empirical relation system and the numerical relation system that maps entities into values and empirical relations among entities into formal relations among values.

Axiomatic approaches formally define desirable properties of the measures for all the attributes to be measured and these are the properties are properties of the numerical relation system. However, they indirectly affect the empirical relation system due to the nature of axioms. For example, if a measure for an attribute does not satisfy axioms that have been stated for numerical relation system, the numerical relation system has to be examined for faults and, in addition, the empirical relation system has to be checked to ensure that it really captures the meaning of the attribute to be measured. It must be remembered that while axioms are stipulated to be true for the purpose of supporting a chain of reasoning, they do not necessarily represent truth.

In the field of business management there have been very few applications of measurement theory. In an era of measurement and management driven by formal accounting systems, there was little need. In the recent spectacular company failures such as *Enron*, the accounting problems seem to be concerned with the application of accounting principles. That is, the empirical relation system allowed for interpretations which, if not encouraged, were not proscribed. The numerical relation system simply used the values suggested but was not in itself at fault. Thus while traditional accounting systems are generally compliant with measurement theory, they are not immune from problems.

The pragmatic requirements of an empirical relation system suited to modern business systems have been set out by *M'Pherson* and *Pike* (2001). They list three key axioms for an empirical relation system which is an expansion of an overarching context. These are completeness, distinctness and independence. The completeness and distinctness of the empirical relation system with respect to the context collectively provide for the full meaning of the "value" defined by the context, and they do not overlap in meaning which eliminates double-counting. The measures must be scale-independent so that a change in one value does not affect any other value reading. There is a fourth point which concerns minimality and is aimed at avoiding huge measurement systems but this is set aside for the time being.

Mapping from the empirical to the numerical involves several further considerations and there has been considerable debate in the literature over the last 30 years on what the axiomatic requirements need to be. The key issue is permissible transformations and the nature of the scales used in the measurement. Put simply, permissible transformations are transformations of a scale of measurement that preserve the relevant relationships of the measurement process. The commonly used terminology categorises scales as follows in *Figure 5*:

The resources used in businesses and the interactions between them are varied and complex. Any measurement scheme used to measure the performance of a business is therefore a multi-attribute system. Immediately, this raises the issue of the commensurability of the measurement space, i.e. the various primary scales must be projected onto a common measurement space that may have many dimensions but only one common scale on each dimension. Commensurability is usually achieved by normalisation, but it also means that a ratio scale is the minimum acceptable as measures being compared or combined from different areas will otherwise be arbitrary. All measures must have a meaningful zero.

Name	Typical description	Transformations	Allowed statistics
Nominal	A classification of the objects	Only those that preserve the fact that objects are different	(Descriptive) Frequencies, mode, information content. (Associative) Chi-square.
Ordinal	A ranking of the objects	Any monotonic increasing transformation, although a transformation that is not strictly increasing loses information.	(Descriptive) Median, quantiles and quartiles. (Associative) Spearman's rank-order correlation coefficient, Kendall's tau, rho.
Interval	Differences between values are meaningful, but not the values of the measure itself	Any affine transformation t(m) = c * m + d, where c and d are constants; the origin and unit of measurement are arbitrary	As above plus arithmetic mean, standard deviation
Ratio	There is a meaningful "zero" value and the ratios between values are meaningful	Any linear (similarity) transformation t(m) = c*m, where c is a constant; the unit of measurement is arbitrary.	As above plus geometric mean
Absolute	All properties reflect the attribute	Only 1-to-1 transformations	All

Fig. 5: Measurement Scales

3.2 Requirements for Business Measures

The five conditions suggested for a business measurement scheme capable of measuring business performance and derived from measurement theory are: completeness, distinctness, independence, agreeability and scaling:

- **Completeness.** If the system to be measured is the whole company then the attributes of the company which are to be the subject of measurement must completely describe the company. In practical terms, the meanings of the attributes of business performance must be fully defined and their aggregate must reflect all the resources used by the firm and the ways in which they are used. It is arguable that a failure to comply with this requirement was at the heart of the Enron situation in which certain bodies were excluded from the financial documentation thus the measures taken were incomplete with respect to the generally held view of what *Enron* comprised.
- **Distinctness.** This is a simple requirement aimed at eliminating double counting. An attribute is acceptable and an entity to be measured if there is no element of its meaning that is contained within the meaning of any other attribute.
- **Independence.** The test for preference independence requires that there is no instantaneous change in the score for each attribute if the input to any of the others is varied. Suppose all measures in a normalised business measurement scheme are set at 0.5 and then one of the inputs becomes first 0, then 1. If you can show that all the others can remain unchanged during this process then they are independent. If this cannot be shown then the attributes to be measured are ill defined.

- **Agreeability.** The issue of agreeability concerns the mapping from the empirical to the numerical relation system. This means that it must be agreed that the meaning of the attribute in the empirical relation system has been fully reflected in the numerical system where the measurement is actually taken. In other words, the attribute must not be represented in the numerical system by a proxy which has a different meaning. Strictly speaking, the numerical relation system must show the same completeness, distinctness, independence and meaning as the empirical system.
- **Scaling.** To make the measurements and any subsequent aggregation of measures valid, they must be observed using a ratio scale. Failure to do this will render meaningless many of the conclusions drawn from the data. Where measures are readily observable physical measures, such as temperature, the correct scale is easily chosen and applied. Many business measures are not of this kind and ratio scale techniques must be used in data collection.

3.3 Business Measurement Systems

Over the last 10-15 years, a great number of systems have been devised to help managers with business performance and with a special emphasis on non-financial measures. According to *Luthy* (1998) and *Williams* (2000), methodologies may be categorised into four groups. These are:

- **Direct Intellectual Capital methods (DIC).** Estimate the $-value of intangible assets by identifying its various components. Once these components are identified, they can be directly evaluated, either individually or as an aggregated coefficient.
- **Market Capitalization Methods (MCM).** Calculate the difference between a company's market capitalization and its stockholders' equity as the value of its intellectual capital or intangible assets.
- **Return on Assets methods (ROA).** Average pre-tax earnings of a company are divided by the average tangible assets of the company. The result is a company ROA that is then compared with its industry average. The difference is multiplied by the company's average tangible assets to calculate an average annual earning from the Intangibles. Dividing the above-average earnings by the company's average cost of capital or an interest rate, one can derive an estimate of the value of its intangible assets or intellectual capital.
- **Scorecard Methods (SC).** The various components of intangible assets or intellectual capital are identified and indicators and indices are generated and reported in scorecards or as graphs. SC methods are similar to DIS methods, expect that no estimate is made of the $-value of the Intangible assets. A composite index may or may not be produced.

The methods offer different advantages dependent on the use to which they are to be put. In this chapter it is assumed that a comprehensive assessment of the contributions intangibles make to the value of the company is the goal and that this requires proper measurement. Hence, the use of measurement theory as the criteria for suitability is justified. In practice, this means that MCM and ROA methodologies can be ignored from the outset since they are fundamentally based on financial figures, augmented by selected intangible assets. If the context of measurement is the value of the company then they are at once incomplete measures. Furthermore, ROA approaches tend to be based on industry comparisons rather than the company itself and many of the MCM approaches view intellectual capital a separable entity from book value and that the sums of these make up market value. This cannot be the case.

DIC and to a lesser extent SC methods offer the potential to create a more comprehensive picture of an organisation's health than financial metrics and they can be easily applied at any level of an organisation. This is because they are directly aimed at management support and DIC is intended to be holistic. *Sveiby* uses this classification system and has generated the list of methodologies in *figure 6*. The list, although large, is not comprehensive. It should be remembered that the aim of this examination is not an exhaustive assessment of all methodologies but is instead, an illustrated description of the attributes required of a measurement system for soft issues, fit to take its place alongside conventional accounting.

Name	Originator	Description of Measure	Type
Technology Broker	*Brooking (1996)*	Value of intellectual capital of a firm is assessed based on diagnostic analysis of a firm's response to twenty questions covering four major components of intellectual capital.	DIC
Citation-Weighted Patents	*Bontis (1996)*	A technology factor is calculated based on the patents developed by a firm. Intellectual capital and its performance is measured based on the impact of research development efforts on a series of indices, such as number of patents and cost of patents to sales turnover, that describe the firm's patents.	DIC
The Value Explorer™	*Andriessen/Tissen (2000)*	Accounting methodology proposed by KMPG for calculating and allocating value to 5 types of intangibles: (1) Assets and endowments, (2) Skills & tacit knowledge, (3) Collective values and norms, (4) Technology and explicit knowledge, (5) Primary and management processes.	DIC
Intellectual Asset Valuation	*Sullivan (2000)*	Methodology for assessing the value of Intellectual Property.	DIC
Total Value Creation, TVC™	*Anderson/McLean (2000)*	A project initiated by the Canadian Institute of Chartered Accountants. TVC uses discounted projected cash-flows to re-examine how events affect planned activities.	DIC
Accounting for the Future (AFTF)	*Nash (1998)*	A system of projected discounted cash-flows. The difference between AFTF value at the end and the beginning of the period is the value added during the period.	ROA
Tobin's q	*Stewart (1997)*	The "q" is the ratio of the stock market value of the firm divided by the replacement cost of its assets. Changes in "q" provide a proxy for measuring effective performance or not of a firm's intellectual capital.	MCM

Investor assigned market value (IAMV™)	*Standfield (1998)*	Takes the Company's True Value to be its stock market value and divides it inot Tangible Capital + (Realised IC + IC Erosion + SCA (Sustainable Competitive Advantage)	MCM
Market-to-Book Value	*Stewart (1997)* *Luthy (1998)*	The value of intellectual capital is considered to be the difference between the firm's stock market value and the company's book value.	MCM
Economic Value Added (EVA™)	*Stewart (1997)*	Calculated by adjusting the firm's disclosed profit with charges related to intangibles. Changes in EVA provide an indication of whether the firm's intellectual capital is productive or not.	ROA
Human Resource Costing & Accounting (HRCA)	*Johansson (1997)*	Calculates the hidden impact of HR related costs which reduce a firm's profits. Adjustments are made to the P&L. Intellectual capital is measured by calculation of the contribution of human assets held by the company divided by capitalised salary expenditures.	ROA
Calculated Intangible Value	*Stewart (1997)*	Calculates the excess return on hard assets then uses this figure as a basis for determining the proportion of return attributable to intangible assets.	ROA
Knowledge Capital Earnings	*Lev (1999)*	Knowledge Capital Earnings are calculated as the portion of normalised earnings over and above expected earnings attributable to book assets.	ROA
Value Added Intellectual Coefficient (VAIC™)	*Pulic (1997)*	Measures how much and how efficiently intellectual capital and capital employed create value based on the relationship to three major components: (1) capital employed; (2) human capital; and (3) structural capital.	ROA
Human Capital Intelligence	*Jac Fitz-Enz (1994)*	Sets of human capital indicators are collected and bench-marked against a database. Similar to HRCA.	SC
Skandia Navigator™	*Edvinsson/Malone (1997)*	Intellectual capital is measured through the analysis of up to 164 metric measures (91 intellectually based and 73 traditional metrics) that cover five components: (1) financial; (2) customer; (3) process; (4) renewal and development; and (5) human.	SC
Value Chain Scoreboard™	*Lev (2002)*	A matrix of non-financial indicators arranged i three categories according to the cycle of development: Discovery / Learning, Implementation, Commercialisation.	SC
IC-Index™	*Roos, Roos, Dragonetti and Edvinsson (1997)*	Consolidates all individual indicators representing intellectual properties and components into a single index. Changes in the index are then related to changes in the firm's market valuation.	SC

Intangible Asset Monitor	*Sveiby (1997)*	Management selects indicators, based on the strategic objectives of the firm, to measure four aspects of creating value from intangible assets. By: (1) growth (2) renewal; (3) utilisation/efficiency; and (4) risk reduction/stability.	SC
Value Creation Index	*Ittner and others (2000)*	Drivers of value are derived from an extensive literature survey and advanced statistics. Metrics are weighted and combined to give a Value Creation Index. The index is compared and combined with financial data.	SC
Balanced Scorecard	*Kaplan/Norton (1992)*	A company's performance is measured by indicators covering four major focus perspectives: (1) financial perspective; (2) customer perspective; (3) internal process perspective; and (4) learning perspective. The indicators are based on the strategic objectives of the firm.	SC

Fig. 6: Measurement methodologies

With the exception of VAIC™, all MCM and ROA methodologies were not included in the assessment for the reasons given above. In addition to this, DIC and SC methodologies aimed at a single specific area of company operations such as human resource management of knowledge management were also excluded as they are, by definition, incomplete. This leaves 9 methodologies. In all cases, the potential of the methodology rather than the outcome of a real instance of usage were assessed. The results from all the analyses are presented in summary in *Figure 7* (for complete results, see *Pike/Roos*, 2003).

It is clear from *Figure 7* that none of the methodologies are fully compliant although it must be remembered that only a selection have been analysed. While *Sveiby*'s recommendation on the general type of methodology that may prove useful has been followed, there may be other methodologies both in the categories selected and from those ignored which would assess well.

The criteria of compliance with measurement theory are clearly a stern test although if the measurement of intangibles is to achieve the acceptability of traditional accounting, it must pass this test. Nevertheless, it is also clear that while there are no candidates suitable as measurement systems, many of the methodologies provide considerable amounts of useful guidance for managers. An example of this is the Balanced Scorecard where it is estimated (*Rigby*, 2000) that more than half the global largest 1000 companies use the balanced scorecard.

Completeness is clearly a difficult criterion to meet with only the VAIC complying. Several of the other methodologies begin with a complete description of the company but filter or develop attributes in the empirical relation system in such a way as to destroy completeness. From a valuation and disclosure standpoint, completeness is a crucially important starting point; off-balance sheet transactions might be considered to be the financial world's analogue.

Compliance with the requirement for distinctness and independence is simply a question of discipline in design. It is perhaps fortunate that the intellectual capital community has thus far failed to standardise on any particular terminology. This eases the issue of defining attributes so that they are valid in the empirical space since there are no fixed positions yet.

Agreeability of the metrics with the attributes is often difficult with unacceptable proxy measures being a common problem. Three of the methodologies were compliant as, in general, their metric system was developed alongside the attributes to which they were to connect. However, two of the three were definitely non-compliant with respect to the qualities of the attributes. The conclusion to be drawn from this is that attributes and metrics were forced (unsuccessfully) to fit. Had the attributes been developed further, it is entirely possible that acceptable attributes and agreeable measures could have been found. The danger of this type of extension to metric systems is that they become unwieldy. However, none of the metric systems is anything other than tiny when considered against the overhead imposed on company by the current requirements of financial management.

	Complete	**Distinct**	**Ind.**	**Agree**	**Scale**
Technology Broker	Possible	Unlike-ly	No	Context specific	Ordinal
Value Explorer	No	No	No	Yes	Nominal
VAIC™	Yes	No	No	Yes	Ratio
Skandia Market Value Scheme	Unknown	No	No	No	Interval
Value Chain Scorecard	No	Unknown	Unknown	Unknown	Unknown
IC Index	No	Probably	Probably	Probably	Ratio
Intangible Assets Monitor	No	Possible	Possible	Possible	Ratio
Value Creation Index	No	Yes	Probably	Yes	Ratio
Balanced Scorecard	No	No	No	Yes	Interval +

Fig. 7: Summary of results

Finally, most of the methodologies were, or could be made compliant with the requirement to use ratio scales. This is especially easy with properly formed objective metrics. Where metrics are subjective, there is no reason why ratio scales cannot continue to be used. The requirement here is that a data collection method, such as pair-wise comparisons as developed by *Saaty* (1980) is used as they naturally to ratio scaling. There are numerous other methods which lead to ratio scales and numerous methodologies that fail to do this and produce interval scales. Where there is no attempt to process measurements, scaling may not be an issue but where some combination is contemplated either to form an index or to link with financial data, the proper scale use is mandatory.

Although there is widespread non-compliance in the sample of methodologies examined, there is enough compliance against each of the criteria to suggest that it is not impossible to design a single methodology that is compliant everywhere.

3.4 Practical Requirements on Measurement Models

In addition to the above methodological requirements, there are also some more practical criteria for a viable measurement model. These include that the model

- does not impose a large measurement overhead,
- facilitates strategic and tactical management and
- generates the information needed by shareholders and investors.

Measurement Overhead

This is an entirely practical point in that measurement schemes tend to grow uncontrollably. Key attributes that have to be measured, especially the non-financial attributes, should be broken down only to the stage where compliance with the requirement of measurement theory (see above), is satisfied. Two dangers emerge from over-measurement. The first is that the cost of data collection far outweighs the benefits of having it and that its collection also causes considerable irritation amongst those carrying out the measurement and those being measured, especially if the redundancy of the measurement is obvious. The second is that management based on over-measurement leads to micro-management and the tendency to instil unwanted behaviours. This latter point arises since people tend to want to improve performance and may focus on many trivial elements in an over-elaborate measurement system. In doing this, they loose sight of the bigger and more important picture.

Strategic Management

If a measurement and management scheme is to be of any real value, it must afford management a means of translating their strategic intent into appropriate actions and feedback information showing whether these actions are working or not. A report by the *Boston Consulting Group* suggests that managers have only three effective levers in the creation of value namely margin, asset productivity and investment (*BCG*, 1999) however others (*Baum* et al., 2000) give a more extensive list including innovation, attraction of talented people, alliances and customer satisfaction. The difference between these appraisals highlights the difference between a finance-dominated approach to management and measurement and one which includes non-financial measures. What appears to be clear is that managers can affect the performance of their businesses at two levels. The first is at the organisational level in which they affect how the processes of value creation in the company are interconnected, e.g. through strategic alliances and positioning. The second is that they can encourage improvements in individual or groups of processes at an operational level, e.g. through investments in soft assets and conditions.

If measurement is to support management effectively, the measures have to be dominated by those that look forward (*Van Buren*, 1999). This is not a new idea since a balanced approach to leading and lagging indicators was encouraged at the outset of the knowledge era by *Kaplan* and *Norton* (*Kaplan/Norton*, 1996) and earlier in the Tableau de Bord developed in France in the 1950s (*Lauzel/Cibert*, 1962). Here lies one of the principle weaknesses of accounting-based methodologies of intellectual capital management. Accounting is based on historical transactions and is thus dominated by lagging measures. Accounting, while detailed, is also cost-based rather than value-based even though some of the outcomes of past investments will have real-time and future consequences.

In contrast, methodologies for managing intellectual capital based on a business approach should suffer from none of the intrinsic weaknesses of financially based management methods. There are, however, serious deficiencies that can considerably degrade the usefulness of the IC approach. The most common of these is the measurement of the presence of resources in the belief that this constitutes value. A typical example of a resource presence metric is the number or percentage of people with a higher degree in a company. Resources represent the potential to create value and unless the measurement of potential is the specific aim (*Pulic*, 2000) then management will be misguided in using such an approach to manage IC. Value is created, as mentioned previously in this chapter, when the resources are employed, and it degrades when they remain unused. If discerning the importance of the value creation pathways is the goal, the attributes that must be measured are the influences that the elements of intellectual capital exert in the generation of value when they are deployed in the company. Alternatively, if some attempt is to be made at relative or even absolute value, then it will be necessary to quantify not just what is influential in the creation of value but also how large the transformation of these intellectual capital elements is. The important distinction here is that what transforms is not necessarily of value and what is influential in creating value might be a small transformation of something very important; the estimation of value requires knowledge of the volume of transformation as well as the influence that transformation actually exerts. The concept resources and transformations of intellectual capital for value creation is based on the notion of stocks and flows, introduced and described by *Roos* (*Roos/Roos*, 1997).

Shareholder Information

A viable measurement model also needs to generate relevant information for external stakeholders, such as shareholders. In order to communicate with these stakeholders, the form and the language of the information have to be such that the stakeholder can understand and evaluate it.

For a long time, the only information that seems to have been required by stakeholders to be reported in a standardised form in most companies has been company performance in financial terms. The exception to this includes companies or agencies involved in delivering public services. For them, financial stability has been a key performance indicator along with the quality level and extent of the service that has been provided. In the UK, however, government agencies largely started only at the end of the 1980's with the launch of the "Next Steps Agencies". In the knowledge era, the concept of stakeholder or even shareholder value extending beyond simple financial performance measures is the crucial change. To communicate with stakeholders now requires a deeper understanding of the attributes of value from the point of view of the stakeholder groups. Internally in the company, this means that the management's strategy must be more sophisticated, and the levers that management pull to improve their business, much more diverse and complex.

Understanding and being able to derive and communicate the attributes of value requires that management models be founded in axiology (value theory). Axiology is a branch of philosophy that developed in Germany and Austria in the 19th century (*Rescher*, 1969) and, in contrast to logical positivism, states that within a well-defined context; meaningful discussion of value can take place. That is, value is not emotional but can be the subject of rational discussion.

To make a practical tool requires the connection of axiology and measurement theory, which has been achieved by *M'Pherson* (*M'Pherson*, 1996) with the *Inclusive Value Methodol-*

ogy[TM] and *Roos* and *Pike* (2000) with the HVA (*Holistic Value Added*) Methodology and the *SEER*[TM] methodology. This opens the way to the generation of management models that can describe and measure the attributes of value for differing stakeholder groups within an embracing context. Managers then have the potential of a methodology, measurement and management scheme in which the impact of strategic decisions on all stakeholder groups can be assessed in advance of their implementation. This does not mean that the problems are all solved and the future is laid bare. Uncertainty in data and risk introduced by the actions of outsiders such as regulators, suppliers, partners, governments, not to mention the actions competitors all conspire to, make the results less than certain. The art of strategic management remains.

These then are the tests and requirements of a truly comprehensive intellectual capital management and measurement scheme. Although some approaches come close to providing a holistic approach by meeting most of the tests, a truly holistic model will require compliance with all tests. In practical terms this means effecting a union between navigators which describe the practical workings of businesses, axiology and measurement theory which will take measures from the navigator and apply them in a rigorous model of value and financial instruments to link the non-dimensional measures of value to money. Only HVA approaches this level and only SEER[TM] achieves it.

4. Disclosure

Continuing on the subject of information to shareholders, this section deals with the disclosure of intellectual capital. In order for a company to enjoy full benefits from managing intangibles, there is a need for disclosure on their performance to the different stakeholders of the company. Today, however, there is still no standard for disclosure of intangibles even though many an institution and authority have set requirements there for. There are a number of other issues around disclosure, which will be discussed in this section.

The measurement section above dealt with the provision of data and estimates of value from management systems. Disclosure is about the mechanisms of information transfer to outside stakeholders and the extent to which this can take place without compromising the security of the company. *Skandia AFS* and *Celemi AFS*, both Swedish companies, started independent publication of intellectual capital data in the mid-1990s, the former using *Edvinsson*'s methodology and language and the latter using *Sveiby*'s methodology and language. There is a growing body of evidence (*Narayanan* et al, 2000) that careful disclosure of information, even strategically sensitive information concerning research and development intentions, is beneficial in terms of market value. The pioneers have shown that this is feasible although the extent of disclosure has been inadequate for the purposes of the investment analyst. This is a key point. The disclosure of "intellectual capital" in the way pioneered by *Skandia* and *Celemi* may not be that helpful since they describe changes in stock or potential at a high level. The important issue is disclosure about the present and future performance of the value creating elements of the company. The goal of disclosure is suggested in *figure 8*.

Given this goal, the purpose of this section is to examine current thinking on the issue of disclosure and determine the nature of the barriers to be crossed if company information disclosure on value creation is to be meaningful and commonplace.

Disclosure should provide relevant, reliable and timely information to those who need to know it so that they can make decisions concerning their relations with the company. At the same time, the information released by the company must not lead to the compromise of sensitive strategic information that would give unfair advantage to others.

Fig. 8: The goal of disclosure

4.1 Financial vs. IC Reporting

The IC perspective is about value creation, thus IC disclosure must embrace a form that enables reporting on value creation. This may seem obvious, although, when looking at the standard form of corporate reporting of today, namely financial accounting, it only measures value realisation. The challenge hence lies in creating a way to measure and report on value creation, i.e. on value which is potentially available for realisation. Disclosure of IC does not replace traditional reporting but it adds a new dimension to it (see *figure 5*).

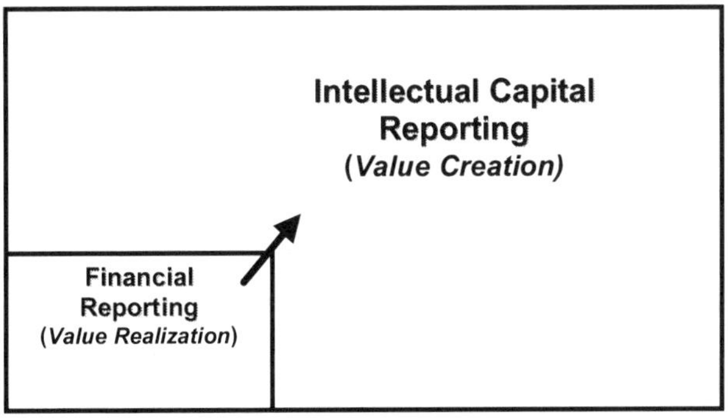

Fig. 9: Financial reporting and the new dimension of IC reporting

Then again, financial reporting has some self-evident advantages. Embedded in this standard is 120 years of experience of modern accounting, building on a 500 year-old bookkeeping tradition. The input numbers are based on 3rd party transactions and financial accounting provides a universal language, e.g. there is no question about what the item 'accounts receivable' is. Finally, since 100% of private sector firms are obliged to use this standard, both management and stakeholders are used to this type of data.

As for IC reporting, there is still no standard and the business community has at most 10 years of experience of managing IC, including the pioneers. The input numbers are based on indicators with varying reliability and relevance, and there is no standardised common language (e.g. there is no formal consensus on what organisational capital is). Lastly, less than 1% of private sector firms have experience from IC reporting. Nevertheless, there are a number of clear benefits of IC reporting. It can for instance address, and to some extent resolve, the following problems associated with traditional disclosure:

- **The risks of insider trading**. Information asymmetries are not as strong in firms that depend more on tangible, observable assets. The higher the reliance on esoteric know-how for competitiveness, the more difficult it becomes for the average outside investor to assess the true value of IC and the larger the gains to be made by insiders.
- **Higher costs of capital.** Companies with more intangible and fewer tangible assets may find it harder to raise capital than companies with more tangible assets, which investors or bankers may regard as security.
- **Overvaluation of intangibles leading to misallocation of capital.** Unregulated disclosure of information about intangible assets makes it possible for companies to manipulate perceptions of their value, leading investors to over-value them.
- **Dulled incentives for knowledge workers and entrepreneurs.** Better information about the value of people and their ideas would reduce the information asymmetry between managers and workers, which managers in theory can exploit to their advantage.
- **Increased volatility.** Inadequate disclosure concerning the quality of intangible assets may feed volatility and uncertainty in capital markets.

There are further incentives specific to institutional investors with IC reporting. Increased transparency and lower information distortions will allow fund managers to reconcile theory and practice to construct portfolios with the defined ideal risk/return trade-off and diversification risk reduction benefits. It also has the potential to allow the establishment of a formalised connection between

- the past (past financial performance, financial policy, accounting changes, corporate governance history, management experience and track record);
- the present (current results, strategic changes, communication skills); and
- the future (management vision and promises, perception of risks and challenges, match between management quality and contingencies)

Figure 10 sets out the financial indicators of most and least importance to investors respectively.

The 10 most valued non-financial indicators, beginning with the most valued indicator	The 10 least valued non-financial indicators, beginning with the least valued indicator
1. *execution of corporate strategy*	1. *compensation ratios*
2. *management credibility*	2. *use of employee teams*
3. *quality of corporate strategy*	3. *environmental and social policies*
4. *innovation*	4. *process quality awards*
5. *ability to attract employees*	5. *product quality awards*
6. *market share*	6. *quality of published materials*
7. *management experience*	7. *quality of customer service departments*
8. *quality of compensation policies*	8. *number of customer complaints*
9. *research leadership*	9. *experience of investor relations personnel*
10. *quality of processes*	10. *employee turnover rates*

Fig. 10: What investors view as important, (Mavrinac/Siesfeld, 1997)

As for management, the incentives lie in creating transparency about the use of private and public funds and explaining the achievements of research and their benefits to stakeholders.

Disclosure of IC also allows for illustrating the development of the intangible assets of the company as well as revealing leverage effects and externalities. Moreover, it provides a vehicle for communicating organisational values and demonstrating their competitiveness. The ultimate benefit is that the company gets valued at a more accurate level than is possible without the disclosure of the performance of intangible assets.

4.2 Progress of IC Disclosure Practice

A major survey of the issues involved in the disclosure of company information on intellectual capital to the investment community was carried out in 1999 (*Rylander/Jacobsen/Roos*, 1999). The principal issues at the time were:

- The information asymmetry gap is growing as the proportion of company value attributable to intangible assets increases.
- Long-term information, particularly on strategic intent and execution, was lacking from company reporting but was considered to be of particular importance to external stakeholders, especially the investors.
- Standards and comparability relating to the disclosure of intellectual capital would remain a major issue.
- Value creation models could provide information to complement traditional reporting required by law.

Since the publication of that report in 1999, there have been a number of developments along the lines foreshadowed in the paper. However, there have been neither solutions nor concrete proposals and hence the problem remains the same.

Regulatory bodies have been pressing and moving towards the mandatory disclosure of certain elements of intangible resources and the recording of elements of goodwill in mergers and acquisition. The most important of them are the FAS 141 and 142, stipulated by FASB (2001).

The UK government commenced a review of UK company law in 1998 with the aim of guiding a significant revision to the laws. The review proposes an evolutionary approach to changes in law centring on three areas. These are changes to the reporting calendar, the extent of information disclosure and the role of auditors. Of them, the second is most interesting. Here, two issues where mandatory disclosure might be considered: 1) a review of the company's last year activities focusing on market changes and positioning and 2) a statement of the company's purpose, strategy and most interesting of all, its principle drivers of performance (*Dickson*, 2000).

Another UK initiative was headed by CEST, *the Centre for Exploitation of Science and Technology*, taking a pragmatic approach to the problem of information asymmetry in a collaborative project: Valuing the New Intangibles (*CEST*, 2000) with the goal of understanding the field of disclosure and the needs of the external and company communities. CEST adopted the approach of running two complimentary and alternating fora, one engaging company representatives, and the other City of London investors. As with the other initiatives, the key areas where more and reliable information is required are the long-term prospects and intentions of the company.

The CICA (*Canadian Institute of Chartered Accountants*) published a report (*Waterhouse*, 1999) presenting an analysis of what is actually published at the time. Although some of the categories of analysis may not align well with the conventional language of intellectual capital, the results are clear. The report shows that both for the CEO and company boards, intellectual capital issues were all considered to be of above average importance in terms of reporting. Operating efficiencies head the list but intellectual capital and corporate learning is a close second and innovativeness and the ability to respond to the market was third.

There is also a major project on intangible assets measurement, monitoring and reporting being undertaken jointly by the *Brookings Institution* and the *Stern School of Business* (*Brookings*, 2000). The project is taking a twin track approach with each track being addressed by a number of groups. The first track deals with trends in company investment practices in intangibles including the effects of regulatory, tax and reporting requirements that influence a company's investment choices. The second is a survey of the laws and regulations that impose barriers or biases.

In Europe, the EU project PRISM is aiming at creating 'a coherent pan-European community of practice out of the fragmented academic and other interest groups currently working on issues relating to business intangibles'. The PRISM action group assembles academics and practitioners from all over Europe facilitating more work and research in the area of IC management and disclosure. *Figure 11* summarises some examples of research projects on IC disclosure.

Country	Research Project
Canada	Canadian Performance Reporting Initiative (CPRI / CICA)
UK	Centre for Exploration of Science and Technology (CEST)
	CIMA and Cranfield University - Understanding corporate value: managing and reporting intellectual capital
EU	PRISM Project
	E*Know-Net Project (MERITUM's successor)
Nordic countries	Nordic Industrial Fund - The Nordika Project, Frame
	Danish Agency for Trade and Industry Guideline for Intellectual Capital Statements
USA	Intangibles Research Project
	FASB - The Financial Accounting Standards Board
	Ross Institute of Accounting Research, Stern School of New York - The Intangibles Research Project
	Brookings Project
Global	OECD
	IASB – The International Accounting Standards Board

Fig. 11: Research projects on IC disclosure

There is little to prevent real progress to be made in improving the information asymmetry between companies and stakeholders. The evidence suggests that there are many benefits to be gained by a more extensive disclosure of information. There would be beneficial effects both in terms of the effects on external reputation, market valuation and the ability to raise capital and internally in the esteem that internal stakeholders (staff) will have in the company and its management. That is not to say that there are no pitfalls, the FASB report highlights where some of them may lie.

Assuming that current work by the FASB, PRISM and the others leads to an equitable and significant closing of the information asymmetry gap then the sticking points that remain are:

- How to provide reliable data that can be used quickly and compared between companies?
- How to address the issue of "information power" inside companies?
- How to collect and deliver up the information that has been subjected to the minimum of internal interpretation but at the same time is not painfully revealing of strategic intent?

None of these three issues even has taken together presents and insurmountable obstacle. The challenge is therefore to find a methodology that does both of these. In addition, it must also constitute a management and measurement methodology that complies with all criteria expressed in the section addressing management.

5. Summary and the Way Forward

The need remains to take another step from the current position in which management and measurement models have progressed to what is commonly called third generation intellectual capital. The debate on disclosure has started to settle on the key issues and at the same time has noted the benefits of careful disclosure.

Analysts need a model that describes the value context of the investors. This is still dominated by financial considerations but intellectual capital, specific ethical positions and the ability of management teams make significant contributions. A model that combines the attributes of value from the standpoint of the investor and analyst will have a limited range of applicability but one that will allow it to be used across a market segment while not being specific to any individual company in that segment. All the analysts require then is input.

The company needs a measurement scheme and model that embodies the management principles they wish to follow. The model must be compliant with the four criteria stated earlier, the most difficult to meet are the last two which require the strategic input and the consequent generation of useful information. However, these requirements align exactly with the needs of the external community. The model and methodology must therefore expose and measure the real routes to value creation in the company and include the areas of uncertainty and risk. As there have to be strategic levers, the model must be forward looking and may include elements of real option theory in estimating values.

As the company would have both the internal model and the external model, management would have the opportunity to see in advance the possible effects of their strategic options on the stakeholder groups as well as the long-term effects on the make-up of their companies. There would inevitably be a tendency to investigate the most favourable solutions but such is the real diversity of stakeholder groups that solutions that please everybody are usually impossible.

There are numerous additions possible to enhance capability especially in the areas of risk and uncertainty and also in leading edge human capital management but the substantial elements all now exist in the field of intellectual capital and its underlying and related sciences. It is now time to build on the work of FASB, CEST, PRISM and others and see whether the vision of people like *Drucker* (1994) can be made a practical reality.

6. References

American Institute of Certified Public Accountants, 2000, (see http://www.xfrml.org/).

Andriessen, D. and *Tissen, R.*, Weightless Weight – Find your Real Value in a Future of Intangible Assets, London 2000.

Bainbridge, A., *Jacobsen, K.* and *Roos, G.*: Intellectual Capital Analysis as a Strategic Tool, in: Strategy and Leadership Journal, Vol. 29 (2001), No. 3.

Barney, J., Firm Resources and Sustained Competitive Advantage, in: Journal of Management, Vol. 17 (1991), No. 1, pp 99–120.

Barney, J. B., Gaining and sustaining competitive advantage. 2nd ed., Prentice-Hall 2002.

Barney, J. B., *Wright P. M.* and *Ketchen, D. J.*, The resource-based view of the firm: Ten years after 1991, in: Journal of Management, Vol. 27 (2001), No. 5.

Baum, G., *Ittner, C.*, *Larcker, D.*, *Low, J.*, *Siedfield, T.*, *Malone, M.*, Introducing the new value creation index, Forbes ASAP (2000), pp. 140-143.

Bontis, N., Assessing knowledge assets: A review of the models used to measure intellectual capital. Working paper, *Queen's Management Research Centre for Knowledge-Based Enterprises*, 2000.

Bontis, N., *Dragonetti, N. C.*, *Jacobsen, K.* and *Roos, G.*, The Knowledge Toolbox; A Review of the Tools Available to Measure and manage Intangible Resources, in: European Management Journal, Vol. 17 (1999), No. 4.

Bontis, N., *Mill, J.*, Web-Based Metrics and Internet Stock Prices, Submitted for review at the International Journal of Electronic Commerce and Business Media (2000).

Boston Consulting Group, The value creators: A study of the world's top performers, 1999, http://www.bcg.com/.

Bradley, S. P. and *Nolan, R. L.*, Capturing Value in the Network Era, in: Harvard Business School Working Paper Series, No. 97-061 (1997).

Brooking, A., Intellectual Capital: Core Assets for the Third Millennium Enterprise, London 1996.

Brooking, A., Intellectual Capital – Core Asset for the Third Millenium Enterprise, Hamilton, Ontario 1999.

CEST (*The Centre for Exploitation of Science and Technology*), Measuring the New Intangibles, 2000, http://www.cest.org.uk/.

Chatzkel, J. B., Conversation with *Göran Roos*, in: Journal of Intellectual Capital, Vol. 3 (2002), No. 2.

Chong, J., Measuring the value of Intangibles for the purpose of Transfer Pricing, Paper based on interview with *Roos, G.*, Canberra, Australia 1999.

Collis, D. J. and *Montgomery, C. A.*, Competing on Resources: Strategy in the 1990's, in: Harvard Business Review, Vol. 73 (1995), No. 4.

Davis, S. and *Meyer, C.*, Future wealth, Addison Wessley, Mass. 2000.

Dickson, M., Survey – Shareholder communication: Overhaul of creeky office, in: The Law Review, Financial Times, 30/6/2000.

Dragonetti, N. and *Roos, G.*, Assessing the Performance of Government Programmes: an Intellectual Capital Perspective, Paper presented at The 18th Strategic Management Society Conference, Nov 1998, Orlando, FL, USA.

Dragonetti, N. C. and *Roos, G.*, Intellectual Capital: A New Language for Strategic Management, Paper presented at the Strategic Management Society Conference, October 1997, Barcelona, Spain.

Dragonetti, N. C. and *Roos, G.*, La evaluación de Ausindustry y el business network programe: una perspectiva desde el capital intelectual, Boletín de Estudios Económicos, Vol. LIII Agosto 1998 Núm. 164.

Dragonetti, N., and *Roos, G.*, Efficiency and Effectiveness in Government Programmes: an Intellectual Capital Perspective, Paper presented at the 2nd World Conference on Intellectual Capital, January 1999, McMaster University, Hamilton, Canada.

Drucker, P. F., Post Capitalist Society, New York 1994.

Edvinsson, L. and *Malone, M. S.*, Intellectual Capital: Realizing your Company's True Value by Finding Its Hidden Brainpower, New York 1997.

Edvinsson, L. and *Malone, M. S.*, Intellectual capital: The proven way to establish your company's real value by measuring its hidden brainpower, Piatkus 1997.

EU/ESPRIT 4, KnowNet: Knowledge management with intranet technologies, Project EP28928. http://www.know-net.org/2000.

Federal Accounting Standards Board, FAS 141, Business Combinations, 2000.

Federal Accounting Standards Board, FAS 142, Goodwill and Other Intangible Assets, 2001.

Fernström, L., Pike, S. and *Roos, G.*, Understanding the Truly Value Creating Resources - The Case of a Pharmaceutical Company, in: International Journal of Learning and Intellectual Capital, Forthcoming December 2003.

Fitz-Enz. J., How to Measure Human Resource Management, McGraw-Hill 1994.

Grant, R. M., Contemporary Strategy Analysis: Concepts, Techniques, Applications. 4th edition, Cambridge, Mass. 2002.

Gupta, O. and *Roos, G.*, Using the Intellectual Capital Perspective to Analyse the Value Creation Potential of Mergers and Acquisitions, Paper presented at The 4th World Congress on Intellectual Capital, 22nd McMaster Business Conference, 2nd World Conference on E-Commerce, Jan 17-19, 2001, Hamilton, Canada. Selected for publication in

Journal of Intellectual Capital Vol. 2 (2001), No. 3, Received highly commended paper award 2001.

Gupta, O. and *Roos, G.*, Using the Intellectual Capital Perspective to Analyse the Value Creation Potential of Mergers and Acquisitions, Paper presented at The 4th World Congress on Intellectual Capital, 22nd McMaster Business Conference, 2nd World Conference on E-Commerce, Jan 17-19, 2001, Hamilton, Canada.

Gupta, O., *Pike, S.* and *Roos, G.*, Intellectual Capital and Improved Corporate Governance, Paper accepted for presentation at The 6th World Congress on the Management of Intellectual Capital and Innovation January 15 - 17, 2003, Hamilton, Canada.

Gupta, O., *Pike, G.* and *Roos, G.*, Evaluating Intellectual Capital and Measuring Knowledge Management Effectiveness" Performance Measurement and Management: Research and Action, Papers from the Third International Conference on Performance Measurement and Management, PMA 2002, Centre for Business Performance, Cranfield.

Guthrie, J., *Steane, P.*, *Fletcher, A.*, *Pike, S.* and *Roos, G.*, Towards an understanding of the value dimensions of ARCBS many diverse stakeholders, submission to ANZAM 2002 - 4th to 6th December 2002, Beechworth, Australia.

Hamel, G. and *Prahalad, C.*, Competing for the Future, Boston 1994.

Helmholtz, H., Zählen und Messen, in: Philosophische Aufsätze, Fues's Verlag, Leipzig, pp. 17-52. Translated by *Bryan, C. L.*, 1930, Counting and Measuring, Van Nostrand 1887.

Hume, D., Original reference: Treatise on human nature, Book III(I)1. see: *Selby-Bigge, L.* (Ed), 1902, An enquiry concerning human understanding, Oxford 1740.

HVA - Holistic Value Added, Intellectual Capital Services Ltd., 2000, http://www.intcap.com/

Investor Relation Society, 2000, http://www.ir-soc.org.uk/

Ittner, C., *Kalafut, P.*, *Larcker, D.*, *Sean Love, S.*, *Low, J.*, *Park, J.*, *Siesfeld, T.* and *Zito, S.*, Measuring the Future: Value Creation Index 2000.

Johansson, U., A model illustration and implications, in: Journal of Human Resource, Costing and Accounting, Vol. 2 (1997), No. 1.

Justice, A., A general treatise on monies and exchanges, London 1707.

Kaplan, R. and *Norton, D.*, The Balanced Scorecard: Translating Strategy into Action, Boston 1997.

Krantz, D., *Luce, R.*, *Suppes, P.* and *Tversky, A.*, Foundations of Measurement, San Diego 1971.

Larsen, H., *Nikolaj, P.*, *Bukh, D.* and *Mouritsen, J.*, Intellectual capital statements – Towards a guideline, Danish Agency for Trade and Industry 2000. (http://www.efs.dk/)

Lauzel, P. and *Cibert, A.*, Des Ratios au Tableau de Bord, Paris 1962.

Lev, B., Seeing is Believing - A Better Approach to Estimating Knowledge Capital, CFO magazine, April 2000.

Lev, B., Intangibles: Management, Measurement and Reporting, Washington 2000.

Lewis, E. and *Lippitt, J.*, Valuing intellectual assets, Journal of Legal Economics, Vol. 9, (1999), No. 1, pp. 31-48.

Lightman, A., The myths that support massive internet market caps, 2000. IntellectualCapital.com. (http://www.intellectualcapital.com/)

Luthy, D. H., Intellectual Capital and its Measurement, Proceedings of the Asian Pacific Interdisciplinary Research in Accounting Conference (APIRA), Osaka, Japan 1998.

Lymer, A., *Debreceny, R.*, *Gray, G.* and *Rahman, A.*, Business reporting on the internet, Submitted to ISAC 1999.

M'Pherson, P. K., *Pike, S.* and *Roos, G.*, Strategic Analysis Adapted For IC Networks and E-Business Networks", Paper accepted for presentation at the Strategic Management Society's 21st Annual International Conference, Oct 21-24, 2001, San Francisco, USA.

M'Pherson, P. K. and *Pike, S.*, Accounting, empirical measurement and intellectual capital, Journal of Intellectual Capital, Vol. 2 (2001), No. 3, pp. 246-260.

M'Pherson, P. K., Business Value Modelling, Globalisation of Information: 48th Conference and Congress of the International Federation for Information and Documentation, Graz, Austria 1996.

Marr, B., *Gupta, O.*, *Pike. S.* and *Roos, G.*, Intellectual Capital and Knowledge Management Effectiveness, Management Decision Special Issue PMA Conference 2002, June 2003.

Narayanan, V., *Pinches, G.*, *Kelm, K.* and *Lander, D.*, The influence of voluntary disclosed qualitative information. In: Strategic Management Journal, Vol. 21 (2000), No. 7, pp. 707-722.

Nash, H., Accounting for the Future: A Discplined Approach to Value-Added Accounting, 1998.

Neely, A., *Marr, B.*, *Roos, G.*, *Pike, S.* and *Gupta, O.*, Towards the Third Generation of Performance Measurement, in: Controlling, 15. Jg. (2003), H. 3/4.

OECD, Measuring and Reporting Intellectual Capital: Experiences, Amsterdam, http://www.oecd.org/, A great many documents at but notable is, Issues and Prospects 1999.

Partanen, T., Intellectual capital accounting, Master's Thesis, Helsinki School of Economics and Business Administration 1998.

Penrose, E. T., The Theory of the Growth of the Firm. New York 1959.

Peteraf, M. A., The Cornerstones of Competitive Advantage: A Resource-Based View." In: Strategic Management Journal, Vol. 14 (1993), No. 3.

Pike, S. and *Roos, G.*, An introduction to intellectual capital, in: Works Institute Journal, Vol. 42 (2000), No. 10, pp. 21-27.

Pike, S. and *Roos, G.*, Measuring the impact of Knowledge Management in Companies, Paper accepted for presentation at The 5th World Congress on Intellectual Capital, 23rd McMaster Business Conference, Jan 16-18, 2002, Hamilton, Canada.

Pike, S. and *Roos, G.*, Mathematics and Modern Business Management, Performance Management Association Symposium, July 2003.

Pike, S. and *Roos, G.*, Mathematics and Modern Business Management, Invited Paper for the Performance Management Association Symposium, INSEAD, July 28-29, 2003.

Pike, S. and *Roos, G.*, Measuring and decision support in the knowledge society, Paper presented at The 4th World Congress on Intellectual Capital, 22nd McMaster Business Conference, 2nd World Conference on E-Commerce, Jan 17-19, 2001, Hamilton, Canada.

Pike, S., *Rylander, A.* and *Roos, G.*, Intellectual Capital Management and Disclosure, in *Bontis, N.* and *Choo, C. W.* (eds.), The Strategic Management of Intellectual Capital and Organizational Knowledge - A Selection of Readings, 2002.

Porter, M. E., Competitive Advantage: Creating and Sustaining Superior Performance, New York 1985.

Pulic, A., An Accounting Tool for IC Management, 2000, http://www.measuring-ip.at/Papers/ham99txt.htm.

Pulic, A., VAICTM – An accounting tool for IC management, University of Graz, 2000, http://www.measuring-ip.at/Papers/ham99txt.htm.

Rescher, N., An Introduction to Value Theory, Prentice Hall 1969.

Rigby, D., Balanced Scorecard Collaborative Announces the Formation of an XML Standards Committee, in Management Tools, Bain and Company 2000, http://www.bain.com.

Roos G. and *Roos, J.*, Measuring your Company's Intellectual Performance, Long Range Planning, Vol. 30 (1997), No. 3, pp. 413-426.

Roos G. and *Jacobsen, K.*, Management in a Complex Stakeholder Organisation; A case study of the application of the IC-Process to a Branch of the Commonwealth Public Service, Monash Mt. Eliza Business Review, Vol. 2 (1999), No. 1.

Roos, G., El capital intelectual desde una perspectiva internacional, Paper presented at Gestión del Conocimiento y Capital Intelectual, Curso de Verano, (Academic Summer Course), 27 July, 1999, EUROFORUM, Escoria, Spain.

Roos, G., Ydinosaaminen Menestystekijänä, in *Ahnger, P.*, *Laakso-Manninen, R.*, *Rantala, S.* and *R. Ritala* (eds.), Markkinalähtöinen Henkilöstöjohtaminen, Vetovalmennus, 1996.

Roos, G. and *Albertsen A.-M.*, Developing a Sense of an Outer Boundary Vision, Paper presented at the Strategic Management Society Conference in Phoenix, Arizona, USA 1996.

Roos, G. and *Lövingsson, G.*, El Proceso CI en el «Nuevo Mondo de las Telecomunicacione»l, in *Güell, A. M.* (ed.), Homo faber, homo sapiens – La gestión del capital intelectual, Ediciones del Bronce, 1999.

Roos, G. and *Roos, J.*, Measuring your Company's Intellectual Performance. Long Range Planning, Vol. 30 (1997), No. 3.

Roos, G. and *Fernström, L.*, Differences in Value Creating Logic and Their Managerial Consequences: The Case of Authors, Publishers & Printers". Paper presented at the Australian International Conference on the Future of the Book, 23rd April 2003, Cairns, Australia.

Roos, G., and *Roos, J.*, Measuring Intellectual Performance, Paper presented at the Knowledge in action Conference in Israel October 1996.

Roos, G., Guthrie, J., Steane, P., Fletcher, A., and *Pike, S.*, Mapping a Transparent Enterprise; The Transparent Enterprise. The Value of Intangibles", Madrid, Spain, November 25-26th, 2002.

Roos, G. and *Jacobsen, K.*, Management in a Complex Stakeholder Organisation, Monash Mt. Eliza Business Review, Vol. 24 (1999), No. 4, pp. 82-93.

Roos, G. and *Lovingsson, F.*, 1999, El Processo CI en EL, Nuevo mundo de las telecomubicaciones: Experiencia de Ericsson Business Consulting Norrkoping, (Ed Güell, A. M.), Homo faber, Homo sapiens: La gestiń del capital intelectual, Ediciones del Bronce, pp. 141-169.

Roos, G. and *Roos, J.*, Measuring your company's intellectual performance, Journal of Long Range Planning, Vol. 30 (1997), No. 3, pp. 413-426.

Roos, G., The Management and Measurement of Intangible resources, Institute Of Chartered Financial Analysts Of India Reader, October 2003.

Roos, J., Roos, G., Dragonetti, N. and *Edvinsson, L.*, Intellectual Capital: Navigating in the New Business Landscape, New York 1997.

Roos, J., Roos, G., Dragonetti, N. C. and *Edvinsson, L.*, Å utvikle kunnskap – veien til fremtidig suksess; Intellektuelkapital, in *Doksrød, H.* (eds), Kunnskap i Arbeid i status og fremtidsbilder, Tano Aschehoug, 1998.

Rumelt, R. P., Theory, Strategy and Entrepreneurship, in: *Teece, D. J.* (ed.), The Competitive Challenge: Strategies for Industrial Innovation and Renewal, Cambridge, Mass. 1987.

Rylander, A., Jacobsen, K. and *Roos, G.*, Towards Improved Disclosure of Intellectual Capital, Paper presented at the 2nd World Conference on Intellectual Capital, January 1999, McMaster University, Hamilton, Canada.

Rylander, A., Jacobsen, K. and *Roos, G.*, Towards Improved Information Disclosure on Intellectual Capital, in: International Journal of Technology Management, Vol. 20 (2000), No. 5-8.

Rylander, A., Jacobsen, K. and *Roos, G.*, Towards improved information disclosure on intellectual capital, International Journal of Technology Management, Vol. 20 (1999), No. 5-8.

Saaty, T., The Analytical Hierarchy Process: Planning, Priority Setting, Resource Allocation, New York 1980.

Scott, D. and *Suppes, P.*, Foundational Aspects of Theories of Measurement, Journal of Symbolic Logic, Vol. 23 (1958), No. 2, pp. 113-128.

Skyrme, D. (Ed.), Business Intelligence report: Measuring the value of knowledge, 1998. http://www.business-intelligence.co.uk/

Stabell, C. B. and *Fjeldstad, Ø. D.*, Configuring value for competitive advantage: on chains, shops, and networks. In: Strategic Management Journal, Vol. 19 (1998), No. 5.

Staff Reports, 1999, FASB looks at non-financial disclosure, Securities Data Publishing. See also http://www.rutgers.edu/Accounting/raw/fasb/tech/index.html

Standfield., K., Extending the Intellectual Capital Framework. 1999. http://www.knowcorp.com/article075.htm

Steering Group Report, Business reporting research project: Electronic distribution of business reporting information, FASB Exposure Draft 2000.

Stewart, T., Intellectual Capital: The New Wealth of Organisations, Doubleday 1997.

Sullivan, P., Value-driven Intellectual Capital: How to convert intangible corporate assets into market value, New York 2000.

Suppes, P. and *Zinnes, J.*, Basic Measurement Theory in *Luce, R.*, *Bush, R.* and *Galanter, E.*, (eds.), Handbook of Mathematical Psychology, New York 1963, pp. 1-76.

Sveiby, K., The New Organizational Wealth: Managing and Measuring Knowledge Based Assets, San Francisco 1997.

The Brookings Institution, Understanding Intangible Sources of Value. 2000. See http://www.brook.edu/es/intangibles/default.htm

Van Buren, M., A yardstick for knowledge management, in: American Society for Training and Development, Vol. 53 (1999), No. 5, pp. 71-78.

Verity, D., Real Options – An analytical approach to aligning risk management and corporate strategy. Corporate Finance Review (Boston), Vol. 3 (1999), No. 7.

Von Krogh, *Roos, G.* and *Kleine, D.*, Knowing in firms, Sage 1998.

Waterhouse, J., Measuring Up, in: CAMagazine, March 1999, pp. 41-43.

Wernerfelt, B., A Resource-Based View of the Firm. In: Strategic Management Journal, Vol. 5 (1984), No. 2.

Williams, M., Is a company's intellectual capital performance and intellectual capital disclosure practices related? Evidence from publicly listed companies from the FTSE 100, Paper presented at McMasters Intellectual Capital Conference, Jan 2001, Hamilton Ontario.

Measuring Intangible Resources for Managerial Accounting Purposes

Thomas Günther [1], *Sandra Kirchner-Khairy* [2]
and Annikka Zurwehme [3]

---

[1] Prof. Dr. *Thomas Günther*, Chair of Business Administration, especially Managerial Accounting and Control, Dresden University of Technology, Germany

[2] *Sandra Kirchner-Khairy*, Dresden University of Technology, Germany

[3] *Annikka Zurwehme*, Dresden University of Technology, Germany

Companies are experiencing an increased influence of intangible resources, such as brands, customer relationships, product and process know-how as well as networks, on their economic success. Therefore, management accounting needs robust methods for the measurement and valuation of such resources. In this chapter an extensive survey of measurement methods currently found in literature is presented and classified. Our evaluation of these methods involves a number of in-depth examples that clarify advantages as well as limitations of application.

1. Introduction

The importance of research on intangible resources has increased during recent years. Attention has been driven by the recognized difference between book value and market value of companies that own generally known intangibles like brands or patents (*Sveiby*, 1997, p. 5). Frequently, these kinds of resources allow a company to excel. They are difficult to duplicate and thus can provide a sustainable competitive advantage. In financial accounting there are strict national and international regulations concerning intangible resources. These regulations improve comparability and reduce fraud and misrepresentation, but they do not necessarily lead to the type of information that would be most useful in internal decision making. Managerial accounting allows a company – without any restrictions - to use its own methods to report intangible resources. We focus, in this chapter, on the needs of managerial accounting. Diverse concepts for the management and evaluation of intangible resources (e.g. *Sveiby*, 1997; *Edvinsson/Malone*, 1997; *Stewart*, 1997; *Lev*, 2001) have been developed in accounting research over the last ten years. Companies and institutions like *Skandia SFA, Canadian Imperial Bank, DuPont, Celemi* and *Deutsche Bank AG* were engaged in managing of and reporting on intangible resources early on. Some consulting companies (e.g. *Arthur D. Little, The Boston Consulting Group, Cap Gemini Ernst & Young, Celemi* and *PriceWaterhouseCoopers*) developed products to support the management of intangible resources.

One reason for the gap between the necessity to manage intangible resources on the one hand and missing management tools on the other might be that only few methods are known. The purpose of this work is to give a structured overview about the manifold methods. We choose seven categories to which individual methods are assigned. For illustration purposes we describe and evaluate one specific method for each category. In order to evaluate the different methods we determine a catalog of evaluation criteria. It is obvious that there is no perfect method that meets all requirements. Nevertheless we think our scheme increases understanding, and make it easier for companies to compare and to choose a suitable method.

2. Intangible Resources within the Management Accounting System

2.1 Intangible Resources, Intangible Assets and Intellectual Capital

There are diverse definitions for relevant terms such as intellectual capital, intellectual property or intangible assets. We consider "intangible resources" to be the non-material and non-financial resources a company can exploit for longer than the current reporting year (distinguishing from current expenses or costs). "Intellectual property" refers to intangible resources that are legally protected, such as brand names, patents or licences. Intangible resources become "intangible assets" if they fulfil the asset definition of the current standards (e.g. *IASC* Framework § 49 and *SFAC* 6 §§ 25 and 26) and legislation (e.g. the regulations in corporate law in countries like Germany or Austria). From our point of view, "intellectual capital" comprises all intangible resources of a company. However, the difference between market value and book value encompasses more than just a firm's intellectual capital, as this would imply e.g. perfectly efficient markets and valuation of all assets in financial statements at market prices.

2.2 A Management Accounting Framework for Intangible Resources

Because of their outstanding relevance for a company's success, intangible resources are analogous to core competencies that have to be carefully developed. In this context the management cycle for intangibles, as can be seen in *figure 1*, can be described as a three-step cycle including recognition, development and exploitation of intangible resources.

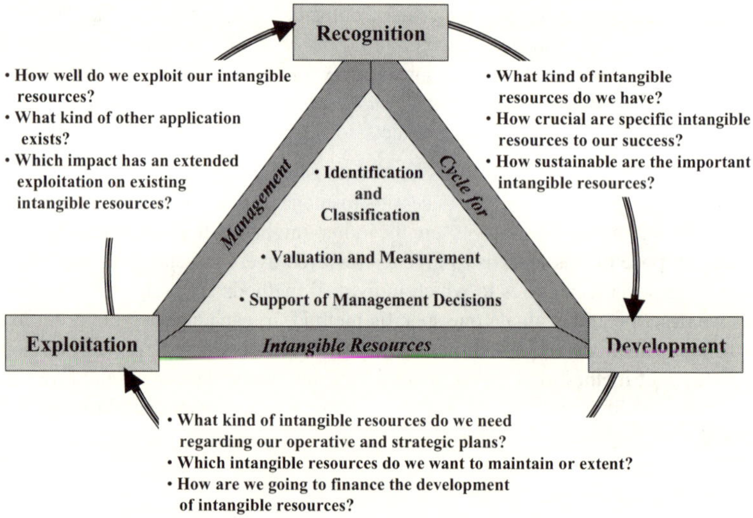

Fig. 1: Management Process for Intangible Resources

As accounting tries to express real flows of goods and services in monetary terms, intangible resources must first be recognised and then evaluated. Based on the mentioned management process the main tasks of managerial accounting are classification (what kind of intangible resources does the company have?), identification (what specific intangible resources exist within a class of intangibles?) and measurement (what is the impact of an intangible resource on the company?). Valuation is a special measurement method that assigns a monetary value to the impact of an intangible resource. These managerial accounting tasks occur in each stage (recognition, development and exploitation) and have to answer within each stage different questions. In the stage of recognition for instance the question "What kind of intangible resources do we have?" has to be answered (see *figure 1*).

In the following, we focus on the measurement and valuation of intangible resources. An intensive discussion of identification and classification can be found in the management literature (e.g. *Sveiby*, 1997 and *Edvinson/Malone*, 1997). In recent years a variety and multitude of measurement techniques has been developed to assess the impact of intangible resources. As mentioned above, the intention of this article is to examine the properties and capabilities of these methods to support managerial accounting. Thus, we must first ask what the general tasks of managerial accounting are and secondly what requirements the methods have to meet to support management. The general tasks of managerial accounting are the co-ordination of different organisational units and their decisions (e.g. budgeting, resource and cost allocation, transfer prices etc.) and the support of management decision-making (e.g. price setting, make or buy, control of efficiency). Managerial accounting is driven by the three formal objectives: shareholder value, profit and liquidity. We see no reason why intangible resources should not be subject to the same framework of accounting and control as tangible or financial resources.

In the long run companies must ensure their own existence by creating the potential for future profits. The long-term objective may be expressed monetarily in the form of shareholder value. Shareholder value can be measured using one of the diverse discounting cash flow techniques. In the short-run companies have to make profit on an accrued annual basis under a solvency restriction (liquidity). Intangible resources contribute to the long-term and the short-term objectives. Their impact on business must also to be measured long-term (e.g. the contribution of a self created brand to future cash flows) as well as short-term (e.g. the contribution of a brand name to the current year's profit). Consequently the purposes that are pursued determine the adequacy of the measurement and valuation method. For instance, a decision about brand transfer can be considered an investment that requires long-term performance measures whereas short-term measures are needed for decisions about the annual marketing budget to promote the brand.

3. Evaluation Criteria for Measurement Methods

Since the purpose of the measurement of intangible resources is to contribute to the general tasks of managerial accounting, these systems can then be regarded as specialized performance measurement systems. Properties that performance measurements should meet have been discussed before (e.g. *Bhargava* et al., 1994, pp. 235; *Siesfeld*, 1998, pp. 11; *Kriegbaum*, 2001, pp. 83; *Arthur Andersen*, 1992, pp. 29; *Global Reporting Initiative*^{TM}, 2002, pp. 16). The catalog of evaluation criteria for the measurement of intangible resources in our opinion consists of measuringability, relevance, completeness, validity, auditability and applicability (see *figure 2*). In the following, let us clarify some aspects of the figure.

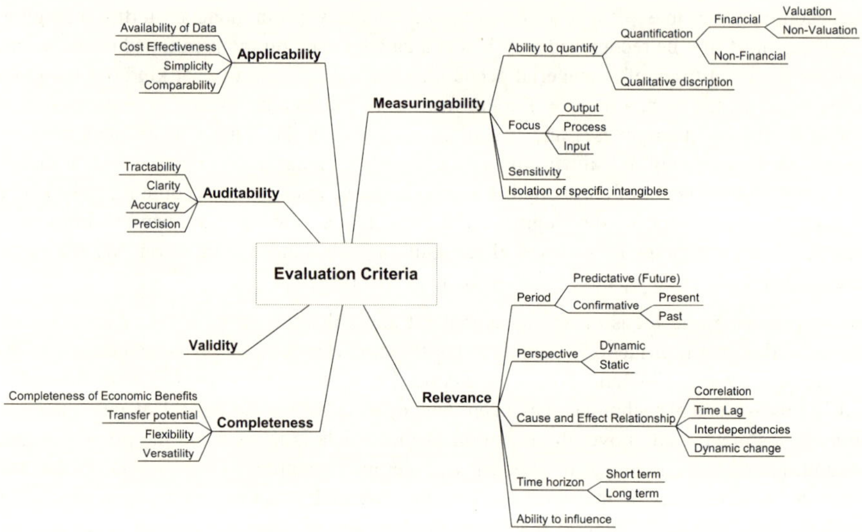

Fig. 2: Evaluation criteria of measurement methods

Measuringability

Measuringability describes how well a tool measures intangible resources. It involves the (i) ability to quantify, (ii) focus, (iii) sensitivity and (iv), the ability to isolate specific intangible resources.

(i:) As highest degree of ability to quantify we consider valuation. It allows the integration into traditional financial and managerial accounting systems. If direct valuation is not possible so-called financial value drivers must be found that highly correlate with the (unknown) values of intangibles. If financial value drivers can not be found non-financial measures must be used. (ii:) There are three possibilities for the kind of measures that can be (focused) on: output, process or input measures. Output measures should be preferred to process measures, and the latter to input measures. Output is the final objective of any business process, but the way the processes are performed and the input into that process may influence its output. (iii:) In order to manage intangible resources it is necessary to determine the consequences of management decisions. Therefore we think that a measurement method should be as sensitive as possible to relevant decision parameters. (iv:) For the measurement and the valuation of individual intangible resources the causal connection between intangibles and other assets (e.g. the *BMW* brand has interdependencies with other assets like customer relationships, process know-how or product quality characteristics) must be separated. That is, we assume an intangible resource has a value unto itself and can be managed independently of the underlying products or services to some extent. For example - with regards to brands - the revenues and costs associated with a no-name product may be compared against those achievable on branded products using conjoint measurement analysis.

Relevance

The selected methodology must be relevant to the requirements of the users. That means that the measures should be appropriate to the decision-making needs. Relevance includes the characteristics (v) period, (vi) perspective, (vii) cause and effect relationship, (viii) time horizon and (ix) the ability to influence.

(v, viii:) Regarding the period, a measurement method can be predictive (focus on future data) or confirmative (focus on present or past data). Value-based management of intangible resources is aimed at both maintenance of and increasing success potential, therefore long-term based measures - and thus predictive data - are most appropriate. Referring to the general tasks of management accounting, short-term as well as long-term measures must be found. (vi:) Thus, the question arises how data from several periods are taken into account. The distinction between dynamic and static perspective comes into play here. The dynamic perspective includes data of several periods and considers therefore the time value of money and the systematic risk reflected by the adjusted discount rate. The static perspective, on the other hand, is limited to data from only one single period. (vii:) A measurement method for a value based management of intangibles must show cause and effect relationships in order to generate value drivers, which can be influenced by actions. The result of these efforts is top-down as well as bottom-up analysis similar to the well-known *DuPont*-Scheme. In the context of intangibles, complex networks of cause and effect relationships must be created where measures have a high correlation to the real value and the manifold interdependencies, time lag and dynamic changes must be considered. (ix:) In order to manage intangible resources a company's influence on value determining factors that are used as measures should be high.

Completeness

A measurement tool fulfills the criterion of completeness if sufficient detail about decision useful matters is considered. From our point of view a method has therefore to take into account (x) all economic benefits, (xi) a transfer potential, (xii) flexibility and has to match (xiii) versatility.

(x:) The value of an intangible resource is based on several sources. This is why it is essential to completely record its benefits and costs. Benefits and costs may include e.g. cash flows derived from additional revenues, reduced costs or extra costs caused by the creation of the intangible resource. Additionally, (xi:) an intangible resource could be also used for other products and services and (xii:) offers future options. (xiii:) Measurement methods for intangible resources should allow to include all types of intangible resources. That means a method should be applicable to all existing intangible resources of a company.

Validity

The validity of a measure determines how well a measurement method measures what it is designed to measure, i.e. if whether or not the measure is an indicator for the desired measurement object. This criterion is referring to systematic error that is embodied in the method itself.

Auditability

Auditability is a criterion that expresses the extent to which a method can be examined for correctness by both internal and external parties. It comprises (xiv) tractability, (xv) clarity, (xvi) accuracy and (xvii) precision.

(xiv:) A measurement method that fulfils the criterion of tractability has to assure, that users are fully informed of the processes, procedures and assumptions embodied in the reported information. Tractability gives the necessary transparency that allows the evaluation of delivered measures. (xv:) Clarity considers the extent to which information is understandable and usable by a user. (xvi:) From our point of view it is essential to achieve a high degree of exactness and low margin of error in reported information necessary for users to make decisions with a high degree of confidence. (xvii:) A measurement method should assure reproducibility of results.

Applicability

In order to assure applicability the methods and the underlying parameters should be clear and relatively easy to apply and communicate in practice. The criteria (xviii) availability of data, (xix) cost effectiveness, (xx) simplicity and (xxi) comparability contribute to the applicability. (xix:) The benefit arising from the valuation should be sufficient to justify the efforts to determine the measurement. (xxi:) In order to compare reported data against prior years and against other companies, the recognition, measurement and presentation of information has to be consistent. Furthermore, comparability will be enhanced if standard approaches can be applied across different intangible resources.

4. Classification of Measuring Methods

Due to the growing importance of intangible resources in the field of management - especially accounting - many measurement concepts have been developed in recent years by researchers, consulting companies and users. Some methods are thoroughly described while others do not reveal details (e.g. concepts of consulting companies). Different suggestions for basic classifications of these methods can be found in the literature (e.g. *Luthy*, 1998 or *Sveiby*, 2001). Here, we further extend upon these classifications by introducing both additional methods and missing categories of methods (such as the real option approach). *Figure 3* shows the classification of diverse measurement methods currently in discussion and gives an overview of the approaches' objectives and characteristics. It should be noted that a very precise classification of the methods is often not possible, as some approaches combine elements of different categories of methods (e.g. in the AFTF approach the value added is a difference of two DCF values). It should also be noted, that the list of methods presented here may not be comprehensive, as the field of measuring intangible resources is developing constantly.

Category/ Methods	Characteristics	References (in chronological order)
	Discounted Cash Flow Methods	
Customer Life Time Value	Value calculated based on customer specific revenues and costs	*Dwyer*, 1989, pp. 8; *Homburg* and *Daum*, 1997, pp. 96
Price Difference	Estimation of cash flows using the price premium of brands	*Arthur Andersen*, 1992; *Mullen*, 1993
Nielsen Brand Performancer	Discounting of future brand profits based on a previously calculated brand power figure	*Hammann*, 1992, pp. 224; *Franzen*, 1994, pp. 1625,
License Income	Valuation of intangibles based on cash flows that might be earned by licensing out the right to exploit the resource	*Mullen* 1993; Reilly and *Schweihs*, 1999, pp. 194; *Smith and Parr*, 2000, pp. 380

Category/ Methods	Characteristics	References (in chronological order)
	Discounted Cash Flow Methods (ctd.)	
Residual Value	Valuation of business with intangible and valuation of business as if it would not have the intangible	*Mullen*, 1993
Brand Valuation	Brand valuation model based on conjoint analysis using DCF approach	*Sattler*, 1997, pp. 55, pp. 215; *Kriegbaum*, 2001, pp. 183
Indicator Model	Brand valuation using the assessment of experts for indicators of the brand product which are subsequently evaluated using conjoint measurement and a DCF approach	*Sattler*, 1997, pp. 215
Allocation of Business Enterprise Value	Enterprise value, that is derived from future cash flows is allocated to the identified resources using the book value; residual is interpreted as value of intellectual property	*Smith* and *Parr*, 2000, pp. 215
Total Value Creation®	Software based measurement and reporting on an organisation's value creation performance using event-based future financial and non-financial value streams	*Total Value Creation International Inc.*, 2000
TechFactor™ Method	Method that is used to calculate the value of technology-based intellectual resources by separation of the technological contribution using one aggregated factor	*Arthur D. Little* (*Hastabacka* and *Khoury, w.Y.*)
Value Explorer®	Evaluation of intangibles as a holistic value based on the evaluation of core competencies using financial (gross profit) and non-financial (sustainability and robustness) data	*Andriessen* and *Tissen*, 2000
Brand Money Value	Value calculated based on brand specific revenues and costs under particular consideration of brand power and brand risk	*GfK* and *PWC Approach* (*Högel* et al., 2002)
	Multiple Methods	
Market Based Brand Valuation	Valuation of brands based on an industry average brand profit and the brand power based on the assessment of external experts	*Bekmeier-Feuerhahn*, 1996, pp. 191
Interbrand Approach	Estimation of representative brand related earnings according to a business specific brand power figure and subsequent multiplication with a derived factor based on empirical data (details not revealed by consulting company)	*Haig* and *Perrier*, 1997, p. 33; *Andrew*, 1997, p. 55
Organizational Intellectual Capital	Calculation of a holistic value through multiplication of some value of Intellectual Capital in dollar I that consists of indicators that are monetary measures and a coefficient of efficiency (i) that results from indicators that are percentages and ratios.	*Edvinsson* and *Malone*, 1997, pp. 177
	Market Derived Measures	
Capital Market based Brand Valuation	Difference between market value and replacement costs as indicator for intangibles; split of the value of intangibles in different categories using regression analysis	*Simon* and *Sullivan*, 1992
Market transaction comparatives	Prices of actual transactions of similar intangibles are used for value estimation	*Mullen*, 1993; *Reilly* and *Schweihs*, 1999, pp. 146; *Smith* and *Parr*, 2000, pp. 165
Market to Book Value	Difference between market value and book value as measure for intangibles	*Nasseri*, 1996
Tobin's q	Difference between market value and replacement costs as measure for intangibles	*Stewart*, 1997

Category/ Methods	**Characteristics**	**References (in chronological order)**
	Real Option Approach	
Real Option	Transfer of Financial Option Pricing Models to Intangible Resources	*Bloxham*, 2000
TRRU®	Transfer of Financial Option Pricing Models to patents	*The Patent and License Exchange Inc.*, 2000
	Excess Returns Approaches	
Premium Profit Method	Isolating profits, that are above 'normal' profits achieved by similar business (using operating margin or RoA difference)	*Mullen*, 1993; *Reilly* and *Schweihs*, 1999, pp. 192
Knowledge Capital	Valuation of Knowledge Capital as a ratio of Management Value Added to Price of Capital	*Strassmann*, 1996
Calculated Intangible Value	Intangible assets value as ability to outperform average competitors	*Stewart*, 1997
VAIC™ - Value Added Intellectual Coefficient	Sum of value added resulting from physical capital, human capital and structural capital employed	*Pulic*, 1998
Accounting for the Future (AFTF)	Value added is calculated as an result of AFTF value at the end of a period minus the AFTF value at the beginning of a period; AFTF value is calculated using the discounted cash flow method	*Nash*, 1998
Knowledge Capital Scoreboard	Performance-based calculation of Knowledge Capital using annual normalised earnings. The normalised earnings are allocated to the broad asset classes' tangible book assets, financial assets and intangible assets using expected returns for the first two asset classes. The residual earnings are interpreted as knowledge earnings that are the basis for the calculation of knowledge capital	*Lev*, 1999
Real Asset Value Enhancer (RAVE™)	Valuation method for customer/human capital; transfer to DCF possible	*Barber* et al., 2000; *Strack* and *Villis*, 2001; *Strack* and *Villis*, 2002
	Cost Approach Methods	
Human Resource Accounting	Origin is within cost approach but was enhanced by the time using non-monetary measures as well as applying discounted cash flow method	*Flamholtz*, 1974 and 1999
Human Resource Costing and Accounting	Development of human resource costing calculations, profit and loss accounts and human resource reports based on financial human resource figures; influenced by human resource accounting and utility analysis	*Gröjer* and *Johanson*, 1996
Historical Costs	Original acquisition or production costs of intangible assets; often used within financial reporting (US GAAP, IAS, German GAAP)	*Arthur Andersen*, 1992; *Smith* and *Parr*, 2000, pp. 193
Cost of Replacement; Reproduction	Current costs for the replacement or reproduction of the existing intangible resource	*Reilly* and *Schweihs*, 1999, pp. 122; *Smith* and *Parr*, 2000, pp. 195

Category/ Methods	Characteristics	References (in chronological order)
	Indicator Approaches	
The Intellectual Capital Audit	Guide for documentation of the existence, the current state and eventually the value of a company's intangibles for specific purposes like planning research & development.	*Brooking*, 1996, pp.82
Two non-monetary indicator classes for measuring human capital	Human Capital measurement through competency test and productivity tests.	*Miller*, 1996, p. 21
Return on Communications	Measurements are divided into five groups (leadership, market, community, employees, finance) and closely linked to a company's profit figures ("value links")	*The Swedish Public Relations Association*, 1996
Skandia Navigator	Indicators in five perspectives (financial, customer, human, process, renewal & development focus)	*Edvinsson* and *Malone*, 1997, pp. 65
Intellectual Capital Accounts	Company specific report showing indicators of the following categories: human resources, customers, technology, processes	*Danish Trade and Industry Development Council*, 1997
Intellectual Capital Navigator	Presentation of different indicators in a management chart without any stringent connection	*Stewart* 1997, pp. 245
Intangibles Assets Monitor	Describing indicators of intangibles in three different categories (external structure, internal structure and competence indicators)	*Sveiby*, 1997
Multidimensional Knowledge Measurement	Definition of four different indicator classes, that are the basis of a complex measurement system	*Probst* et al., 1997
Technology Balance Sheet	Measures the potential and the usage of a company's technologies based on technology attractiveness	*Hartmann*, 1997
IC-Index™	Consolidation of individual indicators into a holistic index	*Roos* et.al., 1998
Liahona - H.O.M.E.S Model	Calculation of a holistic intellectual capital indicator out of several figures	*Bornemann* et al., 1999
Intellectual Capital Report	List of parameters and potential value drivers connected with intangible resources	*Austrian Research Center*, 2000; *Maul*, 2000; *Maul / Menninger* 2000
Intellectual Capital Statement	Composition of indicators for seven different intangible categories	*Arbeitskreis "Immaterielle Werte im Rechnungswesen"*, 2003
Value Creation Index	Scores of value drivers that had in past a statistical proven influence on the market value are combined to a value creation index	*Cap Gemini Ernst & Young Center for Business Innovation*, 2000
ValueNet Works™ Analysis	Analysing value creation through comprehensively mapping value exchange as well through visualisation (HoloMapping™ method) of complex, interwoven and interdependent business processes. Within the value exchange money, knowledge and intangible benefits are considered as a currency	*Allee*, 2000

Category/ Methods	Characteristics	References (in chronological order)
	Indicator Approaches (ctd.)	
Business Logics Model™	Distinction of performance indictors for 21 different specific operations derived from value creation process and called "business logics". The method emphasises on performance measures that people identify with and thus are intended as a guide for the workforce in the short run as well as in the long run.	*Giertz,* 2000
Value Chain Scoreboard™	Indicators for different stages of intellectual capital based on the value chain	*Lev*, 2001
Analytical Accounting System for Intellectual Capital	Model describes and prescribes intellectual capital in a holistic manner under consideration of knowledge management strategy	*Mouritsen* et al., 2001
Management of Intangible Assets (MIA)	Aggregation of various measurements to a value-creating capital, a market capital, a confidence capital and an innovation capital figure	*Sifo Research & Consulting AB*, w. Y.

Fig. 3: Overview on frequently discussed measurement methods for intangible resources

5. Evaluation of Measurement Methods Categories

5.1 Discounted Cash Flow Methods

One of the most common concepts used to determine the value of an intangible resource is the discounted cash flow approach. This assumes that an intangible's value depends on its ability to create future returns over a given period of time. According to this method, periodic cash flows - resulting from revenues minus costs - are projected. Capitalisation of these cash flows using a risk-adjusted discount rate arrives at the net present value of the intangible. The method is not new and has been used for decades for investment appraisal and company valuation. Brands and customer capital are often valued using DCF approaches. The following sections describe the use of DCF approaches in the context of measuring customer capital. Another example for using DCF is brand valuation (*Sattler*, 1997; *Kriegbaum*, 2001)

Customer lifetime value (CLV) applies the net present value method to customer relationships and thus determines the value of a customer or a customer group. CLV is based on a dynamic investigation of customer businesses. A purchase is therefore considered as a part of an ongoing customer relationship, rather than a one time event. The customer relationship follows the customer life-time cycle consisting of four phases: recognition, premature development, expansion and mature development (*Dwyer*, 1989).

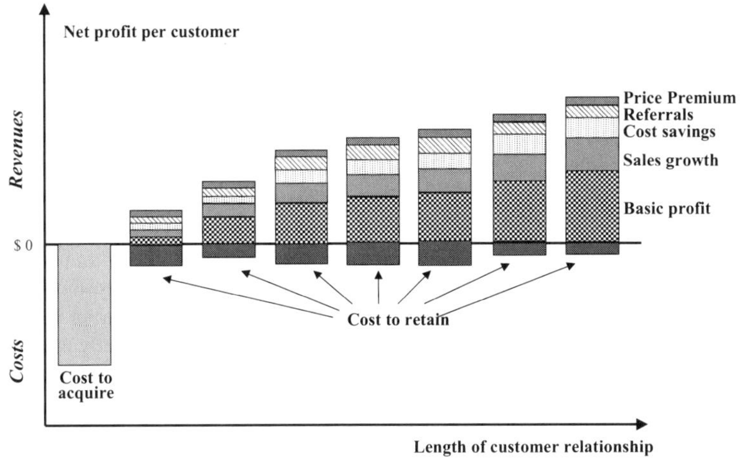

Fig. 4: Calculation of Customer Lifetime Value (Reichheld, 1996)

Evaluation of the customer relationship is based on customer specific revenues and costs. During acquisition, customer revenues are typically deficient and costs relatively high. As the cycle progresses revenues increase and costs decrease. The increased loyalty - enhanced by the length of a customer relationship - opens new profit potentials through cost savings based on less demand for information and additional price premiums from loyalty and switching costs. In addition, satisfied customers make recommendations and increase total units sold. *Figure 4* illustrates an example of a possible cash flow from a customer relationship. In light of scarce resources, management should concentrate on creating and pursuing marketing and selling strategies to selected customers or customer groups based on their CLV.

CLV is calculated as the net present value of all revenues and expenses occurring in conjunction with a given customer relationship during the customer lifetime:

$$CLV_0 = \sum_{t=0}^{T} \frac{E_t - A_t}{(1+r)^t}$$

With

CLV_0:=	Customer Lifetime Value	r:=	discount rate
E_t:=	revenues resulting from the customer relationship during the period t	t:=	period (t = 1, 2, ...,T)
		T:=	length of the customer relationship
A_t:=	expenses resulting from the customer relationship during the period t		

In addition to the monetary dimensions, non-monetary dimensions such as customer satisfaction should be considered. It can be taken into account as a risk component by adjusting the discount rate.

Customer lifetime value analysis is appropriate in industries with long term customer relationships and extensive groups of customers (e.g. retail banking, insurance, telecommunications and leasing). CLV analysis supports company valuation when customer value is relevant. Furthermore, by analysing existing marketing and accounting data, customer group profiles with high CLV can be found. CLV can also be analysed into customer-related value drivers (e.g. churn rates, targets for sales growth or contribution margin) which may help in successfully managing customer value.

All DCF approaches like CLV are long-term measures for intangible resources which automatically fulfil some of the desired evaluation criteria. DCF approaches are financial, future-oriented, dynamic and can measure the output of diverse intangible resources. They are based on logical grounds, but on the other hand are subject to forecast errors for future cash flows. Consequently, it is often appropriate to include some cause and effect analysis which is often not delivered by the DCF approaches listed in figure 3. Validity and tractability are relatively high. Accuracy depends on the quality of the projected data. This frequently requires the use of conjoint measurement or risk analysis which may be costly. DCF approaches may also be used at different levels within the hierarchy of the company. However, as DCF depends heavily on the availability of the data, in models such as for instance in brand valuation not all of the economic benefits or costs of an intangible resource are considered (e.g. restriction on licensing income, neglect of brand related costs or sales volume changes). In some cases intangible resources may not be isolated from their underlying products. For instance, the value of the brand may be set equal to the DCF of its brand-related product profits. In summary, DCF approaches are useful measures for long-term decision support. The crucial but obvious point is how the valuation is performed not what technique is used for it.

5.2 Multiple Methods

Multiple methods are simplified DCF approaches and are based on the assumption that a constant figure (for valuing companies e.g. earnings or cash flow per share) can be related to the value of the intangible resource. Consequently, a multiple must be identified which describes how much revenue has been caused by the investment in the relevant object. Illustrations of this principle can be found in certain "rules of thumb" used by financial analysts. For example, the value of a company's customer base may be expressed as the product of the number of customers and a chosen multiple ($value_C$ = # customers x m_C). A comparable notion on the valuation of human capital can be found in practice as well ($value_E$ = # employees x m_E).

Interbrand, a consulting firm, uses this method to value brands which had been used several times to publish a ranking of the most valuable brands in the world (e.g. Business Week, August 4th 2003, pp. 72). It includes a financial analysis where business earnings are identified, a market analysis where the proportion of earnings attributable to the brand is determined, a brand analysis where brand power is analysed, and a legal analysis (*Haigh/Perrier*, 1997, p. 33).

The brand value is the product of the net earnings of a brand and a multiple. The net earnings of a brand are separated from the estimated earnings of the business depending on the importance of the brand for the business using a scoring analysis. The multiple is a function of brand power which is seen as the primary determinant of the risk profile of the brand. Low brand power will increase the risk that projected earnings will not be realised. Therefore, two brands can have different values even if they have identical earnings just because their brand power and consequently their risk profile is different. The function is an empirically derived S-curve that is not disclosed by *Interbrand.* To determine brand power, seven criteria groups with 80 to 100 sub-criteria that influence brand power are evaluated. The seven criteria groups involve market, stability, leadership, trend, internationality, marketing support and legal protection. Using a scoring model, the criteria are weighted according to their influence on brand power and then summed up to one score (*Andrew*, 1997, pp. 56).

The method suggests that it is possible to separate brand returns from product returns, but it remains unclear how brand returns are isolated. The method also implies that the seven criteria groups directly influence brand power but it is not obvious where the criteria and the criteria weights come from, how or by whom brand power criteria are evaluated. The criteria also seem to be interdependent (e.g. marketing support and internationality are influencing the growth potential of the brand). In addition, the criteria groups and sub-criteria mix the cause of brand power (e. g. marketing support) and their effect (e. g. market leadership). Moreover, the *Interbrand* method does not consider the value of potential brand extensions. Similar to other consulting approaches the measurement procedure is not disclosed in detail, so that the assessment can only be limited, and leaves the impression of a black box, which does not support the acceptance of the method.

Brand valuations using this method are not suitable for long-term performance measures because it might provide the wrong incentives to brand managers. For instance, increases in market share, sales, etc. increase brand value under the *Interbrand* method. But, some activities designed to achieve these goals, e. g. price reductions, may destroy a brand's future value even if they increase sales and market share in the short run. In this case, brand value as a performance measure then even works against long-term brand value (*Roeb*, 1994, p. 114).

The *Interbrand* model as an example of multiple valuation methods reveals the main problems with these concepts. Although brand value can be calculated in different periods and the time value of money can be taken into account as well, it is not a dynamic or future-oriented model because changing cash flows and net present value are completely neglected.

Thus, although the multiple concept is financial in nature and measures output, its future orientation and use as dynamic measurement are limited. Cause and effect analysis can be performed to some extent. Validity is also limited as the connection between multiples and values are often (as in the *Interbrand* case) not empirically proven. Moreover, the specific weights and the assessment of the influencing criteria often seem to be chosen arbitrarily. One doubts if all the necessary economic impacts of the measured intangibles are considered. Nevertheless the utilisation of multiples can be economically useful, especially where searching for a rough estimate of the value of an intangible resource.

5.3 Market Derived Measures

Markets may be useful in evaluating different kinds of assets that are traded. Capital market derived measures based on the market to book ratio (e.g. *Nasseri*, 1996) or the *Tobin*'s Q (e.g. *Stewart*, 1997) use this valuation process for the valuation of intangibles. Essentially, the value of intangibles equals the difference between the market capitalisation and the book value or replacement costs. For this to work, capital markets must be efficient and the difference between market and book value must be solely due to intangibles (intangible resources = goodwill). Unfortunately, all too often factors like hidden reserves, interest rates or political changes are also influencing the market valuation.

Another approach using market evaluation is based on transaction prices for intangible resources (e.g. *Mullen*, 1993; *Smith/Parr*, 2000). For example, *Patent and License Exchange Inc.* (pl-x.com) provides a virtual marketplace for technology. Numerous brand brokers also try to establish a secondary market for brands. However, experience from the valuation of business units using transaction prices indicates that transaction prices often reflect the specific situation of a transaction. Efficient markets do not exist, data is often publicly not available and generally derived multiples vary tremendously from industry to industry and over time.

The advantage of market derived measures lies in their tie to financial reality. In efficient markets validity, accuracy and tractability of the measure are well supported. Unfortunately, in some capital markets (and especially in Europe) the efficiency hypothesis does not hold. The disadvantage is that the cause and effect analysis is missing because the share price is a highly aggregated figure. Furthermore, one doubts if it is possible to disaggregate the goodwill measured at the corporate level into different parts at the business level. This approach is relatively simple and efficient, and may provide a good beginning for valuing intangibles, but we do not believe it provides a proper valuation of one specific intangible.

5.4 Real Options

Option-based valuation approaches have grown in importance in recent years. The major reason for this development is the failure of traditional approaches to consider uncertainty and flexibility in an appropriate way. Many opportunities (real options) of the company are based on know-how, special rights such as patents or licences and, of course, human capital. Thus, the real options approach seems appropriate for valuing intellectual resources (*Bloxham*, 2000). Fundamental examples of real options connected with intangible resources include options to defer investments (e.g. due to flexibility in processes), options to expand (e.g. line extensions using brand transfers or key customer groups), options to shut down and restart operations (e.g. due to flexible workforce), options to abandon for salvage value (e.g. transferring know-how to more attractive fields), options to switch use (e.g. due to built-in flexibility in production due to technological know-how) and corporate growth options (e.g. transfer of know-how according to the core competencies approach) (in general *Trigeorgis*, 1997).

The DCF approach assumes all chosen projects will be realised without any adaptation till the end of the planning horizon. Risk increases the discount rate and thus reduces the value of the project. Because the real options approach forces managers to think about opportunities, even if the approach is not able to value every part of intangible resources in practice, it may help to manage it.

In addition to the present value of expected returns and the investment's cost, option price models involve three additional parameters. They include the length of time the decision may be deferred, the risk-free rate of return and the riskiness of the underlying asset expressed by its distribution. In contrast to the conventional DCF, risk increases the value of options. Two of the most important option pricing models are the continuous model of *Black/Scholes* (*Black* et al., 1973) and the discrete time model of *Cox/Ross/Rubinsteins*' binomial approach (*Cox* et al., 1979).

However, option pricing models are difficult to use in the real world. Textbooks show mostly the calculations for unrealistically simplified problems in the case of real options. The relevant parameters (e.g. volatility) for these option pricing models are hard to estimate.

Real options are normally nested within each other. This complicates their clear identification, not to mention modelling and calculating. Furthermore, real options interact with each other and appear as options on other options (compound options). Thus, a simple accumulation of their values is not possible.

Except for these limitations, the real options approach seems to be very effective in terms of the formerly established criteria. It is capable at measuring flexibility in monetary terms and it is output-oriented. Because it stresses uncertainty and flexibility, the real options approach is future-oriented and dynamic. The treatment of cash flows and risk is hardly to improve upon. Real options can evaluate a variety of economic benefits and costs connected with the intangible. Because some of the most important parameter (e.g. standard deviation) must be estimated in the real world, validity, accuracy and tractability are often not as high as one would like. In addition, the efficiency of the concept may be doubtful due to its complexity and analytical demands.

The real options approach is suitable for valuing specific intangible resources such as patents or special R&D projects (*Bloxham*, 2000). Examples of practical usage are mostly found in these fields. *The Patent and License Exchange Inc.* (pl-x.com) provides a virtual marketplace for technology and uses the real option method TRRU® to value intellectual property. TRRU® is the Technology Risk/Reward Unit Metrics valuation method of *The Patent & License Exchange Inc.* (see e.g. *The Patent & License Exchange*, TRRU® IP Valuation Report).

5.5 Excess Return Approaches

Excess return approaches assume intangible resources are earning a premium return in addition to an average return. The average return is either calculated based on the average rate of return (RoA) earned by specific competitors or by the whole industry (e.g. *Mullen*, 1993 and *Stewart*, 1997). As an alternative an ordinary rate of return on tangible and financial assets as shown in the financial statements is used (e.g. *Lev*, 1999). From our point of view, there may be other reasons for excess returns including competitive advantages in the management of processes, differences in the competitive environment between companies in the same industry or varying percentages of capacity utilisation. Unfortunately, this approach can not be applied to single intangible resources as the company's or the business' performance as a total is evaluated.

One excess return approach for intangibles is the RAVE™ (Real Asset Value Enhancer) concept of *The Boston Consulting Group (BCG)* (*Strack and Villis*, 2001). The RAVE™ concept is a modification of residual income approaches like Economic Value Added™ (e.g. *Stewart*, 1990) (EVA™) or the Cash Value Added concept of *BCG* (CVA) (*Lewis*, 1994). Residual income measures the amount of money created on an annual basis in excess of the charge for capital costs on the capital employed:

$$\begin{aligned} residual\ income &= profit - CC \bullet CE \\ &= (RoI - CC) \bullet CE \end{aligned}$$

With

CC:=	Costs of Capital	CE:=	Tangible Capital Employed	RoI:=	Return on Investment

Only the amount of profit above capital costs embodies a real value creation. The RAVE™ method argues that value management methods should include not only capital invested in tangible assets but also the impact of human capital or customer capital. The RAVE™ concept allows to concentrate on the dominant capital in a particular business, e.g. human capital in personnel-intensive companies or customer capital in marketing-based companies.

BCG adjusts its CVA approach to incorporate key human resource indicators, e.g. personnel costs, number of employees and value added per person (Workonomics™ approach for the intangible resource human capital):

$$CVA = (VAP - AC) \bullet P$$

With					
CVA:=	Cash Value Added	VAP:=	Value Added per Person	AC:=	Average (personnel) costs per person
P:=	Number of employees				

Value added per person is defined as the difference between sales and all other costs except personnel costs and equals profits before personnel cost:

$$VAP = (S - MC - D - CC \bullet CE) / P$$

With S:= Sales MC:= Material Costs and others D:= Depreciation

Using this transformation it is possible to replace figures based on tangible capital with human resource oriented metrics allowing a direct comparison between value creation and major value-generating factors in a personnel-intensive company. Analogous to the residual income, value is created if the value added per person exceeds the costs per person. Growth in the head count is only value creating if this condition holds (VAP > AC). Within the human capital view value can be increased (Δ CVA) first by the improvement of value added per person (see arrow (a) in *figure 5*). A second possibility is a human capital growth if value added per person is bigger than average (personnel) costs per person (arrow (b) in *figure 5*). Finally a decrease of average personnel costs leads also to value increase (arrow (c) in *figure 5*). *Strack* and *Villis* propose a comparable approach related to customer capital, called the Custonomics™ concept.

The main contribution of the RAVE™ concept is the adjustment of the standard CVA model for the requirements of personnel-intensive or marketing-based companies. It is one plausible way of measuring the impact of different intangible resources as an entity on corporate value.

All excess return approaches are monetary-based and intend to measure the output of intangibles. However, they are limited to a static, short-term view and only focus on annual or interim figures. They are based on profit figures, do not include changing future cash flows, and neglect the time value of money. Consequently, they do not completely record a company's potential for success. The methodology assumes a going concern approach and constant returns. Nevertheless, the discounting of future CVA provides not only a net present value figure, but also allows for the adjustment of risk via the discount rate.

With regard to cause and effect relationships, the excess return approaches contain only a few monetary figures. Cause and effect relationships are not explicitly stated, but further indicators can easily be implemented in a periodic monitoring system. Assessment of the validity and accuracy is depending upon the underlying assumption of intangible resources as the "only" reason for excess returns. Because only monetary figures are used in explaining the value of intangibles, the approach seems to be quite tractable. Finally, the approach can be very easily communicated within the company and to external stakeholders. It can also be used within and across business units when adequate comparables can be found.

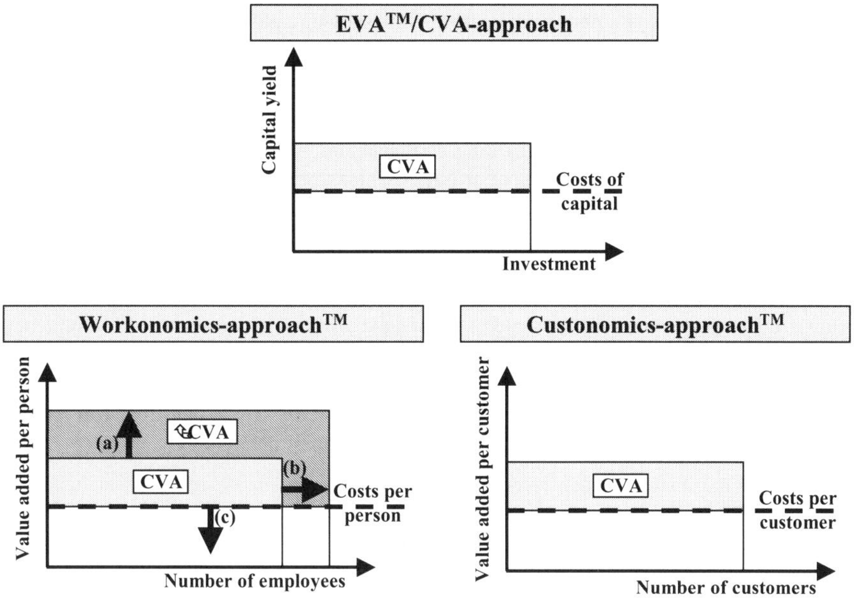

Fig. 5: Residual income methods – the Workonomics™ and Custonomics™ approach (Strack and Villis, 2001, p. 72 and p. 77)

One evident advantage of the RAVE™ method is obviously the clear distinction between different forms of intangibles like human and customer capital. A further separation e.g. into product or process know-how seems to be far more difficult, however. From our point of view, it does not make sense to allocate all the profit to a specific intangible resource since profit is part of value added per employee as well as value added per customer. The sum of the values for both intangibles will therefore overestimate the real value of both resources. The authors also do not state clearly enough how customer-specific costs may be separated from "other costs". For other excess return approaches the distinction in the valuation of different intangible resources seems to be far more complicated.

The excess return concepts are auditable and economic useful because of their simple structure and ability to facilitate intra- and inter-company comparisons, but are limited because of the missing long-term approach.

5.6 Cost Approach

There are two main problems associated with valuing intangible resources using costs. These are based on limitations of financial accounting and on problems of clear allocation for intangibles in order to support internal decision making processes. In either case costs are defined from different points of view.

If intangible resources are recorded as assets on a company's balance sheet, they are entered at historical cost. In the case of purchased intangibles, this refers to the amount that was paid at the time the intangible assets were acquired. Allowable capitalisation of internally generated intangible assets under GAAP requires production costs to be shown. In today's financial accounting often only a portion of costs may be shown (e.g., for patents only legal costs may be booked; R&D costs for gaining the patented know-how are not capitalised). Thus, a firm's financial statements do not completely reflect all of a firm's intangible assets or the costs associated with their creation.

For internal purposes, the costs of creating certain intangible resources are relevant in the decision making process, e. g., in make or buy decisions or transfer pricing. For these purposes it is very important to cover all relevant costs, including imputed costs and opportunity costs (see e.g. the approaches in Human Resource Accounting *Flamholtz*, 1999 and *Gröjer/Johanson*, 1996). Besides historical costs there are the two basic approaches reproduction costs and replacement costs. Furthermore, in our brand management survey we discovered tremendous practical problems associated with the determination of production costs for brands (*Günther/Kriegbaum-Kling*, 2001). It is very difficult to distinguish brand-related from product-related costs (For instance, is a special kind of packaging part of the brand-related costs? Are investments in quality brand-driven?). These questions can only be answered in a specific context and complicate the valuation of intangibles.

By definition the cost approach is not future-oriented as costs are recorded at historical prices. Nevertheless the cost approach can have a dynamic component because the creation process can take a long time and costs should be accumulated up to the decision point. However, the value of an intangible is determined by future returns (future benefits minus future costs) and not by past costs. The cost approach only specifies cause and effects analysis for "inputs" (= costs) but not for outputs (= contributions from future sales). As shown by our performance measurement study (*Günther/Grüning*, 2002), traditional cost measures are associated with high values of validity, accuracy and tractability. The efficiency of this approach occurs only if costs for intangibles can be separated easily from other costs (which is generally not the case). In conclusion, the cost approach tells us something about historical expenditures for the creation of an intangible resource, but is not capable of measuring all the economic benefits which might be derived from a specific intangible resource. Thus, it only provides a second best estimate for value. A common use of this method can be seen in human resource accounting, where specific labour costs are cited as evidence of the potential of personnel.

5.7 Indicator Models

Due to problems concerning the monetary valuation of intangible resources indicator methods concentrate on the creation of complex measurement systems behind monetary terms. These methods try to quantify key drivers of companies' intangible resources and establish cause and effect chains among the major indicators. Although these methods do not allow the explicit calculation of a particular monetary figure, the cause and effect relationships provide information about the impact and the influence of a particular indicator on corporate value. Examples of these models are the *Skandia* Navigator (*Edvinson/Malone*, 1997) and the Intellectual Capital Statement (*Arbeitskreis "Immaterielle Werte im Rechnungswesen"*, 2003).

As one representative indicator method we describe the Value Chain Scoreboard (VCS) developed by *Lev* (*Lev*, 2001). He takes an approach similar to the balanced scorecard by incorporating non-financial and financial measures of different perspectives. Measurements are classified into nine different groups (see *Figure 6*). Each of the nine measurement groups belong to one of the corporate value chain's three stages: discovery, implementation or commercialisation.

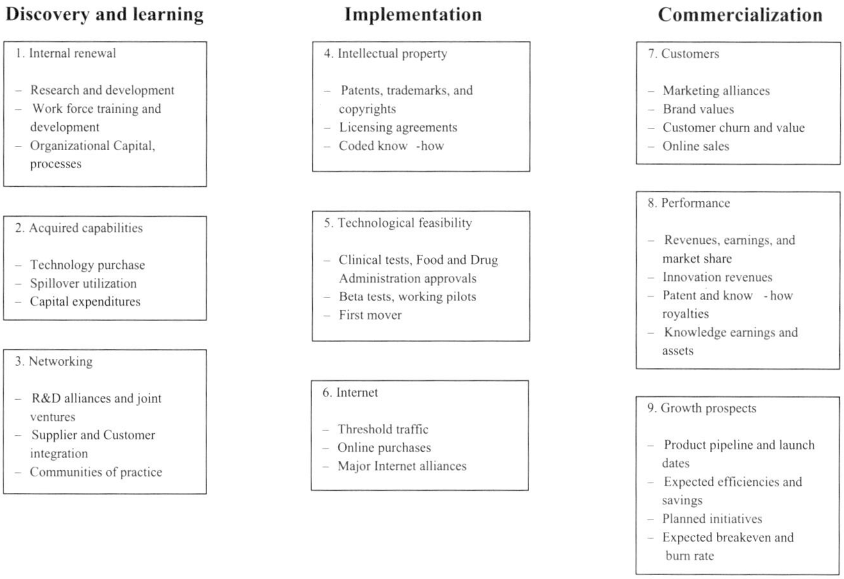

Fig. 6: The Value Chain Scoreboard (Lev, 2001, p. 184)

The VCS provides a broad selection of measurements for different types of intangibles which can be easily adapted through the selection of a subset to the company's specific requirements.

Within the first phase of the value chain "discovering and learning" *Lev* distinguishes between internal renewal, acquired capabilities and networking as major sources of new ideas for products, services, or processes. The second stage, "implementation", considers measurements referring to the transformation of new ideas into working products, services or processes. Readiness for commercialisation reduces an intangible's risk and allows for better estimation of the future development of a "new idea". The third phase, "commercialisation", presents past and potential financial results for the new ideas' outcome.

The VCS is helpful for both internal decision making and as a communications tool for investors. To assure maximal usefulness *Lev* suggests the specific indicators should be quantitative, standardised and confirmed by empirical evidence as relevant to users.

Quantitative indicators can be completed through qualitative indicators in an annex to the scoreboard. Standardisation of the figures enables comparisons across companies. Empirical studies - executed on a large scale by *Lev* and his colleagues - assure the chosen indicators are linked to value and that these linkages are scientifically robust.

The VCS was developed to enhance the information content of financial reports by looking at all intangibles, rather than to serve as a specific tool for measurement or valuation of a chosen intangible. Therefore, it might not fulfil our requirements for the measurement of specific intangibles very well because of *Lev*'s different intent. However, since it contains such a large amount of information, it is by far one of the most developed approaches for assessing the value of intangibles for the entity as a whole.

Indicator approaches have been developed to look behind financial figures in assessing the value of intangibles and to show cause and effect relationships for measures that are often not part of the reporting or performance measurement system of a company. These performance measurement approaches are by definition mostly non-financial and do not intend to evaluate a specific intangible resource. They can deliver a lot of information helpful in forecasting future cash flow (e.g. market growth, product pipeline or expected restructuring impact in the VCS) and assessing risk adjusted discount rates but it is not their primary intent.

Indicator approaches assess the economic benefits from an intangible in a comprehensive manner. *Lev* for example, focuses only on indicators with "confirmation through empirical studies". Consequently, the indicators are limited to those constructed from accessible data. Moreover, the Scoreboard shows measures at different stages of the value chain. This could allow for observation of an intangible's development in addition to cause and effect relationships. Although, indicator systems themselves do not provide detailed relationships supporting the management of an intangible, individual adjustments may be made by companies.

The indicator approaches establish the value of all intangibles of a company compared to other tangible assets. However, seldom do they provide information on a specific intangible.

6. Conclusions

The preceding analysis of existing methods for the measurement of intangible resources results in the following conclusions:

- Monetary valuation for isolated intangible resources is possible for some (e.g. brands and customer capital) but not all intangible resources.
- Monetary valuation of specific intangibles requires a substantial analysis due to the need for separation of product and intangible resource together with the complexity and multitude of influencing factors.
- Multiple and excess return approaches are efficient in use but are based on a static view. For proper measurement of intangible resources, dynamic models like DCF should be additionally regarded.
- Measures derived from markets have the advantage of integrating the market in the valuation process. Unfortunately the highly aggregated market measures are appropriate for the corporate level but may not be used at the business level. In addition, the market approach has difficulties in separating different intangible resources.
- The real option approach is interesting because options are often connected with intangible resources. However, it is complicated and may face limited acceptance by non-financial practitioners.
- Indicator systems may be used if a solid monetary valuation is not possible, as they focus on the drivers for intellectual capital and if cause and effect analysis are necessary to support management decision making. However, improper selection of indicators, scoring and/or weighting systems may limit validity, accuracy and tractability.
- To assess the value creation process properly the indicators must be carefully selected. Following Lev, only indicators shown to have a significant empirical impact on the

company's value should be included. Output measures should be preferred to process and purely input measures.

- The measurement of intangible resources is closely related to the classification and identification of intangible resources. Measurement and especially valuation is facilitated if an intangible resource fulfils the requirements of an asset (e.g. identifiably, controllability and provision of future economic benefits). Often intangible resources are mutually related and are in addition combined with tangible assets.
- Because the measurement of intangible resources is a precondition for their control and management users should be aware of the possibilities as well as the limitations and assumptions of a given method.
- Accordingly different targets in managerial accounting require different measurement methods. Long-term performance (e.g. shareholder value) can only be measured using long-term methods like DCF or real options. Excess return approaches may be adequate to support the short term-orientation within managerial accounting. If cause and effect analysis is needed indicator models should be used.

In conclusion, an optimal method meeting all the requirements of our analysis could not be found. All the methods have their specific strengths and weaknesses. Thus far, the measurement of intangible resources is characterised by a lack of experience, lack of relevant data in companies and few empirical studies on adequate performance measures (see e.g. the studies of *Lev* and colleagues at *NYU Stern School*). Future success will likely depend on better co-ordination between multiple disciplines (e.g. marketing and accounting into brand valuation; engineering, marketing and accounting in the measurement of technical know-how).

7. References

Arbeitskreis „Immaterielle Werte im Rechnungswesen" der Schmalenbach-Gesellschaft für Betriebswirtschaft e. V., Freiwillige externe Berichterstattung über immaterielle Werte (Voluntary disclosure on intangible resources), in: Der Betrieb, Vol. 56 (2000), No. 23, pp. 1233-1237.

Allee, V., Reconfiguration the Value Network, in: Journal of Business Strategy, Vol. 21 (2000), No. 4, pp. 36-39.

Andrew, D., Brand Strength Analysis, in: *Perrier, R.* (ed.,) Brand Valuation, 3rd ed., London 1997, pp. 55-60.

Andriessen, D. and *Tissen, R.,* Weightless Wealth: Find your real value in a future of intangible assets, London 2000.

Arthur Andersen, The Value of Intangible Assets, Economist Intelligence Unit and *Arthur Andersen*, Special Report 1992.

Austrian Research Center, Intellectual Capital Report 1999, research report of the *Austrian Research Center*, Seibersdorf 2000.

Barber, F., Strack, R. and *Villis, U.,* Valuing the Assets of the New Economy. http://www.bcg.com/publications/ search_view_ofas.asp?pubID=566, 2000.

Bekmeier-Feuerhahn, S., Marktorientierte Markenbewertung - eine marktpreisorientierte Bewertung von Marken aus konsumenten- und unternehmensorientierter Perspektive

(Market based brand valuation – a market price based valuation of brands from the customer and company perspective), Wiesbaden 1996.

Bhargava, M., Dubelaar, C. and *Ramaswami, S.,* Reconciling Diverse Measures of Performance: A Conceptual Framework and Test of a Methodology, in: Journal of Business Research, Vol. 31 (1994), No. 4, pp. 235-246.

Black, F. and *Scholes, M.,* The Pricing of Options and Corporate Liabilities, in: Journal of Political Economy, Vol. 81 (1973), No. 3, pp. 637-659.

Bloxham, E., Intangibles, Human Capital, and Options Value Measurement, in: Journal of cost management, Vol. 14 (2000), No. 6, pp. 26-31.

Bornemann, M., Knapp, A., Schneider, U. and *Sixl, K. I.,* Holistic Measurement of Intellectual Capital, International Symposium, Proceedings Measuring and Reporting Intellectual Capital: Experience, Issues and Prospects, Amsterdam (see also http://www.oecd.org//dsti/sti/industry/ indcomp/act/ams-conf/symposium-old.htm#holistic), 1999.

Brooking, A., Intellectual Capital - Core Assets for the third Millennium Enterprise, London, 1996.

Cap Gemini Ernst & Young Center for Business Innovation, Measuring the Future – The Value Creation Index, www.cbi.cgey.com/pub/docs/ New_VCI.pdf, 2000.

Cox, J., Ross, S. and *Rubinstein, M.,* Option Pricing: A simplified approach, in: Journal of Financial Economics, Vol. 7 (1979), No. 3, pp. 229-263.

Danish Trade and Industry Development Council, Intellectual Capital Accounts – Reporting and Managing Intellectual Capital, www.efs.dk/publikationer/ rapporter/engvidenregn/all.html, 1997.

Dwyer, F. R., Customer lifetime valuation to support marketing decision making, in: Journal of Direct Marketing, Vol. 3 (1989), No. 4, pp. 8-15.

Edvinsson, L. and *Malone, M. S.,* Intellectual Capital – Realizing your company´s true value by finding its hidden brainpower, London 1997.

Flamholtz, E. G., Human Resource Accounting, Encino 1974.

Flamholtz, E. G., Human Resource Accounting. Advances in Concepts, Methods and Applications, 3rd Edition, Boston 1999.

Franzen, O., Markenbewertung mit Hilfe von Ertragswertansätzen (Market Valuation based on DCF approaches), in: Deutsches Steuerrecht, Vol. 32 (1994), No. 44, pp. 1625-1629.

Giertz, E., Measuring Success: Identifying Performance Indicators, *Celemi International AB*, Malmö 2000.

Global Reporting Initiative™: Sustainability Reporting Guidelines, www.global reporting.org/GRIGuidelines/2002/gri_2002_guidelines.pdf, 2002.

Gröjer, J.-E. and *Johanson, U.*: Human Resource Costing and Accounting, 2nd ed., Stockholm, 1996.

Günther, T. and *Grüning, M.,* Performance Measurment-Systeme im praktischen Einsatz (Performance Measurement Systems in practical application, in: Controlling, Vol. 14 (2002), No. 1, pp. 5-13.

Günther, T. and *Kriegbaum-Kling, C.*, Brand Valuation and Control: An Empirical Study, in: Schmalenbach Business Review, Vol. 53, October (2001), pp. 263-294.

Haigh, D. and *Perrier, R.*, Valuation of Trade Marks and Brand Names, in: *Perrier, R.* (ed.), Brand Valuation, 3rd ed., London 1997, pp. 25-34.

Hammann, P., Der Wert einer Marke aus betriebswirtschaftlicher und rechtlicher Sicht (The Value of a Brand seen from the business and legal point of view), in: *Dichtl, E.* and *Eggers, W.* (ed.), Marke und Markenartikel als Instrumente des Wettbewerbs (Brand and Brand Product as an Instrument of Competition), München 1992, pp. 205 - 245.

Hartmann, M., Technologie-Bilanzierung: Instrument einer zukunftsorientierten Unternehmensbeurteilung (Technology Balance Sheet: An approach for future-based Corporate valuation), Göttingen 1997.

Högel, S., *Hupp, O.*, *Maul, K.-H.* and *Sattler, H.*, Der Geldwert der Marke als Erfolgsfaktor für Marketing und Kommunikation (Brand money value as success factor for Marketing and Communication), *Gesamtverband Kommunikationsagenturen GWA e.V.* (ed.), Der Geldwert der Marke (Brand money value), Edition Erfolgsbeiträge der Werbung, No. 7, Frankfurt am Main 2002, pp. 37 – 80.

Homburg, Ch. And *Daum, D.*, Marktorientiertes Kostenmanagement: Kosteneffizienz und Kundennähe verbinden (Market based Cost Management: Combining Cost Efficiency and Customer Relationship), Frankfurt am Main 1997.

Hastsbacka, M. A. and *Khoury, I. A.*, Technology Valuation – A Milestone on the Path from Art to Science, http:www.adltechnology.com /servicedocs/value3.pdf, w.Y.

Kriegbaum, C., Markencontrolling (Management Accounting for Brands), München 2001.

Lev, B., Seeing is believing - a better approach to estimating knowledge capital, in: CFO, Vol. 2 (1999), No. 15, pp. 29–37.

Lev, B., Intangibles: Management, Measurement, and Reporting, Washington D.C. 2001.

Lewis, T. G., Steigerung des Unternehmenswertes – Total Value Management (Increasing the corporation value – Total Value Management), Landsberg/Lech 1994.

Luthy, D. H., Intellectual capital and its measurement, Proceedings of the Asian Pacific Interdisciplinary Research in Accounting Conference (APIRA), Osaka, Japan, August 1998.

Maul, K.-H., Die Wissensbilanz als Teil des handelsrechtlichen Jahresabschlusses – Wissensbilanzen dargestellt am Beispiel des Jahresabschlusses von Hochschulen (The knowledge balance sheet as a part of the financial statements – knowledge balance sheets presented as the example of the financial statements of universities), in: Deutsches Steuerrecht, Vol. 47 (2000), No. 38, pp. 2009-2016.

Maul, K.-H. and *Menninger, J.*, Das Intellectual Property Statement - eine notwendige Ergänzung des Jahresabschlusses? (The intellectual property statement – a necessary addition to annual report?), in: Der Betrieb, Vol. 53 (2000), No. 11, pp. 529-533.

Miller, R., Measuring what people know. Human Capital Accounting for the Knowledge Economy, OECD, Paris 1996.

Mouritsen, J., *Larsen, H. T.* and *Bukh, P. N.*, Intellectual capital and the 'capable firm': narrating, visualising and numbering for managing knowledge, in: Accounting, Organizations and Society, Vol. 26 (2001), No. 7-8, pp. 735-762.

Mullen, M., How to value Intangibles, in: Accountancy, Vol. 112 (1993), No. 1, pp. 92-94.

Nash, H. H., Accounting For The Future - a disciplined approach to value-added accounting, http://home.sprintmail.com/~humphreynash/ Draft_Proposal.htm, 1998.

Nasseri, T., Knowledge Leverage: The ultimative Advantage, www.magi.com./ ~godbout/Kbase/kognos11, 1996.

Probst, G., *Raub, St.* and *Romhardt, K.*, Wissen managen – Wie Unternehmen ihre wertvollste Ressource optimal nutzen (Managing Knowledge – How Companies optimally use their most valuable resource), 2nd ed., Wiesbaden 1998.

Pulic, A., Measuring the performance of intellectual potential in knowledge economy, http://www.measuringip.at/OPapers/Pulic/Vaictxt/ vaictxt.html, 1998.

Reichheld, F. R., The Loyalty Effect: Growth, Profits, and Lasting Value, Boston 1996.

Reilly, R. F. and *Schweihs, R. P.*, Valuing Intangible Assets, New York et al. 1998.

Roeb, T., Markenwert: Begriff, Berechnung, Bestimmungsfaktoren (Brand Value: Terminology, Calculation and Drivers), Mainz / Aachen 1994.

Roos, J., *Roos, G.*, *Dragonetti, N. C.* and *Edvinsson, L.*, Intellectual Capital: Navigating in the New Business Landscape, New York 1997.

Sattler, H., Monetäre Bewertung von Markenstrategien für neue Produkte (Monetary Valuation of Brand Strategies for New Products), Stuttgart 1997.

Siesfeld, G. A., Measuring Learning and Knowledge, in: Journal of Strategic Performance Measurement, Vol. 2 (1998), No. 3, pp. 11-17.

Sifo Research and Consulting AB, People and Customer Relations - Management of Intangible Assets, www.sifo.se/html/eng/ba.htm, w.Y.

Simon, C. and *Sullivan, M.*, A Financial Approach to Estimating Firm-Level Brand Equity and Measuring the Impact of Marketing Events, Working Paper, Marketing Science Institute, Report No. 92-116, Cambridge / Massachusetts 1992.

Smith, G. V. and *Parr, R L.*, Valuation of Intellectual Property and Intangible Assets, 3rd ed., New York et al. 2000.

Stewart, G. B., The Quest for Value – The EVA™ Management Guide, New York 1990.

Stewart, T. A., Intellectual Capital: The New Wealth of Organizations, London 1997.

Strack, R. and *Villis, U.*, Die nächste Generation im Shareholder Value Management (The Next Generation in Shareholder Value Management), in: Zeitschrift für Betriebswirtschaft, Vol. 71 (2001), No. 1, pp. 67-84.

Strack, R. and *Villis, U.*, RAVE™: Integrated Value Management for Customer, Human, Supplier and Invested Capital, in: European Management Journal, Vol. 20 (2001), No. 2, pp. 147-158.

Strassmann, P. A., The Value of Computers, Information and Knowledge, www.strassmann. com/pubs/cik/ cik-value.shtml, 1996.

Sveiby, K. E., The New Organizational Wealth: Managing and Measuring Knowledge-Based Assets, San Francisco 1997.

Sveiby, K. E., Methods for Measuring Intangible Assets, http://www.sveiby.com.au/IntangibleMethods.htm, April 2001.

The Patent & License Exchange, TRRU® IP Valuation Report (Marketing Sample), http://www.pl-x.com/media/pdf/sample_report.pdf, 2000.

The Swedish Public Relation Association, Return on Communication, Stockholm 1996.

Total Value Creation International Inc., Total Value Creation™, http://www.totalvalue creation.com, 2000.

Trigeorgis, L., Real options: Managerial Flexibility and Strategy in Resource Allocation, 2nd printing, Cambridge 1997.

Management und Controlling von Intangibles auf Basis der immateriellen Werttreiber des Unternehmens

Roman Stoi [1]

---

[1] Prof. Dr. *Roman Stoi*, Berufsakademie Stuttgart

Intangibles gehorchen anderen Gesetzen als materielle Vermögenswerte und erfordern deshalb neue Ansätze der Planung, Steuerung und Kontrolle. Das Management des intellektuellen Kapitals stellt auch an die Unterstützungsfunktion des Controllings neue Anforderungen. Gegenstand und Natur der Intangibles sowie die notwendigen Anpassungen, die dadurch im Controlling und bei der Unternehmensführung erforderlich sind, werden in diesem Beitrag behandelt.

1. Einführung

In der heutigen Wirtschaft verlieren materielle Vermögensgegenstände zur Differenzierung im Wettbewerb und zur Steigerung des Unternehmenswertes zunehmend ihre Bedeutung. Im Vordergrund stehen die immateriellen Vermögenswerte, die so genannten „Intangibles". Während man sich im skandinavischen und nachfolgend auch im angloamerikanischen Raum schon seit längerer Zeit mit diesem Phänomen auseinander setzt, beginnen deutsche Unternehmen erst langsam die Bedeutung von Intangibles zu erkennen.

Doch was verbirgt sich hinter dem schillernden Begriff „Intangibles"? Wie bei jeder neuen betriebswirtschaftlichen Thematik existiert noch keine einheitliche Definition – für die praktische Verbreitung und Akzeptanz nicht gerade förderlich. Die Begrifflichkeiten reichen von Wissenskapital (z.B. bei *Teece*, 2000) bis zu allen Vermögenswerten, die einen Wert für das Unternehmen besitzen und keine materiellen Güter oder Finanzanlagen darstellen (z.B. *Lev*, 2001). Während die Eingrenzung auf das Wissen der Mitarbeiter zu eng ist und nur einen Bruchteil der immateriellen Vermögenswerte abdeckt, bleibt die letztgenannte Definition für eine konkrete Auseinandersetzung mit dem Thema zu vage.

Für eine Annäherung an den Begriff „Intangibles" kann man zunächst festhalten, dass Finanzanlagen und physisch greifbare, materielle Vermögensgegenstände nicht in diese Kategorie fallen. Sie werden im Folgenden unter dem Begriff Finanzkapital zusammengefasst. Der Unterschied zwischen dem Finanzkapital und dem Marktwert des Unternehmens wird gemeinhin auf immaterielle Ressourcen zurückgeführt, die einen Beitrag zum Unternehmenswert leisten. Die deutsche Begrifflichkeit, die das weite Spektrum dieser immateriellen Ressourcen am ehesten beschreibt, ist die des „intellektuellen Kapitals". Für eine weitere Konkretisierung ist eine Differenzierung in Human-, Kunden-, Partner-/Allianz- und Strukturkapital (in Erweiterung der Einteilung von *Skandia* gem. *Edvinsson/Brünig*, 2000, und in Anlehnung an *Daum*, 2002) sinnvoll.

Das Humankapital umfasst das Wissen und die Kompetenz der Mitarbeiter, das Kundenkapital besteht aus Kundenstamm und Kundenbeziehungen, das Partner-/Allianzkapital umfasst die Anzahl und das Potenzial von Partnerschaften innerhalb der Wertschöpfungskette und das Strukturkapital bezieht sich auf die Leistungsfähigkeit der internen Organisation und das Image des Unternehmens. Das Strukturkapital nimmt dabei eine Sonderstellung ein, da es die Basis und Voraussetzung für den Aufbau aber insbesondere die Nutzung der anderen Formen des intellektuellen Kapitals darstellt. Ohne entsprechende Unternehmenskultur und Führungsstruktur kann beispielsweise das Wissenskapital der Mitarbeiter nur unzureichend genutzt und vor Verlust z.B. durch Kündigung geschützt werden. Ohne entsprechende Geschäftsprozesse z.B. der Auftragsabwicklung lässt sich kein loyaler Kundenstamm mit hoher Wiederkaufrate aufbauen. Für ein Unternehmen mit hohem Bekanntheitsgrad und Ansehen in Wirtschaft und Gesellschaft ist es beispielsweise auch einfacher einen potenten Geschäftspartner zu finden. Darüber hinaus befindet sich nur das Strukturkapital im Eigentum des Unternehmens, während es an seinen Mitarbeitern, Kunden und Geschäftspartnern und dem damit verbundenen intellektuellen Kapital kein Eigentum erwerben kann. Die Verfügungsrechte an diesen Kapitalbestandteilen sind deshalb begrenzt und nur im Rahmen vertraglicher Regelungen durchsetzbar. Deshalb kann man hier zusätzlich in externes intellektuelles Kapital (Human-, Kunden-, Partner/Allianzkapital) und internes intellektuelles Kapital (Strukturkapital) unterscheiden. Die Unterteilung des intellektuellen Kapitals veranschaulicht *Abb. 1*.

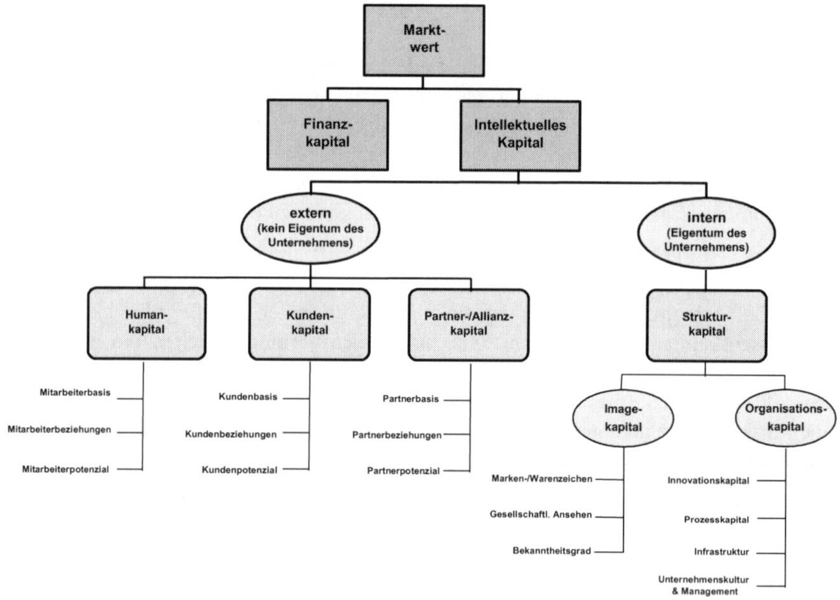

Abb. 1: Differenzierung des intellektuellen Kapitals

Während der durch den Aufbau des intellektuellen Kapitals verursachte Aufwand in der Kostenrechnung und Bilanz erfasst wird, erscheinen immaterielle Werte dort nur, wenn sie entgeltlich erworben wurden oder als Entwicklungsaufwand im Rahmen der Herstellkosten für Gegenstände des Umlaufvermögens aktiviert werden können. Der überwiegende Teil der immateriellen Vermögenswerte und des ihnen gegenüberstehenden intellektuellen Kapitals wird in der Bilanz hingegen nicht ausgewiesen und bleibt damit sowohl dem externen Betrachter als auch häufig der internen Analyse und Steuerung verborgen. Deshalb spricht man in diesem Zusammenhang auch vom „unsichtbaren Vermögen". *Abb. 2* veranschaulicht diesen Sachverhalt.

Investitionen in den Aufbau immateriellen Vermögens erhöhen aber die Ertragskraft, d.h. das Potenzial zur Erwirtschaftung zukünftiger Einzahlungsüberschüsse und damit den Wert des Unternehmens. Während sich diese Potenziale im Buchwert nicht widerspiegeln, werden sie an der Börse – trotz des Einflusses spekulativer und marktgegebener Veränderungen des

Aktienkurses – langfristig durch steigende Kurse und damit einem höheren Marktwert des Unternehmens ausgedrückt. Die Bewertung von Unternehmen erfolgt aber immer noch auf Basis des bilanziellen Anlage- und Umlaufvermögens. Für viele Unternehmen scheint ein solches Vorgehen heute zu keinen sinnvollen Ergebnissen mehr zu führen, denn der Marktwert börsennotierter Unternehmen übersteigt deren Bilanzwert häufig um ein Vielfaches. Neue Ansätze der Unternehmensbewertung, aber auch -steuerung scheinen erforderlich zu sein.

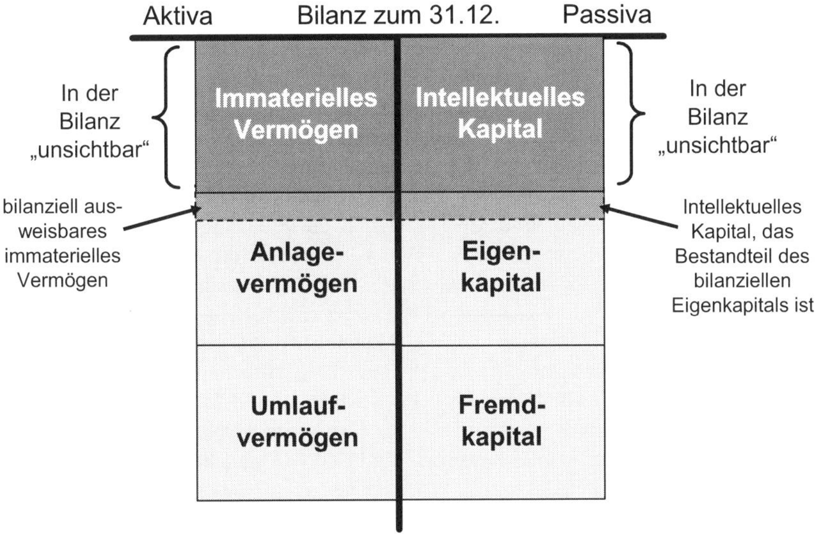

Abb. 2: Immaterielles Vermögen und intellektuelles Kapital bleiben in der Bilanz verborgen

2. Betriebswirtschaftliche Besonderheiten

Intangibles weisen eine Reihe betriebswirtschaftlicher Besonderheiten auf. Inwieweit diese in der heutigen Controlling- und Führungspraxis Berücksichtigung finden, soll im Folgenden näher betrachtet werden.

2.1 Kostenstruktur und Wertschöpfung

Während der Aufbau intellektuellen Kapitals sehr hohe Investitionen erfordert, verursacht dessen Nutzung (Verwendung eines Markennamens, Gebrauch und Vervielfältigung von Software, Verwendung des Wissens eines Mitarbeiters) dagegen kaum Kosten. Durch die Nicht-Rivalität der Nutzung, d.h. der gleichzeitigen und mehrfachen Verwendbarkeit für verschiedene Zwecke ohne resultierende Wertminderung, entstehen auch keine Opportunitätskosten. Die Skalierbarkeit der in den Produkten und Dienstleistungen enthaltenen Intangibles wird nur durch das Marktvolumen beschränkt. Daraus resultiert ein hoher Anteil an Fixkosten (i.d.R. sunk costs) bei geringen proportionalen Kosten und gleich bleibende bzw. steigende Skalenerträge. Bei immateriellen Produkten wie beispielsweise einer Software bedeutet dies, dass jedes zusätzlich verkaufte Produkt annähernd in Höhe des Verkaufserlöses zur Steigerung des Gewinns beiträgt. Aus diesem Grund steht für die Amortisation der produktbezogenen immateriellen Ressourcen die schnelle Erreichung eines hohen Marktanteils im Vordergrund. Hier spielen häufig auch Netzwerkeffekte bei den Kunden (Etablierung von Standards) eine wichtige Rolle (vgl. hierzu ausführlich *Stoi*, 2002). Das Controlling muss in der Lage sein, diese veränderte Kosten- und Erlösstruktur abzubilden und dadurch für die Unternehmensführung gestaltbar zu machen.

Controllingverfahren und Führungstechniken, die auf die effiziente Nutzung der materiellen Vermögenswerte und die optimale Auslastung der vorhandenen Kapazitäten abzielen, sind für Intangibles wenig geeignet. Während materielle Produktionsfaktoren – (manuelle) Arbeit, Betriebsmittel und Werkstoffe – bei der Herstellung von Gütern und Dienstleistungen verbraucht werden, ist dies bei Intangibles nicht der Fall. Ihr Wert wird durch deren Nutzung nicht gemindert, sondern teilweise sogar erhöht. Beispielsweise steigt das Wissen eines Mitarbeiters, dass im Rahmen eines Projektes genutzt wird, durch die im Projekt gemachten Erfahrungen. Andere Beispiele sind der Bekanntheitsgrad eines Produktes oder der Markenwert eines Unternehmens. Bei der Schaffung materieller Vermögenswerte (z.B. der Herstellung eines physischen Produktes) lässt sich die Produktionsmenge und damit der Wert durch Steigerung der Anzahl der Produktionsfaktoren (in diesem Fall: der eingesetzten Rohstoffe, Maschinen, Arbeiter) erhöhen. Zur Schaffung immaterieller Vermögenswerte (z.B. der internen Prozessexzellenz durch Einführung einer ERP-Software) ist jedoch nicht die Anzahl der Produktionsfaktoren oder die Höhe des Investments entscheidend. Vielmehr spielt die Qualität und Adäquanz der investierten (materiellen und immateriellen) Vermögenswerte eine zentrale Rolle. Beispielsweise ist das Ausmaß der Erhöhung der internen Prozessexzellenz durch Investitionen in eine neue Software davon abhängig, ob sich diese für das Unternehmen eignet und in welcher Form deren Implementierung organisiert ist. Eine nicht adäquate Software kann trotz eines hohen Investments zur Minderung der internen Prozessexzellenz und damit zur Senkung des intellektuellen Kapitals führen. Der Unternehmensführung und dem Controlling fehlen weitgehend geeignete Methoden, um Aufbau, Einsatz und Nutzung des intellektuellen Kapitals beurteilen zu können.

2.2 Verfügungsrechte der Unternehmen an Intangibles

Wie erwähnt ist das Ausmaß an Verfügungsrechten an den Komponenten des intellektuellen Kapitals unterschiedlich stark ausgeprägt. Während sich z.B. Marken oder Patente noch relativ gut schützen und auch veräußern lassen, kann ein Unternehmen kein Eigentum an immateriellen Vermögenswerten aus externen Quellen wie Mitarbeitern, Kunden oder Partnerunternehmen erwerben. Für die Unternehmensführung ist deshalb nur eine indirekte Einflussnahme z.B. durch die Gestaltung der Lieferantenbeziehungen oder die Erhöhung der Kundenzufriedenheit mit Hilfe von Qualitätsmaßnahmen möglich. Das Controlling hat hier zunächst die Aufgabe der strukturierten Erfassung des externen immateriellen Vermögens. Im Anschluss übernimmt es die Informationsversorgung des Managements über Ausmaß und Entwicklung der externen Intangibles als Ausgangspunkt für deren Nutzung und Beeinflussung. Bislang ist das Controlling jedoch auf die internen Quellen (Anlagen, Strukturen und Abläufe) fixiert.

Das im Gegensatz zu materiellen Vermögenswerten geringe Ausmaß an Handlungs- und Verfügungsrechten hat ebenfalls zur Folge, dass Intangibles (mit wenigen Ausnahmen) nicht frei handelbar sind und andere Parteien nicht vollständig von deren Nutzung ausgeschlossen werden können (partielle Exkludierbarkeit). In den 80er Jahren waren beispielsweise die japanischen Automobilhersteller dafür berüchtigt, das Design europäischer Fahrzeuge nachzuahmen. Ein anderes Beispiel ist der Wechsel eines erfahrenen Mitarbeiters zu einem Konkurrenten. Diese Spillover-Effekte begrenzen oder gefährden den potenziellen Nutzen einer Investition in Intangibles und erhöhen dadurch deren Risiko. Das Fehlen eines freien Marktes für Intangibles erschwert zudem deren objektive Bewertung.

2.3 Immaterielle Investitionen und Risikomanagement

Investitionen in (aktivierungsfähige!) materielle Vermögenswerte werden in den Unternehmen einer eingehenden Beurteilung unterzogen, wofür dem Controller seit langem eine Reihe an Instrumenten und Verfahren zur Verfügung stehen. Die Bildung immaterieller Vermögenswerte, wie beispielsweise der Aufbau einer Marke oder die Ausbildung eines Mitarbeiters, werden jedoch bilanziell nicht als Investitionen angesehen. Sie werden stattdessen als Aufwand verbucht und entziehen sich dadurch meist einer investitionsrechnerischen Beurteilung. Diese erfordert sowohl die Bewertung des immateriellen Vermögensaufbaus als auch die Prognose der dadurch erzielbaren zukünftigen Einzahlungsüberschüsse. Der Zusammenhang zwischen Investitionen in das intellektuelle Kapital und dem Ergebnis/Unternehmenswert ist jedoch häufig unklar und auch die Wirksamkeit und Wirkungsrichtung einer solchen Investition ist oft nicht eindeutig zu bestimmen.

Zwei Arten von Zeitverzögerungen (time lags) sind zu beachten:

- Investment effectiveness lag: Zeitdauer zwischen der Investition und deren Auswirkung auf den Stand des intellektuellen Kapitals.
- Intangible effectiveness lag: Zeitdauer zwischen der Steigerung des intellektuellen Kapitals und deren Wirkung auf das Ergebnis bzw. den Unternehmenswert.

Traditionelle Investitionsbeurteilungsverfahren führen durch Vernachlässigung des zeitlichen Spielraums und der Alternativen der Investitionsentscheidung in vielen Fällen zur Ablehnung sinnvoller Investitionen. Zur Berücksichtigung der Handlungsflexibilität und des phasenweisen Investitionsprozesses bietet sich der Realoptionsansatz an. Das Ergebnis wäre eine Objektivierung der Investitionsentscheidung und eine Sensibilisierung des Managements für die Bedeutung von Intangibles. Konkrete Anwendungsbeispiele hierfür sind jedoch noch selten. Das Hauptproblem des praktischen Einsatzes ist die Komplexität des Verfahrens und die Schwierigkeit der Datenbeschaffung insbesondere in Bezug auf die Prognose der Rückflüsse und deren Eintrittswahrscheinlichkeiten.

Generell ist das Risiko von Investitionen in Intangibles wesentlich größer als bei materiellen Vermögenswerten, da Intangibles größere und häufigere Wertschwankungen aufweisen. Während der Wert einer Maschine im Rahmen der Nutzungsdauer abnimmt, kann beispielsweise das Wissen eines Programmierers sehr schnell obsolet und damit wertlos werden, wenn sich auf dem Markt eine andere Programmiersprache durchsetzt. Investitionen in Intangibles werden zudem vor allem am Beginn des Innovations- bzw. Produktentwicklungsprozesses durchgeführt. Da in dieser Phase die Unsicherheit über den Erfolg der Investition am höchsten ist, ist auch die Gefahr eines totalen Verlustes wesentlich größer als bei Investitionen in materielle Vermögenswerte, die verstärkt am Ende der Produktentwicklung getätigt werden. Auf der anderen Seite nimmt der Wert des intellektuellen Kapitals aber nicht zwangsläufig durch dessen Nutzung oder im Zeitablauf ab. Der Controller muss dies bei der Investitionsbeurteilung und der Ermittlung der (kalkulatorischen) Abschreibungen berücksichtigen und ein entsprechendes Risikomanagement aufbauen.

2.4 Immaterielle Erlöse

Neben finanziellen Erlösen erzielt ein Unternehmen auch immaterielle Erlöse. Dies kann beispielsweise die Verbesserung des Firmenimage durch Referenzkunden, die Erhöhung der Innovationskraft durch Projekte mit besonders innovativen, kritischen und anspruchsvollen Kunden oder die Steigerung der Kompetenz der Mitarbeiter durch neuartige oder innovative Kundenprojekte sein. Diese immateriellen Erlöse (bzw. im negativen Falle immaterielle Kosten) werden bislang weder qualitativ noch quantitativ durch das Controlling erfasst, führen aber zu einer Änderung des intellektuellen Kapitals. Die fehlende Erfassung hat jedoch zur Konsequenz, dass die Unternehmensführung nicht die Bedeutung immaterieller Kosten und Erlöse erkennen kann. In Folge werden diese weder systematisch geplant noch gesteuert. Die ausschließliche Berücksichtigung monetärer Erlöse ist für eine langfristige und wertorientierte Unternehmensführung aber nicht ausreichend. Monetäre Erlöse sind das Ergebnis der Nutzung von intellektuellem Kapital, das in der Vergangenheit aufgebaut wurde (F&E, Geschäftsprozesse, Markennamen, Mitarbeiterkompetenzen, etc.). Immaterielle Erlöse stellen dagegen die Ertragskraft des Unternehmens und damit die zukünftigen monetären Erlöse sicher. Verfügt das Unternehmen nicht über ausreichende immaterielle Erlöse, kann trotz einer befriedigenden finanziellen Erlöslage der zukünftige Bestand des Unternehmens gefährdet sein.

Kunde	**Umsatz**	Beitrag zum **Finanzkapital** (Gewinn übersteigt die Kapitalkosten)	Beitrag zur **externen Struktur** (Referenzkunden, Kernkompetenzen)	Beitrag zur **internen Struktur** (Wiederverwendbarkeit von Lösungen, Großprojekte)	Beitrag zur **Mitarbeiterkompetenz** (Best-Practice, innovative Projekte)
A	€ 100.000	€ 100.000			
B	€ 90.000		€ 90.000		
C	€ 70.000			€ 70.000	€ 70.000
D	€ 60.000				
E	€ 50.000	€ 50.000			
F	€ 30.000	€ 30.000	€ 30.000	€ 30.000	
Rest	€ 100.000				
Summe	**€ 500.000**	**€ 180.000**	**€ 120.000**	**€ 100.000**	**€ 70.000**
Anteil am Gesamtumsatz*		**36%**	**24%**	**20%**	**14%**

* Die Spaltensummen entsprechen nicht 100 Prozent

Abb. 3: Beispiel zur Einschätzung immaterieller Erlöse (vgl. Sveiby, 2002a)

Einen Ansatz für die Einschätzung immaterieller Erlöse liefert *Sveiby* (2002a). Dabei werden, wie in *Abb. 3* dargestellt, die Kunden zunächst absteigend nach den erzielten Umsätzen sortiert. Anschließend wird für jeden Kunden untersucht, ob er zur Steigerung des Finanzkapitals, der externen bzw. internen Struktur oder der Kompetenz der Mitarbeiter (*Sveiby* unterteilt das intellektuelle Kapital in diese drei Kategorien, vgl. hierzu *Sveiby*, 1998, sowie Punkt 3.2) beiträgt. In jede Spalte, zu der der Kunde einen Beitrag leistet, wird der mit diesem Kunden erzielte Umsatz vermerkt. Die langfristige Zielsetzung dabei ist, neben der absoluten Umsatzhöhe, in allen Spalten möglichst eine prozentuale Gleichverteilung zwischen den materiellen und immateriellen Vermögenswerten zu erreichen. Im Beispiel sind zwar die Kunden profitabel, aber die Mitarbeiter scheinen bei ihren Kundenprojekten zu wenig zu lernen, wodurch langfristig die Innovationskompetenz des Unternehmens gefährdet ist. Die Unternehmensführung sollte sich deshalb in diesem Fall um innovative Kundenprojekte bemühen, die zu einem Ausbau der Mitarbeiterkompetenzen führen. In den Kategorien, zu denen nur wenige Kunden einen Beitrag leisten, bestehen dagegen tendenziell Abhängigkeiten, die langfristig vermieden werden sollten. Bei Kunden, die zu keiner Kategorie beitragen, ist über eine Aufgabe der Geschäftsbeziehung bzw. eine Änderung der Konditionen nachzudenken. Leistet ein Kunde keinen Beitrag zum Finanzkapital, so kann es trotzdem von Vorteil sein diesen weiter zu halten, wenn er das intellektuelle Kapital vermehrt. Gerade diese Sichtweise wird bisher im Controlling häufig vernachlässigt, dessen Schwerpunkt immer noch auf dem (kurzfristigen) monetären Ergebnis liegt.

3. Bewertungsansätze

Bereits frühzeitig hat man sich im Rahmen der Auseinandersetzung mit dem Thema Intangibles (getreu dem Motto: „Only what gets measured, gets done") auch mit deren Messung und Bewertung auseinander gesetzt. Mit einem expliziten Ausweis originären intellektuellen Kapitals in der externen Rechnungslegung ist allerdings in naher Zukunft kaum zu rechnen. Bisherige Bestrebungen zielen vor allem auf qualitative Informationen und weniger auf einen monetären Ausweis. Das SEC hat beispielsweise die Empfehlung ausgegeben, durch einen zusätzlich zum Jahresabschluss zu erstellenden „Supplement report" qualitativ über die Intangibles des Unternehmens und deren Entwicklung zu berichten. In Dänemark wurde bereits eine Rechnungslegungsvorschrift verabschiedet, die zusätzlich zum Jahresabschluss die Erstellung eines „Intellectual Capital Statement" mit ähnlicher Zielsetzung vorschreibt.

3.1 Arten

Die Unternehmen, die sich bereits mit der Thematik der Intangibles auseinander setzen, sind sich jedoch einig, dass für deren Nutzung und Gestaltung eine systematische Erfassung notwendig ist. Tatsächlich existiert bereits eine ganze Bandbreite an Bewertungsverfahren, die wie folgt (vgl. *Sveiby*, 2002b) unterteilt werden kann:

- **Direkte Bewertungsverfahren:** Intangible-Komponenten werden identifiziert und einzeln monetär bewertet, um in Summe den Wert des intellektuellen Kapitals zu erhalten (z.B. Intellectual Asset Valuation Method nach *Sullivan*). Dieser Bewertungsansatz stößt schnell an die Grenzen seiner Praktikabilität, da eine objektive Wertermittlung einzelner immaterieller Vermögensteile wie z.B. einer Kundenbeziehung oder der Managementqualität kaum durchführbar ist. Zur Ermittlung der Summe des intellektuellen Kapitals müssten zumindest die bedeutendsten immateriellen Komponenten identifiziert, bewertet und schließlich addiert werden. Darüber hinaus leisten die direkten Bewertungsverfahren keinen Beitrag zum Management der immateriellen Vermögenswerte.
- **Marktkapitalisierungsverfahren:** Die Differenz zwischen Marktkapitalisierung und Buchwert des Unternehmens wird als Wert des intellektuellen Kapitals interpretiert (z.B. Marktwert-Buchwert-Relationen, Tobin's q). Bei den Marktkapitalisierungsverfahren wird – neben der Kritik an der Annahme, dass der Wert der Intangibles der Differenz zwischen Markt- und Buchwert entspricht – deren unzureichende Steuerungsunterstützung bemängelt, da sie keine Analyse der Ursachen und Zusammenhänge der immateriellen Wertschöpfung beinhalten.
- **ROA-basierte Bewertungsverfahren:** Der Return on Asset des materiellen Vermögens eines Unternehmens wird mit dessen Branchendurchschnitt verglichen. Die Differenz

wird mit dem durchschnittlichen materiellen Vermögen multipliziert, um den Ertrag zu erhalten, der auf Intangibles zurückzuführen ist. Durch Division mit den durchschnittlichen Kapitalkosten erhält man eine Schätzung des Wertes der Intangibles (z.B. Calculated Intangible Value nach *Stewart*). Die ROA-basierten Bewertungsverfahren liefern aufgrund der enthaltenen Annahmen nur vage und teilweise schwer nachvollziehbare Ergebnisse. Sie sind wie die direkten Bewertungsverfahren auch eher ex post rechtfertigend und nicht ex ante potentialorientiert. Die ermittelte Kennzahl repräsentiert bestenfalls die immaterielle Wertschöpfung der Vergangenheit, lässt aber keine Rückschlüsse auf die hierfür verantwortlichen immateriellen Werttreiber zu.

- **Scorecard-Verfahren:** Intangible-Komponenten werden identifiziert, klassifiziert und anschließend mit Hilfe mehrerer Indikatoren gemessen und dargestellt. Die Scorecard-Verfahren erweiterten anfänglich das betriebliche Berichtswesen auf Basis klassischer Finanzkennzahlen um nicht-finanzielle Kenngrößen (z.B. Balanced Scorecard, Tableau de Bord, Quantum Performance Measurement, etc.). Die Verfahren waren jedoch anfangs statisch und enthielten keine Aussagen über die kausalen Beziehungen zwischen den einbezogenen Kenngrößen. Deshalb greifen neuere Ansätze auch die Ursache-Wirkungsbeziehungen zwischen den Dimensionen explizit auf, stellen diese dar und machen sie damit steuerbar (z.B. Strategy Map, Success Map und IC-Navigator). Problematisch dabei ist jedoch, dass die Kausalbeziehungen auf subjektiven Einschätzungen beruhen und nicht empirisch validiert sind. Obwohl die Verfahren finanzielle Kenngrößen beinhalten, eignen sie sich nicht zur Ermittlung eines konkreten monetären Wertes des intellektuellen Kapitals. Sie bestechen vielmehr durch ihre Praktikabilität und fokussieren auf das Management der Intangibles. Deshalb sollen die Scorecard-Verfahren nachfolgend eingehender dargestellt werden.

3.2 Intangibles durch Scorecards steuerbar machen

Das bekannteste Scorecard-Verfahren aus der Praxis ist der in *Abb. 4* dargestellte *Skandia*-Navigator des schwedischen Versicherungsunternehmens *Skandia* – einem der Vorreiter bei der Berücksichtigung der Intangibles in der Unternehmenssteuerung. Die Betrachtung erfolgt aus fünf Perspektiven, die jeweils durch eine Reihe von Indikatoren dargestellt werden. Der Finanzfokus enthält die finanziellen Ergebnisse des Unternehmens und verkörpert die Vergangenheit, da diese das Resultat vergangener Handlungen und Vorgänge sind. Kunden-, Mitarbeiter- und Prozessfokus repräsentieren die gegenwärtigen Aktivitäten des Unternehmens, während sich der Erneuerungs- und Entwicklungsfokus auf strategische Maßnahmen im Hinblick auf das zukünftige Geschäftsfeld des Unternehmens bezieht. Der Prozessfokus bildet zusammen mit dem Erneuerungs- und Entwicklungsfokus das Strukturkapital des Unternehmens ab. Im Mittelpunkt des Navigators steht – sozusagen als Herzstück – der Mitarbeiterfokus, der als aktive Kraft die anderen Dimensionen verbinden soll. Kunden-, Prozess-, Mitarbeiter- sowie Erneuerungs- und Entwicklungsfokus bilden das intellektuelle Kapital ab und sollen es dadurch steuerbar machen.

Ein weiteres Scorecard-Verfahren, das bereits in mehreren skandinavischen Unternehmen (z.B. *Celemi*, *WM-data*, *Ericsson*) angewendet wird, ist der Intangible Asset-Monitor (*Sveiby*, 1997). Das intellektuelle Kapital eines Unternehmens wird dabei unterteilt in die externe Struktur (Beziehungen zu Kunden und Lieferanten sowie Markennamen, Warenzeichen und Image des Unternehmens), die interne Struktur (Patente, Konzepte, Modelle, Computer- und Verwaltungssysteme sowie die Unternehmenskultur) und die Kompetenz der Mitarbeiter. Die drei Kategorien des intellektuellen Kapitals werden mit Hilfe unternehmensspezifischer

Indikatoren nach ihrem Beitrag zu Wachstum/Erneuerung, Effizienz sowie Stabilität des Unternehmens beurteilt. Die Effizienz der Mitarbeiter kann beispielsweise u.a. durch die Wertschöpfung pro Mitarbeiter abgebildet werden und über die Stabilität der externen Struktur gibt z.B. der Anteil an Wiederholungsaufträgen Auskunft. Bei der Auswahl der Indikatoren tritt jedoch häufig das Problem der Abgrenzung zwischen den verwendeten Kategorien und Gesichtspunkten auf. In *Abb. 5* ist beispielhaft der Intangible Asset-Monitor der Firma *Celemi* für das Geschäftsjahr 2000 dargestellt. Zur Veranschaulichung der Zielerreichung werden die Kategorien sowie die einzelnen Indikatoren in den Ampelfarben markiert. Bei grünen Indikatoren (in *Abb. 5* hellgrau) liegt das Ergebnis auf oder über dem Zielwert, gelbe Indikatoren (in *Abb. 5* weiß) zeigen eine leichte und rote (in *Abb. 5* schwarz) eine kritische Abweichung vom Zielwert. Zur Definition der einzelnen Indikatoren sei auf den Geschäftsbericht der Firma *Celemi* (*Celemiab Group AB,* 2001, S. 26 f.) verwiesen.

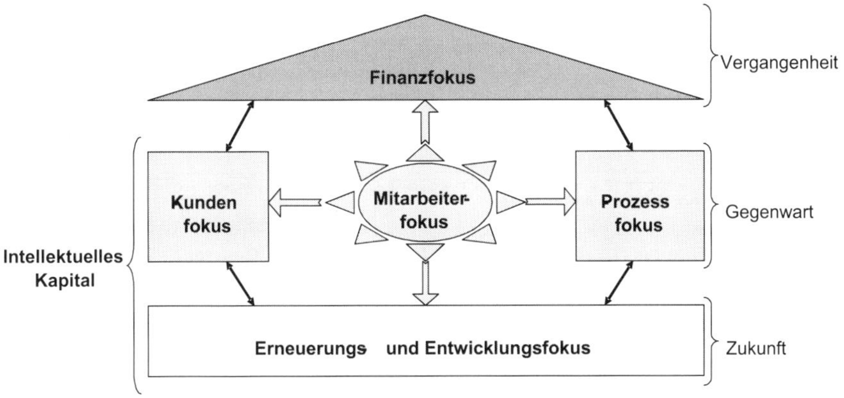

Abb. 4: Skandia-Navigator (Edvinsson/Brünig, 2000, S. 58)

Trotz der bestehenden Kritik insbesondere an den verwendeten Indikatoren zur Beurteilung der Komponenten des intellektuellen Kapitals (vgl. z.B. *North* et al., 1998) stellen die Scorecard-Ansätze für eine wertorientierte Gestaltung und Nutzung der Intangibles bislang das praktikabelste Vorgehen dar. Ähnlich wie bei der Balanced Scorecard nach *Kaplan/Norton* sollte die Auswahl der verwendeten Kategorien und Indikatoren dabei in Abhängigkeit des jeweiligen Unternehmens und seiner Vision und Strategie erfolgen.

Die Ermittlung eines monetären Wertes für das intellektuelle Kapital scheint generell wenig sinnvoll. Dieser wäre sowieso nur bei Veräußerung des Unternehmens bzw. einzelner Intangibles von Relevanz – und dann ist er weniger eine Frage der Bewertung sondern vielmehr des Verhandlungsgeschicks und der Anzahl der Interessenten. Erinnert sei an dieser Stelle an das Gerangel im Herbst 1998 zwischen *VW* und *BMW* um die Marke *Rolls-Royce*, bei der *VW* zwar für 780 Mio. US-Dollar die marode *RR*-Produktionsstätte von *Vickers* kaufte, sich aber *BMW* vom Triebwerkshersteller *Rolls-Royce* für „nur" 66 Mio. US$ die Namensrechte sichern konnte.

Für Unternehmen ist es im laufenden Geschäft entscheidend, in welcher Form das intellektuelle Kapital zur Steigerung des Unternehmenswertes beiträgt und nicht welchen Wert es vermutlich hat!

TANGIBLE ASSETS

Financial Capital	2000	1999	1998
Overall rating year	●	●	●
· Growth/Renewal			
Equity Growth	67% ◉	-3% ●	-3% ●
Net Investment Ratio	8% ●	19% ◉	35% ◉
· Efficiency			
Profit Margin	2% ●	1% ●	0% ●
Net Return on Equity	12% ○	8% ●	1% ●
Profit Capacity % Sales	7% ●	8% ●	12% ○
· Stability			
Solidity	32% ◉	20% ●	29% ○
Liquid Reserves No. Days	19 ●	32 ○	11 ●

INTANGIBLE ASSETS

Customers (External Structure)	2000	1999	1998
Overall rating year	◉	◉	◉
· Growth/Renewal			
Revenue Growth	9% ●	22% ●	8% ●
Image Enhancing Customers	41% ○	54% ◉	59% ◉
· Efficiency			
Revenues per Customer (TSEK)	355 ◉	367 ◉	306 ◉
· Stability			
Customer Satisfaction Index	5 ◉	5 ◉	5,2 ◉
Repeat Orders	78% ◉	68% ◉	66% ◉
5 Largest Customers	39% ○	29% ◉	33% ○

Organization (Internal Structure)	2000	1999	1998
Overall rating year	○	○	○
· Growth/Renewal			
Organization Enhancing Customers	8% ●	21% ◉	51% ◉
Revenues from New Products	26% ○	17% ●	49% ◉
R&D / Revenues	5% ●	14% ◉	12% ◉
Intangible Investments % Value Added	14% ◉	22% ◉	42% ◉
· Efficiency			
Proportion of Admin Staff	13% ◉	20% ◉	25% ●
Revenues per Admin Staff (TSEK)	12727 ◉	9205 ○	6774 ●
· Stability			
Admin Staff Turnover	17% ●	33% ●	13% ○
Admin Staff Seniority (Years)	4,2 ○	3,8 ●	2,6 ●
Rookie Ratio	41% ○	36% ○	41% ○

People (Competence)	2000	1999	1998
Overall rating year	○	◉	○
· Growth/Renewal			
Average Professional Experience (Years)	10,1 ◉	9,2 ◉	8,3 ◉
Competence Enhancing Customers	44% ○	27% ●	59% ◉
Growth in Professional Competence	18% ●	38% ●	8% ●
Experts with Tertiary Degree	75% ◉	80% ◉	67% ○
· Efficiency			
Value Added per Expert (TSEK)	817 ◉	892 ◉	802 ◉
Value Added Margin on Sales	48% ◉	49% ◉	47% ◉
· Stability			
People Satisfaction Index	48% ●	5 ◉	4,6 ○
Expert Turnover	16% ●	14% ●	13% ○
Expert Seniority (Years)	4 ○	4 ○	3,3 ●
Median Age All Employees (Years)	39 ◉	37 ◉	37 ◉

The cells in the Monitor are colored green (◉) if the indicator is equal to or greater than Celemi's strategic plan target. Red cells (●) indicate values less than 0.8% of target. Yellow cells (○) indicate values in between.

Abb. 5: Intangible Asset-Monitor 2000 der Firma Celemi (Celemiab Group AB)

4. Wertgenerierung durch Intangibles

Bei ihren Bestrebungen zur Erhöhung des Unternehmenswertes haben sich Unternehmensführung und Controlling bislang zu sehr auf die materiellen Vermögenswerte konzentriert. Die immateriellen Vermögenswerte als eigentliche Quelle des Unternehmenswerts wurden dagegen weder strukturiert erfasst noch systematisch gesteuert. Bestenfalls findet man Lippenbekenntnisse wie beispielsweise „Unsere Mitarbeiter sind unser wichtigster Vermögenswert“ im Unternehmensleitbild. Über die Wirkungsweise zwischen Investitionen und Intangibles und zwischen Intangibles und dem Unternehmenswert herrscht in den meisten Unternehmen jedoch völlige Unkenntnis.

In Zukunft werden aber nur die Unternehmen langfristig Erfolg haben, die sich der Bedeutung ihrer immateriellen Ressourcen in der Wertschöpfungskette im Hinblick auf die Gestaltung des Unternehmenswertes bewusst sind und diese als Wettbewerbsvorteil nutzen. Während im einen Fall der Markenwert eine zentrale Stellung einnimmt (z.B. wird der Markenwert 2002 von *Coca-Cola* lt. *Interbrand* auf 70 Mrd. US$ und von *Microsoft* auf 64 Mrd. US$ geschätzt), kann bei einem anderen Unternehmen das Wissen der Mitarbeiter (z.B. *SAP* oder *McKinsey*) oder andere Intangible-Komponenten (Netzwerkeffekte wie z.B. bei *Netscape* oder das Organisationskapital wie z.B. bei *Cisco*) ausschlaggebend sein. Sind diese zentralen immateriellen Werttreiber identifiziert, so kann sich das Management auf deren Gestaltung, Steuerung und Ausbau konzentrieren. Die Aufgabe der Bestimmung der wesentlichen immateriellen Werttreiber (Intangible Value Driver) und deren Beziehungen und Wirkungsrichtungen im Hinblick auf den Unternehmenswert wird zu einer herausragenden Aufgabe des Controlling. Denkbar wäre beispielsweise der Aufbau einer Intangible Value

Map, in der die Intangibles des Unternehmens mit ihren Beziehungen sowohl untereinander als auch zum Finanzkapital dargestellt werden. Eine solche „immaterielle Landkarte" wäre nicht nur bei der Formulierung und Überarbeitung der Strategie, sondern auch bei deren operativer Umsetzung von hohem Nutzen für die Unternehmensführung.

Ein praktisches Beispiel für die Identifikation der immateriellen Werttreiber ist der Value Creation Index (VCI) des *Cap Gemini Ernst & Young Center for Business Innovation (vgl. Kalafut/Low*, 2001). Ausgangspunkt war eine vorhergehende Studie aus dem Jahr 1996, bei der eine positive Korrelation zwischen der nicht-finanziellen Unternehmensleistung (Marktpositionierung, Prozessqualität, Innovationsgrad etc.) und der Bewertung durch institutionelle Anleger festgestellt wurde. Auf Basis der Ergebnisse dieser Untersuchung wurden neun maßgebliche Kategorien nicht-finanzieller Leistungsgrößen ermittelt, die den Wert eines Unternehmens bestimmen. Bei diesen immateriellen Werttreiberkategorien handelt es sich um Innovation, Qualität, Kundenbeziehungen, Managementfähigkeiten, Allianzen, Technologie, Markenwert, Mitarbeiter und Unternehmensumfeld. Im Folgenden wurden diese Kategorien für die S&P 500-Unternehmen mit Hilfe mehrerer Indikatoren beurteilt und – gewichtet mit ihrem statistischen Erklärungsgehalt für den Marktwert der Unternehmen – zu einer einzigen Maßgröße für die immaterielle Wertgenerierung – den Value Creation-Index – zusammengefasst. Den Zusammenhang verdeutlicht *Abb. 6.* Als Indikatoren wurden ausschließlich öffentlich zugängliche Unternehmens- und Marktdaten verwendet und bewusst auf subjektive Einschätzungen von Managern und Investoren verzichtet. Dennoch bleibt anzumerken, dass die Auswahl der Indikatoren sowie deren Eignung zur Beurteilung der Werttreiber nicht frei von der subjektiven Einschätzung der Forscher sein können. Der VCI weist eine hohe Korrelation (0,7) zum Marktwert der Unternehmen auf und ermöglicht es, diesen zu 50–90% zu erklären.

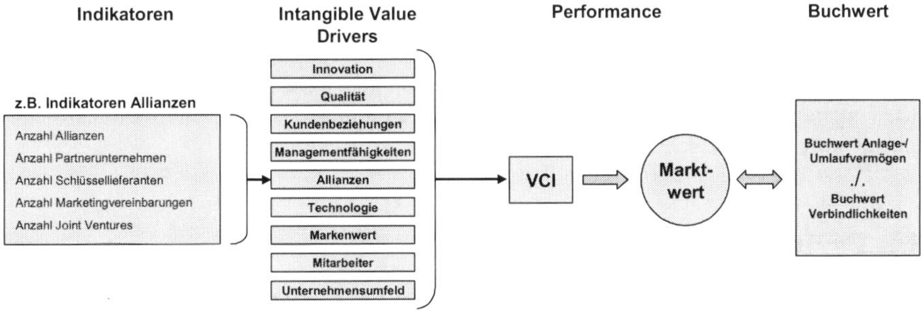

Abb. 6: Value Creation Index-Modell (vgl. Kalafut/Low, 2001)

Im Rahmen der Untersuchung wurde zwischen Konsumgüterherstellern, Gebrauchsgüterherstellern und E-Commerce-Unternehmen unterschieden. Die wichtigsten immateriellen Werttreiber bei Konsumgüterherstellern sind demnach Innovation, Managementfähigkeiten, Mitarbeiter und Qualität, während bei den Gebrauchsgüterherstellern Innovation, Mitarbeiter, Managementfähigkeiten und Allianzen über den Unternehmenswert entscheiden. Bei E-Commerce-Unternehmen haben Allianzen, Innovation, User und Markenwert die höchste Bedeutung. *Abb. 7* zeigt die Konsum- und Gebrauchsgüterhersteller mit dem höchsten immateriellen Wertsteigerungspotenzial auf Basis des VCI-Modells.

US-Top 10 Konsumgüter	US-Top 10 Gebrauchsgüter
1. Procter & Gamble Company	1. Intel Corporation
2. Merck & Co. Inc.	2. Hewlett-Packard Company
3. Coca-Cola Company	3. Sun Microsystems Inc.
4. Johnson & Johnson	4. Lucent Technologies Inc.
5. Pfizer Inc.	5. Cisco Systems Inc.
6. Bristol-Myers Squibb Company	6. EMC Corporation
7. Minnes. Mining & Mfct. Comp.	7. Compaq Computer Corporation
8. DuPont Company	8. Motorola Inc.
9. Time Warner Inc.	9. Ford Motor Company
10. Eli Lilly and Company	10. Corning Inc.

Abb. 7: US-Unternehmen mit dem höchsten intellektuellen Kapital (vgl. Kalafut/Low, 2001)

Da detaillierte Informationen zu Forschungsdesign und -ablauf der empirischen Untersuchung sowie zur Berechnung des VCI nicht publiziert sind, ist eine abschließende Beurteilung der Qualität der Untersuchung an dieser Stelle nicht möglich. Kritisch anzumerken ist die verwendete Ermittlung des immateriellen Wertes durch das Marktkapitalisierungsmodell, das von der einfachen Annahme ausgeht, dass der Wert der Intangibles der Differenz zwischen Markt- und Buchwert entspricht.

Das Modell des Value Creation-Index geht jedoch in die richtige Richtung. Es liefert den Unternehmen auf Basis statistischer Zusammenhänge Hinweise für die Bestimmung der wesentlichen immateriellen Werttreiber und deren Einfluss auf den Unternehmenswert. Der Controller erhält damit wichtige Informationen zur Ermittlung und Einschätzung der unternehmensspezifischen Werttreiber. Zur Ermittlung für das eigene Unternehmen verfügt er darüber hinaus über wesentlich detailliertere und umfangreichere Informationen als die Verfasser dieser Studie. Interessant wäre an dieser Stelle die Bestimmung des VCI für deutsche Unternehmen, da die Ergebnisse aus den USA aufgrund kultureller und wirtschaftlicher Unterschiede nur eingeschränkt übertragbar sind.

5. Fazit

Bei der Planung, Steuerung und Kontrolle von Intangibles steht für die Unternehmen, die zunehmend von immateriellen Vermögenswerten abhängen, nicht die Frage im Vordergrund, wie hoch der monetäre Wert des intellektuellen Kapitals einer Organisation ist. Eine objektive Ermittlung ist kaum möglich und deren Nutzen unter dem Gesichtspunkt der Unternehmensführung ohnehin fraglich. Viel wichtiger ist doch die zielgerichtete Gestaltung des intellektuellen Kapitals zur Steigerung des Unternehmenswertes und damit im Endeffekt die Transformation von Intangibles in monetäre Resultate! Der Schwerpunkt des Controllings sollte deshalb auf der Steuerbarkeit und (qualitativen) Messung und weniger auf der monetären Bewertung des intellektuellen Kapitals liegen. Das Controlling ist bisher auf materielle Ressourcen, physische Produktion und monetäre Ergebnisse fokussiert. Es wird sich dieser Herausforderung stellen und seinen Fokus auf das intellektuelle Kapital mit all seinen Besonderheiten legen müssen. Nur dann kann es seiner Aufgabe der Unterstützung der Unternehmensführung bei der zielorientierten Planung und Steuerung auch in Zukunft nachkommen. Im Vordergrund steht das Verständnis für dessen Natur und seine Auswirkung auf den Unternehmenswert. Nur wenn die Unternehmensführung weiß, wie die Intangibles zusammenwirken und welche Rolle ihnen für die Steigerung des Unternehmenswerts zukommt, wird Sie in der Lage sein die unternehmerische Herausforderung des 21. Jahrhunderts zu meistern.

6. Literatur

Celemiab Group AB (Hrsg.), Annual Report 2000, Malmö 2001.

Daum, J., Intangible Assets, Bonn 2002.

Edvinsson, L. und *Brünig, G.*, Aktivposten Wissenskapital, Wiesbaden 2000.

Lev, B., Intangibles. Management, Measurement, and Reporting, Washington 2001.

Kalafut, P. und *Low, J.*, The value creation index: quantifying intangible value, in: Strategy & Leadership, 29. Jg. (2001), H. 5, S. 9–15.

Stewart, T., Intellectual capital, New York 1997.

Stoi, R., New Economy Controlling, in: *Gleich, R.*, *Möller, K.*, *Seidenschwarz, W.* und *Stoi, R.* (Hrsg.), Controllingfortschritte, München 2002, S. 149–170.

Stoi, R.: Controlling von Intangibles, in: ControllerNews, o. Jg. (2002), H. 3, S. 103-105.

Stoi, R.: Controlling von Intangibles – Identifikation und Steuerung der immateriellen Werttreiber, in: Controlling, 15. Jg. (2003), H. 3/4, S. 38-46.

Sveiby, K. E., Wissenskapital, Landsberg/Lech 1998.

Sveiby, K. E., Measuring Intangible Revenues, www.sveiby.com/articles/Irquantifying.htm, 2002a.

Sveiby, K. E., Methods for Measuring Intangible Assets, www.sveiby.com/articles/ IntangibleMethods.htm, 2002b.

Teece, D., Managing Intellectual Capital, New York 2000.

North, K., *Probst, G.* und *Romhardt, K.*, Wissen messen – Ansätze, Erfahrungen und kritische Fragen, in: ZFO, 67. Jg. (1998), H. 3, S. 158-166.

Integriertes Wertmanagement: Steuerung des Mitarbeiter-, Kunden- und Lieferantenkapitals

Rainer Strack [1] *und Ulrich Villis* [2]

---

[1] *Dr. Rainer Strack*, The Boston Consulting Group, Düsseldorf

[2] *Ulrich Villis* The Boston Consulting Group, München

Viele Unternehmen haben neue wertbasierte Methoden wie EVA™ oder CVA zur strategischen und operativen Steuerung eingeführt. Diese Steuerungskonzepte sind allerdings noch immer stark allein auf das investierte Kapital hin ausgerichtet und damit kaum geeignet, zu erklären oder zu steuern, wie in unserer zunehmend wissensbasierten und dienstleistungsorientierten Gesellschaft Wert geschaffen wird. Dieser Beitrag stellt einen integrierten Wertmanagement-Ansatz (RAVE™) vor, und zwar zur Steuerung von Humankapital (Workonomics™), Kundenkapital (Custonomics™) und Lieferantenkapital (Supplynomics™), allesamt wertorientiert und quantitativ. Die einzelnen Ansätze sind alle mit dem CVA/EVA verknüpft und münden hierdurch in dieselbe zentrale Steuerungskennzahl. Nach kurzer Einführung in das Konzept veranschaulichen die Autoren die Anwendung und den praktischen Nutzen anhand von spezifischen Beispielen.

1. Einleitung

Seit der Geburtsstunde der Bilanz- und Jahreserfolgsrechnung steht das investierte Kapital im Fokus eines jeden Unternehmens und die wesentlichen Steuerungsgrößen wie z. B. der ROI (Return on Investment) beziehen sich auf dieses.

In der alten Industriegesellschaft war das auch richtig. Der Unternehmenserfolg von *Rockefeller's Standard Oil* oder der Stahlwerke der Familie *Krupp* hing in erster Linie von strategischen Investitionen in Kapitalgüter ab. Grundstücke, Gebäude, Produktionsanlagen, Lagerbestände usw. waren die zentralen Vermögensgegenstände. Doch heute ist die Leistung eines Unternehmens mehr ein Resultat von Investitionen in eine neue Art von Vermögenswerten: Mitarbeiter, Kunden und Lieferanten (vgl. *Siegert,* 1999, S. 18 ff.). Die Fähigkeit, die besten dieser „Vermögenswerte“ zu gewinnen, zu halten und zu entwickeln, wird mehr und mehr zum entscheidenden Erfolgsfaktor.

Reagieren Unternehmen angemessen auf diesen Wandel? Letztendlich fordert eine neue Art von Vermögenswerten auch neue Steuerungssysteme. Bereits *Rappaport*, der den Grundstein für die Shareholder-Value-orientierte Unternehmensführung legte (vgl. *Rappaport,* 1981), bemerkt, dass mittlerweile „selbst in einem besonders kapitalintensiven Unternehmen wie beispielsweise *General Motors* ... die Investitionsausgaben nur circa 10 %“ betragen“ (*Rappaport,* 1995, S. 2). Die aus dem Shareholder-Value-Gedanken abgeleiteten Steuerungskonzepte wie zum Beispiel die in vielen Unternehmen verwendeten EVA- oder CVA-Verfahren berücksichtigen dies jedoch nicht ausreichend. Auch sie verwenden weiterhin auf das investierte Kapital bezogene Metriken wie Kapitalrenditen als zentrale Steuerungsgrößen.

Die fehlende Berücksichtigung und Einbindung der wesentlichen Vermögenswerte in die Steuerung des Unternehmens hat erhebliche Folgen für die Qualität der Unternehmensführung. Dabei leidet vor allem die Qualität strategischer Entscheidungen unter dem Mangel adäquater Mess- und Steuerungsinstrumente. Das Potenzial der Wertschöpfungsfaktoren Mitarbeiter, Kunden und Lieferanten wird in vielen Unternehmen nicht ausgeschöpft, weil die wertsteigernden oder -mindernden Wirkungen im Umgang mit diesen „Vermögenswerten“ außerhalb des Blickfelds liegen.

Ähnlich wie das investierte Kapital seit den 1980er-Jahren an der Shareholder-Value-Perspektive ausgerichtet wird, müssen daher auch die „neuen" Kapitalfaktoren Mitarbeiter, Kunden und Lieferanten konsequent in die wertorientierte Unternehmenssteuerung integriert werden.

Das hier vorgestellte Wertmanagementkonzept RAVE (Real Asset Value Enhancer) von *The Boston Consulting Group* bietet dazu einen einfachen Satz von Kennzahlen, die mit der wertorientierten Spitzenkennzahl der Unternehmenssteuerung verbunden sind. Sie bilden den Ausgangspunkt, um die neuen Kapitalfaktoren in die wertorientierten Steuerungsprozesse von Unternehmen zu integrieren (vgl. auch *Barber/Strack/Villis*, 2000; *Strack/Villis*, 2001; *Strack/Villis*, 2002; *Strack/Villis,* 2003).

RAVE ermöglicht es hierdurch, das Unternehmen spezifisch entlang den vorhandenen Vermögenswerten zu steuern. Dabei spielt das klassische Investment-Capital zur Steigerung von EVA bzw. CVA eine immer geringere Rolle, da selbst in traditionellen Industrieunternehmen die kapitalbezogenen Kosten (Abschreibungen und Kosten für das investierte Kapital) nur noch eine vergleichsweise untergeordnete Bedeutung haben (vgl. *Abb. 1*).

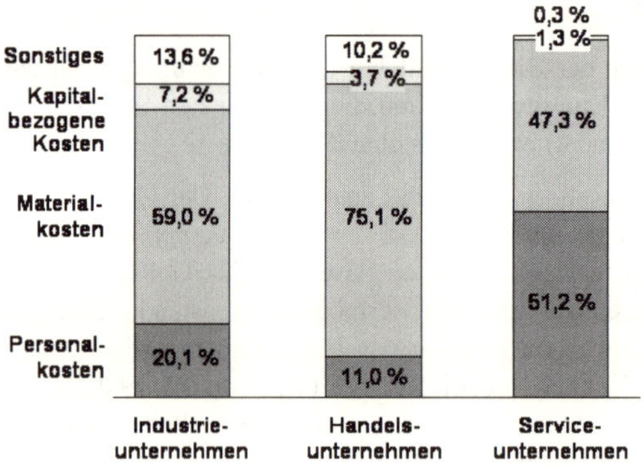

Abb. 1: Kostenstrukturen verschiedener Unternehmen

Im Vergleich von Handels- und Serviceunternehmen zeigt sich, dass ihre Kostenstruktur erwartungsgemäß sehr unterschiedlich ausfällt: Während bei Handelsunternehmen auf die Materialkosten der größte Anteil entfällt, bilden bei den Serviceunternehmen die Personalkosten die größte Position. Dies muss sich auch in der Unternehmenssteuerung niederschlagen. Handelsunternehmen müssen zur Wertsteigerung in erster Linie Produktauswahl, Lieferanten und Materialkosten optimieren, während bei Serviceunternehmen das Management der Mitarbeiter in den Vordergrund rückt. Die effiziente Steuerung des eingesetzten Kapitals ist demgegenüber von untergeordneter Bedeutung. Selbst für Industrieunternehmen sind die Personalkosten oft ein Vielfaches der kapitalbezogenen Kosten. Dieses Bewusstsein ist in vielen Unternehmen bereits vorhanden, jedoch fehlt bisher in den meisten Fällen die konkrete Übersetzung in die Steuerungssysteme.

Neben diesen Input-Faktoren ist als zusätzlicher Output-Faktor die Kundenseite in der Steuerung zu berücksichtigen, und zwar insbesondere für Unternehmen mit hohen Vertriebs- bzw. Marketingkosten. RAVE ermöglicht es, das „Kapital" eines Unternehmens an Kunden ebenfalls in Bezug auf dessen Wertschaffung zu steuern, vor allem mit dem Ziel, wertschaffende Kunden zu akquirieren und zu binden.

Den Ausgangspunkt der neuen Kunden-, Mitarbeiter- und Lieferantenperspektive bildet derselbe EVA/CVA wie im klassischen Wertmanagement. Jedoch wird dieser EVA/CVA nicht wie bisher durch kapitalbasierte Größen, sondern durch mitarbeiter-, kunden- und lieferantenbezogene Kenngrößen ausgedrückt. Die entsprechenden Steuerungskonzepte bezeichnen wir als Workonomics, Custonomics und Supplynomics.

Dabei geht es nicht darum, das klassische kapitalbasierte Wertmanagement zu ersetzen. Vielmehr lassen sich die neuen Steuerungsgrößen als „Spiegelbild" der klassischen Wertmanagementgrößen verstehen. Sie eröffnen hierdurch wichtige ergänzende Perspektiven auf das Wertschaffungspotenzial des Unternehmens. Eine kurze Betrachtung des klassischen Wertmanagementkonzepts erläutert die Parallelen.

2. Das klassische Wertmanagement zur Steuerung des investierten Kapitals

Grundlage des klassischen Wertmanagements ist eine kapitalorientierte Steuerung des Unternehmens. Ausgehend von zunächst statischen Erfolgsgrößen wie Gesamtkapital- oder Eigenkapitalrendite sind Unternehmen in den vergangenen Jahren zu einer stärker dynamisch geprägten – das heißt auf die Wertentwicklung im Zeitverlauf fokussierten – Perspektive übergegangen, die theoretisch auf der Berechnung des Unternehmenswerts basiert. Im Gegensatz zu Renditegrößen wie Return on Investment (ROI), die eine statische Effizienzbetrachtung von Unternehmen darstellen, ermöglichen wertorientierte Steuerungsgrößen eine Berücksichtigung von profitablem Wachstum als Quelle von Wertschaffung. Von diesen so genannten Übergewinn- bzw. Wertbeitragsverfahren sind in der Praxis vor allem das EVA-Konzept (Economic Value Added) von *Stern Stewart & Co.* (vgl. *Stewart*, 1990) und das CVA-Konzept (Cash Value Added) (vgl. *Stelter*, 1999, S. 233 ff.; *Strack/Villis*, 2001, S. 69 ff.) von *The Boston Consulting Group* bekannt geworden. Beide finden heute in vielen Unternehmen Anwendung.

Als Übergewinngröße im EVA- bzw. CVA-Konzept fungiert ein Ergebnis, das um die vom Kapitalmarkt geforderten Kapitalkosten bereinigt ist.

$$\text{Übergewinn} = \text{Ergebnis} - \text{Kapitalkostensatz (KK)} \times \text{investiertes Kapital (IK)}$$

$$\text{bzw. Übergewinn} = (\text{ROI} - \text{KK}) \times \text{IK}$$

Es zeigt sich, warum ein positives Ergebnis nicht ausreicht: Auch die Kapitalkosten müssen erwirtschaftet werden. Anders formuliert: Die Rendite auf das investierte Kapital muss größer sein als der Kapitalkostensatz.

Im Rahmen des EVA-Konzepts ist als Ergebnisgröße der Net Operating Profit after Taxes (NOPAT) definiert; das investierte Kapital wird zu Buchwerten betrachtet. Problematisch bei dieser Betrachtungsweise ist der Einfluss von Abschreibungsmethodik und Anlagenalter auf die Steuerungsgrößen EVA bzw. die Kapitalrendite. Trotz gleich bleibender operativer Leistungsfähigkeit signalisieren die Steuerungsgrößen durch veränderte Abschreibungen eine Wertsteigerung.

Diese mögliche Fehlinterpretation korrigiert der CVA. Im Gegensatz zum EVA basiert dieser auf Cashflow-Größen. Verzerrungen beim investierten Kapital durch unterschiedlich alte Vermögenswerte werden durch Ansatz von historischen Anschaffungs- und Herstellungskosten eliminiert. Die entsprechende Kapitalrendite bezeichnen wir als Cashflow Return on Investment (CFROI).

Aus der Sicht des klassischen Wertmanagements gibt es zwei Wege, den Wertbeitrag (EVA/CVA) zu steigern:

- Verbesserung der Kapitalrendite (ROI/CFROI) oder
- profitables Investitionswachstum (IK), d. h. Wachstum mit einem ROI oberhalb der Kapitalkosten (KK).

Klassisches Wertmanagement setzt bei diesen Zielgrößen an und versucht, durch weiteres Herunterbrechen der Spitzenkennzahlen (Werthebelmanagement) Wertsteigerungspotenziale zu identifizieren (vgl. *Strack/Dörr/Hansen*, 2001, S. 64 ff.).

Die Stellgrößen der Übergewinnansätze (CVA und EVA) beziehen sich auf das Kapital (Kapitalrendite, Kapitalkosten, investiertes Kapital), messen dieses und richten die Unternehmenssteuerung, z. B. Kapitalallokation, daran aus. Auch die wesentlichen Hebel zur Wertsteigerung wie Kapitalrenditeverbesserung und profitables Kapitalwachstum sind rein kapitalorientiert. Daher wird dieser Blickwinkel im Folgenden als „Capital-View" bezeichnet. Kunden, Mitarbeiter und Lieferanten kommen hier nicht explizit vor.

3. Workonomics zur Steuerung des Humankapitals

Workonomics stellt – im Gegensatz zu klassischen Controlling- und Finanzkennzahlen – den Input-Faktor „Mensch" in den Mittelpunkt und ergänzt somit die Finanzperspektive um diesen zentralen „Kapitalfaktor". Bisher fehlen in vielen Unternehmen geeignete Kenngrößen, die eine logische Verbindung zwischen Kapital und Personal zulassen und Transparenz für sinnvolle Steuerungsimpulse schaffen. Diese Vereinheitlichung der Perspektiven gelingt den im Folgenden vorgestellten Workonomics-Spitzenkennzahlen durch den gemeinsamen Fokus auf EVA oder CVA.

3.1 Die Workonomics-Kennzahlen

Die Spitzenkennzahlen ergeben sich dabei durch eine einfache Transformation. Ausgehend von der klassischen kapitalbasierten Definition des EVA bzw. CVA kann dieser durch eine einfache mathematische Umformung allein durch personalbezogene Größen ausgedrückt werden (vgl. *Abb. 2*).

Umformung	**Legende**		
	ROI	=	Kapitalrendite (return on investment)
	KK	=	Kapitalkosten (%)
	IK	=	Investiertes Kapital
	U	=	Umsatz
	PK	=	Personalaufwand
	MK	=	Materialaufwand + sonstige Aufwendungen und Anpassungen
	AFA	=	Abschreibungen
	VAP	=	Value added per person
	ACP	=	Average cost per person
	P	=	Anzahl Mitarbeiter (Vollzeitäquivalente)

$$\begin{aligned} \text{EVA/CVA} &= [\text{ROI} - \text{KK}]\ \text{IC} \\ &= \text{Ergebnis} - \text{KK*IK} \\ &= \text{U} - \text{PK} - \text{MK} - \text{AFA} - \text{KK*IK} \\ &= \Bigg[\underbrace{\frac{\text{U} - \text{MK} - \text{AFA} - \text{KK*IK}}{\text{P}}}_{\text{VAP}} - \underbrace{\frac{\text{PK}}{\text{P}}}_{\text{ACP}}\Bigg]\ \text{P} \\ &= [\text{VAP} - \text{ACP}]\ \text{P} \end{aligned}$$

Abb. 2: Herleitung der Workonomics-Kennzahlen

Die Kennzahlen im Einzelnen (vgl. *Strack/Villis*, 2001, S. 70 ff.; *Strack/Villis*, 2002, S. 149 ff.; *Strack*, 2002, S. 77 ff.; *Strack/Villis,* 2003):

- **Value Added per Person (VAP):** Der Value Added per Person entspricht der durchschnittlichen Wertschöpfung pro Mitarbeiter und kann auch als Mitarbeiterproduktivität interpretiert werden. Der VAP korrespondiert mit der Effizienzgröße ROI auf der Kapitalseite. Die klassische Definition des Value Added als Umsatz minus Materialkosten wurde ergänzt um den Term Abschreibungen und Kapitalkosten. Damit ist sichergestellt, dass durch eine mitarbeitersubstituierende Investition der VAP nicht automatisch erhöht wird. Die Berücksichtigung der Abschreibungen und Kapitalkosten wirkt also als eine Art „Strafterm".
- **Average Cost per Person (ACP):** Hierbei handelt es sich um die durchschnittlichen Personalkosten pro Mitarbeiter. Auf der Kapitalseite entspricht dies den Kapitalkosten.
- **Anzahl Mitarbeiter (P):** Entspricht der Gesamtzahl der Mitarbeiter im Unternehmen als Vollzeitäquivalente und stellt als Wachstumsgröße den Ressourceneinsatz aus Mitarbeiterperspektive dar (analog dem investierten Kapital aus Kapitalperspektive).

Vergleichbar der Kapitalperspektive wird auch bei Workonomics der EVA/CVA durch drei Spitzenkennzahlen beschrieben (vgl. *Abb. 3*): Der EVA bzw. CVA ist die gemeinsame Basis für den Capital-View und den Human-Resources-View. Die Form der Gleichungen ist im Capital-View und im Human-Resources-View dieselbe, sodass hieraus direkt Analogien abgeleitet werden können.

Es korrespondieren:

- die Kapitalkosten (KK) mit den durchschnittlichen Personalkosten (ACP),
- die Kapitalrendite (ROI) mit dem Value Added per Person (VAP) und
- das investierte Kapital (IK) mit der Anzahl Mitarbeiter (P); d.h., der Ressourceneinsatz im Capital-View entspricht dem Ressourceneinsatz im Human-Resources-View.

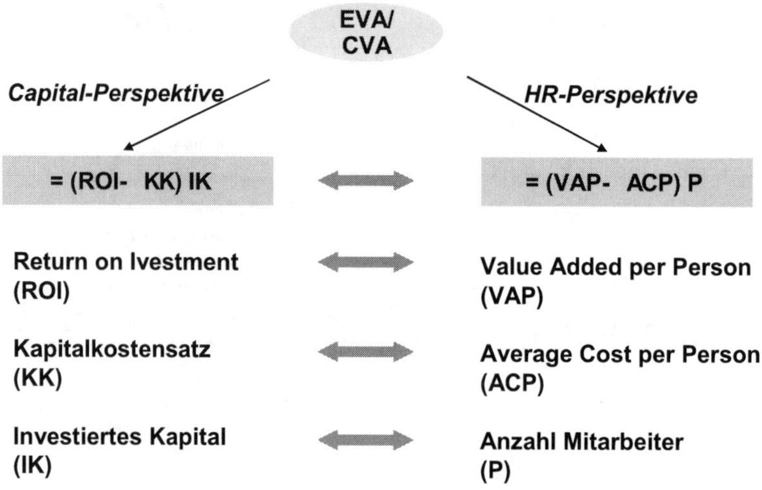

Exakte Definition der Kenngrößen CVA/EVA:
VAP = VA/P mit VA = U - MK - AfA - WACC × IK und ACP = PK/P.

Abkürzungen: VA (€): Wertschöpfung; P: Anzahl Mitarbeiter (Vollzeitäquivalente); U (€): Umsätze; MK (€): Materialkosten plus weitere Aufwendungen plus Anpassung (in Abhängigkeit der konkreten Definition von CVA/EVA); AfA (€): Abschreibungen; WACC (%): gewichtete durchschnittliche Kapitalkosten; IK (€): Investiertes Kapital; PK (€): Personalaufwand (für eine umfassendere Ableitung und Definition der Kenngrößen vgl. *Strack, Villis*, 2001, S. 70-73).

Abb. 3: Die Workonomics-Spitzenkennzahlen

Entsprechend der in *Abb. 3* dargestellten Berechnung des EVA/CVA wird aus HR-Perspektive immer dann Wert geschaffen, wenn der Value Added per Person (VAP) größer ist als die Average Cost per Person (ACP). Dadurch bieten sich drei konkrete Ansatzpunkte zur Wertsteigerung an (vgl. *Strack/Villis*, 2001, S. 73 ff.; *Strack*, 2002, S. 78; *Strack/Villis*, 2003):

- **Steigerung des VAP:** Die Produktivität pro Mitarbeiter kann vor allem durch Prozessverbesserungen, Weiterbildung sowie durch den gesamten Bereich der Personalentwicklung gesteigert werden.
- **Reduzierung der ACP:** Vorstellbar sind alle Maßnahmen, die zur Senkung der Personalkosten beitragen, sei es durch die Reduzierung einzelner Gehaltsbestandteile oder auch durch eine Bindung variabler Vergütungsbestandteile an die Entwicklung des VAP.
- **Profitables Wachstum:** Sobald der VAP größer ist als die ACP, schafft jeder weitere Beschäftigte Wert für das Unternehmen. Einstellungen von Mitarbeitern, deren potenzieller VAP über den Personalkosten liegt, schaffen Wert. Dies bedeutet allerdings auch, dass der Abbau von Mitarbeitern, deren VAP kleiner ist als die ACP, sich ebenfalls wertsteigernd auswirkt. Der Zusammenhang zwischen Produktivität, Höhe des Entgelts und Beschäftigungsniveau wird transparent.

Im Gegensatz zu einer Perspektive, die den Faktor Personal auf eine Kostensicht reduziert, bietet Workonomics eine umfassendere Betrachtung. Die Kostensicht wird ergänzt um Hebel zur Steigerung der Wertschöpfung sowie profitables Beschäftigungswachstum.

Die vorgestellten Kennzahlen können sowohl zur strategischen als auch zur operativen Steuerung herangezogen werden.

Bei der strategischen Steuerung bilden sie die Basis, um auf der Gesamtunternehmensebene das Humankapital in die strategischen Steuerungs- und Planungsprozesse zu integrieren. Die Kennzahlen können dabei für alle Geschäftsbereiche und Segmente berechnet werden. Die Analyse der Istzahlen unterstützt Portfolioentscheidungen und erlaubt es, Fokusfelder für das Management zu identifizieren, konkrete quantitative Zielwerte zu vereinbaren und diese mit Maßnahmen und Projekten zu hinterlegen.

Während die Kennzahlen auf der strategischen Ebene tendenziell eher in der „Breite" für viele unterschiedliche Einheiten erhoben werden, nutzt man sie bei der operativen Steuerung dazu, in einzelnen Geschäftsfeldern die Treiber und Ursachen in der Tiefe zu analysieren und zu steuern. Dazu werden die Spitzenkennzahlen weiter in ihre operativen Bestandteile heruntergebrochen.

Im Folgenden soll die Anwendung von Workonomics zur strategischen Steuerung in den Vordergrund gestellt werden.

3.2 Die Anwendung zur strategischen Steuerung

Beim Personalcontrolling vieler Unternehmen handelt es sich heutzutage meist um eine quantitative Berichterstattung vorwiegend von Bestands- oder Kostengrößen; d.h., nur der Personaleinsatz wird berücksichtigt, nicht aber die Wertschaffung. Darüber hinaus werden qualitative Kenngrößen kaum erhoben und die wenigsten Kenngrößen sowie deren Interpretation sind unmittelbar mit Maßnahmen verknüpft. Eine strategische Steuerung im Sinne eines umfassenden Controllingansatzes findet in der Regel also noch nicht statt.

Workonomics ermöglicht es, das Personal und den Personalbereich des Unternehmens systematisch und auf verschiedene Fristigkeiten hin zu steuern: sowohl aus quantitativer wie auch aus qualitativer und aus Prozesssicht (vgl. *Abb. 4*).

Aus quantitativer Sicht steht dem Personalbereich mit Workonomics ein umfassendes Steuerungsinstrumentarium zur Verfügung, allen voran die drei Spitzenkennzahlen VAP, ACP und P. Diese lassen sich noch weiter herunterbrechen: Kostencontrolling durch Aufspalten der durchschnittlichen Personalkosten (ACP) in die verschiedenen Komponenten sowie Bestandscontrolling durch weitere Differenzierung des Mitarbeiterbestands (P). Die Wertschöpfung (VAP) lässt sich, schrittweise bereinigt um nichtoperative Bestandteile, ebenfalls weiter herunterbrechen. Exemplarisch ist dies in *Abb. 5* zu sehen. Hier ist der Bestand (P) heruntergebrochen zunächst in den Bestand der Vorperiode und die Änderung. Die Änderung sind Einstellungen minus Austritte usw. So bekommt man einen Werthebelbaum auf der Personalseite, der quantitativ mit der Spitzenkennzahl im Unternehmen verbunden ist. Je weiter man diesen Baum aufbricht desto operativer wird die Steuerung.

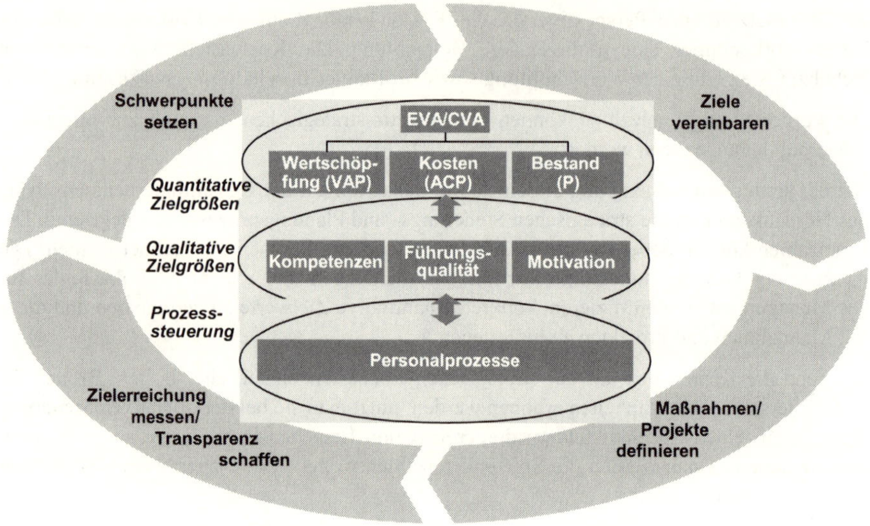

Abb. 4: Strategisches Personalcontrolling

Kennzahlen lassen sich für alle Unternehmensbereiche berechnen. Mit verschiedenen Ausprägungen dieser Kenngrößen lassen sich wiederum entsprechende Maßnahmen verknüpfen. In welcher Tiefe die Kennzahlen analysiert und gesteuert werden, ist dabei von der individuellen Philosophie des Unternehmens abhängig. Es zeigt sich aus der Erfahrung, dass durch die Einführung von Workonomics die Transparenz bereits während der Planungsphase deutlich zunimmt. Produktivitätsveränderungen, die in den klassischen Planzahlen nur implizit vorhanden sind, werden durch die Workonomics-Kennzahlen sofort sichtbar und können adressiert werden.

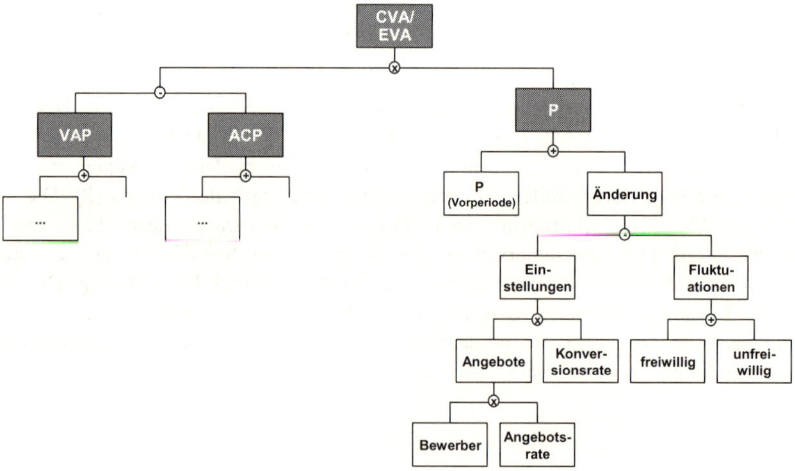

Abb. 5: Der Werthebelbaum auf Personalseite

Personalbezogene Maßnahmen wirken in der Regel nicht unmittelbar auf die Spitzenkennzahlen, sondern individuell über das Arbeitsverhalten der einzelnen Mitarbeiter. Das Gleiche gilt für die Kompetenzen der Mitarbeiter und Führungskräfte. Diese wirken sich jedoch maßgeblich auf die Wertschaffung im Unternehmen aus. Daher werden im Rahmen von Workonomics auch qualitative Größen erhoben und mit Maßnahmen verknüpft: Kompetenzen der Mitarbeiter, deren Motivation sowie die Führungsqualität der Führungskräfte. Auch diese Kennzahlen lassen sich entlang den unternehmensspezifischen strategischen Zielen weiter differenzieren und mit entsprechenden Maßnahmen verbinden.

Neben quantitativen und qualitativen Aspekten werden im Rahmen von Workonomics außerdem personalbezogene Prozesse wie zum Beispiel Besetzung und Entwicklung mit Kennzahlen hinterlegt und mit Maßnahmen verknüpft, um ihre Wirkung auf die quantitativen und qualitativen Zielgrößen aktiv steuern zu können.

Es wird deutlich, dass ein solches Controllingsystem alle wesentlichen Aspekte des Personals sowie des Personalbereichs im Unternehmen in die Steuerung einbezieht. Damit stellt Workonomics eine bedeutende Fortentwicklung des derzeit verbreiteten Personalcontrollings dar.

4. Custonomics zur Steuerung des Kundenkapitals

In der Custonomics-Perspektive rückt das Kundenkapital in den Vordergrund. Ziel ist es, das Wertpotenzial des bestehenden und potenziellen Kundenstamms auf einer quantitativen Basis systematisch und wertorientiert zu steuern und bestmöglich zu realisieren.

Auch in der Kundenperspektive sind die Spitzenkennzahlen direkt mit dem EVA bzw. CVA quantitativ verbunden. Dabei wird der EVA bzw. CVA durch drei kundenrelevante Größen ausgedrückt. Er ergibt sich aus der Differenz von Value Added per Customer (VAC) und den Vertriebs- und Marketingkosten per Customer (ACC) multipliziert mit der Anzahl Kunden (C). Da auch hier der CVA bzw. EVA den Ankerpunkt bildet und die Struktur der Gleichungen den bereits beschriebenen Ansätzen entspricht, ist eine dem Workonomics-Ansatz vergleichbare Analogie ableitbar (vgl. *Abb. 6*).

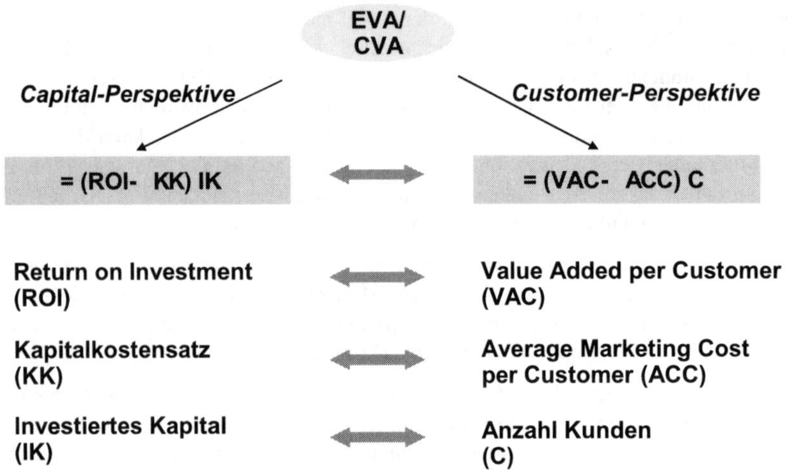

Exakte Definition der Kenngrößen
CVA/EVA: VAC = VA*/C mit VA* = U - MK* - PK* - AfA - WACC × IK* und ACC = SMC/C.

Abkürzungen: VA* (€): Wertschöpfung; C: Anzahl Kunden; U (€): Umsatz; MK* (€): Materialkosten plus weitere Aufwendungen plus Anpassungen ohne den Anteil von Verkauf und Marketing; PK* (€): Personalkosten ohne den Anteil von Verkauf und Marketing; AfA* (€): Abschreibungen ohne den Anteil von Verkauf und Marketing; WACC (%): gewichtete durchschnittliche Kapitalkosten; IK* (€): Investiertes Kapital ohne den Anteil von Verkauf und Marketing; SMC (€): Aufwendungen für Verkauf und Marketing (vgl. auch *Strack/Villis*, 2001, S. 76-77 für eine Ableitung der Kennzahlen).

Abb. 6: Custonomics-Spitzenkennzahlen

Aus der Customer-Capital-Perspektive wird Wert geschaffen (positiver Delta-CVA/EVA) durch Vergrößerung des Value Added per Customer (VAC) (z. B. durch Cross-Selling), durch Senkung der Marketing- und Vertriebskosten (ACC) (z. B. durch Variabilisierung der Vertriebskosten) oder durch profitables Wachstum der Kundenbasis (C) (z. B. durch spezielle Customer-Akquisitionsprogramme).

Für viele kundenorientierte Unternehmen (z. B. Handel, Telekommunikation, Versicherungen, Banken etc.) wird ein geeignetes quantitatives Steuerungsinstrumentarium geschaffen.

Auch das Custonomics-Konzept kann sowohl als strategisches Instrument zur Steuerung von Unternehmen, Unternehmensbereichen oder einzelnen Kundensegmenten genutzt werden als auch zur operativen Steuerung von Vertriebsstandorten bzw. im Extremfall zur Steuerung von Entscheidungen einzelner Vertriebsmitarbeiter.

Auf der strategischen Ebene werden ähnlich wie im Workonomics-Konzept die quantitativen Spitzenkennzahlen um qualitative und prozessorientierte Kennzahlen ergänzt (vgl. *Abb. 7*).

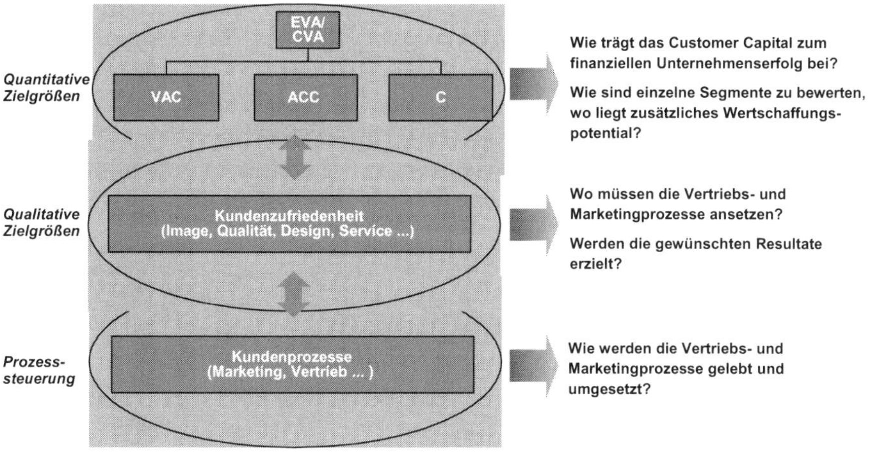

Abb. 7: Strategische Steuerung des Kundenkapitals

Somit werden Wirkungszusammenhänge zwischen quantitativen Ergebnisgrößen, qualitativen Kundeneinschätzungen und Marketing- und Vertriebsprozessen transparent. Durch eine Integration dieser drei Ebenen in den strategischen Steuerungs- und Planungskreislauf des Unternehmens können so gezielt Schwachstellen aufgedeckt, langfristige Ziele vereinbart und konkrete Projekte und Maßnahmen eingeleitet werden. Neben dieser strategischen Anwendung, bei der man sich in der Regel auf wenige Spitzenkennzahlen und Segmente beschränkt, kann das Konzept auch zu einem operativen Steuerungsinstrument ausgeweitet werden. Dabei erweist sich Custonomics als mehr als ein reines Controllingtool. Es liefert beispielsweise auch Anhaltspunkte für eine neue strategische Kundensegmentierung.

5. Supplynomics zur Steuerung des Lieferantenkapitals

Ergänzend zur mitarbeiter- und kundenbezogenen Perspektive können im Rahmen des Supplynomics-Konzepts auch Material bzw. Lieferanten in die wertorientierte Steuerung einbezogen werden. Hierbei wird die durchschnittliche Wertschöpfung pro Lieferant (Value Added per Supplier, VAS) den durchschnittlichen Kosten pro Lieferant (ACS) gegenübergestellt. Die Anzahl der Supplier (S) wird analog den anderen Ansätzen verwendet.

Exakte Definition der Kenngrößen CVA/EVA: VAC = VA**/S mit VA** = U - PK** - AfA** - WACC × IK** und ACS = MK**/S.

Abkürzungen: VA** (€): Wertschöpfung; S: Anzahl Lieferanten; U (€): Umsatz; PK** (€): Personalaufwendungen ohne Anteil von Einkaufsbereichen; AfA** (€): Abschreibungen ohne Anteil von Einkaufsbereichen; WACC (%): gewichtete durchschnittliche Kapitalkosten; IK** (€): Investiertes Kapital ohne Anteil von Einkaufsbereichen; MK** (€): Materialkosten inklusive Anteil von Einkaufsbereichen.

Die Lieferanten können in diesem Ansatz auch durch Produkte oder Produktgruppen (Supplies) ersetzt werden, was zu einer wertorientierten Produkt-Deckungsbeitragsrechnung führt. Supplynomics ermöglicht es damit, eventuell vorhandene Produkt-Deckungsbeitragsrechnungen an die Spitzenkennziffern EVA bzw. CVA anzubinden.

Wie gestaltet sich in diesem Ansatz der Zusammenhang zwischen den Kennzahlen und einer Wertsteigerung? Analog zu Workonomics bzw. Custonomics kann Wert gesteigert werden,

- indem die durchschnittliche Wertschöpfung pro Lieferant/Produkt (VAS) erhöht wird,
- indem die durchschnittlichen Kosten pro Lieferant/Produkt (ACS) gesenkt werden oder
- durch eine Ausweitung der Lieferantenbasis bzw. zusätzliche Produkte (S), vorausgesetzt, dass die durchschnittliche Wertschöpfung größer ist als die durchschnittlichen Kosten des Lieferanten bzw. Produkts.

Eine Steigerung der durchschnittlichen Wertschöpfung (VAS) kann beispielsweise durch produktgruppenspezifisches Pricing oder Ausweitung des Umsatzes pro Lieferant erreicht werden, die Senkung der durchschnittlichen Kosten (ACS) zum Beispiel durch Verringerung der Logistik- und Lagerhaltungskosten oder auch der Einkaufspreise.

Der Supplynomics-Ansatz eignet sich insbesondere für Unternehmen mit geringer Wertschöpfungstiefe, bei denen der Wertbeitrag für einzelne Lieferanten bzw. Produkte ermittelt werden kann, wie z. B. Handels- oder Konsumgüterunternehmen. Über das Kennzahlensystem wird es möglich, den Wertbeitrag einzelner Lieferanten oder Produkte unternehmensweit zu vergleichen und zu steuern.

6. Zusammenfassung: Integriertes Wertmanagement

Grundidee des vorgestellten Wertmanagementkonzepts ist es, die Unternehmenssteuerung auf die wesentlichen Vermögenswerte zu fokussieren. Die Haupthebel zur Steigerung des Shareholder-Value liegen immer weniger in der Optimierung des investierten Kapitals im klassischen Sinne. Sie liegen vielmehr in einer zielgerichteten Entwicklung und Steuerung des Kunden-, Mitarbeiter- und Lieferantenkapitals (vgl. *Abb. 8*).

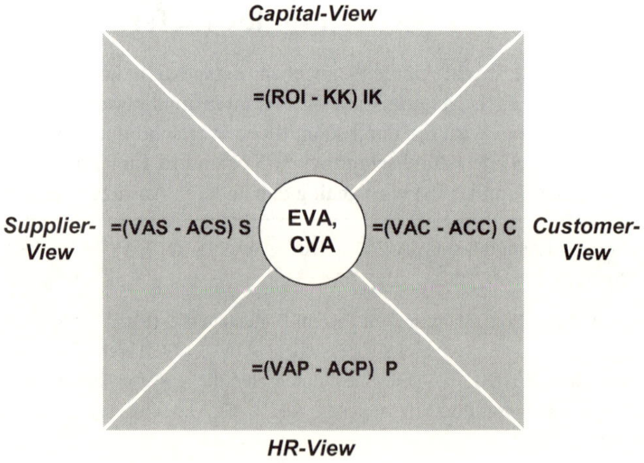

Abb. 8: Spitzenkennzahlenintegriertes Wertmanagement

Der entscheidende Unterschied zur traditionellen Sichtweise ist der Wechsel der Perspektiven: Die ehemaligen Kostenpositionen Kapitalkosten, Personalkosten, Materialkosten und kundenbezogene Kosten werden nun zu den Kapitalpositionen Investment-Capital, Human Capital, Supplier-Capital und Customer-Capital. Aus dem veränderten Verständnis ergibt sich ein neuer Handlungsrahmen: Kosten versucht man zu minimieren, Kapital versucht man zu entwickeln.

Die Vermögenswerte, die in Mitarbeitern, Kunden oder Lieferanten schlummern, müssen gezielt identifiziert und gesteuert werden. Wir schlagen vor, das Mitarbeiter-, Kunden- und Lieferantenkapital ähnlich wertorientiert und quantitativ zu steuern, wie es seit vielen Jahren mit dem investierten Kapital üblich ist. Die verwendeten Kennzahlen müssen dabei die Verbindung zum Shareholder-Value sicherstellen. Ausgangspunkt sind Spitzenkennzahlen, die alle direkt mit bestehenden Übergewinngrößen wie zum Beispiel dem EVA oder CVA verbunden sind.

Durch die Tatsache, dass ein EVA oder CVA bereits in vielen Unternehmen als oberste Steuerungsgröße verwendet wird, werden unserer Erfahrung nach die Einführung und die Akzeptanz beim Management und den Mitarbeitern zusätzlich erleichtert. Auch sollen die vorgestellten neuen Perspektiven nicht das bereits vorhandene kapitalbasierte Steuerungssystem ersetzen, sondern können dieses insbesondere auf Grund der Verbindung zur gleichen Spitzenkennzahl sinnvoll ergänzen. Vor allem bei Konglomeraten mit einer Vielzahl von Geschäftsbereichen ist es damit möglich, das Steuerungssystem geschäftsfeldspezifisch auf das jeweils dominierende Kapital auszurichten, gleichzeitig jedoch durch die Verbindung zur einheitlichen Spitzenkennzahl ein konsistentes und vergleichbares Gesamtbild zu bewahren. So können beispielsweise besonders mitarbeiterintensive Bereiche durch Workonomics, kundengetriebene Bereiche durch Custonomics, lieferantenabhängige Bereiche durch Supplynomics und schließlich anlagenintensive Bereiche durch die klassischen kapitalbasierten Finanzkennzahlen gesteuert werden. Auch können die häufig operativ eingesetzten Balanced-Scorecard-Systeme in das vorgestellte wertoriente Steuerungssystem integriert werden, sodass Value-Managment und Balanced Scorecard zu einem integrierten System verbunden werden.

Die vorgestellten Kennzahlen sind dabei nur die Basis für einen tiefer gehenden Wandel. Denn es geht nicht um das pure Messen oder Kontrollieren von Mitarbeitern, Kunden oder Lieferanten, sondern um das Scharfstellen der gesamten Unternehmenssteuerung auf diese entscheidenden Kapitalfaktoren. Auch organisatorisch wird sich eine stärkere Orientierung am Mitarbeiter-, Kunden- und Lieferantenkapital in einer neuen Rolle des HR-, Marketing- und Einkaufsdepartments niederschlagen. So wird zum Beispiel das HR-Department aus seiner selbst definierten Rolle des Dienstleisters schrittweise zum Steuerer des Humankapitals avancieren.

Durch die notwendige Integration in den kontinuierlichen Controllingprozess rückt das neue Kapital darüber hinaus stärker ins Zentrum der Diskussionen des Topmanagements – Diskussionen, die nicht nebulös auf der Basis verstreuter qualitativer Beobachtungen oder mehr oder weniger unsystematisch gesammelter Kennzahlen stattfinden, sondern die sich auf Fakten stützen, deren Relevanz systematisch und quantitativ vom Ziel der Shareholder-Value-Maximierung abgeleitetet wurde. Dass in der Interpretation ein qualitatives Urteil des Managements auch weiterhin zwingend erforderlich ist, versteht sich von selbst.

7. Literatur

Barber, F., *Strack, R.* und *Villis, U.*, Das „Kapital" der New Economy bewerten, in: *Von Oetinger, B.* (Hrsg.), Das Boston Consulting Strategie-Buch – Die wichtigsten Management-Konzepte für den Praktiker, S. 467-473, München 2000.

Rappaport, A., Selecting Strategies that Create Shareholder Value, in: Harvard Business Review, Vol. 59 (1981), No. 5/6, S. 139-149.

Rappaport, A., Shareholder-Value: Ein Handbuch für Manager und Investoren, Stuttgart 1995.

Siegert, T., Humankapital: Erfolgsmessung und Partizipation, in: *Bühler, W.* und *Siegert, T.* (Hrsg.), Unternehmenssteuerung und Anreizsysteme, Stuttgart 1999.

Stewart, G. B. III, The Quest for Value – A Guide for Senior Managers, 1990.

Stelter, D., Wertorientierte Anreizsysteme für Führungskräfte und Mitarbeiter, in: *Bühler, W.*, und *Siegert, T.* (Hrsg.): Unternehmenssteuerung und Anreizsysteme, Stuttgart 1999.

Strack, R., *Hansen, J.* und *Dörr, T.*, Wertmanagement: Implementierung auf operativer Ebene und Erweiterung zur quantitativen Steuerung des Human und Customer-Capitals, in: Kostenrechnungspraxis, Sonderheft 1/2001, S. 63-72.

Strack, R. und *Villis, U.*, RAVE™: Die nächste Generation im Shareholder-Value-Management, in: Zeitschrift für Betriebswirtschaft, 71. Jg. (2001), H. 1, S. 67-84.

Strack, R. und *Villis, U.*, RAVE™: Integrated Value Management for Customer, Human, Supplier and Invested Capital, in: European Management Journal, Vol. 20, (2002), No. 2, S. 147-159.

Strack, R. und *Villis, U.*, Steuerung des „neuen" Kapitals: Kunden, Mitarbeiter und Lieferanten in der wertorientierten Unternehmensführung, in: *Wiedmann, K.* und *Heckemüller, C.* (Hrsg.), Ganzheitliches Corporate Finance Management, Wiesbaden 2003, S. 383-401.

Strack, R., Workonomics: Wertorientierte Steuerung des Humankapitals, in: *Klinkhammer, H.* (Hrsg.), Personalstrategie, Neuwied 2002.

3. Rechnungslegung und Reporting

Die Notwendigkeit einer Berücksichtigung von Intangible Assets in der externen Unternehmensrechnung und im Reporting ist ein wesentlicher Treiber für den Bedeutungszuwachs des Themenfeldes. Im Zuge des New Economy-Booms in den 1990er Jahren kam es zu zahlreichen Unternehmensakquisitionen, bei denen der Kaufpreis das Nettovermögen der betreffenden Firmen bei weitem übertraf. Die übernehmenden Unternehmen reagierten mit einem erheblichen Aufbau der Goodwillpositionen in den Bilanzen. Die Aktivierungsfähigkeit des Goodwills ist abhängig von den jeweils anzuwendenden internationalen Rechnungslegungsstandards (HGB, IAS/IFRS, US-GAAP) und den Inhalten. Ein einheitliches Verständnis zum externen Reporting von Intangibles fehlt bisher. Dem steht ein zunehmendes Anwachsen von Ausgaben für immaterielle Vermögenswerte, insbesondere in den Hochtechnologieindustrien und -ländern gegenüber. Für die Erhaltung bestehender Wettbewerbsvorteile bzw. für die Schaffung neuer kommt insbesondere der Forschung und Entwicklung eine zentrale Bedeutung zu. Im Sinne einer entscheidungsnützlichen Unternehmenspublizität gilt es, Informationen zu diesem Intangible Asset strukturiert aufzubereiten und zu berichten.

Erfassung immaterieller Werte in der Unternehmensberichterstattung vor dem Hintergrund handelsrechtlicher Rechnungslegungsnormen

Arbeitskreis "Immaterielle Werte im Rechnungswesen"[1]

---

[1] *Arbeitskreis "Immaterielle Werte im Rechnungswesen" der Schmalenbach Gesellschaft für Betriebswirtschaft e.V.*

Dem *Arbeitskreis "Immaterielle Werte im Rechnungswesen" der Schmalenbach-Gesellschaft für Betriebswirtschaft e.V.* gehören momentan die folgenden Personen an: Prof. Dr. *Walther Busse von Colbe*, Ruhr-Universität Bochum; Dr. *Rolf Uwe Fülbier*, Ruhr-Universität Bochum; Dr. *Torsten Ganske* (Arbeitskreisleiter), *Deutsche Telekom AG*; Prof. Dr. *Thomas Günther*, Technische Universität Dresden; Prof. Dr. *Axel Haller*, Johannes Kepler Universität Linz/Österreich; Dr. *Matthias Hendler*, *KPMG Deutsche Treuhand-Gesellschaft AG*; Prof. Dr. *Isabel von Keitz*, Fachhochschule Münster; *Guido Kerkhoff*, *Deutsche Telekom AG*; Dr. *Harald Köster*, *Henkel KGaA*; Prof. Dr. *Hans Peter Möller*, *RWTH Aachen*; Prof. Dr. *Bernhard Pellens* (Arbeitskreisleiter), Ruhr-Universität Bochum; Dr. *Achim Reinhold*, *ZDF*; WP/StB *Rüdiger Reinke*, *KPMG Deutsche Treuhand-Gesellschaft AG*; Dr. *Wolfgang Ritter*, *SAP AG*, und *German Weinberger*, *Deutsche Post AG*.

Die externe Rechnungslegung ist im Wesentlichen auf die Erfassung und Abbildung materieller Werte gerichtet. Diese Vorstellung entspricht jedoch bereits seit geraumer Zeit aufgrund des Wandels von der Industrie- zur Dienstleistungs- und Hochtechnologiegesellschaft nicht mehr der Realität. Immer wichtiger werden daher immaterielle Werte. Als solche werden allgemein physisch nicht greifbare wirtschaftliche Vorteile, wie Rechte, Humanressourcen, Kundenbeziehungen, Prozess- und Produkttechnologien-Know-how, Marken und Marktpositionen, verstanden. Vor diesem Hintergrund der erkannten mangelnden Informationsqualität von publizierten Abschlüssen im Hinblick auf immaterielle Vermögenswerte hat der Arbeitskreis Empfehlungen für die Darstellung von immateriellen Werten in der handelsrechtlichen Rechnungslegung sowie Unternehmensberichterstattung erarbeitet. Diese betreffen die Kategorisierung, die Bilanzierung dem Grund und der Höhe nach sowie die Berichterstattung. Im Bezug auf die Berichterstattung sind die Empfehlungen des Arbeitskreises insbesondere für Unternehmen konzipiert, die von öffentlichem Interesse sind (z.B. große bzw. kapitalmarktorientierte Unternehmen) oder bei denen immaterielle Werte für die wirtschaftliche Lage eine wesentliche Bedeutung besitzen. Dabei sind sie nicht nur auf nach HGB Rechnung legende Unternehmen, sondern auch auf solche gerichtet, die international anerkannte Standards (IAS/IFRS und US GAAP) anwenden.

1. Einführung

Die gesellschafts- bzw. handelsrechtlichen Rechnungslegungsvorschriften sind trotz der Reformen von 1937, 1965 und 1985 immer noch wesentlich von der Vorstellung geprägt, dass materielle Werte wie Grundstücke, Gebäude, Produktionsanlagen oder Vorräte die entscheidenden Werttreiber von Unternehmen darstellen und deshalb die externe Rechnungslegung im Wesentlichen auf die Erfassung und Abbildung dieser Werte gerichtet sein sollte. Diese Vorstellung entspricht jedoch bereits seit geraumer Zeit nicht mehr der mikro- und makroökonomischen Realität, deren Entwicklung aufgrund des Wandels von der Industrie- zur Dienstleistungs- und Hochtechnologiegesellschaft wesentlich von der Schaffung und Erhaltung immaterieller Werte abhängig ist. Als solche werden allgemein physisch nicht greifbare wirtschaftliche Vorteile, wie Rechte, Humanressourcen, Kundenbeziehungen, Prozess- und Produkttechnologien-Know-how, Marken und Marktpositionen, verstanden. Dieser mangelnde „Realitätsbezug" in den Abbildungsregeln des deutschen Bilanzrechts zeigt sich insbesondere an dem 1965 kodifizierten, pauschalen Ansatzverbot für alle nicht entgeltlich erworbenen immateriellen Vermögensgegenstände des Anlagevermögens gemäß § 248 Abs. 2 HGB, dessen Rechtfertigung bis heute aus der primär auf Zahlungsbemessung ausgerichteten, von Vorsichts- und Objektivierungsgesichtspunkten geprägten Tradition des deutschen Bilanzrechts abgeleitet wird. Obgleich anzumerken ist, dass auch die vorrangig auf die Informationsvermittlung ausgerichteten internationalen Rechnungslegungssysteme (IAS/IFRS und US-GAAP) der zunehmenden Bedeutung von selbst erstellten immateriellen Werten nur wenig gerecht werden. Denn auch hier besteht nur für eine spezifische Teilmenge dieser Werte bei Erfüllung bestimmter Voraussetzungen eine Aktivierungspflicht, die darüber hinaus aufgrund der „weichen“ Ansatzkriterien den Bilanzierenden einen großen Auslegungs- und Gestaltungsspielraum eröffnet.

Das Ausmaß der nicht aktivierten immateriellen Werte kann u.a. an der Höhe der bei Unternehmensübernahmen gebildeten Goodwill zum Ausdruck kommen (so ergab sich z.B. bei der *Mannesmann*-Übernahme durch *Vodafone* im Jahre 2000 ein Goodwill von ca. 150 Mrd. Euro; bei der Übernahme von *Voicestream Wireless* durch die *Deutsche Telekom* im Jahre 2001 waren es etwa 40 Mrd. Euro; zur Diskrepanz von Buch- und Börsenwert des Eigenkapitals einiger Kapitalgesellschaften vgl. auch *Kubin*, 1998, S. 538.). Wenngleich ein Goodwill nicht mit den immateriellen Vermögenswerten gleichzusetzen ist, so enthält er regelmäßig eine Vielzahl von immateriellen Werten, die vom übertragenden Unternehmen selbst erstellt und von dem übernehmenden Unternehmen erworben wurden, aber nicht separat aktiviert werden. Dies obwohl sich aus der handelsrechtlichen Generalnorm und der im Rahmen der GoB etablierten „Einzelerwerbsfiktion“ sowohl für den Einzel- als auch den Konzernabschluss ein Gebot einer möglichst weitgehende Disaggregation des Sammelpostens „Goodwill“ („derivativer Firmenwert“) durch die separate Aktivierung einzelner im Rahmen des Unternehmenserwerbs erworbener immaterieller Werte ableiten lässt (vgl. z.B. *Adler/Düring/Schmaltz* 1995, § 255, Tz. 265; ebenda § 301, Tz. 80; *Knop/Küting*, 1995, § 255 Tz. 453).

Vor diesem Hintergrund der erkannten mangelnden Informationsqualität von publizierten Abschlüssen im Hinblick auf immaterielle Vermögenswerte hat der Arbeitskreis Empfehlungen für die Darstellung von immateriellen Werten in der handelsrechtlichen Rechnungslegung sowie Unternehmensberichterstattung erarbeitet (wesentliche Teile der hier vorgestellten Empfehlungen sind bereits an anderer Stelle publiziert worden, siehe: *Arbeitskreis „Immaterielle Werte im Rechnungswesen“ der Schmalenbach-Gesellschaft für Betriebswirtschaft e.V.*, 2001, S. 989ff. und Derselbe, 2003, S. 1233ff.). Dabei widmete er sich folgenden Teilfragen:

- Wie lassen sich immaterielle Vermögenswerte definieren bzw. kategorisieren (**Kategorisierung**)?
- Welche immateriellen Werte sind unter welchen Voraussetzungen in der Bilanz zu aktivieren (**Bilanzierung dem Grunde nach**)?
- Wie sind die aktivierten immateriellen Werte im Zugangszeitpunkt sowie in den Folgejahren zu bewerten (**Bilanzierung der Höhe nach**)?
- Welche Angaben sollten über die aktivierten und die nicht aktivierten immateriellen Vermögenswerte in Anhang und Lagebericht offen gelegt werden (**Berichterstattung**)?

Die nachfolgend vorgestellten Empfehlungen zielen darauf ab, den Aussagewert handelsrechtlicher Abschlüsse in Bezug auf immaterielle Werte deutlich zu erhöhen, indem einerseits entsprechend der Konvergenzbestrebungen auf dem Gebiet der externen Rechnungslegung innerhalb der EU auf die bestehenden internationalen Rechnungslegungsstandards Bezug genommen, gleichzeitig jedoch andererseits an den fundamentalen Grundsätzen des handelsrechtlichen Systems (vgl. zu den Systemgrundsätzen zur Ermittlung und Aufrechterhaltung der Grundsätze ordnungsmäßiger Buchführung (GoB) z.B. *Baetge/Kirsch*, 2002, Tz. 76 ff.; *Leffson*, 1987, S. 179), insbesondere dem Grundsatz der Pagatorik, und den europäischen Bilanzrichtlinien als Rahmenbedingungen festgehalten wird. Im Vordergrund steht die Frage, ob und wie immaterielle Werte in einem handelsrechtlichen Abschluss im Sinne der Generalnorm (§ 238 Abs. 1 HGB, § 264 Abs. 2 HGB und § 297 Abs. 2 HGB) abgebildet werden können und welche Darstellungsform sowie Darstellungsinhalte bei der Berichterstattung von immateriellen Werten außerhalb der Bilanz am zweckmäßigsten erscheinen. Im Bezug auf die Berichterstattung sind die Empfehlungen des Arbeitskreises insbesondere für Unternehmen konzipiert, die von öffentlichem Interesse sind (z.B. große bzw. kapitalmarktorientierte Unternehmen) oder bei denen immaterielle Werte für die wirtschaftliche Lage eine wesentliche Bedeutung besitzen. Dabei sind sie nicht nur auf nach HGB Rechnung legende Unternehmen, sondern auch auf solche gerichtet, die international anerkannte Standards (IAS/IFRS und US GAAP) anwenden.

Unerlässliche Voraussetzung für die Entwicklung dieser Empfehlung ist die Konkretisierung des Abbildungsgegenstandes, d.h. die Umschreibung und Kategorisierung immaterieller Werte, die nachfolgend vorgenommen wird.

2. Konkretisierung des Begriffs „immaterielle Werte"

2.1 Kategorisierung immaterieller Werte

Bei dem Unterfangen, das Phänomen „immaterielle Werte" für Rechnungslegungszwecke grundsätzlich, d.h. unabhängig von ihrer bilanziellen Erfassung, konkretisieren zu wollen, stellt man fest, dass sich bisher weder in Deutschland noch international eine einheitliche Definition oder zumindest Umschreibung für diesen Begriff etabliert hat. Ebenso wird deutlich, dass verschiedene Begriffe, wie „immaterielle Güter", „immaterielle Vermögensgegenstände", „knowledge based assets", „intellectual property", „intangible assets", „intellectual capital", häufig als Synonyme bzw. inhaltsähnliche Begriffe für „immaterielle Werte" Verwendung finden und ein Konsens bezüglich der inhaltlichen Abgrenzung dieser Begriffe nicht besteht. Lediglich hinsichtlich der Bezeichnungen „immaterielle Wirtschaftsgüter" und „immaterielle Vermögensgegenstände" im Steuer- bzw. Handelsrecht lässt sich von einer Konkretisierung der Begrifflichkeiten ausgehen. Sie stellen abstrakt aktivierungsfähige Werte dar (vgl. zur abstrakten Aktivierungsfähigkeit im Handels- und Steuerrecht *Baetge/Kirsch/Thiele*, 2002, S. 124 ff.) und sind mithin enger definiert als der hier relevante Begriff „immaterielle Werte".

Sowohl im deutschen als auch im internationalen Rechnungslegungsschrifttum werden immaterielle Werte von anderen Werten i.d.R. negativ abgegrenzt (vgl. *von Keitz*, 1997, S. 5f. m.w.N), indem sie als solche umschrieben werden, denen im Gegensatz zu den materiellen Werten die „körperliche Substanz" fehlt. Werte, die sich sowohl aus immateriellen als auch aus materiellen Komponenten zusammensetzen, sind dann als immateriell zu klassifizieren, wenn die materielle Komponente nur untergeordnete Bedeutung hat und vornehmlich Transport-, Dokumentations-, Speicherungs- oder Lagerungszwecken dient. Demnach zählt z.B. auf einem Datenträger gespeicherte Software zu den immateriellen Werten, wenngleich der Datenträger physisch greifbar ist. Finanzielle Werte, wie Geldforderungen, zeichnen sich zwar ähnlich wie immaterielle Werte durch die fehlende körperliche Substanz aus. Gleichwohl lassen sie sich von den immateriellen Werten durch das Kriterium „monetär" abgrenzen. Insgesamt können immaterielle Werte mithin als nicht monetäre Werte ohne körperliche Substanz negativ umschrieben werden.

Im Gegensatz zur negativen Abgrenzung hat sich eine international anerkannte positive Definition von immateriellen Werten bisher nicht etabliert. Vielmehr finden sich im Rechnungslegungsschrifttum verschiedene, wenn auch z.T. ähnliche Umschreibungen. Gemeinsam ist den meisten Umschreibungen, dass sie den Begriff „immaterielle Werte" nicht durch eine abstrakte Angabe des Begriffsinhalts definieren, sondern vielmehr in Form einer Kategorisierung beschreiben. So haben im deutschen Schrifttum u.a. *Mutze* (*Mutze*, 1960, S. 23 f.), *Husemann* (*Husemann*, 1976, S. 246f.), *Reuleaux* (*Reuleaux*, 1987, S. 48 ff.), *von Rütte/Hoenes* (*Rütte/Hoenes*, 1995, S. 39 ff.) und im internationalen Schrifttum z.B. *Hendriksen/van Breda* (*Hendriksen/Van Breda*, 1992, S. 633), *Roos* et al. (*Roos* et al., 1998, S. 57), *Edvinsson* und *Malone* (*Edvinsson/Malone*, 1997, S. 73) und *Sveiby* (*Sveiby*, 1998) solche Kategorisierungen vorgenommen. Diese sind allerdings z.T. sehr grob und damit wenig operational, z.T. nicht umfassend und z.T. aufgrund ihrer Mehrstufigkeit für Abbildungszwecke zu komplex. Unter Berücksichtigung dieser Kritikpunkte und dem Strukturierungskonzept von *Edvinsson* und *Malone*, schuf der Arbeitskreis folgende einstufige, von der Bilanzierung unabhängige, umfassende Kategorisierung für immaterielle Werte (vgl. *Abb. 1*), deren sieben einzelne Kategorien sich wie folgt beschreiben lassen:

Abb. 1: Kategorisierung der immateriellen Werte

- **Innovation Capital:** Immaterielle Werte im Bereich der Produkt-, Dienstleistungs- und Verfahrensinnovationen eines Unternehmens. Hierzu zählen z.B. neue Software, Patente, Filme oder Rezepturen eines Unternehmens.
- **Human Capital:** Immaterielle Werte eines Unternehmens im Personalbereich. Hierzu zählen das im Personal und Management inhärente Wissen (z.B. Ausbildung und Experten-know-how der Mitarbeiter), deren Kompetenz (z.B. Führungsqualität) sowie sonstige immaterielle Werte im Personalbereich, wie etwa ein gutes Betriebsklima oder eine Knowledge-Datenbank.
- **Customer Capital:** Immaterielle Werte eines Unternehmens im Absatzbereich. Beispiele hierfür sind Kundenlisten, Marktanteile, Kundenzufriedenheit, Marken und Abnahmeverträge.
- **Supplier Capital:** Analog zum Customer Capital handelt es sich hier um immaterielle Werte eines Unternehmens im Beschaffungsbereich. Hierzu können z.B. Verträge über den Bezug von bestimmten Rohstoffen oder Entwicklungskooperationen mit Zulieferanten zählen.
- **Investor Capital:** Immaterielle Werte im Finanzbereich eines Unternehmens, die sich in günstigen Konditionen für die Eigen- und Fremdkapitalbeschaffung niederschlagen. So kann ein Unternehmen, das ein Bonitäts-Rating durchführen lässt, ggf. seine Kreditkonditionen verbessern.

- **Process Capital:** Immaterielle Werte im Organisationsbereich, insbesondere im Zusammenhang mit der Ablauf- und Aufbauorganisation. Hierzu zählen z.B. ein funktionierendes Vertriebsnetz, eine hochwertige Qualitätssicherung und ein gutes Kommunikationsnetz.
- **Location Capital:** Immaterielle Werte eines Unternehmens, die sich aus dessen Standort ergeben. Dies sind z.B. Standortvorteile, die auf einer günstigen Verkehrsanbindung beruhen, oder lokale Steuervorteile.

Durch die sieben Kategorien werden die „immateriellen Werte" eines Unternehmens umfassend beschrieben. Jeder einzelne immaterielle Wert ist mindestens einer dieser sieben Kategorien zuzuordnen, wobei manche auch mehreren Kategorien zugerechnet werden können, da die Kategorien nicht überschneidungsfrei sind. So kann z.B. ein Vertriebssystem sowohl als Customer Capital als auch als Process Capital klassifiziert werden.

2.2 Abgrenzung vom Goodwill

Wenngleich die immateriellen Werte hier bewusst weit verstanden werden, besteht keine Identität mit einem Goodwill. Denn der Goodwill als Differenz aus dem Ertragswert, bzw. im Fall eines Unternehmenserwerbs dem Kaufpreis abzüglich dem Nettosubstanzwert eines Unternehmens, enthält neben den nicht separat aktivierten immateriellen Werten auch weitere Komponenten, die nicht zu den immateriellen Werten zählen. So bildet z.B. der Mehrpreis aufgrund des Verhandlungsgeschicks des Verkäufers einen Teil des derivativen Goodwills, stellt aber keinen immateriellen Wert dar. Die immateriellen Werte enthalten hingegen neben den nicht einzeln aktivierten immateriellen Werten auch die nicht dem Goodwill zuzuordnenden separat aktivierten immateriellen Werte. Insofern kann mit der anschließenden Diskussion über die Abbildung von immateriellen Werten der Sammelposten „Goodwill" zwar zum Teil reduziert bzw. erklärt werden, einzelne Bestandteile des Goodwill werden von dieser Diskussion indes nicht erfasst. Die Beziehungen zwischen den immateriellen Werten und einem Goodwill lassen sich anhand der *Abb. 2* veranschaulichen:

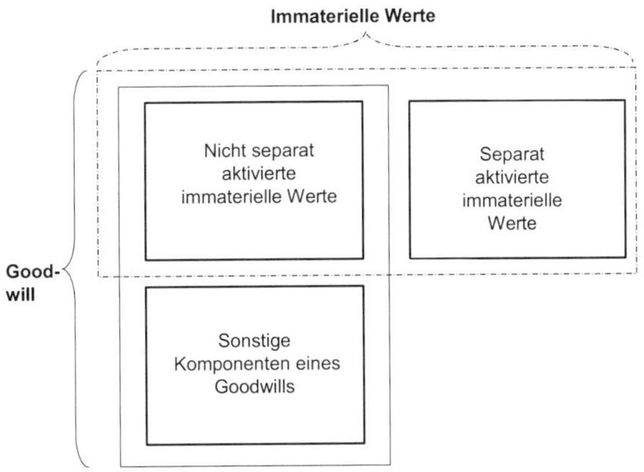

Abb. 2: Verhältnis von immateriellen Werten zum Goodwill

3. Erfassung immaterieller Werte in der Bilanz

3.1 Bilanzansatz (Bilanzierung „dem Grunde nach")

Ausgaben, wie für Forschungs- und Entwicklungs- oder Weiterbildungsmaßnahmen etc., stellen selbst keine immateriellen Werte dar. Vielmehr führen – wie in *Abb. 3* an verschiedenen Beispielen veranschaulicht – diese Ausgaben möglicherweise zu immateriellen Werten, z.B. zu Forschungsergebnissen oder auch nur zu Teilergebnissen. Im Rahmen eines pagatorisch abgesicherten Rechnungslegungssystems stellt sich die Frage, ob die Ausgaben zum Erwerb, zur Schaffung oder zur Verbesserung von immateriellen Werten zu aktivieren oder sofort als Aufwand zu erfassen sind. Grundsätzlich bestehen für diese Ausgaben die folgenden fünf bilanziellen Behandlungsmöglichkeiten, die nachfolgend im Hinblick auf ihre Anwendbarkeit näher untersucht werden:

- Aktivierung als immaterieller Vermögensgegenstand,
- Aktivierung als Rechnungsabgrenzungsposten,
- Aktivierung als Bilanzierungshilfe,
- Aktivierung als Goodwill im Rahmen des Unternehmenserwerbs oder
- sofortige Verrechnung als Aufwand.

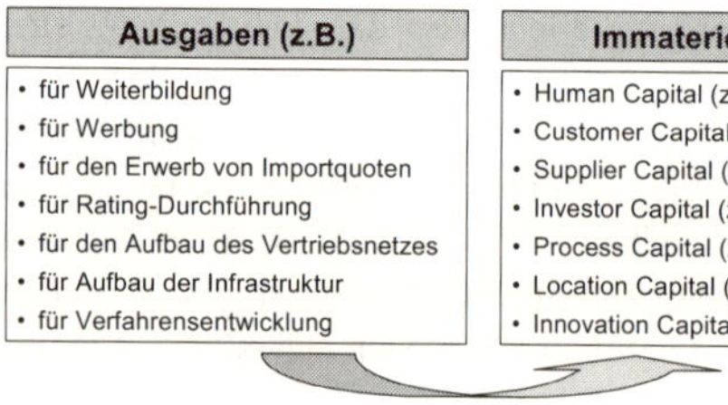

Abb. 3: Zusammenhang von Ausgaben und immateriellen Werten

Aktivierung als immaterieller Vermögensgegenstand

Ansatzkriterien: Abstrakte und konkrete Aktivierungsfähigkeit

Entsprechend dem einleitend angeführten Bekenntnis des Arbeitskreises zu den Systemgrundsätzen des bestehenden deutschen Bilanzrechts, wird bei dem entwickelten Vorschlag an den im Rahmen der GoB etablierten Ansatzkriterien festgehalten und nicht – obwohl sich dies im Zuge der Internationalisierungsentwicklungen der Rechnungslegungsnormen innerhalb Europas auch anbieten würde – auf den stärker dynamisch ausgerichteten „asset"-Begriff der IAS/IFRS zurückgegriffen.

Gemäß der traditionellen deutschen Rechnungslegungskonzeption müssen immaterielle (wie materielle) Werte abstrakt und konkret aktivierungsfähig sein, um als Vermögensgegenstand in der Handelsbilanz aktiviert zu werden. Hinsichtlich der abstrakten Aktivierungsfähigkeit wird ein Vermögensgegenstand im Wesentlichen durch seine selbstständige Verwertbarkeit charakterisiert, die gegeben ist, wenn ein Gegenstand, zumindest dem Wesen nach außerhalb des Unternehmens monetär verwertet werden kann und damit auch einzeln bewertbar ist (vgl. hierzu *Baetge/Kirsch/Thiele*, 2002, S. 127ff. und *Adler/Düring/Schmaltz,* 1998,

§ 246 Tz. 15ff.). In der Regel dürfte dies durch Veräußerung geschehen, so dass hier einschränkend auch das Kriterium der Einzelveräußerbarkeit genannt wird (vgl. zur abstrakten Aktivierungsfähigkeit im Sinne der Einzelveräußerbarkeit u.a. *Freericks*, 1976, S. 141ff.). Eine solche selbstständige Verwertbarkeit oder Einzelveräußerbarkeit ist bei den als Rechte konkretisierten immateriellen Werten (z.B. gewerbliche Schutzrechte, Konzessionen) regelmäßig erfüllt. Für immaterielle Werte, bei denen eine solche Konkretisierung durch ein Recht nicht bzw. noch nicht stattgefunden hat (z.B. ungeschützte Erfindungen oder noch nicht patentierte Innovationen), ist die Erfüllung der allgemeinen Ansatzkriterien bei Ermangelung einer Veräußerung häufig sehr fraglich.

Hinsichtlich der konkreten Aktivierungsfähigkeit kann bisher von einer Gleichbehandlung materieller und immaterieller Vermögensgegenstände keine Rede sein. Der Gesetzgeber hat Vorsichts- und Objektivierungsgesichtspunkte durch das in § 248 Abs. 2 HGB postulierte Aktivierungsverbot für immaterielle Vermögenswerte des Anlagevermögens überbetont und insofern eine Diskriminierung selbst erstellter immaterieller Vermögensgegenstände des Anlagevermögens hingenommen. § 248 Abs. 2 HGB steht wohl mit dem Vollständigkeitsgebot im Sinne des § 246 Abs. 1 HGB in Konflikt und reduziert insofern auch die Rechenschaftspflicht. Wegen des Wandels von den materiellen hin zu immateriellen Werttreibern wird aufgrund des Aktivierungsverbots immer weniger „Vermögen" in der Bilanz sichtbar. Vor diesem Hintergrund wäre die Abschaffung von § 248 Abs. 2 HGB u.a. mit dem Vollständigkeitsgebot und den ordnungsmäßigen Rechenschaftspflichten zu begründen (so auch z.B. *Busse von Colbe*, 2000, S. 662; *Pellens/Fülbier*, 2000, S. 58). Immaterielle Werte hätten folglich - wie materielle Werte - ausschließlich der abstrakten Aktivierungsfähigkeit zu genügen und immaterielle Vermögensgegenstände des Anlagevermögens würden wie solche des Umlaufvermögens behandelt werden, auf die das Ansatzverbot des § 248 Abs. 2 HGB eindeutig nicht anzuwenden ist.

Die Abschaffung von § 248 Abs. 2 HGB und mit ihr die Aktivierung selbst erstellter immaterieller Werte im Anlagevermögen ist nach der 4. EG-Richtlinie (78/660/EWG) zulässig, da Art. 9 bzw. 10 Buchstabe C Abs. I Nr. 2b hier ein Mitgliedstaatenwahlrecht vorsieht. Zudem würde diese Abschaffung einen Schritt in Richtung international üblicher Vorgehensweisen darstellen und wird deshalb auch im Rahmen des DRS 12 dem Gesetzgeber empfohlen. Ein § 248 Abs. 2 HGB vergleichbares Aktivierungsverbot existiert nicht im IAS/IFRS- oder US-GAAP-System (vgl. ausführlich zur Bilanzierung immaterieller Vermögenswerte nach IAS und US-GAAP *von Keitz*, 1997; *Pellens/Fülbier*, 2000, S. 35 ff.).

Kriterien zur Objektivierung der abstrakten Aktivierungsfähigkeit

Bei Abschaffung des § 248 Abs. 2 HGB stellt sich die Frage, ab wann Ausgaben zur innerbetrieblichen Erstellung immaterieller Vermögensgegenstände aktiviert werden müssen. Analog zu den „Anlagen im Bau", unter denen unfertige, selbst erstellte Gegenstände des Sachanlagevermögens subsumiert werden, existieren auch „immaterielle Vermögensgegenstände in der Entstehung". Der Erstellungs- und Entstehungsprozess ist hier zum Bilanzstichtag noch nicht abgeschlossen. Bei der Frage nach dem genauen Aktivierungszeitpunkt hilft die Analogie zum Sachanlagevermögen nur bedingt weiter. Ausgaben werden dort sofort aktiviert, sofern keine Zweifel daran bestehen, dass die im Bau befindliche Anlage bei Fertigstellung aktivierungsfähig ist (vgl. *Adler/Düring/Schmaltz,* 1997, § 266 Tz. 66ff.; *Ellrott/Schmidt-Wendt* 2003, § 247 Tz. 561). Diese Zweifel scheinen bei materiellen Anlagegegenständen in der Regel kaum aufzutreten, zumal die angesetzten Ausgaben im Rahmen eines nachgelagerten Niederstwerttests auf ihre Amortisierbarkeit hin zu überprüfen sind. Im Gegensatz hierzu sind solche Zweifel im Zusammenhang mit „immateriellen Vermögenswerten in der Entwicklung" aufgrund deren mangelnden Greifbarkeit und Flüchtigkeit eher gerechtfertigt.

Für auf die Schaffung solcher Vermögenswerte bezogene Ausgaben sind somit konkrete Charakteristika festzulegen, deren Erfüllung mit genügend großer Objektivität einen immateriellen Vermögenswert in Entstehung annehmen lässt. Vergleichbar zu den in IAS 38.40 ff. und den US-GAAP (z.B. SFAS 86, SOP 98-1 hinsichtlich der Aktivierung selbsterstellter Software) entwickelten Anforderungen (gemäß IAS 38.45 muss der Bilanzierende u.a. die technische Fähigkeit, die Ressourcen und auch die Absicht besitzen, das aus der Entwicklung resultierende Produkt der Marktreife zuzuführen, wobei die Existenz eines Marktes im Fall externer Verwendung ebenfalls nachzuweisen ist; vgl. dazu ausführlich *Fülbier/Honold/Klar*, 2000, S. 836 ff.) schlägt demnach der Arbeitskreis vor, für die handelsrechtliche Bilanzierung von „immateriellen Vermögensgegenständen in der Entstehung" „de lege ferenda" konkretisierende Anforderungen zu formulieren, die auf die (abstrakte) Aktivierungsfähigkeit des durch entsprechende Ausgaben geschaffenen immateriellen Vermögensgegenstandes bei Fertigstellung hindeuten. Hierdurch soll gewährleistet werden, dass für immaterielle Werte in der Entstehung die gleichen Aktivierungskriterien Anwendung finden wie für materielle Vermögenswerte, nur werden sie hier, dem besonderen Charakter dieser Werte Rechnung tragend, weiter konkretisiert.

Da sich der Entstehungsprozess meist im Rahmen von Projekten vollzieht, soll zur objektivierten Abbildung realer Gegebenheiten der Projektbezug gewählt werden. Bei der Aktivierung projektbezogener Ausgaben ist darauf zu achten, dass die Abgrenzung der Vermögensgegenstände nicht verschwimmt. So ist z.B. vorstellbar, dass bestimmte Projektausgaben zur Schaffung materieller Rahmenbedingungen (Laboraustattungen etc.) dem Projekt und damit direkt dem immateriellen Vermögensgegenstand zugerechnet werden. Alternativ – insbesondere dann, wenn die materiellen Vermögensgegenstände nicht ausschließlich in einem Projekt eingesetzt werden – ist der Vermögensgegenstand zunächst dem Sachanlagevermögen zuzurechnen und dann ggf. über die Abschreibung dem immateriellen Vermögensgegenstand anteilig zuzuordnen.

Ein immaterieller Vermögensgegenstand in der projektbezogenen Entstehung sollte demnach „de lege ferenda" aktiviert werden, wenn folgende Anforderungen erfüllt sind:

- Projekt ist initiiert worden. Dies hat auf der Basis eines dokumentierten Geschäftsführungsbeschlusses zu geschehen und die Budgetfreigabe muss erfolgt sein.
- Projektabgrenzung und -beschreibung ist möglich. Das Projekt kann hinsichtlich seiner sachlichen, zeitlichen und finanziellen Dimension (z.B. Projektstrukturplan) hinreichend präzise abgegrenzt werden. Zur Erfüllung des Grundsatzes der Einzelbewertung müssen die projektbezogenen Ausgaben dem Projekt prinzipiell zurechenbar sein.
- Möglicher Projektnutzen ist darstellbar. Das Projekt muss einen Nutzen im Sinne der selbstständigen Verwertbarkeit besitzen und dieser Nutzen muss potenziell darstellbar sein.
- Aktive weitere Projektverfolgung ist sichergestellt.

Genügt das Projekt diesen Anforderungen, sind die projektbezogenen Ausgaben zu aktivieren (siehe später Punkt 3.2). Bei der Überprüfung der Anforderung ist nach dem so genannten „Management Approach" vorzugehen, d.h. die im internen Rechnungswesen festgehaltenen Sachverhalte und verwandten Daten sind für die Entscheidung über die Aktivierungsfähigkeit relevant (der Begriff wurde primär im Rahmen der Segmentberichterstattung nach US-GAAP geprägt; vgl. *Haller,* 2000, S. 768).

Problematisch bleiben Auslegungs- und Gestaltungsspielräume bei den Anforderungskriterien. Um Ausgestaltungsmöglichkeiten im Vorfeld, insbesondere bei den Dokumentationspflichten, weitgehend zu reduzieren und ein daraus resultierendes De-facto-Wahlrecht bei der Aktivierung „immaterieller Vermögensgegenstände in der Entstehung" zu reduzieren, stellen die jeweiligen Kriterien bereits auf die bloße Dokumentationsmöglichkeit ab. Ähnlich wie beim Umfang der Herstellungskosten darf für die Frage der Aktivierung von selbsterstellten immateriellen Werten nicht die Qualität der Unternehmensorganisation maßgebend sein. Ist diese Möglichkeit der Dokumentation gegeben, muss aktiviert werden, so dass die bilanzierenden Unternehmen in diesen Fällen letztlich zur Projektbeschreibung und Dokumentation (z.B. des Projektnutzens) verpflichtet sind.

Wie bei materiellen Anlagen im Bau gilt auch für die Aktivierung „immaterieller Vermögensgegenstände in der Entstehung" das (gemilderte) Niederstwertprinzip. Begründete Zweifel an dem Projektnutzen, insbesondere auch in den Folgeperioden, erfordern außerplanmäßige Abschreibungen ggf. aller Projektausgaben.

Die Aktivierung „immaterieller Vermögensgegenstände in der Entstehung" ist europarechtlich zulässig, sofern das Mitgliedstaatenwahlrecht in Art. 9 bzw. 10 Buchstabe C Abs. I Nr. 1 der 4. EG-Richtlinie (78/660/EWG) nicht nur die Aktivierung von Forschungs- und Entwicklungskosten für Innovation Capital erlaubt, sondern auch (Entwicklungs-)Ausgaben für z.B. Customer Capital.

Aktivierung als Rechnungsabgrenzungsposten

Sofern Ausgaben zur Schaffung immaterieller Werte nicht als Vermögensgegenstand aktiviert werden können, weil es an dem Vermögensgegenstandscharakter im Sinne der abstrakten Aktivierungsfähigkeit mangelt, ist (auch bereits in Rahmen einer „de lege lata"-Betrachtung) zu prüfen, ob die Ausgaben den Charakter eines aktiven transitorischen Rechnungsabgrenzungspostens aufweisen.

In der handelsrechtlichen Bilanzierung haben Rechnungsabgrenzungsposten den Charakter von Korrekturposten, die Vermögensgegenstände und Schulden ergänzen. Sie dienen der Periodisierung bestimmter Zahlungsgrößen und damit der periodengerechten bzw. nutzennäheren Erfolgsermittlung im Sinne des Grundsatzes der Abgrenzung der Sache und der Zeit nach (vgl. dazu u.a. *Coenenberg*, 2003, S. 378 ff.; *Trützschler*, 2002, § 250 Tz. 1 ff.). Gemäß § 250 Abs. 1 HGB sind aktive (transitorische) Rechnungsabgrenzungsposten Ausgaben vor dem Abschlussstichtag, die Aufwand erst für eine bestimmte Zeit nach diesem Abschlussstichtag darstellen.

Von einer späteren Erfolgswirksamkeit der Ausgaben ist zweifelsfrei auszugehen, wenn ihnen ein Rechtsverhältnis zugrunde liegt, das im Gegenzug eine spätere Leistungsgewährung vorsieht. Der Zusammenhang zwischen Ausgaben und späterer Erfolgswirkung ist hier vertraglich eindeutig identifiziert. Diese Sichtweise entspricht langjähriger BFH-Rechtsprechung (vgl. grundlegend BFH I 208/63 vom 31.5.1967, BStBl. III 1967, S. 607 (608); dazu auch *Hartung*, 1994, S. 220 m.w.N.). Im Kontext der von ihm behandelten Themenstellung sieht es der Arbeitskreis als zweckmäßig, gerechtfertigt und gesetzeskonform an, im Sinne einer erweiterten Interpretation auch einen mittelbaren, aber statistisch eindeutig nachweisbaren Zusammenhang zwischen der Ausgabe und ihrer späteren Erfolgswirkung als ausreichende Bedingung für den Ansatz eines Abgrenzungspostens zu fordern. So ist es z.B. nicht ungewöhnlich, bei Direktwerbemaßnahmen statistisch verlässliches Datenmaterial über den künftigen Werberücklauf in Form von Erträgen zu besitzen und damit einen eindeutigen Zusammenhang zu den getätigten Werbeausgaben herstellen zu können. Ähnliches kann z.B. auch für die Preisvergünstigung eines Handys als Gegenleistung für den Mobilfunk-Vertragsabschluss unterstellt werden.

Der Rückgriff auf statistisch verlässlich berechnete Zusammenhänge wird bei der Auslegung der „bestimmten Zeit" schon jetzt akzeptiert. Diesem unbestimmten Rechtsbegriff wird entsprochen, wenn ein Zeitpunkt oder -raum eindeutig kalendarisch bestimmt werden kann. Dabei wird die Zeit nicht auf das folgende Geschäftsjahr begrenzt, kann sich demnach auf einen späteren Zeitraum beziehen oder mehrere Geschäftsjahre umfassen (vgl. *Baetge/Kirsch/Thiele*, 2002, S. 466; von einem "strengen Zeitraumbezug" spricht *Coenenberg*, 2003, S. 379; ähnlich *Schildbach*, 2000, S. 169). Nach herrschender Literaturmeinung wird aber auch auf „statistisch verlässlich berechnete Durchschnittsgrößen umfangreicher Datenkollektive" (*Sieben/Ossadnik*, 1986, S. 50) zurückgegriffen, um den zeitlichen Zusammenhang zwischen der Ausgabe und ihrer späteren Erfolgswirksamkeit zu belegen. Statistisch zweifelhafte (z.B. bei statistisch nicht signifikanten Korrelationen), insbesondere subjektive Schätzungen sind aber grundsätzlich ausgeschlossen. Entsprechend sind Werbeausgaben, deren spätere Erfolgswirksamkeit nicht anhand von statistischen Daten nachgewiesen werden kann, sofort als Aufwand zu erfassen.

Der Rückgriff auf mittelbare, statistisch abgesicherte Zusammenhänge unter völligem Verzicht auf nicht fundierte Schätzungen ist demnach schon heute gesetzeskonform. Angesichts der Übereinstimmung von § 250 Abs. 1 Satz 1 HGB mit Art. 18 Satz 1 der 4. EG-Richtlinie (78/660/EWG) liegt diese Konformität auch europarechtlich vor. Die Charakteristika eines Rechnungsabgrenzungspostens sind in der 4. EG-Richtlinie sogar weniger restriktiv ausgestaltet als im deutschen Handelsrecht. So ist das Zeitbestimmtheitspostulat nur in der deutschen Fassung von Art. 18 Satz 1 der 4. EG-Richtlinie enthalten, während es in der offiziellen englischen und auch französischen Fassung nicht auftaucht (vgl. bereits *Schulze-Osterloh*, 1992, S. 664; dazu auch *Beisse*, 1995, S. 70; *Hartung*, 1994, S. 216 f.). In der Literatur mehren sich daher die Stimmen, dem Kriterium der „bestimmten Zeit" keine größere Bedeutung mehr beizumessen, sofern nur ein „objektiv nachprüfbares bilanzrechtliches Synallagma zwischen Ausgabe und Aufwand" (*Hartung*, 1994, S. 222) besteht. Dem steht jedoch die von Vorsichts- und Objektivierungsgesichtspunkten geprägte Argumentation des Gesetzgebers bei der Aktiengesetzreform 1965 entgegen, der dieses Kriterium in das reformierte AktG integrierte, um die damals ausufernde Ausgabenaktivierung zu begrenzen (vgl. dazu *Trützschler*, 2002, § 250 Tz. 9 ff.). Dieser Argumentation soll hier nicht widersprochen werden. Vielmehr wird das Zeitbestimmtheitspostulat des § 250 Abs. 1 Satz 1 HGB hier nicht in Frage gestellt, da es dem Rückgriff auf mittelbare, statistisch eindeutig nachweisbare Zusammenhänge zwischen der Ausgabe und ihrer späteren Erfolgswirkung in sachlicher und auch zeitlicher Hinsicht nicht entgegen steht, gleichzeitig aber wesentlich zur Objektivierung der als Abgrenzungsposten zu erfassenden Sachverhalte dient.

Aktivierung als Bilanzierungshilfe

Sofern Ausgaben zur Schaffung immaterieller Werte weder unter den Vermögensgegenständen noch unter den Rechnungsabgrenzungsposten subsumiert werden können, kommt möglicherweise eine Aktivierung als Bilanzierungshilfe in Frage. Das Handelsrecht verwendet diesen Begriff explizit nur im Zusammenhang mit Ingangsetzungsaufwendungen (§ 269 HGB) und aktiven latenten Steuern (§ 274 Abs. 2 HGB). Filtert man bei diesen beiden Posten die Gemeinsamkeiten heraus, so handelt es sich hier um gesondert auszuweisende Aktivposten, die keine Vermögensgegenstände darstellen, die nur von Kapitalgesellschaften per Ansatzwahlrecht gebildet werden können, für die eine Ausschüttungssperre existiert und bei denen spezielle Auflösungsvorschriften greifen (vgl. *Busse von Colbe*, 1986, S. 88). In diesem engen handelsrechtlichen Sinne sieht der Arbeitskreis keine Möglichkeiten zur Aktivierung von Ausgaben zur Schaffung immaterieller Werte unter den Bilanzierungshilfen. Auch der enge Rahmen der 4. EG-Richtlinie (78/660/EWG) steht hier entgegen (Art. 37 Abs. 1 der 4. EG sieht allenfalls die Möglichkeit vor, Forschungs- und Entwicklungsaufwendungen im Sinne einer Bilanzierungshilfe zu behandeln).

Aktivierung als Goodwill

Ausgaben für immaterielle Werte können auch im Rahmen eines Unternehmenserwerbs angefallen sein. Sind diese Werte abstrakt aktivierungsfähig, müssen sie in der Einzel- bzw. Konzernbilanz des erwerbenden Unternehmens angesetzt werden (Auswirkungen auf den Einzelabschluss ergeben sich im Rahmen einer Übernahme i.S.d. § 255 Abs. 4 HGB). Sie gelten als erworbene immaterielle Vermögensgegenstände und fallen somit schon nach geltender Rechtslage nicht unter das Aktivierungsverbot des § 248 Abs. 2 HGB und damit nicht unter den Goodwill bzw. derivativen Firmenwert. Diese Ansatzpflicht gilt auch für immaterielle Vermögensgegenstände, die in der Bilanz des erworbenen Unternehmens nicht angesetzt worden sind (vgl. zu § 255 Abs. 4 HGB z.B. *Knop/Küting*, 1995, § 255 Tz. 453; zur Abbildung im Konzernabschluss z.B. *Adler/Düring/Schmaltz* 1996, § 301, Tz. 80, entsprechend auch die Forderung in IAS 22.26 ff., SFAS 141.39, DRS 4.17).

Erfüllen die erworbenen immateriellen Werte hingegen nicht die Kriterien der abstrakten Aktivierungsfähigkeit, werden sie als Bestandteil des Goodwill bzw. derivativen Firmenwerts bilanziert. Der Goodwill (bzw. derivative Firmenwert) bezeichnet den bei der Bilanzierung von Unternehmenserwerben im Rahmen eines so genannten *share deal* (bzw. *asset deal*) verbleibenden positiven Unterschiedsbetrag zwischen Gesamtkaufpreis und Summe der Zeitwerte aller vorhandenen Aktiva abzüglich der Schulden zu Zeitwerten. Er ist damit ein „verfahrensbedingter, also technischer Differenzbetrag" (*Moxter*, 1979, S. 743), der in seiner Rechtsnatur uneinheitlich beurteilt wird: als Vermögensgegenstand (vgl. z.B. *Biener/Berneke*, 1986, S. 117; *Ellrott/Aicher*, 2003, § 255 Tz. 511), als Bilanzierungshilfe (vgl. z.B. *Knop/Küting,* 1995, § 255 Tz. 432) oder auch als Wert eigener Art („aliud") (vgl. *Adler/Düring/Schmaltz*, 1995, § 255 Tz. 272; *Baetge/Kirsch/Thiele,* 2002, S. 263).

Um im Rahmen von Unternehmensakquisitionen die weitgehende Erfassung erworbener immaterieller Werte im Goodwill im Sinne der Generalnorm (§ 264 Abs. 2 und § 297 Abs. 2 HGB) zu vermeiden, wird unter Bezugnahme auf die Fiktion des entgeltlichen Einzelerwerbs die Forderung nach weitgehender Aufteilung des Unterschiedsbetrages und Ansatz aller aktivierungsfähigen immateriellen Werte abgeleitet (vgl. *Adler/Düring/Schmaltz* 1995, § 255, Tz. 265; ebenda § 301, Tz. 80). Die Bilanzierungspraxis zeigt jedoch, dass die Aufteilung des Unterschiedsbetrages auf immaterielle Werte immer noch die Ausnahme ist. Im Rahmen der Konzernrechnungslegung mag dies an der Möglichkeit der erfolgsneutralen Goodwillverrechnung liegen (§ 309 Abs. 1 Satz 3 HGB), die zumindest in der Vergangenheit von Unternehmensseite gegenüber der erfolgswirksamen Abschreibung immaterieller Vermögensgegenstände häufig präferiert wurde. Wie schon vielfach in der Literatur wird auch hier gefordert, die erfolgsneutrale Goodwillverrechnung gemäß § 309 Abs. 1 Satz 3 HGB abzuschaffen. Ein Verbot der erfolgsneutralen Goodwillverrechnung entspricht nicht nur internationalen Grundsätzen (SFAS 141, IAS 22), sie ist auch sachgerecht, da die im Kaufpreis entgoltenen Werte und Vorteile bei erfolgsneutraler Verrechnung mit keinem Aufwand korrespondieren und damit dem Grundsatz periodengerechter Erfolgsermittlung, insbesondere dem Prinzip der Abgrenzung der Sache und der Zeit nach, widersprechen (vgl. ähnlich *Weber/Zündorf*, 1998, § 309 Rn. 34). Da die 7. EG-Richtlinie (83/349/EWG) hier in Art. 30 Abs. 2 ein Mitgliedstaatenwahlrecht vorsieht, wäre die Abschaffung auch richtlinienkonform. Im Übrigen hat auch das DRSC die erfolgsneutrale Goodwillverrechnung in DRS 4.28 negativ beurteilt und untersagt sie sogar explizit in DRÄS 1, Art. 7, Ziff. 13.

Zwischenergebnis

Abb. 4 zeigt die gemäß des Vorschlags des Arbeitskreises bei Bilanzansatzentscheidungen in Zusammenhang mit selbst erstellten immateriellen Werten zu vollziehenden Schritte in Form eines Ablaufschemas zusammenfassend auf.

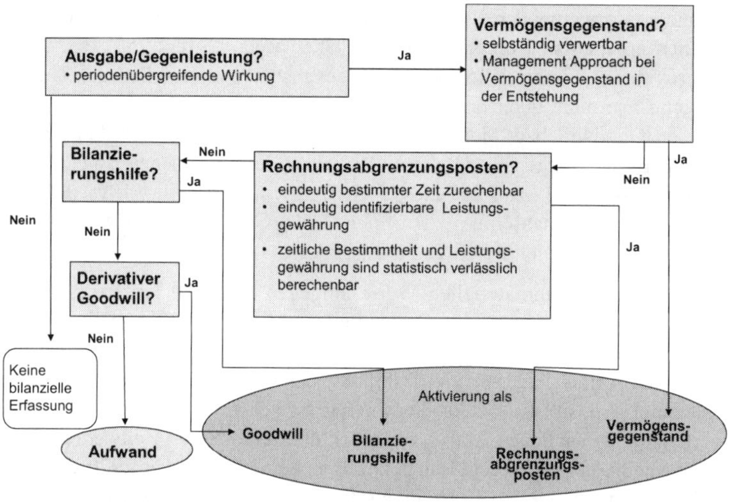

Abb. 4: Ablaufschema zur bilanziellen Erfassung immaterieller Werte

3.2 Bilanzbewertung (Bilanzierung „der Höhe nach")

Grundsätzliches

Wie bei materiellen so ist auch bei immateriellen Vermögensgegenständen hinsichtlich des Bewertungszeitpunktes zwischen Zugangs- und Folgebewertung zu differenzieren. Ebenso bestimmt die Form des Zugangs des Vermögensgegenstandes den relevanten Wertmaßstab, nämlich Anschaffungskosten bei Fremdbezug und Herstellungskosten bei Selbsterstellung. Darüber hinaus ist zwischen dem Zugang eines einzelnen Vermögensgegenstandes und dem Zugang eines Bündels von Vermögensgegenständen zu unterscheiden. Grundsätzlich verursacht die durch den Arbeitskreis empfohlene Aufgabe des Aktivierungsverbots gemäß § 248 Abs. 2 HGB keine Änderung oder gar Erweiterung der „de lege lata" angewandten Bewertungsprinzipien und -regeln, insbesondere aufgrund des eingangs erwähnten Bekenntnisses des Arbeitskreises zu den tradierten Systemgrundsätzen, vor allem dem Grundsatz der Pagatorik. Somit wird von dem Vorschlag einer regelmäßigen Neubewertung oder einer grundsätzlichen Bewertung zum Zeitwert über die ursprünglichen Anschaffungs- oder Herstellungskosten hinaus – analog zu DRS 12.A9, jedoch im Gegensatz zu IAS 38.64 – abgesehen. Die folgenden Ausführungen beschränken sich deshalb auf eine kurze Darstellung der Anwendung der bestehenden Bewertungsregeln auf nach den oben dargelegten Grundsätzen aktivierungsfähige, selbst erstellte und entgeltlich erworbene immaterielle Vermögensgegenstände sowie die Diskussion von dabei auftretenden spezifischen Detailproblemen.

Zugangsbewertung beim Erwerb

Nach § 253 Abs. 1 HGB ist ein entgeltlich erworbener immaterieller Vermögensgegenstand im Zeitpunkt des Zugangs mit seinen Anschaffungskosten anzusetzen, die alle dem Gegenstand einzeln zurechenbaren Ausgaben enthalten, die geleistet werden, um den Gegenstand zu erwerben und ihn in einen betriebsbereiten Zustand zu bringen; d.h. Nebenkosten und nachträgliche Anschaffungskosten sind zu addieren und Anschaffungspreisminderungen sind abzusetzen (§ 255 Abs. 1 HGB; vgl. hierzu im Detail *Adler/Düring/Schmaltz,* 1995, § 255, Tz. 21 ff.; dies entspricht den Bestimmungen des IAS 38.22 ff.). So umfassen die Anschaffungskosten erworbener Software z.B. neben deren Kaufpreis auch die Ausgaben für Installation, Customizing etc., die dieser einzeln zurechenbar sind.

Wird ein immaterieller Vermögensgegenstand im Rahmen eines Tausches erworben, so gelten hier die GoB, die ein Wahlrecht gewähren, den Buchwert des hingegebenen Vermögensgegenstandes fortzuführen oder den neuen Gegenstand zum beizulegenden Wert des hingegebenen (alten) Gegenstandes zu bewerten (vgl. *Adler/Düring/Schmaltz*, 1995, § 255, Tz. 89 ff.; konkrete Bestimmungen für die Anwendung der ergebniswirksamen oder -neutralen Methode enthält IAS 38.34f.).

Wird ein Vermögensgegenstand nicht einzeln, sondern im Verbund mit anderen Gegenständen erworben (sog. „Güterbündel"), so ist gemäß dem Grundsatz der Einzelbewertung jeder Gegenstand für sich zu aktivieren und zu bewerten. Folglich hat der Erwerber die Anschaffungskosten des Güterbündels (inkl. Nebenkosten, nachträglichen Anschaffungskosten und Anschaffungspreisminderungen) auf die einzelnen Vermögensgegenstände des Bündels aufzuteilen. Als Wertmaßstab hat hierbei der beizulegende Zeitwert (*fair value*) des jeweiligen Vermögensgegenstandes zu gelten. Dabei handelt es sich um den Wert, der sich bei sachkundigen und unabhängigen Parteien zu marktgerechten Konditionen als Kaufpreis ergeben würde, wenn der Gegenstand einzeln erworben werden würde. Zur Bestimmung dieses Wertes ist bei Fehlen von Marktpreisen und Preisen vergleichbarer Transaktionen auch die Anwendung von Ertrags- bzw. Cashflow-Barwerten angebracht. Eine Bewertung der Gegenstände über ihrem beizulegenden Zeitwert ist aufgrund des Vorsichtsprinzips nicht zulässig. Übersteigt somit der für das Bündel bezahlte Kaufpreis die Summe der erworbenen Zeitwerte, so ist die Differenz im Einzelabschluss als sonstiger betrieblicher Aufwand zu erfassen. Liegt der Kaufpreis unter der Summe der Zeitwerte, so sind bei der Festlegung der Anschaffungskosten der einzelnen Vermögensgegenstände die Zeitwerte abzustocken. Bei Erwerb von immateriellen Vermögensgegenständen im Rahmen eines Güterbündels erscheint es aufgrund der mit der Zeitwertbestimmung dieser Gegenstände i.d.R. verbundenen größeren Unsicherheiten sachgerecht, diese bei der Abstockung relativ stärker abzuwerten als materielle Gegenstände (in Analogie zu *Adler/Düring/Schmaltz*, 1995, § 255, Tz. 106 f. u. § 248, Tz. 20).

Ein typisches Beispiel eines Bündels, das aus einem immateriellen und einem materiellen Vermögensgegenstand besteht, stellen diejenigen Mobiltelefonverträge dar, bei denen der Vertrag neben der Leistung der Netznutzung auch die Lieferung eines Telefons umfasst. Soweit die beiden Elemente des Vertrages, die Netzleistung und das Telefon, getrennt verwertet werden können, handelt es sich um zwei Vermögensgegenständc, deren einzelne Anschaffungskosten aus Erwerbersicht einzeln zu Zeitwerten zu bewerten sind.

Ein häufiger Spezialfall des Erwerbs eines Güterbündels ist der Erwerb eines ganzen Unternehmens oder eines Unternehmensteils. Im Rahmen eines so genannten *asset deal*, d.h. dem tatsächlichen Erwerb der Vermögensgegenstände und Schulden (unmittelbare Rechtsnachfolge) durch das Käuferunternehmen, werden die (materiellen und immateriellen, auch wenn letztere beim Bilanzierenden nicht bilanziert wurden) Vermögensgegenstände und Schulden des erworbenen Unternehmens direkt in die Bilanz des Erwerbers übernommen und einzeln zum beizulegenden Wert (Zeitwert) bewertet. Weicht der Kaufpreis von der Summe der Zeitwerte der erworbenen Vermögensgegenstände und Schulden (= Nettovermögen zu Zeitwerten) ab, so stellt die Differenz einen so genannten „derivativen Firmenwert“ (Goodwill) dar, der positiv oder negativ sein kann und wie zuvor dargestellt bilanziert werden sollte. Bei einem *share deal* ist im Rahmen der Konzernabschlusserstellung analog zu verfahren. Auch hier gilt trotz des faktischen Erwerbs von Unternehmensteilen gem. § 300 HGB die so genannte „Einzelerwerbsfiktion“ im Hinblick auf immaterielle und materielle Vermögensgegenstände und Schulden (vgl. *Coenenberg*, 2003, S. 588 f.).

Zugangsbewertung aktivierungsfähiger immaterieller Vermögensgegenstände bei Herstellung

Gemäß § 253 Abs. 1 HGB sind auch selbst erstellte immaterielle Vermögensgegenstände mit ihren Herstellungskosten anzusetzen, wobei hierfür entweder ein Teilkosten- oder ein Vollkostenansatz gewählt werden kann (während IAS 38.53-55 momentan lediglich einen Vollkostenansatz zulässt, sieht das IASB für die Zukunft einen Teilkostenansatz vor; siehe Exposure Draft of Proposed Amendments to IAS 38, ED-IAS 38.58).

Die Herstellungskosten eines immateriellen Vermögensgegenstandes sind auf die gleiche Art zu ermitteln wie diejenigen eines materiellen Gutes. So ergäben sich die Herstellungskosten der oben bereits erwähnten Software im Falle der Selbsterstellung aus den direkten Kosten, i.e. den Gehaltskosten der Programmierer, Kosten für Material etc., aus den anteiligen Kosten der Computernutzung während der Entwicklung, aus den Personalkosten für die Installation und aus den Kosten, die für die Anpassung an die Unternehmensbedürfnisse anfallen.

Bei der Bewertung der „immateriellen Vermögensgegenstände in der Entstehung“ müssen die einbezogenen Kosten einen eindeutigen Projektbezug aufweisen, um das Ermessen des Bilanzierenden möglichst gering zu halten. Diesbezüglich empfiehlt der Arbeitskreis, § 255 Abs. 2 HGB um eine entsprechende Klarstellung zu ergänzen (ein klarer Projektbezug wird auch in IAS 38.39 ff. gefordert). So würde man z.B. bei selbst entwickelter Software nur die Löhne der direkt mit der Programmierung, dem Test, der Installation und dem Customizing betrauten Personen in die Herstellungskosten einrechnen, sofern das Projekt insoweit abgegrenzt wurde.

Die für die Ermittlung der Herstellungskosten erforderliche zeitliche Bestimmung des Herstellungsvorgangs muss sich ebenfalls aus der konkreten Projektsituation ergeben. Folglich sind nur die Herstellungskosten zu aktivieren, die ab dem Zeitpunkt anfallen, in dem alle oben erläuterten Ansatzkriterien erfüllt sind, d.h. ab dem von einem „immateriellen Vermögenswert in Entstehung“ gesprochen werden kann. Eine solche Festlegung des Zeitraumes entspricht sowohl § 255 Abs. 2 S. 5 HGB als auch IAS 38.53 und dient ebenfalls der Eindämmung von Willkür bei der Zurechnung von Ausgaben zu den Herstellungskosten immaterieller Vermögensgegenstände.

Ausgaben, die nach Abschluss der Herstellung im Zusammenhang mit einem immateriellen Vermögensgegenstand getätigt werden und nicht dessen Vermarktung dienen, sind daraufhin zu prüfen, ob es sich um Erhaltungs- oder um nachträgliche Herstellungskosten handelt. Nachträgliche Herstellungskosten sind dem Buchwert des Gutes zuzurechnen. Sie werden in der Folge so behandelt wie der Buchwert des Gutes. Dieses Vorgehen entspricht § 255 Abs. 1 HGB und IAS 38.60-62. So wären die Kosten für die Behebung von Programmierfehlern sicherlich als Erhaltungsaufwand einzustufen, die Kosten des Umbaus der Software zu einer neuen Version dagegen als Herstellungsaufwand der neuen Version.

Sollten in einem einzigen Herstellungsprozess mehrere Vermögensgegenstände gleichzeitig entstehen und lassen sich die Herstellungskosten der einzelnen Vermögensgegenstände nicht klar separieren, so sind die Kosten des Herstellungsprozesses grundsätzlich nach sinnvoll erscheinenden Schlüsselgrößen bzw. Verfahren auf die Vermögensgegenstände zu verteilen (vgl. *Adler/Düring/Schmaltz*, 1995, § 255, Tz. 244).

Folgebewertung aktivierungsfähiger immaterieller Vermögensgegenstände

Auch bei immateriellen Vermögenswerten ist im Hinblick auf die Bewertung zwischen solchen zu differenzieren, die dazu bestimmt sind, dem Unternehmen voraussichtlich über einen längeren Zeitraum Nutzen zu stiften (d.h. dem Geschäftsbetrieb „dauernd" zu dienen; § 247 Abs. 2 HGB), und die deshalb ins Anlagevermögen aufzunehmen sind, und solchen, die diese Bestimmung nicht aufweisen, und die folglich im Umlaufvermögen auszuweisen sind. Ist bei den ersteren das Nutzungspotenzial begrenzt, d.h. nimmt es im Laufe der Zeit durch Gebrauch oder lediglich den Zeitablauf ab, so ist der Nutzenverzehr durch planmäßige periodische Abschreibungen über die geschätzte Nutzungsdauer in der Bilanzbewertung nachzuvollziehen (vgl. hierzu auch DRS 12.14). Folglich sollte die planmäßige Abschreibung nutzungsorientiert sein und versuchen, dem tatsächlichen Verzehr des wirtschaftlichen Nutzens pro Periode möglichst zu entsprechen (analog DRS 12.19). Bei immateriellen Vermögensgegenständen wird die Nutzungsdauer i.d.R aus den zugrunde liegenden Rechtsverhältnissen bzw. wirtschaftlichen Faktoren bestimmt (siehe auch DRS 12.15ff.).

Soweit sich ein Nutzenverzehr nicht konkret identifizieren und jeder Periode verursachungsgerecht zuordnen lässt, sollte – entsprechend den Anforderungen nach IAS 38.88 – die lineare Abschreibungsmethode Anwendung finden. Da dies bei immateriellen Vermögenswerten die Regel sein dürfte, bewirkt die Forderung der Anwendung der linearen Abschreibung eine Reduktion des Ermessensspielraums bei der Bestimmung der Abschreibungsmethode und eine Erhöhung der zwischenbetrieblichen Vergleichbarkeit.

Lässt sich die Nutzungsdauer nicht konkret bestimmen, so ist die „voraussichtliche Nutzungsdauer" bestmöglich zu schätzen (dies entspricht auch § 253 Abs. 2 HGB und (derzeit) IAS 38.79). Dabei ist die „betriebsindividuelle Nutzbarkeit" relevant, die sowohl von einer allgemeinen wirtschaftlichen Nutzungsdauer sowie rechtlichen Nutzungsdauer (d.h. einem rechtlich abgesicherten Nutzungszeitraum) abweichen kann (vgl. DRS 12.15; siehe auch *Adler/Düring/Schmaltz*, 1995, § 253, Tz. 369). Ist die Schätzung des Nutzungspotentials mit beträchtlichen Risiken verbunden, so sollte die Unsicherheit gemäß dem Vorsichtsprinzip durch eine relativ kurz angesetzte Nutzungsdauer Berücksichtigung finden. Für eine Festlegung einer maximal zulässigen Nutzungsdauer (wie z.B. 20 Jahre in IAS 38.79 und DRS 12.18) sieht der Arbeitskreis keine sachliche Rechtfertigung.

Im Sinne des Grundsatzes der sachlichen Abgrenzung sollte ein immaterieller Vermögensgegenstand nur in den Abrechnungszeiträumen planmäßig abgeschrieben werden, während denen er tatsächlich genutzt wird. Planmäßige Abschreibungen sollten daher nicht schon ab dem Erwerbszeitpunkt, sondern erst ab dem Zeitpunkt der Inbetriebnahme vorgenommen werden. Solange der Gegenstand noch nicht in Betrieb genommen wurde, liegt keine Nutzung vor und es gibt keine Rechtfertigung, eine planmäßige Abschreibung zur Berücksichtigung der Abnahme des Nutzungspotenzials vorzunehmen. Dies ist anders zu beurteilen, wenn trotz einer Nichtnutzung des Vermögensgegenstandes dessen Wert kontinuierlich aufgrund technischer und wirtschaftlicher Veralterung abnimmt (z.B. Veralterung einer genutzten Software oder des einer Lizenz zugrunde liegenden Verfahrens- oder Produkt-Know-how). Hierbei sorgt eine stetige Abbildung des periodischen Wertverzehrs in Form einer planmäßigen Abschreibung für eine bessere Entsprechung der Generalnorm als eine ebenfalls denkbare Vornahme einer außerplanmäßigen Abschreibung zum jeweiligen Bilanzstichtag.

Für immaterielle Vermögensgegenstände, deren Nutzenpotenzial zeitlich nicht begrenzt ist, besteht nach Auffassung des Arbeitskreises für eine planmäßige Abschreibung keine Rechtfertigung (ebenso DRS 12.20; a.A. z.B. *Adler/Düring/Schmaltz*, 1995, § 253, Tz. 356). Beispiele hierfür sind Wegerechte, dauerhafte Frequenzrechte ohne zeitliche Begrenzungen, Luftrechte für Wolkenkratzer, Markennamen oder Rechte an Literaturveröffentlichungen nach Ablauf von Autorenrechten. Ändert sich die Nutzungsdauereinschätzung und erweist sich eine zeitliche Begrenzung als realistischer, so ist ab dem Zeitpunkt der geänderten Einschätzung eine planmäßige Abschreibung entsprechend der voraussichtlichen zeitlich begrenzten Nutzung vorzunehmen (ebenso DRS 12.21).

Alle immateriellen Vermögenswerte unterliegen dem Niederstwertprinzip, wie es in § 253 Abs. 2 u. 3 HGB zum Ausdruck kommt. Danach ist zu jedem Bilanzstichtag der Buchwert mit dem jeweiligen beizulegenden Zeitwert (falls vorhanden mit dem Börsen- und Marktpreis) zu vergleichen. Hierbei ist regelmäßig der Wiederbeschaffungszeitwert des Vermögensgegenstandes zugrunde zu legen. Insbesondere bei Patenten oder anderen Nutzungsrechten ergibt sich der beizulegende Wert zumeist aus dem Ertragswert (vgl. auch *Adler/Düring/Schmaltz*, § 253 Tz. 464). Bei immateriellen Gegenständen des Umlaufvermögens dürfte häufig eine Relevanz der Absatzpreise bestehen. Während immaterielle Vermögensgegenstände des Umlaufvermögens bei einem Unterschreiten des Zeitwertes unter den Buchwert immer außerplanmäßig abzuschreiben sind (strenges Niederstwertprinzip; § 253 Abs. 3 HGB), ist bei Kapitalgesellschaften eine solche Abschreibung von immateriellen Vermögensgegenständen des Anlagevermögens nur vorzunehmen, wenn die Wertminderung voraussichtlich von Dauer ist (§ 253 Abs. 2 HGB; im Gegensatz zu DRS 12.22 und IAS 38.97-102, die von einem strengen Niederstwertprinzip für alle immateriellen Vermögenswerte ausgehen). Vorübergehende Wertminderungen dürfen im Bereich immaterieller Vermögensgegenstände des Anlageve-rmögens nicht zu außerplanmäßigen Abschreibungen führen.

Gemäß § 280 Abs. 1 HGB ist für Kapitalgesellschaften eine Wertaufholung vorzunehmen, soweit der Grund für die in einem vorangegangenen Geschäftsjahr vorgenommene außerplanmäßige Abschreibung weggefallen ist und die Voraussetzungen des § 280 Abs. 2 HGB nicht erfüllt sind (entspricht auch DRS 12.23). Aufgrund des Anschaffungskostenprinzips ist eine Wertzuschreibung nur maximal bis zu den fortgeführten Anschaffungs- und Herstellungskosten (d.h. Anschaffungs- und Herstellungskosten abzüglich etwaiger planmäßiger Abschreibungen) zulässig.

3.3 Bewertungsrelevante Angaben

Teilweise in Erweiterung der Bestimmungen nach IAS 38.107 und DRS 12.24 ff. hält der Arbeitkreis im Hinblick auf den Ausweis und die Bewertung die nachfolgenden Angaben für sachgerecht.

Im Rahmen des von Kapitalgesellschaften zu erstellenden Anlagespiegels gem. § 268 Abs. 2 HGB sollte hinsichtlich der zu gewährenden Daten für sämtliche ausgewiesenen Kategorien von immateriellen Vermögensgegenständen zwischen solchen, die entgeltlich erworben, und solchen, die selbst erstellt wurden, differenziert werden. Es sollte deutlich werden, welche Beträge auf erworbene und welche auf selbst erstellte immaterielle Vermögensgegenstände entfallen und welche Abschreibungsbeträge jeweils planmäßig und außerplanmäßig verrechnet wurden.

Um den Abschlussadressaten eine bessere Einschätzung der Bedeutung von immateriellen Vermögensgegenständen für ein Unternehmen im allgemeinen und innerhalb einer Periode im speziellen zu ermöglichen, sollte der Wert der bilanzierten immateriellen Vermögensgegenstände und seine Veränderungen in der Bilanz, im Anlagespiegel, in der Gewinn- und Verlustrechnung sowie in einer Kapitalflussrechnung (soweit erstellt) gesondert angegeben werden. Darüber hinaus sollten im Anhang sowohl für erworbene als auch für selbst erstellte immaterielle Vermögensgegenstände die entsprechenden Zeitwerte dargestellt werden, wenn sie verlässlich bestimmbar und für die Beurteilung der wirtschaftlichen Lage des Unternehmens wesentlich sind. Hierfür kommen zumindest solche immateriellen Vermögensgegenstände in Frage, für die – analog zu IAS 38.64 ff. – ein aktiver funktionsfähiger Markt vorliegt. Wesentliche Unterschiede zwischen Buchwerten und Zeitwerten sowie wesentliche Veränderungen der Buchwerte innerhalb einer Periode sollten erläutert werden. Ferner sind ihrem Wert nach als wesentlich zu beurteilende immaterielle Vermögensgegenstände zu erläutern. Hilfreich wäre es auch, die bei den planmäßigen Abschreibungen unterstellten Nutzungsdauern zu nennen.

4. Freiwillige externe Berichterstattung über immaterielle Werte

4.1 Intellectual Capital Statement als Bestandteil des Lageberichts

Aufgrund der Tatsache, dass immaterielle Werte auch bei Anwendung der oben dargestellten Bilanzierungsregeln (aber auch der international üblichen Standards IAS/IFRS und US-GAAP) wegen ihrer häufig mangelnden Identifizierbar- und Bewertbarkeit sich oft einer umfassenden bilanziellen Erfassung entziehen und darüber hinaus die Angabepflichten zu solchen Werten in den bestehenden Normen äußerst rudimentär sind, entwickelte der Arbeitskreis Grundsätze und einen Gestaltungsvorschlag, wie ein Unternehmen außerhalb der Bilanz freiwillig über seine „immaterielle Vermögenslage“ Bericht erstatten sollte. Hierbei entspricht der Arbeitskreis Intentionen des DRSC (siehe Empfehlung in DRS 12.33), aber auch jenen des FASB in den USA (siehe *FASB* 2003 u. *Haller/Dietrich* 2001), die Unternehmen zu einer verstärkten Berichterstattung über immaterielle Werte motivieren, um das in der Berichterstattungspraxis diesbezüglich feststellbare Informationsdefizit zu überwinden (vgl. hinsichtlich des Informationsdefizits z. B. *PriceWaterhouseCoopers/Günther/-Beyer*, 2003; allerdings gibt es im Gegensatz zu Deutschland in skandinavischen Ländern eine Reihe von Unternehmen, die umfangreich über immaterielle Werte Bericht erstatten, z. B. *Skandia*, *Celemi International*, *WM-data AB*, *KREAB*, *Jacobson & Widmark*, *Carl Bro a/s*, *Coloplast a/s).*

Das vom Arbeitskreis entwickelte Berichterstattungskonzept ist als Empfehlung gedacht, wie ein (Konzern-)Lagebericht um Informationen zu immateriellen Werten erweitert werden kann. Er konkretisiert damit einerseits die gemäß DRS 12.33 f. geforderte „Berichterstattung über das intellektuelle Kapital des Konzerns" und zeigt andererseits auch eine Möglichkeit auf, wie die im Rahmen des Value Reporting u.a. durch den *Arbeitskreis „Externe Unternehmensrechnung" der Schmalenbach-Gesellschaft* geforderte Berichterstattung über immaterielle Werte ausgestaltet werden kann (vgl. *Arbeitskreis „Externe Unternehmensrechnung" der Schmalenbach-Gesellschaft,* 2002, S. 2338 ff.). Die Empfehlung richtet sich vor allem an Unternehmen bzw. Konzerne, die von öffentlichem Interesse sind (z.B. große und kapitalmarktorientierte Unternehmen) oder bei denen immaterielle Werte für die wirtschaftliche Lage eine wesentliche Bedeutung besitzen, unabhängig davon, ob sie ihren (Konzern-) Abschluss nach HGB oder nach international anerkannten Grundsätzen erstellen (denn für Mutterunternehmen, die die Befreiung des § 292a HGB in Anspruch nehmen, und auch für alle nach dem 1.1.2005 nach IFRS bilanzierende Unternehmen besteht eine Pflicht zur Lageberichtserstellung).

Für die Bezeichnung des Berichterstattungskonzepts wird der Begriff „Intellectual Capital Statement" gewählt, da „Intellectual Capital" im Gegensatz zu „Intellectual Property" (vgl. zum Begriff z. B. *Maul/Menninger*, 2000, S. 529 ff.), das lediglich rechtlich geschützte immaterielle Werte, wie Patente oder Marken, umfasst, sämtliche immateriellen Werte, unabhängig von ihrer rechtlichen Schützbarkeit und Bilanzierungsfähigkeit beinhaltet (vgl. zum Begriff des „Intellectual Capital" (IC) *Edvinsson/Malone*, 1997, S. 3ff.).

Die Entwicklung der Konzeption und der konkreten Ausgestaltung des Vorschlags des Arbeitskreises wurde durch verschiedene Quellen gespeist. Zum einen wurden in der Literatur bzw. von anderen Institutionen vorgestellte und diskutierte Ansätze ausgewertet, insbesondere die "Guideline for Intellectual Capital Statements" der *Danish Agency for Trade and Industry* (*Danish Agency for Trade and Industry*, 2000). Zum anderen flossen die praktischen Erfahrungen der Mitglieder zum derzeit „Mach- und Umsetzbaren" ein. Der Arbeitskreis entschied sich für die Umsetzung eines sog. Indikatormodells. Prominente Beispiele für ein solches Modell sind der *Skandia*-Navigator (vgl. *Edvinsson/Malone*, 1997, S. 65ff.), das Liahona – H.O.M.E.S. Modell (Vgl. *Bornemann/Knapp/Schneider/Sixl*, 1999), der Intangible Assets Monitor (vgl. *Sveiby*, 1997, S. 11), der Intellectual Capital Navigator (vgl. *Stewart*, 1998, S. 236ff.), das Value Chain Scoreboard (vgl. *Lev*, 2001), das Intellectual Property Statement (vgl. *Maul/Menninger*, 2000, S. 529ff.) oder die Wissensbilanz (vgl. *Austrian Research Center*, 2002 und *Maul*, 2000, S. 2009ff.). Hierdurch können am ehesten unterschiedliche Sichten (z. B. monetär als auch nicht-monetär; Input-, Prozess- als auch Output-Betrachtungen) auf immaterielle Werte in der gewählten Klassifikation immaterieller Werte gewährleistet werden. Zudem ermöglicht der Indikatorenansatz eine Quantifizierung der Berichterstattung über immaterielle Werte auch für Sachverhalte, die nicht monetär gemessen werden können. Bei der Ausgestaltung des Vorschlags fanden auch wesentliche Indikatoren, die sich in empirischen Kapitalmarktstudien als wertrelevant erwiesen haben, Berücksichtigung (vgl. z. B. stellvertretend die Studien von *Aboody/Lev*, 1998, S. 161ff.; *Deng/Lev/Narin*, 1999, S. 20ff. und *Lev/Sougiannis*, 1996, S. 107ff.).

4.2 Berichtsgrundsätze

Nach Auffassung des Arbeitskreises sollten bei Erstellung eines „Intellectual Capital Statement" (ICS) folgende Berichtsgrundsätze Anwendung finden:

- Ziel der gesamten Berichterstattung ist die Darlegung von Strategien zum Management immaterieller Werte und der Identifikation von Werttreibern. Dabei soll sich die Berichterstattung sowohl auf aktivierte immaterielle Vermögenswerte als auch auf solche immaterielle Werte erstrecken, die (noch) keinen zahlenmäßigen Niederschlag im Abschluss gefunden haben bzw. nicht als immaterielle Vermögenswerte aktivierbar sind.

- Die Berichterstattung sollte anhand der unter Punkt 2.1 dargestellten sieben Intellectual Capital-Kategorien Innovation Capital, Human Capital, Customer Capital, Supplier Capital, Investor Capital, Process Capital und Location Capital erfolgen.
- Im ICS sollte das Management Ziele und Maßnahmen des Unternehmens im Hinblick auf immaterielle Werte konkret benennen sowie begründen, und zwar sowohl allgemein als auch jeweils innerhalb der einzelnen Kategorien.
- Der Management Approach fungiert als Ausgangspunkt, d.h. Auswahl und Zusammensetzung der im ICS dargestellten relevanten Indikatoren obliegen allein der Unternehmensleitung. Management Approach bedeutet jedoch auch, dass diejenigen Größen ausgewählt werden, die für die interne Unternehmensteuerung relevant sind. Hinsichtlich der periodisch dargestellten Indikatoren gilt der Grundsatz der Stetigkeit, wobei fakultative Erweiterungen bei Relevanz (z.B. branchentypische Spezifika) möglich sind.
- Es obliegt den Adressaten, die Indikatoren letztlich aus ihrem Blickwinkel und auf der Basis ihres Informationsstandes zu interpretieren. So sind z.B. Angaben zur Altersstruktur der Mitarbeiter nur unternehmens- oder branchenindividuell von Adressatenseite interpretierbar. Eine jüngere bzw. eine ältere Mitarbeiterstruktur ist damit niemals per se gut oder schlecht. Primat des ICS ist daher die Bereitstellung der Informationen, die Analyse und Einschätzung obliegt dem jeweiligen Adressaten.
- Die Aufnahme des ICS im Lagebericht stellt die sachgerechteste Lösung dar (vergleichbar dazu DRS 12.32ff und *Arbeitskreis „Externe Unternehmensrechnung" der Schmalenbach-Gesellschaft,* 2002, S. 2340). Gegen eine Aufnahme in den Anhang sprechen mehrere Gründe. Erstens entzieht sich der Inhalt des ICS dem Prüfungsziel der Abschlussprüfung. Zweitens wäre bei einer Aufnahme in den Anhang mit einer größeren Zurückhaltung der Unternehmen bei der Berichterstattung zu rechnen. Darüber hinaus liegt der Zweck des Anhangs primär in der Erläuterung der Zahlen und Inhalte des Abschlusses, und erstreckt sich gerade eben nicht auf Bereiche, die in den Abschlussinstrumenten keinen oder nur einen eingeschränkten Niederschlag (z.B. nur als laufender Aufwand) finden.
- Das ICS sollte sich nicht nur auf wesentliche immaterielle Werte beschränken, sondern zur besseren Vergleichbarkeit alle sieben IC-Kategorien umfassen. Falls ein Unternehmen in einer bestimmten Kategorie aus Managementsicht keine immateriellen Werte hält, ist bewusst eine Leermeldung zu erklären.
- Alle immateriellen Werte sind – unabhängig von ihrer Aktivierung – (entsprechend Tz. 37 von DRS 14) zu erläutern: Dabei ist allerdings, soweit dies sinnvoll ist, der Bezug zu Bilanz und GuV-Rechnung herzustellen. Auch auf die Nutzung von Werten, die in Vorperioden bereits in den Aufwand gegangen sind, ist einzugehen. Soweit möglich sollen (vorhandene) beizulegende Zeitwerte (z.B. Schätzungen von Markenwerten durch externe Dritte wie *Interbrand*, *GfK* oder *Nielsen*) mit angegeben werden.
- Es wird empfohlen, die Berichterstattung zur besseren Klarheit strukturell zu differenzieren. Die Angaben sind gemäß der Struktur der Segmentberichterstattung zu gliedern (vgl. *Arbeitskreis „Externe Unternehmensrechnung" der Schmalenbach-Gesellschaft,* 2002, S. 2340). Daneben sind weitere Strukturierungen wie die Darstellung der Fortbildungstage getrennt nach *white/blue collar*-Mitarbeitern oder nach Funktionsbereichen (z. B. Forschung und Entwicklung, Produktion, Verwaltung, Service etc.) denkbar.

- Allgemeine GoB (u.a. Klarheit, Richtigkeit, Willkürfreiheit, Vollständigkeit, Stetigkeit) sind auch für die freiwillige Berichterstattung über immaterielle Werte wesentlich (vergleichbar *Arbeitskreis „Externe Unternehmensrechnung" der Schmalenbach-Gesellschaft,* 2002, S. 2339f.). Die Berichtskontinuität ist insoweit wichtig, als unangenehme Tatsachen nicht verschwiegen werden dürfen, z. B. durch Weglassen von Indikatoren, bei denen die im Vorjahr verkündeten Sollwerte verfehlt wurden. Andererseits erscheinen angesichts der Tatsache, dass die Diskussion über sinnvolle Berichtsinhalte noch sehr stark im Fluss ist, Änderungen eher begründbar als im traditionellen Abschluss.

4.3 Darstellungsvorschlag eines Intellectual Capital Statement

Grundlegender Berichtsaufbau

Ein ICS beginnt mit einem allgemeinen Teil, in dem eine generelle Erläuterung der Strategie des Unternehmens im Umgang mit immateriellen Werten und des Zusammenhangs immaterieller Werte mit dem langfristigen Unternehmenserfolg des betrachteten Unternehmens erfolgt. Daran schließt sich für jede der sieben IC-Kategorien eine detaillierte Darstellung der für die jeweilige Kategorie als aussagefähig betrachteten Indikatoren an. Dabei ist für jede der Kategorien auf folgende vier Punkte einzugehen:

- **Strategie**
  Hierbei sind im Wesentlichen nachfolgende Fragen zu erläutern: Wo steht das Unternehmen in dieser Kategorie? Wo will es hin? Welchen Bezug, welche Bedeutung hat die IC-Kategorie für die Unternehmensstrategie? Denkbar sind z.B. Ausführungen zur Bedeutung von Marken bei Konsumgüterunternehmen oder Kundenbeziehungen bei Finanzdienstleistern oder gar Darstellungen als Ursache-Wirkungs-Diagramme analog zur Balanced Scorecard.

  Ob die Indikatoren, über die berichtet wird, mit der Strategie bzw. den Werttreibern sinnvoll zusammenhängen, lässt sich anfangs von Externen nur schwer beurteilen. Hier muss man (vorerst) dem Management folgen. Bei Unplausibilität kann es aber konstruktive Diskussionen geben. Außerdem lenkt die Berichterstattungspflicht die Aufmerksamkeit auf die Entwicklung der Indikatoren im Zeitablauf.

- **Katalog der Indikatoren**

  Auflistung der Indikatoren unter Angabe des aktuellen Werts der Berichtsperiode, Werte vergangener Berichtsperioden (soweit verfügbar) und wenn möglich des Zielwerts für zukünftige Perioden.

- **Definition und Wechselwirkung der Indikatoren**

  Wie kann die Entwicklung einer IC-Kategorie gemessen werden? Welcher Zusammenhang besteht zwischen mehreren Indikatoren? Es empfiehlt sich auf genau spezifizierte Definition der Indikatoren i.d.R. über exakte Berechnungsformeln zu achten, da die Interpretation der Indikatoren von deren Definition abhängt. Zum Beispiel kann die Neuproduktrate wie folgt definiert werden:

  $$\text{Neuproduktrate} = \frac{\text{Umsatz der in den letzten 3 Jahren neu eingeführten Produkte}}{\text{Gesamtumsatz}}$$

- **Kommentar**
  Wie sind die Indikatoren der aktuellen Berichtsperiode zu beurteilen? Die Werte der In-

dikatoren sind bezüglich ihrer vergangenen Entwicklung vom Ersteller zu diskutieren und im Zeitablauf zu beurteilen. In diesem Zusammenhang sollten Soll-Ist-Diskrepanzen erklärt werden, sofern das Unternehmen in den Vorperioden Soll-Werte angegeben hat.

Nach den kommentierten Kennzahlen zu den sieben IC-Kategorien folgt ein abschließender Überblick bzw. eine Zusammenfassung zu einem Gesamtbild in verbaler Form. *Abb. 5* stellt den Gesamtaufbau des Intellectual Capital Statement zusammenfassend dar.

Einführung:	Generelle Strategie des Managements immaterieller Werte; Zusammenhang zum langfristigen Unternehmenserfolg
Für jede Kategorie:	1. **Strategie** (spezifisch für die jeweilige Kategorie) 2. **Katalog der Indikatoren** - Aktueller Wert der Berichtsperiode - Aktuelle(r) Wert(e) vergangener Berichtsperioden - wenn möglich Sollwerte für zukünftige Periode(n) 3. **Definition und Wechselwirkung der Indikatoren** - Erläuterung und exakte Definition des Indikators und dessen erfolgte Differenzierung 4. **Kommentar** (Stand und Entwicklung der Kategorie)
Zusammenfassung:	Bewertung der immateriellen Werte als Einheit

Abb. 5: Aufbau des Intellectual Capital Statement

Als Darstellungsform erscheint eine tabellarische Auflistung, wie im nachfolgenden Vorschlag zu den einzelnen IC-Kategorien aufgezeigt, ergänzt um eine einleitende Ziel- und Maßnahmenbeschreibung und einen abschließenden Überblick sowie eine Würdigung sinnvoll. Diese kann fakultativ durch Diagramme (z.B. Entwicklung der Indikatoren über die Zeit), Netzdarstellungen (z.B. Ursache-Wirkungsketten), Strukturierung der Analyseebenen (z.B. die IC-Kategorien als Betrachtungsobjekte analog zu den Fenstern einer Balanced Scorecard) o.ä. bereichert werden. Zu einem besseren Verständnis erscheint die Orientierung an einer „Input-Prozess-Output-Logik“ und damit die (interne) Dreiteilung der Indikatoren in Input (*resource mix*), Prozess als Veränderung im Jahresverlauf (*upgrade, downgrade*) und Output (*impacts*) hilfreich. Dabei sind Output-Kennzahlen (z.B. Neuproduktrate) gegenüber Prozess-Kennzahlen (z.B. Anzahl laufender Entwicklungsprojekte (Produkt-Pipeline)) und Prozess-Kennzahlen gegenüber Input-Kennzahlen (z.B. F&E-Aufwand in % des Umsatzes) zu präferieren.

Zwischen einzelnen IC-Kategorien gibt es aufgrund der Natur der immateriellen Werte Überschneidungen. Zum Beispiel kann der Erfolg von Markenartikelunternehmen sowohl durch gute Kundenbeziehungen (Customer Capital), hohe und permanente Produktinnovationen (Innovation Capital) als auch gute und exklusive Prozess- und Produktqualität (Process Capital) bedingt sein. Eine eindeutige Identifizierbarkeit und damit Zuordenbarkeit zu einzelnen Positionen wie bei einem Vermögenswert wird bei immateriellen Werten nicht immer möglich sein. Zudem können einige immaterielle Werte je nach Fall unterschiedlichen IC-Kategorien zuordenbar sein. So können z.B. Netzwerke und Kooperationen in fallweise dem Innovation Capital (z.B. F&E-Kooperationen), dem Process Capital (z.B. Joint Ventures in der Produktion) oder dem Customer Capital (z.B. gemeinsame Erschließung von Kundennutzen) zugerechnet werden.

Berichterstattung zu einzelnen Kategorien immaterieller Werte

Nachfolgend wird für jede einzelne IC-Kategorie die Struktur der grundsätzlich als aussagekräftig erachteten Indikatoren zur Messung der immateriellen Werte dargestellt und erläutert:

Indikator	**Erläuterung/Differenzierung**
F&E-Ausgaben	Angabe der F&E-Ausgaben; F&E-Ausgaben im Verhältnis zum Umsatz; Informationen zur Streuung/Konzentration der F&E-Ausgaben
Portfolio von Patenten und ähnlichen Schutzrechten	Zahl, Zusammensetzung und (Rest-)Laufzeiten der Schutzrechte/Patente Vorstellbar wäre eine Art Anlagespiegel analog zum Sachanlagevermögen (Anfangsbestand, Zugänge, Abgänge, Umbuchungen, Endbestand) nur mit Stückzahlen, ohne Werte
Angemeldete Patente und ähnliche Schutzrechte	Zahl, Zusammensetzung der angemeldeten Schutzrechte/Patente
Patentklagen u. sonstige Schutzrechtsklagen	Zahl und Bedeutung aktueller Patent-/Schutzrechtsklagen/-ansprüche
Neuproduktrate	Umsatz der in den letzten drei Jahren eingeführten Produkte zum Gesamtumsatz

Abb. 6: Kategorie 1: Innovation Capital

Indikator	**Erläuterung/Differenzierung**
Altersstruktur der Mitarbeiter	Klassifizierung nach Altersgruppen (in Jahren): [<25] [25-39] [40-54] [>54]
Unternehmenszugehörigkeit	Klassifizierung der Zugehörigkeitsdauer (nach Jahren) [<5] [5-15] [>15]
Fluktuation	Zahl der Mitarbeiter, die in der Berichtsperiode das Unternehmen verließen, im Verhältnis zu Gesamtmitarbeiterzahl
Mitarbeiterqualifikation	Einteilung in % der Gesamtmitarbeiterzahl: Lehre/Ausbildung im Haus bzw. bei einem anderen Unternehmen ▪ Hochschulabschluss (Letztlich jedoch Management Approach, d.h. unternehmens- oder branchenindividuelle Auflistung von Abschlüssen; z.B. Sparkassenakademie, Bankakademie etc. bei Kreditinstituten)
Weiterbildung	▪ Angabe der Ausgaben (pro Mitarbeiter) Angabe der Weiterbildungstage (pro Mitarbeiter)
Mitarbeiterzufriedenheit	Wichtig ist insbesondere die Angabe und Erläuterung der Methode, auf deren Basis die Mitarbeiterzufriedenheit ermittelt wurde.
Fehlzeiten	Angabe der Tage pro Mitarbeiter
Wertbeitrag	Explizit berechnet als: (Wertschöpfung pro Mitarbeiter – Personalkosten pro Mitarbeiter) x Mitarbeiterzahl

Abb. 7: Kategorie 2: Human Capital

Die Mitarbeiterzahl wird bereits im Anhang als Information zur Verfügung gestellt (§ 285 Nr. 7 und § 314 Abs. 1 Nr. 4 HGB). Zudem wird für die Segmentberichterstattung die Angabe der Mitarbeiterzahl pro Segment empfohlen. Bei der Mitarbeiterzahl ist eine aussagefähige Definition zu wählen, die einen Vergleich zwischen den Unternehmen und deren verschiedenen Geschäftssystemen erlaubt (z.B. Umrechnung auf Vollzeitäquivalente).

Die Angabe zum Wertbeitrag folgt der Grundidee einer Zerlegung des Residualgewinnes (z.B. gemessen als EVA™ oder CVA) gemäß dem Workonomics™-Ansatz der *Boston Consulting Group* in die Wertschöpfung der Mitarbeiter und die Personalkosten (vgl. *Strack/Villis*, 2001, S. 67ff.). Der Ansatz ist insbesondere bei personalintensiven Geschäftssystemen (z.B. bei Filialunternehmen oder vertriebsstarken Unternehmen) von Interesse.

Indikator	**Erläuterung/Differenzierung**
Kundenzufriedenheit	Wichtig ist insbesondere die Angabe und Erläuterung der Methode, auf deren Basis die Kundenzufriedenheit ermittelt wurde. Eventuell ist ein Zugriff auf den sog. Kundenbarometer sinnvoll, der von relativ vielen Unternehmen standardisiert genutzt wird und damit Branchenvergleiche erlaubt.
Kundenqualität	Kundenbindungsdauer Wiederkaufrate (%-Satz der Kunden, die bereits im Vorjahr Umsätze tätigten) Großkunden (ABC-Analyse = %-Satz der Kunden, die 50 % bzw. 90% des Umsatzes ausmachen)
Marktanteil	Umsatz- bzw. stückzahlenbezogene Anteile pro Produkt oder Produktgruppe bzw. Markt (Wichtig ist insbesondere die Angabe zur Abgrenzung des Marktes)
Auflistung (wesentlicher) Marken	Denkbar wäre – wie bei den Patenten – eine Art Anlagespiegel (Anfangsbestand, Zugänge, Abgänge, Umbuchungen, Endbestand) nur mit Stückzahlen ohne Werte. Unter Umständen Angabe von Umsatz pro Marke
Wertbeitrag	Explizit berechnet als (Wertschöpfung pro Kunde – Kosten pro Kunde) x Kundenzahl, analog zum Wertbeitrag pro Mitarbeiter in der Kategorie 2

Abb. 8: Kategorie 3: Customer Capital

Analog zum Wertbeitrag beim Human Capital kann auch der Wertbeitrag beim Customer Capital kundenbezogen ermittelt werden. Nach dem Custonomics™-Ansatz der *Boston Consulting Group* (vgl. *Strack/Villis*, 2001, S. 67 ff.) lässt sich der Residualgewinn in eine Wertschöpfung pro Kunde und in die Kosten pro Kunde zerlegen. Hierdurch lässt sich insbesondere der Wert von Kundenbeziehungen in Unternehmen darstellen, die systematisch Kundenbeziehungen aufbauen und entwickeln (z. B. Finanzdienstleister, Energieversorgungs- oder Telekommunikationsunternehmen).

Indikator	**Erläuterung/Differenzierung**
Lizenzen	Zahl und Struktur erhaltener Lizenzen als Lizenznehmer
Schlüssellieferanten	Lieferantenbindungsdauer
Wertschöpfungstiefe	Umsatz – Materialaufwand – zugekaufte Dienstleistungen (Materialaufwand wird zwar selbst beim Umsatzkostenverfahren i.d.R. angegeben, zugekaufte Dienstleistungen gehen jedoch häufig im „sonstigen betrieblichen Aufwand" unter)

Abb. 9: Kategorie 4: Supplier Capital

Indikator	**Erläuterung/Differenzierung**
Aktionärsstruktur	Zahl/Anteilsquote in- und ausländischer Aktionäre Zahl/Anteilsquote privater und institutioneller Aktionäre Zahl/Anteilsquote der Belegschaftsaktionäre Free Float in %
Bedeutung bei Analysten	Zahl der Roadshows/Analystentreffen Durchschnittliche Teilnehmerzahl an den Analystentreffen Zahl der Analystenberichte
β-Faktor (EK-Markt)	Angaben zur Art der Ermittlung (z.B. gewählter Index und Zeitraum)
Bonität (FK-Markt)	Rating Zinsaufschlag bei emittierten Anleihen gegenüber risikolosem Zins
Ergebnisse von IR- und Geschäftsberichts-wettbewerben o.ä.	Freiwillige Angabe, sofern Daten vorhanden und bekannt

Abb. 10: Kategorie 5: Investor Capital

Indikator	**Erläuterung/Differenzierung**
Schnelligkeit der Prozessabläufe	Zeit für Auftragsabwicklung (Durchlaufzeit) Lieferzeit Liefertreue
Prozessqualität	Kennzahlen zur Messung der Prozessqualität (z.B. First Pass Yield, Liefergenauigkeit) Darstellung des Qualitätsmanagementsystems Wichtig ist insbesondere die Angabe und Erläuterung der Methode, auf deren Basis die Prozessqualität ermittelt wurde.
Produktqualität	Rückweisquote pro Produkt Beschwerdequoten Gewährleistungsaufwendungen / Umsatzerlöse Bewertung im Kundenbarometer

Abb. 11: Kategorie 6: Process Capital

Indikator	Erläuterung/Differenzierung
Standortqualität	Standorte der Produktion Standorte der F&E Standorte der Verwaltung
Medienpräsenz	Z.B. Verlinkungsdichte im Internet
Arbeitsmarktattraktivität (Attraktivität für „Key Personnel")	Z.B. Ranking als potentieller Arbeitgeber bei Hochschulabsolventen Eventuell Hinweis auf Ausbildungsstätten vor Ort (z.B. Halbleiterindustrie und TU Dresden)

Abb. 12: Kategorie 7: Location Capital

5. Zusammenfassung

Die Empfehlungen des Arbeitskreises wurden unter der Prämisse des Festhaltens an tradierten Systemgrundsätzen der handelsrechtlichen Rechnungslegung entwickelt. Sie lassen sich wie folgt zusammenfassen:

- Immaterielle Werte lassen sich in folgende sieben Kategorien einteilen: Innovation Capital, Human Capital, Customer Capital, Supplier Capital, Investor Capital, Process Capital, Location Capital.
- Das pauschale Aktivierungsverbot des § 248 Abs. 2 HGB für nicht entgeltlich erworbene immaterielle Vermögensgegenstände des Anlagevermögens erweist sich vor dem Hintergrund der wachsenden Bedeutung immaterieller Werte im Hinblick auf die Generalnorm und den Grundsatz der Vollständigkeit als nicht mehr haltbar. „De lege ferenda" sollten für immaterielle Vermögensgegenstände die gleichen Kriterien der abstrakten Aktivierungsfähigkeit gelten wie für materielle Vermögensgegenstände des Anlagevermögens sowie für immaterielle Vermögensgegenstände des Umlaufvermögens.
- Für „immaterielle Vermögensgegenstände in der Entstehung" bedarf es weiterer Anforderungen zur Objektivierung der abstrakten Aktivierungsfähigkeit. Nach Maßgabe einer die Daten des internen Rechnungswesens berücksichtigenden Vorgehensweise (Management Approach) sollen immaterielle Vermögensgegenstände in der (projektbezogenen) Entstehung „de lege ferenda" aktiviert werden, wenn das Projekt bereits initiiert worden, die Projektabgrenzung und -beschreibung möglich, der mögliche Projektnutzen darstellbar und die weitere aktive Projektverfolgung sichergestellt ist sowie die darauf entfallenden Kosten zuverlässig ermittelbar sind.
- Soweit eine Erfassung der Ausgaben für immaterieller Werte unter den Vermögensgegenständen aber auch nach Abschaffung des § 248 Abs. 2 HGB unzulässig ist, sollten die betreffenden Ausgaben künftig verstärkt dahingehend überprüft werden, ob sie unter den aktiven Rechnungsabgrenzungsposten zu aktivieren sind.
- Im Rahmen einer Unternehmensakquisition erworbene aktivierungsfähige immaterielle Werte sind einzeln zu aktivieren, soweit sie die Aktivierungskriterien erfüllen. Alle übrigen immateriellen Werte finden sich im derivativen Goodwill wieder, der zu aktivieren und nicht erfolgsneutral zu verrechnen ist.

- Aufgrund der Prämisse, am pagatorischen Prinzip festzuhalten, sind immaterielle Vermögensgegenstände grundsätzlich wie materielle zu bewerten. Im Speziellen sollten in einem Güterbündel erworbene Gegenstände zum beizulegenden Zeitwert angesetzt, die Herstellungskosten von „immateriellen Vermögensgegenständen in Entstehung" in sachlicher und zeitlicher Hinsicht Projekt bezogen ermittelt und Vermögensgegenstände, die keinem Nutzenverzehr unterliegen, nicht planmäßig abgeschrieben werden.
- Aufgrund ihrer beschränkten Abbildungsmöglichkeit im Abschluss sollte der Bedeutung immaterieller Werte insbesondere durch eine standardisierte, freiwillige Berichterstattung im Sinne eines Business Reporting innerhalb des Lageberichts Rechnung getragen werden. Zu diesem Zweck wurde die Erstellung eines „Intellectual Capital Statement" vorgeschlagen, das strukturierte und damit leichter vergleichbare Informationen zu immateriellen Werten in einer für die Adressaten nützlichen und für die Unternehmen in Bezug auf die damit entstehenden Kosten der Datengenerierung und -publizität zumutbaren Weise gewährt.

6. Literatur

Aboody, D. and *Lev, B.*, The Value-Relevance of Intangibles: The Case of Software Capitalization, in: Journal of Accounting Research, Supplement 1998, S. 161-191.

Adler, H., Düring, W. und *Schmaltz, K.*, Rechnungslegung und Prüfung der Unternehmen, bearbeitet von *Forster, K.-H.* u.a., 6. Aufl., Stuttgart 1995-1998.

Arbeitskreis „Externe Unternehmensrechnung" der Schmalenbach-Gesellschaft für Betriebswirtschaft e.V., Grundsätze für das Value Reporting, in: DB, 55. Jg. (2002), H. 45, S. 2337-2340.

Arbeitskreis „Immaterielle Werte im Rechnungswesen" der Schmalenbach-Gesellschaft für Betriebswirtschaft e.V., Kategorisierung und bilanzielle Erfassung immaterieller Werte, in: DB, 54. Jg. (2001), H. 19, S. 989-995.

Arbeitskreis „Immaterielle Werte im Rechnungswesen" der Schmalenbach-Gesellschaft für Betriebswirtschaft e.V., Freiwillige externe Berichterstattung über immaterielle Werte, in: DB, 56. Jg. (2003), H. 23, S. 1233-1237.

Austrian Research Center, Wissensbilanz 2002, http://www.arcs.ac.at/publik/fulltext/wissensbilanz/arc_wissensbilanz_2002.pdf.

Baetge, J. und *Kirsch, H.-J.*, Grundsätze ordnungsmäßiger Buchführung, in: *Küting, K.* und *Weber, C.-P.* (Hrsg.), Handbuch der Rechnungslegung, Bd. 1, 5. Aufl., Stuttgart 2002.

Baetge, J., Kirsch, H.-J. und *Thiele, S.*, Bilanzen, 6. Aufl., Düsseldorf 2002.

Beisse, H., Wandlungen der Rechnungsabgrenzung, in: *Förschle, G., Kaiser, K.* und *Moxter, A.* (Hrsg.), Rechenschaftslegung im Wandel, Festschrift *Budde*, München 1995, S. 67-85.

Biener, H. und *Berneke, W.*, Bilanzrichtlinien-Gesetz, Düsseldorf 1986.

Bornemann, M., Knapp, A., Schneider, U. and *Sixl, K.*, Holistic measurement of intellectual capital; International Symposium "Measuring and Reporting Intellectual Capital: Experience, Issues, and Prospects", Amsterdam 1999.

Busse von Colbe, W., Bilanzierungshilfe, in: *Leffson, U., Rückle, D.* und *Großfeld, B.*, Handwörterbuch unbestimmter Rechtsbegriffe im Bilanzrecht des HGB, Köln 1986, S. 86–94.

Busse von Colbe, W., Ausbau der Konzernrechnungslegung im Lichte internationaler Entwicklungen, in: ZGR 2000, S. 651–673.

Coenenberg, A. G., Jahresabschluss und Jahresabschlussanalyse, 19. Aufl., Stuttgart 2003.

Danish Agency for Trade and Industry: A Guideline for Intellectual Capital Statements, Copenhagen 2000.

Deng, Z., Lev, B. and *Narin, F.*, Science & Technology as Predictors of Stock Performance, in: Financial Analysts Journal, Vol. 55 (1999), No. 3, S. 20-32.

Edvinsson, L. and *Malone, M.*, Intellectual capital: realizing your company's true value by finding its hidden roots, New York 1997.

Ellrott, H. und *Aicher, H.-P.*, Kommentierung zu § 255 HGB, in: *Berger, A* . et al. (Hrsg.), Beck'scher Bilanzkommentar, 5. Aufl., München 2003.

Ellrott, H. und *Schmidt-Wendt, D.*, Kommentierung zu § 247 HGB, in: *Berger, A.* et al. (Hrsg.), Beck'scher Bilanzkommentar, 5. Aufl., München 2003.

Freericks, W., Bilanzierungsfähigkeit und Bilanzierungspflicht in Handels- und Steuerbilanz, Köln u.a. 1976.

FASB, Disclosure about intangible assets, http://accounting.rutgers.edu/raw/fasb/project/intangibles.shtml (Stand: August 2003).

Fülbier, R.., Honold, D. und *Klar, A.*, Bilanzierung immaterieller Vermögenswerte, Möglichkeiten und Grenzen der Bilanzierung nach US-GAAP und IAS bei Biotechnologieunternehmen, in: RIW, 46. Jg. (2000), S. 833-844.

Haller, A., Segmentberichterstattung, in: *Haller, A., Raffournier, B.* und *Walton, P.* (Hrsg.), Unternehmenspublizität im internationalen Wettbewerb, Stuttgart 2000, S. 755-805.

Haller, A. und *Dietrich, R.*, Freiwillige Unternehmensberichterstattung in den USA, in: KoR, 1. Jg. (2001), H. 2, S. 206-211.

Hartung, W., Rechnungsabgrenzungsposten und richtlinienkonforme Auslegung, in: *Ballwieser, W.* et al. (Hrsg.), Bilanzrecht und Kapitalmarkt, Festschrift für *A. Moxter*, München 1994, S. 213-225.

Hendriksen, E. and *Van Breda, M.*, Accounting theory, 5. Aufl., Homewood 1992.

Husemann, K., Grundsätze ordnungsmäßiger Bilanzierung für Anlagegegenstände, 2. Aufl., Düsseldorf 1976.

Knop, W. und *Küting, K.*, Kommentierung zu § 255, Anschaffungs- und Herstellungskosten, in: *Küting, K.* und *Weber, C.-P.* (Hrsg.), Handbuch der Rechnungslegung, Bd. 1a, 4. Aufl., Stuttgart 1995.

Kubin, K. W., Der Aktionär als Aktienkunde – Anmerkungen zum Shareholder Value, zur Wiedervereinigung der internen und externen Rechnungslegung und zur globalen Verbesserung der Berichterstattung, in: *Möller, H.* und *Schmidt, F.* (Hrsg.), Rechnungswesen als Instrument für Führungsentscheidungen, Festschrift für *Coenenberg*, Stuttgart 1998, S. 525-558.

Leffson, U., Die Grundsätze ordnungsmäßiger Buchführung, 7. Aufl., Düsseldorf 1987.

Lev, B., Intangibles: Management, Measurement, and Reporting, Washington 2001.

Lev, B. and *Sougiannis, T.*, Penetrating the Book-to-Market Black Box: The R&D-Effect, in: Journal of Business, Finance and Accounting, Vol. 23 (1996), No. 4-5, S. 107-138.

Maul, K.-H., Die Wissensbilanz als Teil des handelsrechtlichen Jahresabschlusses, in: DStR, 38. Jg. (2000), H. 47, S. 2009-2016.

Maul, K.-H. und *Menninger, J.*, Das "Intellectual Property Statement" - eine notwendige Ergänzung des Jahresabschlusses?, in: DB, 53. Jg. (2000), H. 11, S. 529-533.

Moxter, A., Immaterielle Anlagewerte im neuen Bilanzrecht, in: BB, 34. Jg. (1979), S. 1102-1109.

Mutze, O., Aktivierung und Bewertung immaterieller Wirtschaftsgüter nach Handels- und Steuerrecht, Berlin 1960.

Pellens, B. und *Fülbier, R. U.*, Ansätze zur Erfassung immaterieller Werte in der kapitalmarktorientierten Rechnungslegung, in: *Baetge, J.* (Hrsg.), Zur Rechnungslegung nach International Accounting Standards (IAS), Düsseldorf 2000, S. 35-77.

PriceWaterhouseCoopers, Günther, T., Beyer, D., Immaterielle Werte und andere weiche Faktoren in der Unternehmensberichterstattung – eine Bestandsaufnahme, Frankfurt 2003

Reuleaux, S., Immaterielle Wirtschaftsgüter: Begriff, Arten und Darstellung im Jahresabschluss, Wiesbaden 1987.

Roos, J., *Edvinsson, L.* and *Roos, G.*, Intellectual capital: navigating the new business landscape, Basingstoke u.a. 1998.

Rütte, M. und *Hoenes, R..*, Rechnungslegung immaterieller Werte, Bamberg 1995.

Schildbach, T., Der handelsrechtliche Jahresabschluss, 6. Aufl., Herne u.a. 2000.

Schulze-Osterloh, J., Der Ausweis von Aufwendungen nach dem Realisations- und dem Imparitätsprinzip, in: *Moxter, A.* et al. (Hrsg.), Rechnungslegung, Festschrift *Forster*, Düsseldorf 1992, S. 653-670.

Sieben, G. und *Ossadnik, W.*, Aufwand für eine bestimmte Zeit nach dem Abschlußstichtag, in: *Leffson,U.*, *Rückle, D.* und *Großfeld, B.* (Hrsg.), Handwörterbuch unbestimmter Rechtsbegriffe im Bilanzrecht des HGB, Köln 1986, S. 46-53.

Stewart, T., Der vierte Produktionsfaktor: Wachstum und Wettbewerbsvorteile durch Wissensmanagement, München/Wien 1998.

Strack, R. und *Villis, U.*, RAVE™: Die nächste Generation im Shareholder Value Management, in: ZfB, 71. Jg. 2001, H. 1, S. 67-84.

Sveiby, K.-E., Measuring Intangibles and Intellectual Capital - An Emerging First Standard, Internet Version: http://www.sveiby.com/articles/emergingstandard.html, 1998.

Sveiby, K.-E., The new organizational wealth: managing and measuring knowledge-based assets, San Francisco 1997.

Trützschler, K., Kommentierung zu § 250, in: *Küting, K.* und *Weber, C.-P.* (Hrsg.), Handbuch der Rechnungslegung, Bd. 1, 5. Aufl., Stuttgart 2002.

Weber, C.-P. und *Zündorf, H.*, Kommentierung zu § 309 HGB, in: *Küting, K.* und *Weber, C.-P.*, Handbuch der Konzernrechnungslegung, Band II, 2. Aufl., Stuttgart 1998.

von Keitz, I., Immaterielle Güter in der internationalen Rechnungslegung, Düsseldorf 1997.

Die Bilanzierung von immateriellen Vermögenswerten in der nationalen und internationalen Rechnungslegung

Hans-Georg Bruns [1], *Markus Zeimes* [2] *und Michael G. Thuy* [3]

---

[1] Prof. Dr. *Hans-Georg Bruns*, International Accounting Standards Board, London, UK

[2] *Markus Zeimes*, KPMG Deutsche Treuhand-Gesellschaft AG

[3] *Michael G. Thuy,* Philip Morris International Management S.A., Lausanne, Schweiz

Immaterielles Vermögen wird zunehmend zum entscheidenden Erfolgspotenzial für die Zukunft von Unternehmen. Ursächlich hierfür sind im Wesentlichen zwei Faktoren: Zum einen befindet sich die Wirtschaft in den Industrieländern auf dem Weg in eine Hochtechnologiegesellschaft, bei der immaterielle Vermögenswerte wichtiger und die Bedeutung von materiellen Produktionsfaktoren, wie Grundstücken, Gebäuden und Produktionsanlagen, merklich geringer wird. Zum anderen war insbesondere in den Jahren des „New Economy Boom" ein erheblicher Anstieg von Unternehmensakquisitionen zu verzeichnen. Dabei wurden Kaufpreise bezahlt, die deutlich über dem Nettovermögen der Unternehmen lagen und somit zu einem erheblichen Aufbau von Goodwillpositionen in den Bilanzen der übernehmenden Gesellschaften führten. Aufgrund der Anwendung unterschiedlicher Rechnungslegungssysteme werden immaterielle Vermögenswerte sowie Goodwill jedoch nicht einheitlich bilanziert. Durch die im Sommer 2002 getroffene Entscheidung des Europäischen Parlaments und des Rats zur verpflichtenden Anwendung der International Accounting Standards (IAS; künftig: IFRS – International Financial Reporting Standards) für europäische kapitalmarktorientierte Konzerne ab 2005 hat die unterschiedliche Bilanzierung von immateriellen Vermögenswerten des Anlagevermögens für europäische Unternehmen weiter an Bedeutung gewonnen.

1. Vorschriften zur Bilanzierung immaterieller Vermögenswerte und Goodwill

1.1 Handelsgesetzbuch (HGB) und Deutsche Rechnungslegungs-Standards (DRS)

Im Rahmen der handelsrechtlichen Rechnungslegung immaterieller Vermögensgegenstände des Anlagevermögens (der im Rahmen dieses Beitrags verwendete Begriff des „Vermögenswertes" entspricht dem englischen Begriff des „Assets", wie er in den IFRS (F.49a) und US-GAAP (CON 6.25) definiert ist) gelten neben den allgemeinen Ansatz- und Bewertungsvorschriften insbesondere das pauschale Aktivierungsverbot gemäß § 248 Abs. 2 HGB für selbst erstellte Werte sowie der vom *Deutschen Standardisierungsrat* (*DSR*) verabschiedete Deutsche Rechnungslegungsstandard DRS 12 *Immaterielle Vermögenswerte des Anlagevermögens* (vgl. *DRS*).

Die Bilanzierung eines derivativen Geschäfts- oder Firmenwert (im Folgenden „Goodwill") als Vermögensgegenstand erfolgt nach Maßgabe des § 255 Abs. 4 HGB. Der im Rahmen von Unternehmenserwerben entstehende Goodwill ist gemäß § 301 Abs. 3 Satz 1 HGB zu bilanzieren. Die Folgebilanzierung ist gemäß § 309 HGB vorzunehmen. Zudem gelten die Regelungen des DRS 4 *Unternehmenserwerbe im Konzernabschluss.*

1.2 US-Generally Accepted Accounting Principles (US-GAAP)

Das amerikanische *Financial Accounting Standards Board* (*FASB*) als Standardsetter für die US-Generally Accepted Accounting Principles (US-GAAP) hat 2001 die Rechnungslegungsvorschriften für immaterielle Vermögenswerte und Goodwill überarbeitet und einen grundlegend neuen Ansatz eingeführt. Danach ist bei immateriellen Vermögenswerten zu unterscheiden, ob sie im Rahmen eines Unternehmenserwerbs zugegangen sind (Statement of Financial Accounting Standard – SFAS 141: *Business Combinations*) oder separat erworben wurden (SFAS 142: *Goodwill and Other Intangible Assets*). Bei derivativem Goodwill ist zudem keine planmäßige Abschreibung mehr vorgesehen (siehe dazu Abschnitt 3.8).

Darüber hinaus enthält das stark auf Einzelfallregelungen aufgebaute System der USA eine Vielzahl weiterer Vorschriften zur Bilanzierung immaterieller Vermögenswerte, wobei die Vorschriften in einer Hierarchie („House of GAAP") zueinander stehen (vgl. *Pellens*, 2001, S. 123). Neben dem Statement des Accounting Principles Board (APB) No. 17 *Intangible Assets* aus dem Jahre 1970, SFAS 2 *Accounting for Research and Development Costs* und den genannten SFAS 141 und 142 bestehen eine Vielzahl weiterer, auch branchenspezifischer Vorschriften und Detailregelungen (vgl. *Küting/Ulrich*, 2001, S. 953; *KPMG*, 2002, S. 32).

1.3 International Financial Reporting Standards (IFRS)

Die International Financial Reporting Standards (IFRS, bis 2001: IAS - International Accounting Standards) regeln die Bilanzierung immaterieller Vermögenswerte in IAS 38 (1998) *Intangible Assets* sowie in den Interpretationen des Standing Interpretations Committee (SIC) SIC 6 *Costs of Modifying Existing Software* und SIC 32 *Intangible Assets – Web Site Costs*. Zur Bilanzierung eines erworbenen Goodwills ist IAS 38 (1998) in Verbindung mit IAS 22 (1998) *Business Combinations* anzuwenden.

Ein durchzuführender Werthaltigkeitstest der immateriellen Vermögenswerte und des Goodwill ist analog der allgemeinen Vorschrift des IAS 36 (1998) *Impairment of Assets* vorzunehmen. Am 7. Dezember 2002 hat der IASB einen Standardentwurf ED3 *Business Combinations* veröffentlicht, der in weiten Teilen an die US-GAAP Regelungen anknüpft. Dieser Standard ersetzt IAS 22 und geht einher mit der Überarbeitung der Standards IAS 36 und IAS 38 (siehe Abschnitt 4.3).

2. Der Bilanzansatz von immateriellen Vermögenswerten und Goodwill

2.1 Immaterielle Vermögenswerte

Bislang hat sich weder in Deutschland noch international eine einheitliche Definition für immaterielle Vermögenswerte etabliert (vgl. auch *Küting/Ulrich*, 2001, S. 955 ff.). Im deutschen Handelsrecht haben immaterielle Vermögensgegenstände den gleichen Kriterien zu genügen wie materielle Vermögensgegenstände. Nach herrschender Meinung ist hierbei die ständige Verwertbarkeit das wesentliche Bilanzierungskriterium. Demnach sind nur solche immateriellen Vermögensgegenstände aktivierungsfähig, die in Geld transformierbar sind und somit zur Schuldendeckung eines Unternehmens beitragen können. Das Kriterium der selbständigen Verwertbarkeit wird in einem zweiten Schritt durch die Forderung nach einer objektivierten Werthaltigkeit ergänzt. Da der Wert immaterieller Vermögensgegenstände des Anlagevermögens in der Regel schwer schätzbar und deshalb unsicher ist, hat der Gesetzgeber festgelegt, dass nur Vermögensgegenstände aktivierungsfähig sind, welche entgeltlich erworben wurden. Als Aktivierungsgrundlage werden die Anschaffungskosten herangezogen. Eine Aktivierung selbst erstellter immaterieller Vermögensgegenstände scheidet – mangels sicherer Bewertungsbasis - demnach gemäß § 248 Abs. 2 HGB aus.

Diese allgemein gültigen Ansatzgrundsätze stellen sicher, dass vergleichbare Sachverhalte bilanziell gleich abgebildet werden.

DRS 12 orientiert sich bei der Definition eines immateriellen Vermögenswertes – zur Erreichung einer Konvergenz mit IFRS und US-GAAP – am dynamischen Kriterium des wahrscheinlichen künftigen Nutzens. Dennoch sind immaterielle Vermögenswerte nach DRS 12.7 und immaterielle Vermögensgegenstände gemäß den handelsrechtlichen Vorschriften gleichzusetzen, da sich die DRS innerhalb der kodifizierten Regelungen bewegen. Gleichwohl hat sich DRS 12 in seinem als „Empfehlungen an den Gesetzgeber" verstandenen de lege ferenda Teil für die Aufhebung des Bilanzierungsverbots in § 248 Abs. 2 ausgesprochen.

Als „Intangible Assets" gemäß US-GAAP werden nach *Larson* und *Miller* (vgl. *Larson/Miller*, 1995, S. 118) immaterielle Vermögenswerte bezeichnet, „...welche Rechte oder andere Privilegien beinhalten." Sie sind dadurch definiert, dass sie klar identifizierbar, ohne physische Substanz und nicht monetär sind. Die Werte sind bilanzierungspflichtig, wenn sie entweder aus vertraglichen oder anderen Rechten entstanden und separierbar, d.h. veräußerbar oder übertragbar, sind. Dabei ist es grundsätzlich unerheblich, ob die Werte selbst erstellt oder erworben wurden. Für eine Aktivierung ist allein auf die Voraussetzungen der SFAS 141 und 142 abzustellen.

Nach IFRS sind immaterielle Vermögenswerte klar identifizierbare nicht monetäre Werte ohne physische Substanz, die in der Verfügungsmacht des Unternehmens stehen. Sie müssen ein Ergebnis von Ereignissen der Vergangenheit darstellen, von denen ein künftiger Zufluss wirtschaftlichen Nutzens erwartet wird (IAS 38.7). Für die tatsächliche Aktivierungsfähigkeit wird in einem zweiten Schritt gefordert, dass die entsprechenden Anschaffungs- oder Herstellungskosten verlässlich bestimmbar sind (vgl. IFRS Framework F 89).

Darüber hinaus gelten gemäß IAS 38.51 explizite Aktivierungsverbote für selbst geschaffene Markennamen, Drucktitel, Verlagsrechte, Kundenlisten und ihrem Wesen nach ähnliche Sachverhalte. Ferner gilt ein explizites Bilanzierungsverbot für Aufwendungen aus Aus- und Weiterbildungsaktivitäten, Werbekampagnen und Verkaufsförderungsmaßnahmen, Gründungsaufwendungen und Kosten für den Anlauf des Geschäftsbetriebs sowie Kosten im Zusammenhang mit der Verlegung oder Reorganisation des Unternehmens (IAS 38.57).

2.2 Goodwill

Hinsichtlich der Bilanzierung eines Goodwill ist zwischen dem originären (selbst geschaffenen) und derivativen (entgeltlich erworbenen) Goodwill zu unterscheiden. Ein originärer Goodwill ist mangels Objektivierbarkeit des Wertes nach keinem der drei Rechnungslegungssysteme aktivierbar. Nach den Vorschriften des deutschen Handelsrechts ist ein derivativer Goodwill aktivierungsfähig, da sich durch den entgeltlichen Erwerb der entsprechende Vermögenswert weitestgehend konkretisiert. Das handelsrechtliche Wahlrecht des § 255 Abs. 4 HGB ermöglicht einen gewissen Bilanzierungsspielraum; streng abzugrenzen davon ist jedoch der zwingende Ansatz von sog. „geschäftswertähnlichen Vermögensgegenständen", die nach den allgemeinen Grundsätzen des § 253 HGB anzusetzen und zu bewerten sind (vgl. *Beck'scher Bilanzkommentar*, 2003, § 255 Rd.518). § 301 HGB regelt die Bilanzierung eines aus der Kapitalkonsolidierung entstehenden Goodwill. Zudem regelt

§ 309 HGB zusätzlich zur Bilanzierung nach § 255 Abs. 4 HGB, das ein aus der Kapitalkonsolidierung entstandener Goodwill anstelle einer Aktivierung mit den Rücklagen verrechnet werden kann.

Ermittelt wird der derivative Goodwill als Residualgröße. Er ergibt sich als Unterschied zwischen dem Betrag, den der Erwerber eines Unternehmens aufwendet (Kaufpreis) und der Differenz aus den zum Zeitwert bewerteten Vermögensgegenständen und Schulden. Aus wirtschaftlicher Sicht besteht der Goodwill aus diskontierten Gewinnen (Überschüsse der zukünftigen Einnahmen über dem im Rahmen des Unternehmenszusammenschluss erworbenen Reinvermögen). Diese Gewinne resultieren aus vielerlei Ursachen, die nicht unbedingt auf konkrete, zu aktivierende Vermögenswerte zurückzuführen sind. Beispielhaft zu nennen wären hier u.a. Produktions- und Management Know-how, Standortvorteile, Kundenstamm etc. (vgl. *Beck'scher Bilanzkommentar*, 2003, § 309 Rd.5).

Im Gegensatz zum handelsrechtlichen Wahlrecht, den Goodwill mit dem Eigenkapital zu verrechnen, ist nach US-GAAP ein aus der Konsolidierung entstehender, positiver Unterschiedsbetrag (die Anschaffungskosten übersteigen das beteiligungsproportional oder voll zu Zeitwerten bewertete Reinvermögen) gemäß SFAS 141.43 zwingend zu aktivieren. Eine eventuelle Abschreibung wird auf Basis des regelmäßig vorzunehmenden Impairmenttests berechnet (siehe 3.8).

3. Die Bewertung immaterieller Vermögenswerte und Goodwill

3.1 Allgemeine Bewertungsregeln für immaterielle Vermögenswerte

Immaterielle Vermögensgegenstände sind handelsrechtlich gemäß § 253 Abs. 1 S. 1 HGB zu Anschaffungskosten zu bewerten. Der laufenden Wertminderung wird durch eine planmäßige Abschreibung über die erwartete Nutzungsdauer Rechnung getragen (§ 253 Abs. 2 S. 1, 2 HGB). Eine außerplanmäßige Abschreibung ist nach § 253 Abs. 2 S. 3 HGB nur dann vorzunehmen, wenn der beizulegende Wert des Gegenstands unter seinem Buchwert liegt, dem Bilanzersteller also bereits Indizien hinsichtlich einer etwaigen Wertminderung vorliegen. Nach DRS 12 ist festgelegt, dass immaterielle Vermögenswerte, deren Nutzung zeitlich unbegrenzt ist, keiner planmäßigen Abschreibung unterliegen (DRS 12.20). Außerdem fordert DRS 12.22, dass alle immateriellen Vermögenswerte, unabhängig von ihrer Nutzungsdauer, jährlich einem Werthaltigkeitstest zu unterziehen und ggf. auf den niedrigeren Wertansatz abzuschreiben sind.

Nach US-GAAP sind immaterielle Vermögenswerte ebenfalls zu Anschaffungs- oder, falls selbst erstellt, zu Herstellungskosten zu bewerten, soweit es sich um Vermögenswerte handelt, die nicht im Rahmen eines Unternehmenszusammenschlusses erworben wurden (SFAS 142.9). Die Bewertung in der Folgezeit richtet sich danach, ob es sich um einen immateriellen Vermögenswert mit begrenzter oder unbestimmter Nutzungsdauer handelt. Eine planmäßige Abschreibung hat nur zu erfolgen, wenn die festgestellte Nutzungsdauer begrenzt ist. Als Nutzungsdauer wird der Zeitraum definiert, über den der Vermögenswert voraussichtlich direkt oder indirekt zum Cashflow eines Unternehmens beiträgt. Eine Obergrenze für die Nutzungsdauer existiert demnach nicht. Liegen Anzeichen dafür vor, dass der Buchwert immaterieller Vermögenswerte über der Summe der künftigen nicht abgezinsten Cashflows liegt, ist ein Werthaltigkeitstest (Impairmenttest) durchzuführen. Vermögenswerte, deren Nutzungsdauer unbestimmbar ist, werden nicht planmäßig abgeschrieben. Zur Überprüfung ihrer Werthaltigkeit sind sie regelmäßig, d.h. nicht nur bei bestehendem Verdacht, einem Impairmenttest zu unterziehen.

Immaterielle Vermögensgegenstände die im Rahmen eines Unternehmenszusammenschlusses (durch Erwerb eines Teilbetriebs oder einer zu konsolidierenden Beteiligung) erworben wurden, sind zu ihren Zeitwerten (Fair Values) zu erfassen (vgl. *Dawo*, 2003, S. 139).

Nach IFRS sind Vermögenswerte, die nicht im Rahmen eines Unternehmenszusammenschlusses erworben wurden (vgl. *Baetge/v. Keitz*, 2002, S. 19), im Zugangszeitpunkt ebenfalls mit den Anschaffungs- bzw. Herstellungskosten zu aktivieren.

Die Herstellungskosten eines selbst geschaffenen immateriellen Vermögenswertes umfassen gemäß IAS 38.54 sämtliche Kosten, die der Schaffung, Herstellung und Vorbereitung eines Vermögenswertes auf seinen beabsichtigten Gebrauch direkt zugerechnet werden können (vgl. *Hayn/Graf Waldersee*, 2002, S. 59).

Gemäß IFRS ist eine Folgebewertung grundsätzlich mit fortgeführten Anschaffungs- oder Herstellungskosten durchzuführen. Dabei hat sich die Abschreibung am tatsächlichen Werteverzehr des Vermögenswertes zu orientieren. Gemäß IAS 38.79 besteht die widerlegbare Vermutung, dass dessen Nutzungsdauer auf 20 Jahre begrenzt ist. Am Ende jeden Geschäftsjahres ist die Abschreibungsmethode sowie die Nutzungsdauer zu überprüfen (IAS 38.94). Außerplanmäßige Abschreibungen sind bei Vorliegen der Voraussetzungen gemäß IAS 36 vorzunehmen. Der immaterielle Vermögenswert darf anstelle der fortgeführten Anschaffungs- oder Herstellungskosten in der Folgezeit unter bestimmten Voraussetzungen alternativ mit dem beizulegenden Zeitwert neubewertet werden (Allowed Alternative Treatment: IAS 38.64). Der beizulegende Zeitwert ist unter Rückgriff auf einen aktiven Markt zu ermitteln, welcher die nach IAS 38.7 dargestellten Kriterien (im Wesentlichen homogene Produkte, vertragswillige Parteien, transparenter Markt) kumulativ erfüllen muss.

Im Rahmen eines Unternehmenszusammenschlusses erworbene immaterielle Vermögenswerte sind mit dem fair value anzusetzen (IAS 38.27). Kann ein Wert jedoch nicht mit hinreichender Sicherheit ermittelt werden, wird kein eigenständiger immaterieller Vermögenswert bilanziert. Vielmehr fließen diese immateriellen Werte dann in den Goodwill ein.

Im Folgenden sind wesentliche Einzelfallregelungen zur Bilanzierung immaterieller Vermögenswerte aufgeführt.

3.2 (Auftragsunabhängige) Forschungs- und Entwicklungskosten

Aufwendungen für Forschung und Entwicklung sind bilanziell getrennt zu behandeln.

Forschungskosten sind nach allen drei Rechnungslegungssystemen (HGB, US-GAAP, IFRS) unmittelbar als Aufwand zu erfassen. Nach HGB greift das allgemeine Aktivierungsverbot des § 248 Abs. 2 HGB für selbst erstellte immaterielle Vermögensgegenstände des Anlagevermögens. Nach US-GAAP und IFRS gelten Forschungsaufwendungen als noch unzureichend zielgerichtete Suche nach neuen Erkenntnissen, die nicht zu gesicherten Zahlungsmittelzuflüssen führen und deswegen keine Aktivierung zulässt (SFAS 2.12 bzw. IAS 38.42).

Das handelsrechtliche Aktivierungsverbot des § 248 Abs. 2 HGB erstreckt sich auch auf Entwicklungskosten. Ein allgemeines Aktivierungsverbot gilt ebenfalls nach US-GAAP (SFAS 2.12; Emerging Issues Task Force (EITF) 99-5). Zu dieser grundsätzlichen Regelung bestehen allerdings zahlreiche Ausnahmen. So regeln die SFAS 19, 25, 50, 53, 69 und weitere Standards und Verlautbarungen branchenspezifische Sachverhalte, wie den Abbau von Bodenschätzen oder die Behandlung von Aufnahmekosten der Film- und Musikindustrie. Weitere Ausnahmen bestehen für (die nachfolgend angesprochenen) Fälle zur Entwicklung von Software zur internen Nutzung, selbst entwickelter Software zur Vermarktung, der Entwicklung von Internet-Seiten oder für bestimmte Werbekampagnen. Kosten für Auftragsforschung, die innerhalb einer Periode abgeschlossen wird, werden gemäß SFAS 68 „Research and Development Arrangements" als Aufwand erfasst.

Im Gegensatz zu HGB und US-GAAP besteht für Entwicklungskosten nach IFRS (vgl. die Definition in IAS 38.7) eine Aktivierungspflicht, wenn das Unternehmen kumulativ folgende Nachweise erbringen kann (IAS 38.45):

- Die technische Realisierbarkeit der Fertigstellung des immateriellen Vermögenswertes, damit er zur internen Nutzung oder zum Verkauf zur Verfügung stehen wird;
- die Absicht, den immateriellen Vermögenswert fertig zu stellen sowie ihn zu nutzen oder zu verkaufen;
- die Fähigkeit, den immateriellen Vermögenswert zu nutzen oder zu verkaufen;
- wie der immaterielle Vermögenswert einen voraussichtlichen künftigen wirtschaftlichen Nutzen erzielen wird. Nachgewiesen werden muss von dem Unternehmen u.a. die Existenz eines Marktes für die Produkte des immateriellen Vermögenswertes oder den immateriellen Vermögenswert an sich oder, falls er intern genutzt werden soll, der Nutzen des immateriellen Vermögenswertes;
- die Verfügbarkeit adäquater technischer, finanzieller und sonstiger Ressourcen, um die Entwicklung abschließen und den immateriellen Vermögenswert nutzen oder verkaufen zu können; und
- die Fähigkeit, die dem immateriellen Vermögenswert während seiner Entwicklung zurechenbaren Ausgaben zuverlässig zu bewerten.

Sind Aufwendungen nicht eindeutig in eine Forschungs- und Entwicklungsphase trennbar, so sind die gesamten Kosten als Forschungsaufwendungen einzuordnen und aufwandswirksam zu erfassen (IAS 38.41). Eine nachträgliche Aktivierung von bereits als Aufwand erfassten Entwicklungskosten ist gemäß IAS 38.59 unzulässig.

In jedem Fall sind zur Erfüllung der Aktivierungsvoraussetzungen ein umfangreiches Projektcontrolling, eine gut strukturierte Kostenrechnung und eine ausreichende Dokumentation seitens des Unternehmens erforderlich. Die Erfüllung der oben genannten Kriterien bietet den Unternehmen jedoch einen Ermessensspielraum, wodurch Aktivierungspflicht für Entwicklungsaufwendungen einem faktischen Wahlrecht gleichkommt (vgl. *Dawo*, S. 205).

3.3 Implementierung von Software

Gemäß § 255 Abs. 1 HGB sind nur durch Unternehmensfremde entstandene Softwarekosten zu aktivieren, die den Vermögensgegenstand in einen betriebsbereiten Zustand versetzen (Anschaffungsnebenkosten). Für entstandene eigene Aufwendungen scheidet eine Aktivierung nach § 248 Abs. 2 HGB aufgrund der bekannten Bewertungsproblematik aus.

Sowohl nach US-Statement of Position (SOP) 98.1-21 als auch IFRS (IAS 38.24) sind alle direkt zurechenbaren Kosten, demnach auch Personalkosten für unmittelbar beteiligte Mitarbeiter und weitere erbrachte Leistungen zu aktivieren, welche die Software auf deren

Nutzung vorbereiten, und die erwarteten Zahlungsmittelüberschüsse aus dem Vermögenswert sicherstellen oder erhöhen. Es erfolgt keine Unterscheidung zwischen extern angefallenen und unternehmensintern entstandenen Kosten.

3.4 Software zur Eigennutzung, Modifikation bestehender Software

Wie bei der Implementierung von Software sind nach deutschem Handelsrecht nur derivative Vermögenswerte bzw. erworbene Leistungen anzusetzen. Für Eigenleistungen des Unternehmens besteht ein Aktivierungsverbot.

Nach US-GAAP sind die Kosten für Entwicklung von intern genutzter Software zu aktivieren. Handelt es sich dagegen um die Modifikation einer bereits bestehenden Software, darf nur dann eine Aktivierung der Kosten erfolgen, wenn die Modifikation mit großer Sicherheit zusätzliche Funktionalität liefert (SOP 98-1.24). Wartungsaufwand ist demgegenüber sofort erfolgswirksam zu erfassen (EITF 97-13).

IFRS folgt dem Gedanken der US-GAAP Regelung und sieht eine Aktivierung von Kosten ebenfalls nur dann vor, wenn dadurch der wirtschaftliche Nutzen der Software erhöht wird (SIC 6.4) und somit zusätzliche Zahlungsmittelzuflüsse generiert werden.

Eine Trennung von Erhaltung, Modifikation und Verbesserung eines aktivierten Vermögenswerts dürfte sich in der Praxis als regelmäßig schwierig herausstellen. Gute Dokumentationsverfahren sind hier unerlässlich.

3.5 Software zur Vermarktung

Eine weitere US-GAAP spezifische Detailregelung besteht für die Behandlung von EDV-Forschungs- und Entwicklungskosten (SFAS 86 *Accounting for the Costs of Computer Software to be Sold, Leased, or Otherwise Marketed*). Die Vorschrift stellt klar, dass die angefallenen Kosten (Planung, Design, Programmierung und Testläufe) bis zur Sicherstellung der technischen Durchführbarkeit des Projekts (Technical Feasibility) Entwicklungskosten darstellen, aus welchen kein aktivierbares Nutzenpotenzial erwächst. Die im Anschluss anfallenden Kosten für die Produktion der Software-Masterprogramme sind hingegen zu aktivieren. Wie auch im Falle eigengenutzter Software sind hier die Grenzen zwischen erfolgswirksamer Erfassung und Aktivierung als Vermögenswert sicherlich fließend; eine aussagekräftige Kostenrechnung zur Erfassung und Abgrenzung der Kosten ist in jedem Fall unabdingbar.

Weder nach HGB noch nach IFRS findet sich eine konkrete Bilanzierungsvorgabe zu diesem Sachverhalt, dennoch ist zu unterstellen, dass das allgemeine Bilanzierungsverbot nach HGB bzw. die Aktivierungsvoraussetzungen des IAS 38.45 Anwendung finden.

3.6 Kosten der Entwicklung von Internet-Seiten

Weder das Handelsgesetzbuch noch DRS 12 behandelt die spezielle Thematik der bilanziellen Behandlung von Kosten der Entwicklung von Internet-Seiten.

Nach US-GAAP werden Entwicklungskosten eigener Internet-Seiten, welche nicht vermarktet oder anderweitig veräußert werden, analog den Vorschriften zu selbst erstellter Software behandelt; somit sind nur Kosten ausgewählter Projektstufen (ab Erreichung der „Technical Feasibility“) aktivierungspflichtig. Die EITF-Statements 00-2 und 00-20 der Emerging

Issues Task Force regeln weitergehende Sachverhalte zur Behandlung von Web-Hosting oder der Bilanzierung von Inhalten von Internet-Seiten.

Das damalige SIC (zwischenzeitlich ersetzt durch das International Financial Reporting Interpretations Committee, IFRIC) des IASC hat in SIC 32 festgelegt, dass die extern oder intern angefallenen Kosten für die Entwicklung der eigenen Inter- oder Intranetseite einen selbsterstellten immateriellen Vermögenswert darstellen, der bei Erfüllung der Voraussetzungen des IAS 38.19 und .54 zu aktivieren ist.

Dabei ist zu beachten, dass der Erstellungs- und Unterhaltsprozess der Internetseite in Phasen aufzuteilen ist. Nur Kosten, die vergleichbar zu der allgemeinen Entwicklungskostendefinition des IAS 38.7 sind, werden aktiviert. Kosten anderer vor- oder nachgelagerter Phasen („Planning, Content Development, Operating") sind als Aufwand zu behandeln.

Die Erfüllung der Voraussetzungen u.a. des IAS 38.45 d, wonach der Nachweis des voraussichtlichen wirtschaftlichen Nutzens des Vermögenswerts erbracht werden muss, ist nach dem Wortlaut des SIC 32 als streng auszulegen. So kann dieser Nachweis wohl erbracht werden, wenn über die Unternehmens-Web-Seite beispielsweise Bestellungen ausgeführt werden können, die zu einem Geldzufluss beim Unternehmen führen. Ein wahrscheinlicher ökonomischer Nutzenzufluss beim Unternehmen und damit eine Aktivierung der Kosten wären dahingegen wohl zu verneinen, wenn die Internet Seite nur als Werbeträger fungiert.

3.7 Werbekampagnen

Kosten für Werbekampagnen sind sowohl nach deutschem Recht als auch nach IFRS (IAS 38.57c) sofort aufwandswirksam zu erfassen.

IAS 38.51 stellt diesbezüglich klar, dass diese Art von Kosten – auch wenn von der Entstehung eines „Intangible Assets" ausgegangen werden kann – nicht von den Kosten der Entwicklung des Unternehmens als Ganzes unterschieden werden kann; infolgedessen scheidet eine Aktivierung als immaterieller Vermögenswert aus.

Die US-amerikanische Rechnungslegung sieht bezüglich der Behandlung von Werbeausgaben eines ihrer wenigen Wahlrechte vor. SOP 93-7.26 legt fest, dass Kosten für Werbung entweder sofort als Aufwand verbucht – oder alternativ – als Vermögenswert aktiviert werden dürfen. Der Aktivposten ist jedoch nach erstmaliger Präsentation der Werbung in der Öffentlichkeit sofort vollständig aufwandswirksam abzuschreiben. Schreiber (vgl. *Schreiber*, 2003, S. 175) untersucht in ihrem Beitrag die Systematik dieses Wahlrechts vor dem Hintergrund der Konsistenz mit dem Conceptual Framework und weiteren US-GAAP Vorschriften. Dabei ist die von der Autorin geäußerte Kritik u.E. gerechtfertigt, denn ein aufgrund eines Wahlrechts gewährter Bilanzierungsspielraum führt gerade nicht zu einer Erhöhung des Informationszieles des Abschlusses.

Eine explizite Ausnahme zu dem allgemeinen Wahlrecht des SOP 93-7.26 wird hingegen für Ausgaben, die in den Bereich der Direktwerbung fallen, gewährt. So ist für Kosten bezüglich so genanntem „Direct-Response Advertising" eine Aktivierung verpflichtend, wenn der Hauptzeck dieser speziellen Werbemaßnahmen darin zu sehen ist, Kunden zu generieren, die nachgewiesener Weise auf diese Direktwerbung reagiert haben und sichergestellt ist, das die Werbemaßnahme einen wahrscheinlichen künftigen Nutzen hervorruft (SOP 93-7.33 a, b). Während letztere Voraussetzung unproblematisch erscheint, gilt für Erstere gemäß SOP 93-7.34 eine besondere Dokumentationspflicht.

3.8 Derivativer Goodwill

Der derivative Goodwill ist eine Residualgröße aus dem Kaufpreis, der für das Unternehmen bezahlt wurde und dem Zeitwert der Vermögenswerte und Schulden des Unternehmens (Nettosubstanzwert). Er enthält regelmäßig immaterielle Werte, die in der originären Bilanz des erworbenen Unternehmens nicht angesetzt wurden. Ist ein immaterieller Vermögenswert jedoch vom Goodwill separierbar, besteht eine Aktivierungspflicht in der in den Konzernabschluss einzubeziehenden Bilanz. Aus diesem Grund sind im Rahmen eines Unternehmenserwerbs übertragene immaterielle Werte einer besonderen Betrachtung zu unterziehen. Eine Identifikation und Separierung solcher Werte stellt sich in der Praxis jedoch regelmäßig als schwierig dar (vgl. *Davis*, 2002, S. 697).

In Deutschland findet beispielsweise für in einem erworbenen Unternehmen angefallene Entwicklungskosten zum Zeitpunkt des Erwerbs das Aktivierungsverbot des § 248 Abs. 2 HGB keine Anwendung, da die Vermögensgegenstände nicht selbst erstellt, sondern im Rahmen des Unternehmenskaufs erworben wurden. Trotzdem erfolgt eine separate Aktivierung nur selten, da die Bilanzierungsregeln für Goodwill mehr bilanzpolitischen Spielraum bieten als die für immaterielle Vermögensgegenstände (vgl. Arbeitskreis „Immaterielle Werte im Rechnungswesen" der *Schmalenbach Gesellschaft für Betriebswirtschaft e.V.*, 2001, S. 989).

Das deutsche Handelsrecht bietet den Unternehmen vielfältige Wahlrechte zur Goodwillbilanzierung. So kann ein Goodwill beispielsweise erfolgsneutral mit den Rücklagen verrechnet werden. Wird der Goodwill hingegen aktiviert, kann er entweder pauschal über maximal vier Jahre oder planmäßig über seine voraussichtliche Nutzungsdauer abgeschrieben werden (§ 309 Abs. 1 S. 3 HGB). Aufgrund des geltenden Niederstwertprinzips sind auf den Goodwill gegebenenfalls außerplanmäßige Abschreibungen vorzunehmen. Da Kriterien für eine Abschreibung jedoch nicht definiert sind, erfolgt eine außerplanmäßige Wertberichtigung eher aus bilanzpolitischen Erwägungen. Eine Wertaufholung wird nach herrschender Meinung abgelehnt (vgl. *Pellens/Sellhorn*, 2001, S. 713). Der stark an den Regelungen der IFRS orientierte DRS 4 verbietet eine erfolgsneutrale Verrechnung des Goodwills und schreibt eine Aktivierung mit anschließend planmäßiger Abschreibung vor. Dabei gilt die widerlegbare Vermutung, dass die Nutzungsdauer 20 Jahre nicht überschreitet (DRS 4.27, 4.30).

Die Goodwillbilanzierung nach US-GAAP ist durch die bereits angesprochenen Vorschriften SFAS 141 und SFAS 142 im Jahre 2001 umfassend novelliert worden. Während ein Goodwill bis dahin planmäßig entsprechend dem unterstellten, wirtschaftlichen Werteverzehr abgeschrieben wurde, ist eine planmäßige Abschreibung nach der neuen Regelung nicht mehr erlaubt. Vielmehr ist ein im Rahmen eines Unternehmenserwerbs entstehender Goodwill auf Unternehmensteilbereiche („Reporting Units") aufzuteilen. Mögliche Wertminderungen dieses bereits auf Ebene der Reporting Units „segmentierten" Goodwill sind auf der Grundlage eines mindestens jährlich durchzuführenden Werthaltigkeitstests zu berücksichtigen (vgl. *Nestler/Thuy*, 2002, S. 169). Dieser zweistufige Test beginnt mit der Ermittlung des Zeitwertes („Fair Value") der Reporting Unit (vgl. *Küting/Weber/Wirth*, 2001, S. 189).

Der Wert ist sowohl bei der erstmaligen Aufteilung des Goodwills auf die Unternehmensteilbereiche zum Zeitpunkt des Unternehmenserwerbes als auch beim durchzuführenden Impairmenttest die zentrale Größe. In einem weiteren Schritt wird durch einen Vergleich des ermittelten Fair Value mit den Buchwerten der Vermögenswerte der Reporting Unit die Höhe des Abwertungsbedarfes festgestellt. Ein ausführliches Prüfungsschema zur Feststellung und Erfassung einer Wertminderung nach US-GAAP bietet *Dawo* (vgl. *Dawo,* 2003, S. 171).

Hintergrund dieses umfangreichen wie revolutionären Projekts war die Überarbeitung der Konsolidierungsvorschriften in den USA und die daraus resultierende Abschaffung der Pooling of Interests-Methode der Kapitalkonsolidierung. Im Zuge der Schaffung von SFAS 141 und 142 war das FASB der Meinung, dass die meisten der in den Unternehmensbilanzen aktivierten Goodwillpositionen ihren Wert nicht regelmäßig verlieren würden, insbesondere sei nicht davon auszugehen, dass eine planmäßige lineare Abschreibung den tatsächlichen Werteverzehr des Goodwill widerspiegelt (SFAS 142 B.79). Die planmäßige Abschreibung der Goodwillpositionen wurde daraufhin durch eine Wertminderungsprüfung (Impairment) ersetzt. Im Schrifttum wird dies als „Impairment-Only-Approach" bezeichnet (vgl. *Pellens/Sellhorn*, 2001, S. 713).

Nach IFRS ist ein Goodwill zwingend zu aktivieren und planmäßig über seine voraussichtliche Nutzungsdauer abzuschreiben. Der Abschreibungszeitraum soll den Zeitraum umfassen, in welchem dem Unternehmen aus dem Goodwill ein Nutzen zufließt. Vom Abschreibungszeitraum wird widerlegbar vermutet, dass dieser 20 Jahre nicht überschreitet (IAS 22.44). Die Nutzungsdauer und die Abschreibungsmethode sind gemäß IAS 22.54 mindestens einmal jährlich auf ihre Angemessenheit hin zu überprüfen und bei wesentlicher Abweichung anzupassen. Neben der planmäßigen Abschreibung sieht IAS 22.55 ff. Regelungen zur außerplanmäßigen Abschreibung des Goodwills gemäß IAS 36 vor. Wird ein Goodwill über weniger als 20 Jahre abgeschrieben, erfolgt nur bei begründeten Annahmen einer Wertminderung ein Werthaltigkeitstest. Liegt die Nutzungsdauer darüber, hat der Impairmenttest jährlich zu erfolgen (IAS 36.79 ff). Wertaufholungen sind nur unter restriktiven Voraussetzungen möglich (IAS 36.109).

4. Neue Entwicklungen zur Bilanzierung immaterieller Werte und Goodwill

4.1 Handelsgesetzbuch (HGB) und Deutsche Rechnungslegungs-Standards (DRS)

Das zum 26. Juli 2002 in Kraft getretene Transparenz- und Publizitätsgesetz (TransPuG) hat durch die Modifikation des HGB unter anderem zur Aufhebung der Anschaffungskostenrestriktion im Rahmen der Konsolidierung nach der Neubewertungsmethode geführt (Aufhebung des § 301 Abs. 1 S. 4 HGB). Somit ist auch DRS 4.24, der eng an die im HGB verankerte Anschaffungsrestriktion gebunden war, obsolet. Auswirkungen ergeben sich bei künftigen Unternehmenserwerben genau dann, wenn das Eigenkapital des erworbenen Unternehmens dessen Kaufpreis übersteigt. In diesen Fällen müssen die stillen Rücklagen nunmehr nicht in ihrem Gesamtbetrag gekürzt werden, um die Obergrenze einzuhalten. Stattdessen wird in Höhe der Differenz ein passivischer Unterschiedsbetrag (negativer Goodwill) ausgewiesen (vgl. *Busse von Colbe*, 2002, S. 1585).

In DRS 12.33 wird empfohlen, im Konzernlagebericht über das intellektuelle Kapital zu berichten. Beispielhaft wird in DRS 12.34 eine entsprechende Kategorisierung des intellektuellen Kapitals vorgestellt, ohne auf die konkrete Ausgestaltung einzugehen. Kürzlich wurde ein Vorschlag des Arbeitskreises „Immaterielle Werte im Rechnungswesen" für die Gestaltung einer freiwilligen ergänzenden Berichterstattung veröffentlicht. Demnach sind für jede Kategorie die Strategie, ein Katalog der Indikatoren, Definition und Wechselwirkung der Indikatoren und eine entsprechende Kommentierung zu erläutern (vgl. Arbeitskreis „Immaterielle Werte im Rechnungswesen" der *Schmalenbach-Gesellschaft für Betriebswirtschaft e.V.*, 2003, S. 1233 – 1237). Die Berücksichtigung dieses Vorschlages soll die praktische Umsetzung des DRS 12.34 ermöglichen und somit auch in der Berichterstattung der wachsenden Bedeutung immateriellen Vermögens Rechnung tragen.

4.2 US-Generally Accepted Accounting Principles (US-GAAP)

Nach einer umfänglichen Überarbeitung der Vorschriften zur Bilanzierung immaterieller Vermögenswerte und des Goodwill durch SFAS 141 und 142 im Jahre 2001, hat sich das FASB im Anschluss daran mit der Darstellung von weiteren Anhangangaben zu immateriellen Vermögenswerten, die im Unternehmen selbst geschaffen wurden und nicht Gegenstand eines Unternehmenserwerbs waren, befasst.

Das Projekt beschäftigt sich mit der Schaffung von Erfassungs- und Anhangsangaberegelungen zu immateriellen Vermögenswerten, die nicht im Abschluss bilanziert werden. Diese Werte können im Unternehmen selbst geschaffen worden sein (wie Marken, Kundenbeziehungen, Lizenzen und Patente) oder auch im Rahmen von Unternehmenszusammenschlüssen erworben und sofort abgeschrieben worden sein (wie laufende Forschungs- und Entwicklungsarbeiten). Diskutiert werden hierzu sowohl die Angabe quantitativer Größen, wie Zeitwerte oder aufgelaufene Kosten bis zur Entstehung der immateriellen Werte, als auch diverse qualitative Informationen. Dabei ist das Projekt darauf ausgerichtet, anstelle von möglichen freiwilligen Angaben konkrete Anforderungen für Angaben über nicht ausgewiesene immaterielle Vermögenswerte zu formulieren (*www.fasb.org/project/intangibles.shtml* S. 2). Dieses Projekt ruht zurzeit jedoch.

Vor dem Hintergrund gesteigerter Unternehmenstransparenz wäre eine Weiterverfolgung dieses Projekts sicherlich zu begrüßen; nicht zu vernachlässigen ist jedoch die Tatsache der erschwerten und ungenauen Ermittlung der offen zu legenden Informationen. Auch dürfte das eine oder andere Unternehmen der Einbeziehung von intern generierten Vermögenswerten in den Jahresabschluss – wenn auch nur als Anhangsangabe – nicht besonders positiv gegenüberstehen.

4.3 International Financial Reporting Standards (IFRS)

Im Rahmen des IASB-Projekts „Business Combinations" wurden im Dezember 2002 neben dem Standardentwurf für Unternehmenszusammenschlüsse (ED 3 Business Combinations) auch überarbeitete Standardentwürfe zu IAS 36 und IAS 38 veröffentlicht. Daraus gehen grundlegende Änderungen der Bilanzierung von immateriellen Vermögenswerten (ED-IAS 38) und der Regelung von außerordentlichen Abschreibungen aufgrund des Impairmenttests (ED-IAS 36) hervor. Im Folgenden werden ausgewählte Änderungen dargestellt.

Zunächst wird in ED-IAS 38.7 eine geringfügig abgeänderte Definition für einen immateriellen Vermögenswert vorgeschlagen. Es wird nunmehr nicht auf die Verwendung des Vermögenswertes abgestellt. Zudem wird in ED-IAS 38.11 das Kriterium der Identifizierbarkeit konkretisiert. Dieses Kriterium gilt zum einen als erfüllt, wenn ein immaterieller Vermögenswert separierbar ist, d.h. einzeln oder im Zusammenhang mit anderen Vermögenswerten, Schulden oder Verträgen vom Unternehmen getrennt werden kann. Zum anderen gilt die Identifizierbarkeit als erfüllt, wenn der immaterielle Vermögenswert aus vertraglichen oder gesetzlichen Rechten resultiert, unabhängig von deren Übertragbarkeit oder Separierbarkeit. Dadurch soll erreicht werden, dass im Rahmen eines Unternehmenszusammenschlusses erworbene immaterielle Vermögenswerte weitgehend vom Goodwill getrennt bilanziert werden. Der umfassendere Einzelansatz von immateriellen Vermögenswerten spiegelt nach Meinung des IASB deren zunehmende Bedeutung wider (vgl. ED-IAS 38 B7).

Die nach IAS 38 geltenden Ansatzkriterien haben weiterhin Bestand. Ergänzend stellt ED-IAS 38 klar, dass für den Einzelerwerb (ED-IAS 38.22) die Voraussetzung des wahrscheinlichen zukünftigen Nutzenzuflusses immer gegeben ist. Gleiches gilt für den Erwerb von immateriellen Vermögenswerten durch Unternehmenszusammenschluss (ED-IAS 38.29). Die zuvor erwähnten expliziten Aktivierungs- und Bilanzierungsverbote nach IAS 38.51 beispielsweise für selbst geschaffene Markennamen und IAS 38.57 für die Gründung und den Anlauf eines Geschäftsbetriebes bleiben weiterhin bestehen.

Hinsichtlich der Folgebewertung stellt das IASB mit dem ED-IAS 38 neue Regelungen vor, die sich an den US-GAAP Regelungen des SFAS 142 orientieren. Zwar bleiben die unter Abschnitt 3.1 dargestellten Fortführungskonzeptionen (Benchmark Treatment und Allowed Alternative Treatment) grundsätzlich bestehen, allerdings ergeben sich grundlegende Veränderungen für die Nutzungsdauer- und Abschreibungssystematik. Die bislang geltende widerlegbare Annahme der maximalen Nutzungsdauer von 20 Jahren entfällt. Stattdessen ist zu prüfen, ob die Nutzungsdauer eines immateriellen Vermögenswertes begrenzt oder unbestimmbar ist. Eine unbestimmbare Nutzungsdauer ist anzunehmen, wenn die Nettozahlungszuflüsse aus dem Vermögenswert keinen vorhersehbaren zeitlichen Beschränkungen unterliegen. Diese immateriellen Vermögenswerte werden zukünftig nicht mehr planmäßig abgeschrieben, sondern mindestens jährlich einem Impairmenttest unterzogen. Immaterielle Vermögenswerte mit begrenzter Nutzungsdauer sind demgegenüber weiterhin planmäßig abzuschreiben. Ein Impairmenttest ist für diese immateriellen Vermögenswerte lediglich bei entsprechenden Anzeichen für eine Wertminderung durchzuführen. Der vorgeschriebene jährliche Impairmenttest für immaterielle Vermögenswerte mit einer Nutzungsdauer von über 20 Jahren entfällt und ist ebenfalls nur bei entsprechenden Anzeichen für einen Wertverlust durchzuführen. Letztlich bewirken die Neuregelungen einen Übergang auf einen Impairment-Only-Approach für immaterielle Vermögenswerte mit unbestimmbarer Nutzungsdauer (vgl. *Langecker/Mühlberger*, 2003, S. 119).

Für den Fall der unbestimmbaren Nutzungsdauer schreibt ED-IAS 38.117 vor, den Buchwert des Vermögenswertes und Faktoren für die Nutzungsdauerbestimmung darzulegen.

Nicht zuletzt aufgrund des geplanten Wegfalls von planmäßigen Goodwillabschreibungen wird der Impairmenttest innerhalb der Rechnungslegung nach IFRS zunehmend bedeutsamer. Daher sollen im Folgenden ausgewählte Änderungsvorschläge für die Werthaltigkeitsprüfung gemäß ED-IAS 36 dargestellt werden. Zunächst fordert ED-IAS 36.8A im Gegensatz zu den geltenden Regelungen einen jährlichen Impairmenttest unabhängig von Wertverlustindikatoren für nicht planmäßig abzuschreibende und noch nicht in Betrieb genommene immaterielle Vermögenswerte als auch für erworbenen Goodwill.

Die Grundkonzeption des Impairmenttests wird beibehalten (vgl. *Kümpel*, 2003, S. 1491). Allerdings konkretisiert ED-IAS 36.25A die Vorgaben für die Berechnung des Nutzungswerts, um den Ermessensspielraum des Bilanzierenden einzuschränken. Zudem werden bei der Nutzungswertermittlung nach ED-IAS 36.26A in Anlehnung an US-GAAP die Eintrittswahrscheinlichkeiten der möglichen Zahlungsmittelzuflüsse berücksichtigt.

Wie bisher ist der Impairmenttest grundsätzlich für den einzelnen Vermögenswert durchzuführen. Muss von diesem Grundsatz abgewichen werden, weil einem Vermögenswert keine von anderen Vermögenswerten unabhängigen Zahlungsströme zugeordnet werden können, sind Zahlungsmittelgenerierende Einheiten (ZGE) zu identifizieren. Für die Bildung von ZGE ist nunmehr die Ebene der Segmentberichterstattung als Zuordnungsobergrenze festgelegt (vgl. ED-IAS 36.74).

Hinsichtlich der Berücksichtigung von Goodwill in den ZGE stellen ED-IAS 36.73 und .74 klar, dass die Zuordnung auf einer vernünftigen und stetigen Basis erfolgen muss. Um dies zu gewährleisten können auch mehrere ZGE für die Zuordnung des Goodwills zusammengefasst werden. Dabei müssen die ZGE (bzw. mehrere ZGE) auf der tiefsten Ebene der Steuerung des return on investment angesiedelt sein. Darüber hinaus wurde in Anlehnung an die US-GAAP ein zweistufiges Verfahren für den Imparimenttest von ZGE, denen Goodwill zugeordnet wurde, vorgeschlagen. Allerdings hat das IASB kürzlich beschlossen (vgl. *IASB* Update Juli 2003, S.2), aufgrund der Kostenintensivität und Komplexität der Durchführung des Tests, die derzeitigen Regelungen zum einstufigen Impairmenttests beizubehalten.

Aufgrund der in ED 3.39 zwingend vorgegebenen Anwendung der Neubewertungsmethode ergeben sich auch Änderungen hinsichtlich der Berücksichtigung von Minderheitenanteilen bei der Goodwill-Bewertung. So ist eine den Minderheitenanteilen entsprechende Hochrechnung des Goodwills notwendig. Eine Wertminderung des Goodwills ist also nur in Höhe des Konzernanteils zu erfassen (vgl. *Kümpel*, 2003, S. 1492). Darüber hinaus sieht ED-IAS 36 ein durchgängiges Zuschreibungsverbot für Goodwill vor, um die Aktivierung von originärem Goodwill auszuschließen (vgl. ED-IAS 36 C63).

Die dargestellten Änderungsvorschläge nebst den entsprechenden Kommentaren werden derzeit im IASB diskutiert. Die Verabschiedung der endgültigen Standards ist für das erste Quartal 2004 vorgesehen.

4.4 Konvergenz von IFRS und US-GAAP

Der IASB und das FASB möchten gemäß einer Vereinbarung von Oktober 2002 bis zum Jahre 2004 IFRS und US-GAAP weitgehend harmonisieren. Die Bilanzierung und Bewertung von immateriellen Vermögenswerten und Goodwill gemäß dem Standardentwurf ED 3 sowie ED-IAS 36 und ED-IAS 38 sind vor diesem Hintergrund als gemeinsames Projekt des IASV mit dem FASB erarbeitet worden. Da Unternehmenszusammenschlüsse in der Regel wertmäßig und hinsichtlich ihrer nachhaltigen Bilanzwirkung sehr bedeutsam sind, ist das Projekt für die Konvergenzziele sehr wichtig (vgl. *Fladt/Feige*, 2003, S. 249).

Gleichwohl muss festgehalten werden, dass die Regelungsvorschläge des IASB nicht deckungsgleich mit denen der US-GAAP sind. Dies gilt insbesondere für die Durchführung des Werthaltigkeitstests (vgl. *Fladt/Feige*, 2003, S. 262). Allerdings ergeben sich bei der Durchführung des Impairmenttests in beiden Rechnungslegungssystemen weit reichende Ermessensspielräume, so dass eine Vergleichbarkeit ohnehin nur schwer erreichbar ist.

5. Zusammenfassung

Die Rechnungslegungsvorschriften nach HGB/DRS, US-GAAP und IFRS zur Bilanzierung immaterieller Werte weisen signifikante Unterschiede auf.

Nach deutschem Handelsrecht werden immaterielle Vermögensgegenstände dem Schuldendeckungsgedanken folgend definiert. Im Unternehmen selbst generierte Vermögenswerte werden – aufgrund der Bewertungsproblematik und dem daraus resultierenden Vorsichtsgedanken – nicht in den Unternehmensbilanzen ausgewiesen. Dies hat zur Folge, dass Branchen, in denen immaterielle Vermögenswerte einen wichtigen Produktionsfaktor darstellen, ihr Erfolgpotenzial nicht im Abschluss darstellen können.

Der Deutsche Standardisierungsrat hat im DRS 12 empfohlen, das HGB-Ansatzverbot für selbsterstellte immaterielle Vermögenswerte aufzuheben, um damit eine Annäherung an die IFRS zu ermöglichen. Eine diesbezügliche HGB-Reform ist derzeit jedoch nicht absehbar. Insbesondere gilt die durch das Maßgeblichkeitsprinzip gewährleistete Verknüpfung von Jahresabschlusses und Steuerbilanz als problematisch.

Demgegenüber wird nach US-GAAP und IFRS die Wahrscheinlichkeit eines künftigen Nutzenzuflusses durch den zu aktivierenden Wert herausgestellt. Neben dem eher prinzipienbasierten Aktivierungsgrundsätzen der IFRS beinhalten die US-GAAP-Vorschriften eine Vielzahl von Einzelregelungen die allgemeine Aktivierungsvoraussetzungen eher vermissen lassen. Dennoch kann auch nach US-GAAP eine allgemeine Systematik in den diversen Vorschriften zur Aktivierung von „Intangible Assets“ gefunden werden.

Die Bilanzierung von Goodwill wurde in den letzten Jahren nahezu vollständig geändert. Der US-Standard SFAS 142 hat die planmäßige Abschreibung von Goodwill abgeschafft und setzt einen zweistufigen Impairmenttest an dessen Stelle. Zudem fordert SFAS 142 die spezifische Identifikation, „Herauslösung“ und separate Aktivierung von immateriellen Werten aus dem im Rahmen eines Unternehmenszusammenschlusses erworbenen Goodwill. Dies führt zu einer gesteigerten Transparenz in den Bilanzen der Unternehmen. Das IASB hat sich in den veröffentlichten Standardentwürfen ED3, ED-IAS 36 und ED-IAS 38 stark den US-GAAP angenähert.

Sowohl bei der Behandlung von immateriellen Werten als auch bei der Goodwill-Bilanzierung wird es in Zukunft starke Unterschiede nur noch zwischen deutscher Rechnungslegung einerseits und US-GAAP und IFRS andererseits geben. Eine Annäherung für alle Unternehmen kann nur durch eine Reform des Handelsgesetzbuches auf der Grundlage der 4. und 7. EG Richtlinie erfolgen.

6. Literatur

Arbeitskreis „Immaterielle Werte im Rechnungswesen“ der *Schmalenbach Gesellschaft für Betriebswirtschaft e.V.*, Kategorisierung und bilanzielle Erfassung immaterieller Werte, in: Der Betrieb, 54. Jg. (2001), H. 19, S. 989–995.

Baetge, J. und v. Keitz, I., IAS 38 – Immaterielle Vermögenswerte (Intangible Assets), in: *Baetge, J., Dörner, D., Kleekämper, H., Wollmert, P. und Kirsch, H.J.* (Hrsg.), Rechnungslegung nach International Accounting Standards (IAS), 2. Auflage, Hannover u.a. 2002.

Beck'scher Bilanzkommentar, 5. Auflage, München 2003.

Brücks, M. und Wiederhold, P., Exposure Draft 3 „Business Combinations“ des IASB, in: Zeitschrift für kapitalmarktorientierte Rechnungslegung, 3. Jg. (2003), H. 1, S. 21–29.

Busse von Colbe, W., Kleine Reform der Konzernrechnungslegung durch das TransPuG – Ein weiterer Schritt zur Internationalisierung und Kapitalmarktorientierung, in: Bilanzrecht und Betriebswirtschaft, 57. Jg. (2002), H. 31, S. 1583–1588.

Davis, M. K., US-GAAP: Rechnungslegung für immaterielle Vermögensgegenstände nach Verabschiedung der FASB Standards 141 und 142, in: Der Betrieb, 55. Jg. (2002), H. 14, S. 697–701.

Dawo, S., Immaterielle Güter in der Rechnungslegung nach HGB, IAS/ IFRS und US-GAAP – Aktuelle Rechtslage und neue Wege der Bilanzierung und Berichterstattung, Herne/Berlin 2003.

DRS – Deutsches Rechnungslegungs-Standards Committee, Deutsche Rechnungslegungs-Standards, Stuttgart, Loseblattsammlung, Stand 9/2002.

Exposure Draft ED 3 – Business Combinations, Download, IASB, *www.iasb.org.uk*.

Fladt, G. und Feige, P., Der Exposure Draft 3 „Business Combinations“ des IASB – Konvergenz mit den US-GAAP?, in: Die Wirtschaftsprüfung, 56. Jg. (2003), H. 6, S. 249–262.

Hayn, S. und Graf Waldersee, G., IAS/US-GAAP/HGB im Vergleich, 3. Auflage, Stuttgart 2003.

IAS – International Accounting Standards Board, International Accounting Standards 2002, London 2002.

IASB, IASB Update, July 2003.

IAS-Verordnung – Amtsblatt der Europäischen Gemeinschaften L 243 vom 11. September 2002, Verordnung (EG) Nr. 1606/2002 des Europäischen Parlaments und des Rates vom 19. Juli 2002 betreffend die Anwendung internationaler Rechungslegungsstandards.

KPMG, Implementing IAS – IAS compared with US-GAAP, 2002.

Kümpel, T., Geplante Änderungen der bilanziellen Behandlung von Wertminderungen bei Vermögenswerten nach ED-IAS 36, in: Betriebs-Berater, 58.Jg. (2003), H. 28/29, S. 1491–1494.

Küting, K. und Ulrich, A., Abbildung und Steuerung immaterieller Vermögenswerte (Teil I), in: Deutsches Steuerrecht, 39. Jg. (2001), H. 23, S. 953–960.

Küting, K., Weber, C.-P. und Wirth, J., Die neue Goodwillbilanzierung nach SFAS 142, in: Zeitschrift für kapitalmarktorientierte Rechnungslegung, 1. Jg. (2001), H. 5, S. 185–198.

Langecker A. und Mühlberger M., Berichterstattung über immaterielle Vermögenswerte im Konzernabschluss: Vergleichende Gegenüberstellung von DRS 12, IAS 38 und IAS 38 rev., in: Zeitschrift für kapitalmarktorientierte Rechnungslegung, 3. Jg. (2003), H. 3, S. 109–123.

Larson, K.D. and Miller, P.B.W., Financial Accounting, 6. Auflage, Chicago 1995.

Nestler A. und Thuy M.G., Verfahren zur Bewertung von Reporting Units im Rahmen des Goodwill-Impairments nach SFAS 142, in: Zeitschrift für kapitalmarktorientierte Rechnungslegung, 2. Jg. (2002), H. 4, S. 169–179.

Pellens B., Internationale Rechnungslegung, 4. Auflage, Stuttgart 2001.

Pellens B. und Sellhorn T., Neue Goodwill-Bilanzierung nach US-GAAP, in: Der Betrieb, 54. Jg. (2001), H. 14, S. 713–720.

Proposed Amendments to IAS 38 – Intangible Assets, Download, IASB, *www.IASB.org.uk.*

Schneider, S., Die grundsätzliche bilanzielle Behandlung von Werbeausgaben gemäß SOP 93-7 sowie Ingangsetzungsausgaben gemäß SOP 98-5, in: Internationales Steuerrecht, 12. Jg. (2003), H. 5, S. 175–180.

TransPuG (Transparenz- und Publizitätsgesetz) – Bundesgesetzblatt Jahrgang 2002, Teil I, Nr. 50, ausgegeben zu Bonn am 25. Juli 2002, Gesetz zur weiteren Reform des Aktien- und Bilanzrechts, zu Transparenz und Publizität (Transparenz- und Publizitätsgesetz).

US-GAAP – American Institute of Certified Public Accountants, AICPA Technical Practice Aids as of June 1, 2002, New York 2002.

US-GAAP – Financial Accounting Standards Board, EITF Abstracts – A Summary of Proceedings of the FASB Emerging Issues Task Force as of September 20–21, 2000, Norwalk 2000.

US-GAAP – Financial Accounting Standards Board, Original Pronouncements as of June 1, 2002, Norwalk 2002.

Ein besonderer Dank gilt Frau *Kati Beiersdorf* für die große Unterstützung im Rahmen der Überarbeitung des Beitrags für dieses Buch.

Fair-Value-Bewertung von Intangibles sowie die bilanzielle Behandlung des Goodwill im Rahmen von Business Combinations

Reinhard Heyd [1]

---

[1] Prof. Dr. *Reinhard Heyd*, Fachhochschule Nürtingen, Universität Ulm.

Die Bilanzierung immaterieller Vermögenswerte einschließlich des Goodwills ist seit einiger Zeit in der Diskussion. Neue Regelungen nach US-GAAP und IFRS zielen auf eine verstärkte Anwendung der Fair Value Bewertung. Der Impairment Only Approach wird allgemeine Praxis für die Goodwill-Bilanzierung. Der Beitrag zeigt theorie- und anwendungsbezogene Aspekte der bilanziellen Behandlung von Business Combinations auf und beurteilt die Qualität der Fair Value Bewertung anhand der Kriterien Relevance und Reliability.

1. Der Fair Value als Bewertungsmaßstab

Mit der Einführung der internationalen Rechnungslegung und der damit verbundenen Ausrichtung auf die Vermittlung entscheidungsrelevanter Informationen sowie die Harmonisierung von internem und externem Rechnungswesen werden vermehrt entscheidungsorientierte Wertansätze zur Bewertung der betrieblichen Erfolgspotenziale verwendet. Das Postulat der Fair Presentation läßt sich nach dem Conceputal Framework von US-GAAP, dem Framework von IAS und dem Framework-Entwurf des Deutschen Standardisierungsrates als Kompromiss zwischen den Anforderungen von Relevance und Reliability beschreiben. Während das traditionelle HGB-Konzept auf retrospektiv ermittelten Anschaffungs- oder Herstellungskosten basiert, gewinnt in IAS/IFRS und US-GAAP der Fair Value (beizulegender Zeitwert) immer größere Bedeutung. Er wird als entscheidungsorientierter Wert betrachtet. Die Bewertung zum Fair Value soll

- der vollständigen Darstellung des realisierbaren Haftungspotenzials dienen, welches ein Asset repräsentiert,
- stille Reserven offenlegen,
- die Entscheidungsorientierung im Rechnungswesen erhöhen (Opportunitätskosten) sowie
- die zeitliche und zwischenbetriebliche Vergleichbarkeit verbessern mit dem Ziel, die Bilanzanalyse zu vereinfachen.

Der Fair Value wird nach IAS 38.7 (gleichlautend IAS 16.6, 17.3, 18.7, 19.7, 20.3, 21.7, 22.8, 25.4, 32.5, 33.9, 39.8, 40.4) definiert als „the amount for which an asset could be exchanged or a liability settled between knowledgeable, willing parties in an arm's length transaction“. Er stellt die bestmögliche Schätzung (Best Estimate) des Preises dar, den ein Unternehmen erzielt hätte, wenn es in einer unter normalen Geschäftsbedingungen ablaufenden Transaktion mit Dritten den Vermögenswert zum Bewertungsstichtag veräußert oder die Schuld beglichen hätte. Der Fair Value ist grundsätzlich gegenwartsbezogen, stellt allerdings eine Markteinschätzung für den Barwert der künftig aus dem Asset erzielbaren Cashflows dar. Als Value in Use beschreibt er einen unternehmensindividuellen Einzelertragswert (vgl. *Mujkanovic*, 2002, S. 113–117); dieser weicht jedoch regelmäßig vom Ertragswert des Gesamtkonzerns bzw. seiner Segmente ab.

Die Ermittlung des Fair Value erfolgt in drei Stufen:

- Zunächst wird ein Marktwert für das in Frage stehende Asset ermittelt; ein öffentlich notierter Marktpreis auf einem aktiven Markt ist der bestmögliche Indikator für den Fair Value.
- Ersatzweise versucht man einen Vergleichswert für das Asset zu finden. Entweder bietet eine zeitnahe Transaktion für dasselbe Asset oder eine zeitgleiche Transaktion eines ähnlichen Assets nützliche Anhaltspunkte für die Bestimmung des Fair Value.
- Ist auch dies nicht zielführend, so wird ein plausibler Schätzwert durch Diskontierung der zukünftigen durch das Asset generierten Ein- und Auszahlungen ermittelt, wobei finanzmathematische Verfahren wie die Discounted-Cashflow-Methode (DCF) herangezogen werden (z.B. IAS 39.100).

Während die ersten beiden Stufen eine marktmäßige Verwertung des Assets unterstellen, geht die dritte Stufe von einer internen Nutzung als Value in Use aus.

Die gegenwärtige Anwendung des Fair Values nach IFRS/IAS bezieht sich auf

- ausgewählte Financial Instruments, insbesondere Trading- und Available-for-Sale-Wertpapiere und Hedges (IAS 39),
- Sachanlagen im Rahmen des Allowed Alternative Treatment (IAS 16),
- Investment Property, d.h. als Kapitalanlagen gehaltene Immobilien (Benchmark Treatment nach IAS 40),
- Immaterielle Vermögenswerte bei Tauschvorgängen und im Rahmen des Allowed Alternative Treatment (IAS 38) sowie
- die Abbildung von Business Combinations.

Bei US-GAAP ist darüber hinaus eine Fair Value Bewertung des Goodwills im Rahmen des mindestens einmal jährlich stattfindenden Impairment Tests geboten (SFAS 142, 143). Dem wird sich die IFRS-Rechnungslegung nach ED-3 voraussichtlich anschließen (siehe unten Kapitel 3.5.1).

2. Bewertung von Intangible Assets nach IAS 38

Intangible Assets sind nach IAS 38.7 identifizierbare, nicht monetäre Vermögenswerte ohne physikalische Substanz, die für die Herstellung von Erzeugnissen oder Erbringung von Dienstleistungen, die Vermietung an Dritte oder Zwecke der eigenen Verwaltung genutzt werden. Sie sind anzusetzen, wenn dem Unternehmen aus dem Vermögenswert der künftige wirtschaftliche Nutzen wahrscheinlicherweise zufließen wird und die Anschaffungs- oder Herstellungskosten des Intangible Assets zuverlässig bemessen werden können. Für Bewertungszwecke lassen sich immaterielle Vermögenswerte wie folgt einteilen:

Kriterien	**Alternativen**		
Nutzungsdauer	Begrenzte, nicht bestimmbare Nutzungsdauer	Bestimmbare Nutzungsdauer	
Bewertungsverfahren	Impairment Only	Fortgeführte Anschaffungs- oder Herstellkosten	Neubewertungsmethode

Sofern die Aktivierungsvoraussetzungen erfüllt sind, sind Intangible Assets bei Erstverbuchung mit ihren Anschaffungs- oder Herstellungskosten zu bewerten. Dies gilt für IAS/IFRS, US-GAAP und HGB gleichermaßen. Wird ein immaterieller Vermögenswert im Tausch gegen Eigenkapitalinstrumente des berichtenden Unternehmens erworben, so entsprechen dessen Anschaffungskosten dem beizulegenden Zeitwert der emittierten Eigenkapitalinstrumente, die wiederum dem beizulegenden Zeitwert des erworbenen immateriellen Vermögenswertes entsprechen (IAS 38.26).

In den Folgeperioden sind Intangible Assets nach dem Benchmark Treatment mit den fortgeführten Anschaffungs- oder Herstellungskosten anzusetzen. IAS 38.84 geht davon aus, dass die Nutzungsdauer eines immateriellen Vermögenswertes immer begrenzt ist. Unabhängig von der planmäßigen – regelmäßig der linearen – Abschreibung ist ein Impairment-Test im Hinblick auf eingetretene Wertminderungen durchzuführen (IAS 38.97). Er hat mindestens einmal jährlich zu erfolgen, wenn die Nutzungsdauer zwanzig Jahre übersteigt (IAS 38.101). Beim Impairment Test ist der Buchwert mit dem erzielbaren Betrag (Recoverable Amount) zu vergleichen (IAS 38.7). Dieser ist definiert als der höhere der beiden Beträge aus Nettoveräußerungspreis (Net Selling Price) und Nutzungswert (Value in Use). Letzterer wird aus den abgezinsten künftigen Cashflows, die der immaterielle Vermögenswert generieren kann, ermittelt (IAS 36.26-56). Ist der erzielbare Betrag für einen einzelnen Vermögenswert nicht zu schätzen, so ist der erzielbare Betrag der zahlungsmittelgenerierenden Einheit zu bestimmen, zu der der Vermögenswert gehört (IAS 36.65). Ist ein Wertminderungsaufwand einer zahlungsmittelgenerierenden Einheit erfolgswirksam in Ansatz zu bringen, so ist dieser zuerst dem Goodwill zuzurechnen, dann anteilig auf die anderen Vermögenswerte der Einheit nach dem Verhältnis der Buchwerte zu verteilen (IAS 36.88). Sind die Gründe, die in früheren Jahren zu einem Impairment geführt haben, in späteren Jahren weggefallen, so besteht nach IAS 36.94 ff. eine grundsätzliche Wertaufholungspflicht. Obergrenze sind die fortgeführten Anschaffungs- oder Herstellungskosten (IAS 36.102). Wird die Wertaufholung bei einer zahlungsmittelgenerierenden Einheit durchgeführt, so erfolgt die Wertaufholung zuerst anteilig auf die Vermögenswerte (ohne Goodwill) entsprechend deren Buchwerten, danach unter den strengen Voraussetzungen des IAS 36.109 auf den Goodwill (IAS 36.107). Obergrenze für den Buchwert der einzelnen Vermögenswerte einer zahlungsmittelgenerierenden Einheit ist der niedrigere Betrag aus erzielbarem Betrag (Recoverable Amount) und fortgeführten Anschaffungskosten.

Als Allowed Alternative Treatment sieht IAS 38.64 für Abschlüsse nach dem erstmaligen Ansatz vor, Intangible Assets mit einem Neubewertungsbetrag fortzuführen, der sich aus dem Fair Value zum Zeitpunkt der Neubewertung ergibt, abzüglich späterer Abschreibungen und Wertminderungen. Neubewertungen sind – sofern das Allowed Alternative Treatment Anwendung findet – hinreichend regelmäßig durchzuführen, wobei die Häufigkeit von Neubewertungen von der Volatilität des beizulegenden Zeitwertes abhängt (IAS 38.68). Sie kommen nur für die Bewertung ganzer Gruppen immaterieller Vermögenswerte in Betracht (IAS 38.70). Bei der Fair-Value-Bewertung von Intangible Assets ist stets auf einen aktiven Markt Bezug zu nehmen (ggf. bei übertragbaren Taxilizenzen, Fischereilizenzen oder Produktionsquoten), andernfalls ist das Benchmark Treatment anzuwenden (fortgeführte Anschaffungs- oder Herstellungskosten). IAS 38.67 nimmt an, dass es für Markennamen, Drucktitel bei Zeitungen, Musik- und Filmverlagsrechten sowie Patenten und Warenzeichen keinen aktiven Markt gibt. Einzelne immaterielle Vermögenswerte einer neubewerteten Gruppe, für die kein aktiver Markt besteht, sind mit fortgeführten Anschaffungskosten anzusetzen. Die Wertänderungen aus der Neubewertung sind wie folgt zu behandeln:

- Fair-Value-Erhöhungen werden in eine Neubewertungsrücklage eingestellt, es sei denn sie machen einen erfolgswirksam verbuchten Neubewertungsverlust früherer Jahre rückgängig.
- Fair-Value-Minderungen werden als Aufwand erfasst, es sei denn, sie machen eine in früheren Jahren erfolgsneutral gebildete Neubewertungsrücklage rückgängig. (IAS 38.63–78).

3. Bilanzielle Behandlung des Goodwill (derivativer Firmenwert)

3.1 Allgemeines

Der derivative Firmenwert ist der Betrag, um den der Kaufpreis eines Unternehmens die Differenz zwischen Aktiva und Schulden, bewertet zu Fair Values, übersteigt. Während für selbst geschaffene Firmenwerte (originärer Goodwill) nach HGB, IAS/IFRS und US-GAAP ein Aktivierungsverbot besteht, existiert für einen derivativen Firmenwert bei Erwerb eines Unternehmens als Sachgesamtheit nach IAS/IFRS und US-GAAP eine Aktivierungspflicht, nach HGB ein Aktivierungswahlrecht (§ 255 Abs. 4 HGB). Tritt ein Goodwill aus Konsolidierungsvorgängen auf nachdem stille Reserven bzw. Lasten aufgeteilt sind, so besteht auch für diesen nach IAS/IFRS und US-GAAP eine Aktivierungspflicht, nach § 309 HGB dagegen ein Wahlrecht

- zur Aktivierung und planmäßigen Abschreibung oder
- zur sofortigen erfolgsneutralen Verrechnung mit den Gewinnrücklagen. Diese Vorgehensweise vermeidet, dass in späteren Jahren Ergebnisse durch Goodwill-Abschreibungen belastet werden. Durch die Reduzierung des Eigenkapitals verbessert sich bei gegebenem Gewinn die Eigen- und Gesamtkapitalrentabilität. Diese Vorgehensweise wird außerhalb des Geltungsbereichs des HGB als problematisch angesehen und findet international keine Akzeptanz. Auch mit DRS 4 ist dieses Vorgehen nicht vereinbar (DRS 4.28).

	Originärer Goodwill	**Goodwill aus Unternehmenserwerben als Sachgesamtheit**	**Goodwill aus Konsolidierungsvorgängen**
HGB	Aktivierungsverbot	Aktivierungswahlrecht (§ 255 Abs. 4 HGB)	Wahlrecht nach § 309 HGB: - Aktivierung und planmäßige Abschreibung Sofortige erfolgsneutrale Verrech nung mit den Gewinnrücklagen
IAS	Aktivierungsverbot	Aktivierungspflicht	Aktivierungspflicht
US-GAAP	Aktivierungsverbot	Aktivierungspflicht	Aktivierungspflicht

3.2 Bilanzielle Behandlung des Goodwill nach HGB und DRS

Der Goodwill ist eine Residualgröße. Seine quantitative Höhe im Jahresabschluss ergibt sich aus der Fair Value-Bewertung aller anderen Vermögensgegenstände und Schulden im Zeitpunkt des Unternehmenserwerbs. Wird der Goodwill aktiviert, so ist er grundsätzlich erfolgswirksam abzuschreiben; lediglich nach § 309 Abs. 1 Satz 3 HGB darf wahlweise eine erfolgsneutrale Verrechnung erfolgen. Die Abschreibungsfrist beträgt fünf Jahre („die folgenden vier Jahre", §§ 255 Abs. 4 Satz 2, 309 Abs. 1 Satz 1 HGB). Die Abschreibung kann aber auch planmäßig auf die Geschäftsjahre verteilt werden, in denen er voraussichtlich genutzt werden kann (§§ 255 Abs. 4 Satz 2, 309 Abs. 1 Satz 2 HGB). Auch Zwischenlösungen und Mischformen sind denkbar (*Pellens/Sellhorn,* 2001, S. 714; *Küting*, 2000, S. 97-106; *Lewicki,* 2002, S. 56-62). Das HGB sieht kein standardisiertes Verfahren zur Ermittlung möglicher Wertminderungen vor, wodurch ein gewisser bilanzpolitischer Freiraum eröffnet wird. Trotz des für Kapitalgesellschaften geltenden Zuschreibungsgebotes nach § 280 Abs. 1 HGB sieht ein Teil der Literatur eine Wertaufholung derivativer Firmenwerte kritisch, da dadurch die Möglichkeit einer Aktivierung orginärer Firmenwerte nicht ausgeschlossen ist (*Pellens/Sellhorn,* 2001, S. 714 m.w.N.). DRS 4 „Unternehmenserwerbe im Konzernabschluss", knüpft grundsätzlich an internationale Vorschriften zur Goodwill-Aktivierung an. Zwar wird darin die nach § 309 Abs. 1 Satz 3 HGB mögliche erfolgsneutrale Behandlung nicht mehr als zulässig erachtet (DRS 4.28) und eine Aktivierungspflicht mit Abschreibung auf eine Nutzungsdauer von (widerlegbar) 20 Jahren gefordert, es wird aber das allgemeine Wertaufholungsgebot auch für den derivativen Firmenwert festgeschrieben (DRS 4.36). DRS 4 benennt keine Konkretisierungen für die Messung wertmindernder Faktoren; lediglich eine jährliche Überprüfung der Werthaltigkeit (Impairment-Test) wird verlangt, wobei die Höhe der außerplanmäßigen Goodwill-Abschreibungen im Konzernanhang anzugeben ist (*Pellens/Sellhorn,* 2001, S. 714).

Um jedoch sicherzustellen, dass eine Goodwill-Bilanzierung nach US-GAAP (SFAS 142 Impairment Only Approach) mit den EU-Richtlinien konform ist, was eine Voraussetzung für die mit befreiender Wirkung durchgeführte Konzernrechnungslegung nach US-GAAP aufgrund von § 292a HGB darstellt, wurde am 6.4.2002 DRS 1a durch das Bundesjustizministerium bekanntgemacht. In ihm wird ausgeführt,

- dass der US-amerikanische Impairment-Only-Approach für den Goodwill nicht in Widerspruch zu den wesentlichen Bewertungsgrundsätzen der EG-Richtlinien steht,
- dass es neben der Einteilung in begrenzte und unbegrenzte Nutzungsdauern auch eine Einteilung in bestimmbare und unbestimmbare Nutzungsdauern gibt und
- dass bei einer unbestimmbaren Nutzungsdauer – wie dies beim Goodwill angenommen wird – der periodische Werthaltigkeitstest an die Stelle der planmäßigen Abschreibung zu treten hat.

Somit bleibt festzuhalten, dass der Impairment Only Approach mit den EU-Richtlinien konform ist. Die Variantenvielfalt zulässiger Darstellungsmöglichkeiten des Goodwill ist somit beträchtlich (kritisch: *Busse von Colbe,* 2001, S. 877-879; *Henselmann,* 2000, S. 283-287; *Kleindiek,* 2001, S.2572-2577; *Krawitz,* 2002, S. 145-148).

3.3 Bilanzielle Behandlung des Goodwill nach IAS

Nach IAS 22.44 ff. (revised 1998) ist der Goodwill planmäßig über die bestmöglich geschätzte Nutzungsdauer abzuschreiben, während der dem Unternehmen zukünftige Nutzenstiftungen zufließen. Es besteht eine widerlegbare Vermutung, dass die Nutzungsdauer regelmäßig 20 Jahre vom Zeitpunkt der Erstverbuchung an nicht überschreitet. Bei Widerlegung dieser Vermutung durch stichhaltige Gründe, also bei Zugrundelegung einer Nutzungsdauer von mehr als 20 Jahren, ist ein jährlicher Impairment-Test nach IAS 36 durchzuführen; bei einer angesetzten Nutzungsdauer von weniger als 20 Jahren ist ein Impairmenttest nur bei Vorliegen bestimmter Indikatoren erforderlich. Diese Indikatoren können sich beispielsweise beziehen auf externe Faktoren, wie z.B. Entwicklungen auf den relevanten Beschaffungs-, Absatz- und Kapitalmärkten wie auch auf unternehmensinterne Entwicklungen, die als negative Soll-Ist-Abweichungen aus dem internen Rechnungswesen ersichtlich sind. Die gewählte Abschreibungsmethode soll der erwarteten Wertentwicklung möglichst entsprechen. Ohne besondere Anhaltspunkte dafür dass eine andere Abschreibungsmethode den Entwertungsverlauf besser abbildet, ist die lineare Methode anzuwenden. Sowohl Restnutzungsdauer als auch Abschreibungsmethode sind jährlich einer Überprüfung zu unterziehen; der Abschreibungsverlauf ist ggf. anzupassen. IAS 8 ist entsprechend anzuwenden. Auch außerplanmäßige Abschreibungen auf den erzielbaren Betrag (Recoverable Amount) kommen in Betracht. Wegen der fehlenden Möglichkeit, den Goodwill isoliert zu bewerten, ist als Bewertungseinheit eine größere Cash-Generating Unit zu wählen (IAS 36.79). Diese entspricht regelmäßig der mit dem Goodwill in Zusammenhang stehenden Tochtergesellschaft, in Ausnahmefällen ist hier der ganze Konzern als Zurechnungseinheit zugrunde zu legen. Im Falle eines Impairments auf die Cash-Generating Unit ist zuerst der Goodwill zu vermindern bevor die übrigen Vermögenswerte der Cash-Generating Unit anteilig gekürzt werden. Nachträgliche Werterhöhungen können nur unter den restriktiven Bedingungen des IAS 36.109 durch eine Wertaufholung zum Ausdruck gebracht werden. Das bedeutet, dass eine Wertaufholung auf Firmenwerte nur zulässig ist, wenn die Wertminderung durch ein spezielles, außergewöhnliches und einmaliges externes Ereignis (z.B. geänderte Rechtsvorschriften etc.) ausgelöst wurde, dass später durch externe Ereignisse in seiner Wirkung umgekehrt wurde. Dadurch soll verhindert werden, dass originäre Firmenwerte aktiviert werden.

3.4 Bilanzielle Behandlung des Goodwill nach US-GAAP

Grundsätzliches zum Impairment-Only-Approach

Nach US-GAAP war bis 2001 eine Goodwill-Aktivierung und Abschreibung auf längstens 40 Jahre zwingend, wobei die lineare Abschreibungsmethode als übliche Norm, andere Methoden nur in Ausnahmefällen angewendet wurden und eine fortlaufende Überprüfung von Abschreibungsdauer und -verfahren notwendig war. Traten Indikatoren einer Wertminderung in Erscheinung, so musste die Notwendigkeit einer außerplanmäßigen Abschreibung überprüft werden. Wurde eine solche durchgeführt, so war der Abschreibungsaufwand im operativen Ergebnis zu zeigen; Gründe, Höhe und Ermittlungsmethode der Abschreibung waren in den Notes anzugeben (SFAS 121.13 f), ferner die Segmente, in denen sich die Goodwill-Bewertung vollzog (SFAS 131); Wertaufholungen waren nicht zulässig (*Pellens/Sellhorn,* 2001, S. 715 ff; *Stauber/Ketterle,* 2001, S. 955-966; *Teitler-Feinberg,* 2001, S. 331-338 *Alvarez/Biberacher,* 2002, S.346-353).

Die dieser Form der Goodwill-Bilanzierung zugrundeliegende Behandlung eines Unternehmenserwerbs (Purchase-Methode) hat aus Sicht der Konzernleitungen den Nachteil, dass aus der Akquisition über viele Jahre hinweg Ergebnisminderungen aufgrund von Goodwill-Abschreibungen die Folge sind. Daher war in der amerikanischen Bilanzierungspraxis die bis 2001 alternativ zulässige Pooling-of-Interest-Methode besonders beliebt. Sie zeichnet sich dadurch aus, dass beim Unternehmenszusammenschluss „unter Gleichen" unterstellt wird, dass die Verbindung schon immer bestanden habe, somit eine Übernahme der Bilanzposten zu Buchwerten erfolgt und ein Unterschiedsbetrag aus der Erstkonsolidierung als Differenz zwischen Kaufpreis und so ermitteltem anteiligem Eigenkapital erfolgsneutral mit den Konzernrücklagen verrechnet wird. Die Pooling-of-Interest-Methode ist allerdings starker nationaler und internationaler Kritik ausgesetzt, insbesondere weil Rentabilitätskennziffern als Quotient von Periodenergebnis zu Eigenkapital bei ihrer Anwendung betragsmäßig zu hoch ausgewiesen werden, da

- die mit dem Goodwill-Erwerb zusammenhängende Ausgabe nie die GuV-Rechnung belastet und somit ein zu hoher Gewinn (Zähler) ausgewiesen wird, und
- die durch die direkte Verrechnung mit dem Eigenkapital verbundene Eigenkapitalminderung zu einem zu geringen Betrag der Bezugsgröße im Nenner führt.

Aufgrund dieser Kritik hat das FASB in einem Exposure Draft „Business Combinations and Intangible Assets" von 1999 die Abschaffung der Pooling-of-Interest-Methode vorgeschlagen. Dieser Vorstoß ist allerdings auf erheblichen Widerstand der US-amerikanischen Wirtschaft gestoßen, sodass in einer Überarbeitung des Exposure Drafts vom 14.2.2001 zwar an der Abschaffung der Pooling-of-Interest-Methode festgehalten, die dann allein zulässige Purchase Methode aber so modifiziert wurde, dass

- zwar die stillen Reserven aus allen Assets und Liabilities aufzudecken, d.h. diese zum Fair Value anzusetzen sind, und
- die identifizierbaren immateriellen Vermögenswerte zu Fair Values aktiviert werden unabhängig davon, ob sie vor der Akquisition angesetzt waren oder nicht, dass aber
- keine planmäßigen, sondern nur noch fallweise außerplanmäßige Abschreibungen auf den Goodwill vorzunehmen sind (Impairment-Only-Approach), welcher den unternehmensinternen Reporting Units zugewiesen wird.

Unter Reporting Units versteht man grundsätzlich ein „Operating Segment" gemäß SFAS 131 oder ein Level darunter in der Unternehmenshierarchie. Besteht ein Operating Segment aus mehreren Teilbereichen, so ist zu prüfen, ob alle Teilbereiche ökonomisch ähnlich sind bzw. ob einer der Teilbereiche selbst eine Reporting Unit ist. Ist dies der Fall, ist das Operating Segment selbst eine Reporting Unit. Ist dies nicht der Fall, ist zu prüfen, ob die Reporting Unit ein Business darstellt. Ein Business liegt vor, sobald finanzielle Informationen verfügbar sind, eine regelmäßige Kontrolle durch das Segment Management erfolgt und der Teilbereich auch selbständig am Markt existieren könnte.

Mit der Aufdeckung stiller Reserven im Erstkonsolidierungszeitpunkt kommt es regelmäßig zu latenten Steuerabgrenzungen, die aber wegen der gleichzeitigen Erhöhung des Goodwills erfolgsneutral dargestellt werden. Erfolgswirksame Differenzen kann es geben, wenn der Goodwill in der Steuerbilanz planmäßig, im US-GAAP-Konzernabschluss aber außerplanmäßig abgeschrieben wird.

Mit SFAS 142 „Business Combinations", der am 29.6.2001 verabschiedet wurde, ist der Impairment Only Approach als geltende US-GAAP-Regelung für Geschäftsjahre, die nach dem 15.12.2001 beginnen, anzuwenden.

Erstbilanzierung

Die Goodwill-Behandlung nach SFAS 141 und 142 läßt sich in die Erst- und Folgebilanzierung einteilen:

Die Erstbilanzierung (im Übernahmezeitpunkt) beinhaltet folgende Einzelschritte:

- Bestimmung des Goodwills aus der Übernahme (Business Combination) im Rahmen der Purchase Methode. Dies setzt voraus, dass die Übernahmekosten für das erworbene Unternehmen auf die akquirierten Assets und Liabilities gemäß den Fair Values der einzelnen Bilanzposten verteilt werden. Anschließend wird die vorläufige Restgröße auf identifizierbare immaterielle Vermögenswerte untersucht. Sie können eingeteilt werden in

- Marketing-related
- Customer-related
- Artistic-related
- Contact-based
- Technology-based Intangible Assets (*Pfeil/Vater,* 2002, S. 68)

▪ Ablösung und gesonderten Ausweis der übernommenen, identifizierbaren Intangible Assets, die bislang nicht bilanziert waren. Sie sind mit ihrem Fair Value in den Konzernabschluss aufzunehmen. Der nach Separierung dieser Intangible Assets verbleibende Restbetrag stellt den Goodwill der Akquisition dar. Durch den erweiterten separaten Ausweis von immateriellen Vermögenswerten als einzelne Bilanzposten umfasst der Goodwill nur noch den Wert der übernahmebedingten Synergien aus dem Unternehmenszusammenschluss.

Übernahmekosten (Kaufpreis) für das erworbene Unternehmen	
-	(Fair Values der übernommenen Assets - Fair Values der übernommenen Liabilities)
=	vorläufige Restgröße
-	Separierung und gesonderter Ausweis identifizierbarer immaterieller Vermögenswerte, die in der Bilanz des übernommenen Unternehmens noch nicht ausgewiesen waren
=	Goodwill der Akquisition

▪ Zuordnung der übernommenen Vermögenswerte und Schulden einschließlich identifizierbarer Intangible Assets auf die einzelnen Reporting Units nach dem betrieblichen Einsatz entsprechend der Management Absicht; gemeinsam genutzte Vermögenswerte sind vernünftig und nachvollziehbar aufzuteilen; diese Aufteilung ist stetig anzuwenden, Corporate Items sind in einer Nebenbuchhaltung zu berücksichtigen.

▪ Verteilung des Gesamtgoodwills aus der Unternehmensakquisition auf die Reporting Units auf der Grundlage von Management Judgement oder relativ zu den Fair Values der den jeweiligen Reporting Units zugeordneten ansatzfähigen Vermögenswerten und Schulden. Der Goodwill kann auch einer Reporting Unit zugeordnet werden, die keine Vermögenswerte und Schulden zugewiesen erhält (SFAS 142.35). In diesem Fall wird der Goodwill als Wertänderung des Fair Values der gesamten Reporting Unit durch den Erwerbsvorgang interpretiert (With-and-Without-Computation). Bei der Goodwill-Allokation auf die Reporting-Units sind große Beurteilungsspielräume vorhanden, welche bilanzpolitisch genutzt werden können (*Kütting/Weber/Wirth,* 2001, S. 188). In der Praxis werden Goodwills vornehmlich auf ertragsstabile Reporting Units verteilt, bei denen ein Goodwill-Impairment in absehbarer Zeit nicht zu erwarten ist.

Durch die Definition von Reporting Units, die Fair Value Bewertung aller Vermögenswerte und Schulden einschließlich immaterieller Assets, die bislang nicht aktiviert waren, die Verteilung der Assets und Liabilities sowie die Goodwillallokation auf die Reporting Units erweitert sich der bilanzpolitische Beurteilungsspielraum in den einzelnen Perioden gegenüber der früheren Regelung.

Folgebilanzierung

Normalerweise ist mindestens einmal jährlich ein Impairmenttest für die Goodwill-Bewertung durchzuführen; bei Auftreten von Umständen, die eine Fair Value Minderung vermuten lassen, ist er auch unterjährig geboten. Er gliedert sich in zwei Teilschritte „Prüfung, ob ein Impairment vorliegt" und „Quantifizierung des Impairments".

Prüfung, ob ein Impairment vorliegt

Beim Impairmenttest erfolgt eine Neubestimmung des Fair Values der Reporting Unit durch Neubewertung aller bedeutsamen Faktoren für eine Gesamtbewertung. Es ist zu prüfen, ob ein Impairment wahrscheinlich ist durch Vergleich des Fair Values der Reporting Unit, ermittelt durch Gesamtbewertung mittels diskontierter Cashflows, mit den auf den Stichtag ermittelten Fair Values der Vermögenswerte, Schulden und des Goodwills der Reporting Unit. Dabei kann es zu einer Vermischung von originärem und derivativem Goodwill kommen; ferner hängt der Wert des Goodwills von den eigens für den Impairmenttest aufgedeckten stillen Reserven in der Reporting Unit ab.

Der Goodwill ist unverändert mit dem in einem früheren Test ermittelten Wert fortzuführen, wenn

- die Vermögenswerte und Schulden sich nicht wesentlich verändert haben,
- der Fair Value des Goodwills beim letzten Impairmenttest seinen Buchwert bei weitem überstiegen hat,
- eine Analyse der seit dem letzten Impairmenttest eingetretenen Ereignisse und Änderungen in den Rahmenbedingungen es unwahrscheinlich erscheinen lassen, dass der aktuelle Fair Value der Reporting Unit ihren Buchwert unterschreitet.

Quantifizierung des Impairments

Sind Anhaltspunkte für eine Wertminderung des Goodwills seit dem letzten Impairmenttest gegeben, so ist die Höhe eines evtl. Impairments zu bestimmen. Für ein Impairment im Rahmen der Folgebilanzierung bildet der Fair Value der Reporting Unit die Ausgangsbasis (Gesamtbewertung) (*Lüdenbach/Schulz,* 2002, S. 489-499). Dabei können Marktquotierungen für die betreffende Unternehmenseinheit bei börsennotierten Gesellschaften eine Rolle spielen; bei nicht-börsennotierten Einheiten ist eine bestmögliche Schätzung mittels der Discounted Cashflow-Methode durchzuführen, welche die Erwartungen der Marktteilnehmer berücksichtigt und auf SFAC No. 7 aufbaut (*Starbatty,* 2001, S. 543-558). In diesem Fall werden aber konzerninterne Faktoren wie Synergieeffekte und strategische Vorteile nicht berücksichtigt. Ferner ist der Unternehmenswert nicht nur von den geschätzten Cashflows und ihrer zeitlichen Verteilung, sondern auch vom zugrundegelegten Marktzins als Diskontierungsfaktor abhängig.

Der Wertberichtigungsbedarfs bestimmt sich durch Gegenüberstellung des Fair Values der Reporting Unit als Unternehmenswert und den Fair Values der Vermögenswerte und Schulden aus Sicht eines virtuellen Erwerbers der Reporting Unit einschließlich der bisher nicht bilanzierten, aber aus Sicht des gedachten Erwerbers bilanzierungsfähigen materiellen und immateriellen Vermögenswerte. Sie werden eigens für den Impairmenttest identifiziert und zum Fair Value bewertet. Der Differenzbetrag wird mit dem Gesamtbuchwert des für die Reporting Unit bilanzierten Goodwills verglichen. Übersteigt der Goodwill-Buchwert den ermittelten Fair Value, so ist in Höhe der Differenz eine Abwertung vorzunehmen. Der um die außerplanmäßige Abschreibung korrigierte Goodwill stellt für zukünftige Geschäftsjahre die „New Basis of Accounting“ dar. Nachträgliche Zuschreibungen sind nicht zulässig (*Küting/Weber/Wirth,* 2001, S. 191 m.w.N).

Fair Value der Reporting Unit (Gesamtbewertung)
- Fair Values der Vermögenswerte und Schulden
= Fair Value des Goodwills der Reporting Unit
- Bisheriger Buchwert des Goodwills der Reporting Unit
= Wertberichtigungsbedarf

Zusammenfassend lassen sich die Arbeitschritte im Erwerbszeitpunkt und bei Durchführung des Impairmenttests (Folgebilanzierung) wie folgt darstellen

Arbeitsschritte im Erwerbszeitpunkt
- Bildung von Reporting Units
- Zuordnung von Vermögensgegenständen und Schulden auf die Reporting Units
- Aufteilung des Goodwills auf die Reporting Units
Arbeitsschritte bei Durchführung des Impairmenttests (Folgebilanzierung)
- Feststellung eines Wertberichtigungsbedarfs (Haben sich die Wertverhältnisse gegenüber dem letzten Impairment-Test geändert?)
- Feststellung der Höhe des Wertberichtigungsbedarfs: Vergleich der Differenz zwischen dem Fair Value der Reporting Unit (Gesamtbewertung) und den Fair Values der Vermögenswerte und Schulden (ohne Goodwill) mit dem bisherigen Buchwert des Goodwills

Anhangangaben

In den Notes ist eine Berichterstattung über die wirtschaftlichen Motive der Aquisition und der methodischen Behandlung des Goodwills erforderlich. Dabei sind Angaben zu machen über

- die Bedeutung der Aquisition sowie der damit verbundenen Ziele und Erwartungen,
- die Bewertungsparameter für die Darstellung des Goodwill im laufenden und in späteren Abschlüssen,
- die Zurechnung des Goodwills auf nachgelagerte dezentrale Konzerneinheiten.

Ziel sind ist eine objektivierte Werthaltigkeitsbeurteilung und eine Begrenzung des mit der Goodwillbilanzierung zusammenhängenden bilanzpolitischen Freiraums.

Ferner sind

- der Goodwillbetrag aus Unternehmenserwerben,
- die außerplanmäßigen Abschreibungen der Berichtsperiode sowie
- die Abgänge aus dem Goodwill, die in dem Gewinn/Verlust aus dem Abgang einer Business Unit verrechnet wurden

für den Konzern insgesamt wie auch für die einzelnen Segmente anzugeben. Schließlich ist über wesentliche Änderungen bei der Goodwillzuordnung auf die Segmente zu berichten.

Hat ein Impairment stattgefunden, so sind Angaben zu machen über

- die Umstände, die zu der außerplanmäßigen Abschreibung geführt haben,
- den Betrag der außerplanmäßigen Abschreibung und die Methode zur Ermittlung dieses Betrages (z.B. Marktpreis, Comparable-Business-Approach, Barwert-Verfahren o.ä.)
- ggf. die Auswirkungen auf die Folgeperioden.

Bilanzpolitik

Als bilanzpolitisches Ziel im Zusammenhang mit der Goodwillbilanzierung kann die Vermeidung bzw. Verlagerung von Impairments in die Zukunft bezeichnet werden. Dies läßt sich damit begründen, dass

- es sich bei Goodwill-Impairments meist um ganz wesentliche Beträge handelt, die sowohl Ertrags- als auch Kapitalstrukturkennzahlen negativ beeinflussen,
- die Leiter der Reporting Units im Falle eines Goodwill-Impairments eine Under-Performance zum Ausdruck bringen müssen und
- die Gründe für ein Goodwill-Impairment für das Konzernmanagement eine Reputationseinbuße darstellen, weil ein Goodwill-Impairment entweder
  - vorgenommen wird, weil der Kaufpreis für das erworbene Unternehmen zu hoch war oder
  - die Synergieeffekte, die sich der Übernehmende von der Akquisition erwartet haben, nicht eingetreten sind.

Als Instrumente zur Realisierung dieses bilanzpolitischen Ziels kommen Beurteilungsspielräume im Zusammenhang mit dem Impairment-Test in Betracht

- bei der Gesamtbewertung der Reporting Unit (Auswahl des Planungshorizontes, des Zinssatzes, der Bestimmung des Cashflows am Ende des Planungshorizontes etc),
- bei der Auswahl von Zeitpunkt und Häufigkeit des Impairments,
- bei der Auswahl der Indikatoren für den Werthaltigkeitstest, SFAS 142 nennt einen nicht abschließenden Katalog von Beispielen für Sachverhalte, die Anlass für einen Impairment-Test geben sollen:
  - Verlustausweise in der Vergangenheit, im laufenden Jahr oder Verlusterwartungen in der Zukunft,
  - Entwicklung der wirtschaftlichen Rahmenbedingungen zu ungunsten des Unternehmens wie z.B. sinkende Umsätze, Marktanteile, Verlängerung der Zeitdauer von der Idee zum Produkt, Fluktuation von Key Employees bzw. Key Employee Groups in Forschung und Entwicklung oder Marketing sowie neu erkannter Restrukturierungsbedarf etc.,
  - absehbare Gesetzesänderungen, die die wirtschaftliche Tätigkeit des Unternehmens negativ beeinflussen,
  - Pläne über die mit mehr als 50%-iger Wahrscheinlichkeit bevorstehende Veräußerung des Berichtsunternehmens (more likely than not),
  - Werthaltigkeitsprüfung für einen wesentlichen Bestandteil des Berichtsunternehmens,
  - erheblich und dauerhaft sinkende Börsenkurse des Unternehmens (Marktkapitalisierung) unter den Buchwert des Eigenkapitals,

  - erheblich und dauerhaft sinkende Börsenkurse des Unternehmens (Marktkapitalisierung) entgegen dem allgemeinen Börsentrend,
  - Zurückstufung von emittierten Wertpapieren des Unternehmens oder einer Berichtseinheit hinsichtlich der Ratingkategorie;
- bei der Fair Value Bewertung der bilanzierten Assets und Liabilities der Reporting Unit zum Impairmentzeitpunkt,
- bei der Auswahl der nur im Impairmentstatus anzusetzenden immateriellen Assets sowie deren Bewertung zum Fair Value.

Bilanzanalyse

Goodill-Ausweise in Abschlüssen nach dem Impairment-Only-Approach sind weder vergleichbar mit Goodwill-Ausweisen bei planmäßiger Abschreibung noch mit Darstellungen von Business Combinations nach der Pooling-of-Interest-Methode.

Im Vergleich zur Pooling-of-Interest-Methode erhöht sich beim Impairment Only Approach durch den Goodwill-Ausweis sowohl die Bilanzsumme als auch das Eigenkapital. In beiden Fällen kommt es nicht zu Goodwillabschreibungen, bei der Darstellung von Unternehmensakquisitionen nach SFAS 141, 142 kommt es aber zu ergebnisminderndem Abschreibungsaufwand auf die aufgelösten stillen Reserven in den Assets und auf die Fair Values der gesondert ausgewiesenen Intangible Assets. Daraus ergeben sich Effekte auf die Eigen- und Gesamtkapitalrentabilität, die nicht allgemein vorab quantifizierbar sind, sowie negative Auswirkungen auf die Ergebnisausweise.

Im Vergleich zur planmäßigen Goodwill-Abschreibung bei der Purchase-Methode (HGB) ist der Ergebnisausweis nach US-GAAP ohne Impairment höher; das Eigenkapital und die Bilanzsumme werden nicht durch planmäßige Goodwillabschreibungen reduziert.

Allerdings beinhalten Goodwill-Beträge nach dem Impairment-Only-Approach ein latentes Verlustpotenzial, da Impairments wegen der innewohnenden Beurteilungsspielräume über einige Zeit hinausgeschoben werden können. Die Nachholung des Goodwill-Impairments hat meist gravierende Auswirkungen auf Ergebnis- und Kapitalstrukturkennzahlen (Big Bath Effect). Zur Abschätzung möglicher Verlustgefahren künftiger Perioden können verbale Erläuterungen zusätzliche Transparenz schaffen. Impairment-Abschreibungen sind zu begründen und hinsichtlich ihrer Rahmenbedingungen in den jeweiligen Segmenten zu würdigen. Sowohl die Impairment-Abschreibung als auch der neue Goodwill-Buchwert innerhalb des jeweiligen Segmentes sind betragsmäßig zu beziffern. Die Goodwillabschreibung wird im Rahmen des Operating Income gezeigt.

Beurteilung des Impairment Only Approachs nach US-GAAP

Die Anwendung des Impairment Only Approaches für die Goodwillbewertung in den Reporting Units läßt sich durch folgende Argumente bewerten:

- Die Beurteilungsspielräume bei der Fair Value Bewertung der übernommenen Vermögenswerte und Schulden im Zeitpunkt der Erstkonsolidierung können so genutzt werden, dass Wertansätze abnutzbarer Vermögensgegenstände zugunsten des Goodwill reduziert werden, um künftige Ergebnisbelastungen durch Abschreibungen zu minimieren.
- Es ist nicht festgelegt, wie der Impairment-Betrag auf die Goodwill-Komponenten der Reporting Unit zu verteilen ist, was bei späteren Endkonsolidierungen von Bedeutung ist.

- Goodwillbeträge sind bei der Ermittlung des Abgangserfolgs nur dann zu berücksichtigen, wenn der ausscheidende Teil der Reporting Unit ein „Business“ darstellt (allein lebensfähiger Teilbetrieb).
- Bei der Fair Value Bewertung der Reporting Units können auch selbsterstellte Intangibles und FuE-Aufwendungen aktiviert werden.
- Die Fair Values der Vermögenswerte und Schulden sind nur für den Goodwill-Impairment relevant; sie beeinflussen die Abschreibungsbasis dieser Objekte selbst nicht, was den für ihre Bemessung angewandten Beurteilungsspielraum erhöht.
- Die Gesamtbewertung einer Reporting Unit durch Abzinsung der mit ihr verbundenen Cashflows stellt eine Abkehr von der Einzelbewertungskonzeption dar.
- Die Gesamtbewertung kann zu einer stillen Aktivierung von originärem Goodwill führen.
- Die nach dem Unternehmenserwerb generierten neuen Zahlungsströme stehen in keinem Bezug zum ursprünglich bilanzierten derivativen Goodwill.
- Es kann sich um eine stille Einbeziehung von Saldierungseffekten zwischen Goodwill aus verschiedenen Erwerbsvorgängen und auch mit einem zum Erwerbszeitpunkt vorhandenen originären Goodwill handeln.
- Es besteht eine erhebliche Subjektivität in der Ermittlung des Unternehmenswertes, insbesondere hinsichtlich
  - der Zahlungsstromprognose, durch die der Beurteilungsspielraum dazu genutzt wird, Goodwillabschreibungen in die Zukunft zu verlagern, bis so gravierende Abschreibungsgründe vorliegen, dass eine Vermeidungsstrategie für außerplanmäßige Abschreibungen nicht mehr durchzuhalten ist,
  - der Möglichkeit, negative Unternehmensentwicklungen zu verschleiern,
  - der Möglichkeit im Falle eines ungünstigen Unternehmensumfeldes durch eine hohe Goodwillabschreibung spätere Jahre von weiterer Aufwandsverrechnung zu entlasten (Big Bath) (*Küting/Weber/Wirth,* 2001, S. 192).
- Wegen bisher hoher Goodwillabschreibungen führt die Neuregelung zu (vorübergehend) höheren Gewinnausweisen; diese Höherbewertung des Unternehmens und seiner Anteile kann genutzt werden, wenn die Anteile als Akquisitionswährung verwendet werden.
- Mit dem Impairment Only Approach wird externes Wachstum begünstigt, da internes Wachstum mit ergebnisbelastenden Aufwendungen verbunden ist.

3.5 Weiterentwicklungen durch das IASB

Um zu einer Konvergenz mit den US-GAAP zu gelangen, wurde ein IASB-Projekt „Business Combinations“ aufgelegt, das in zwei Phasen eingeteilt ist und die Grundlagen von SFAS 141, 142 übernehmen soll. Damit gewinnt der Impairment Only-Approach weltweite Bedeutung (*Brücks/Wiederhold,* 2003, S. 28-29; *Fladt/Feige,* 2003, S. 249-262).

IASB-Projekt Business Combinations Phase I (ED 3)

In Phase I (IASB-ED-3) wird die Darstellung aller Unternehmenszusammenschlüsse nach der Purchase Methode vorgeschrieben und die Pooling of Interest Methode verboten. Ausgenommen sind Joint-Venture-Gründungen und Unternehmenszusammenschlüsse unter gemeinschaftlicher Beherrschung (Common Control). Für Reverse Acquisitions wird die Fresh Start Methode diskutiert, bei der alle Vermögenswerte und Schulden des rechtlichen Erwerbers zum beizulegenden Zeitwert anzusetzen sind.

Bewertung bei Erstkonsolidierung

Bei der Purchase Methode sind alle immateriellen Vermögenswerte getrennt vom Goodwill auszuweisen, wenn sie aus einem Vertrag oder Recht resultieren oder eindeutig identifizierbar sind. Eine sogenannte „Assembled Workforce“ (übernommene Mitarbeiterschaft) erfüllt jedoch nicht die Voraussetzungen zur Aktivierung, weil keine Ansatz-, Bewertungs- und Allokationsmethoden vorhanden sind. Dagegen sollen laufende FuE-Projekte im Gegensatz zur derzeitigen Rechtslage aktivierbar sein, wenn sie separierbar sind oder ein vertraglich oder gesetzlich durchsetzbares Recht darstellen (*Brücks/Wiederhold,* 2003, S. 25). Der neue Standard wird (analog SFAS 141) Beispiele aktivierbarer Intangibles enthalten; sie sind bei erstmaliger Erfassung zum Fair Value zu bewerten. Ansatzpflichtig sind im Rahmen von Business Combinations nach ED 3.45 auch bedingte Schulden (Contingent Liabilities), deren beizulegender Zeitwert verlässlich gemessen werden kann (Unterschied zu IAS 22.26 ff und IAS 37.27). Restrukturierungsrückstellungen dürfen in Zukunft nur noch im Rahmen der Erstkonsolidierung gebildet werden, wenn die betreffende unsichere Verpflichtung bereits vor Begründung des Control-Verhältnisses durch das gekaufte Unternehmen wirtschaftlich verursacht wurde (ED 3.40 f.).

Alle identifizierbaren Vermögenswerte, Schulden und bedingten Verpflichtungen sind im Übernahmezeitpunkt zu ihrem Fair Value am Erwerbsstichtag anzusetzen. Dies bedeutet eine vollständige Neubewertung aller anzusetzenden Vermögenswerte und (bedingten) Schulden auch in Höhe der Minderheitenanteile (Ausnahme: Goodwill, Full-Goodwill-Approach) (*Pellens/Basche/Sellhorn,* 2003, S. 1-4). Das bisherige Benchmark Treatment nach IAS 22.32 zur Neubewertung lediglich in Höhe des Anteils der (neuen) Muttergesellschaft entfällt.

Der Goodwill ist bei der Erstkonsolidierung nachvollziehbar und begründbar auf eine oder mehrere Cash-Generating Units zuzuordnen (ED IAS 36.73 ff.) als kleinstmögliche Gruppe von Vermögenswerten innerhalb eines Unternehmens, die Einzahlungen aus der fortlaufenden Nutzung unabhängig von den Einzahlungen anderer Vermögenswerte generiert (ED IAS 36.5). Die Smallest Group of CGUs kann mit den internen Management Groups übereinstimmen, darf jedoch nicht größer sein als ein Primary Segment nach IAS 14. Bei Aufgabe einer Geschäftseinheit innerhalb einer CGU, zu der ein Goodwill gehört, muss dieser bei der Bestimmung des Aufgabegewinns oder –verlusts einbezogen werden proportional zum Wert des aufgegebenen oder beibehaltenen Teils der CGU. Wenn die Segmentstruktur neu bestimmt wird, muss auch der darin enthaltene Goodwill neu verteilt werden.

Für den Goodwill soll in Zukunft der Impairment Only Approach gelten, d.h. es sollen keine planmäßigen sondern nur noch fallweise außerplanmäßige Abschreibungen stattfinden. Diese sind im Rahmen eines jährlichen Niederstwerttests zu ermitteln (ED 3.53) Sind die Anschaffungskosten für den Unternehmenserwerb niedriger als die Differenz der Fair Values aus anteiligen Vermögenswerten und (bedingten) Schulden (negativer Unterschiedsbetrag) so fordert ED 3.55(a), die Neubewertung der Fair Values nochmals zu überprüfen. Ein danach immer noch bestehender negativer Unterschiedsbetrag ist unmittelbar als Ertrag auszuweisen und nicht mehr in der Bilanz zu passivieren (*Theile/Pawelzik,* 2003, S. 316-324).

Bewertung bei Folgekonsolidierungen

In den Folgeperioden sind bedingte Schulden mit ihrem beizulegenden Zeitwert zu bewerten (ED 3.46). Wertänderungen werden direkt in der GuV-Rechnung erfasst, was eine Abweichung von IAS 37 und dem Framework darstellt.

Für die Bewertung von Intangibles wird in der Folgebilanzierung unterschieden zwischen Intangibles mit „Finite" und „Indefinite Useful Lives":

- Intangibles mit Finite Useful Life sind weiter nach IAS 38 zu bilanzieren, d.h. mit den Anschaffungs- oder Herstellungskosten oder dem „Revalued Amount" (Neubewertung) anzusetzen, jeweils gemindert um kumulierte planmäßige und außerplanmäßige Abschreibungen,
- Intangibles mit Indifinite Useful Life, zu denen auch der Goodwill zählt, sind nicht mehr planmäßig abzuschreiben, sondern unterliegen einem jährlichen Impairment Test. Eine Neubewertung zum Fair Value nach IAS 38 ist auch für diese Intangibles zulässig. Es erfolgt keine planmäßige Abschreibung; der Impairment Test ist in Übereinstimmung mit ED-IAS-36 mindestens einmal jährlich durchzuführen, bei Anzeichen für eine Wertminderung auch unterjährig.

Impairmenttest für den Goodwill

Der Werthaltigkeitstest für den Goodwill ist – ähnlich wie nach US-GAAP – zweistufig angelegt (*Pellens/Sellhorn,* 2003, S. 402 f.; *Frohwein/Lüdenbach,* 2003, S. 65-72; *Wüstemann/Duhr,* 2003, S. 250 f.). Zu vergleichen ist der Buchwert mit dem Recoverable Amount der Cash Generating Unit. Entsprechend der allgemeinen Definition des Recoverable Amount ist dieser der höhere Betrag aus Net Selling Price und Value in Use.

- Der Net Selling Price ist der zwischen unabhängigen Geschäftspartnern festgelegte Preis abzüglich der direkt dem Verkauf zurechenbaren Kosten (bei verbindlichem Verkaufsvertrag) oder ein Marktpreis, wenn die Cash Generating Unit an einem aktiven Markt gehandelt wird oder eine Schätzung.
- Der Nutzungswert wird mit Hilfe von abgezinsten Zahlungsströmen alternativ nach dem traditionellen Ansatz oder dem Expected Cashflow-Ansatz ermittelt. Durch Rückgriff auf betriebsindividuelle Informationen (immaterielle Ressourcen, Synergien, Wachstumsaussichten) werden bei der Ermittlung des Barwertes außerbilanzielle Faktoren berücksichtigt. Die Schätzung der Zahlungsströme ist in einen Detailplanungszeitraum (längstens fünf Jahre) und einen Restzeitraum zu unterteilen (ED IAS 36.28).
  - Beim traditionellen Ansatz werden die zu erwartenden Zahlungsströme mit dem Opportunitätskostensatz diskontiert. Dabei entspricht die anzusetzende Zahlungsreihe gerade der wahrscheinlichsten Ausprägung der zu erwartenden Zahlungen (ED IAS 36.B4). Die mit den Zahlungsvorgängen zusammenhängenden Unsicherheiten werden ausschließlich im Diskontierungsfaktor berücksichtigt, d.h. der risikolose Zins ist mit einem aus dem Markt abgeleiteten Risikofaktor zu multiplizieren. Der Risikofaktor ist ggf. für jede Periode zu ermitteln und entspricht der Prämie für die Übernahme des Risikos aus dem Engagement.
  - Der Expected Cashflow Ansatz übernimmt nicht nur die wahrscheinlichste Variante der künftigen Zahlungsreihe, vielmehr fließen die möglichen Zahlungsstromausprägungen mit ihren jeweiligen Eintrittswahrscheinlichkeiten in die Berechnung ein. Die Erwartungswerte der zu betrachtenden Perioden werden diskontiert. Da die Unsicherheit der zukünftigen Zahlungsströme bereits in die Erwartungswerte Eingang

gefunden hat, berücksichtigt der anzuwendende Zinssatz lediglich den Zeitwert des Geldes (risikoneutraler Zins); eine Risikoanpassung ist nur für die Risiken möglich, die nicht in den Erwartungswerten berücksichtigt wurden.

Ist der Recoverable Amount der Cash Generating Unit niedriger als deren Buchwert einschließlich des anteiligen Goodwills, so ist im zweiten Schritt der Goodwill auf seinen Abschreibungsbedarf hin zu untersuchen (ED IAS 36.85). Dies erfordert einen Vergleich zwischen dem Goodwill der Cash Generating Unit und dem Implied Value des betreffenden Goodwills (ED IAS 36.85(a)). Der Implied Value des Goodwills ermittelt sich als Differenz zwischen dem Recoverable Amount der Cash Generating Unit und den beizulegenden Zeitwerten der Vermögenswerte und (bedingten) Schulden der Cash Generating Unit (ED IAS 36.86). Liegt der Goodwill-Buchwert über dem Implied Value des Goodwills, so ist auf diesen abzuschreiben. Sollte der nach dieser Abschreibung verbleibende Buchwert der Cash Generating Unit immer noch über ihrem Recoverable Amount liegen, so ist der verbleibende Wertminderungsbedarf auf die übrigen Vermögenswerte der Cash Generating Unit im Verhältnis von deren Buchwerten zu verteilen. Erhöht sich der Recoverable Amount einer Cash Generating Units in späteren Jahren wieder nachdem sie vorher abgeschrieben wurde, so sind alle materiellen und immateriellen Vermögenswerte außer dem Goodwill proportional ihrer Buchwerte zuzuschreiben bis höchstens auf die fortgeführten Anschaffungs- oder Herstellungskosten. Für den Goodwill besteht ein Wertaufholungsverbot, sodass außerplanmäßige Goodwillabschreibungen bei Wertsteigerungen in künftigen Perioden nicht mehr rückgängig gemacht werden können.

Für den Werthaltigkeitstest ergibt sich folgendes Ablaufschema:

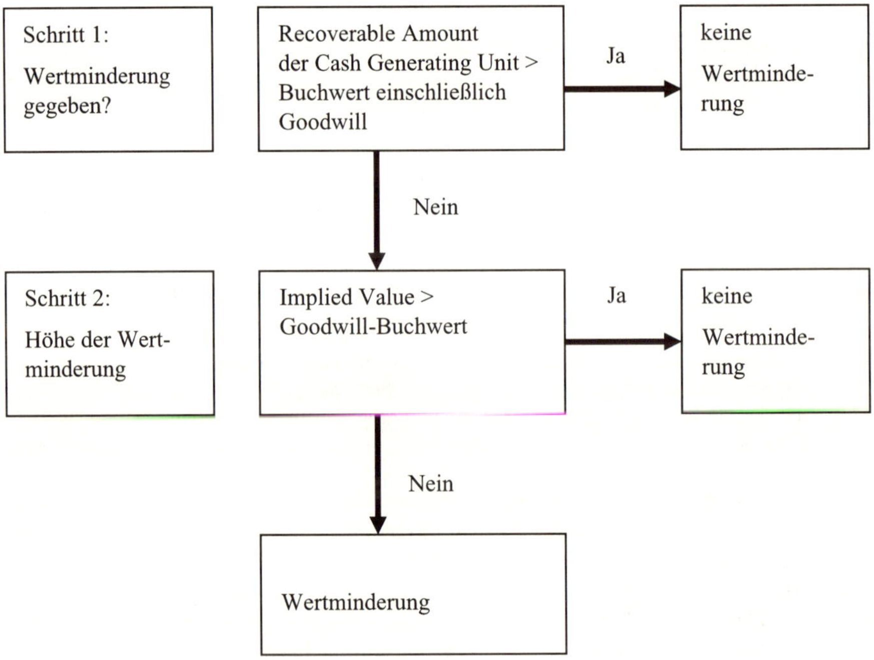

Buchwertanpassungen nach der Erstkonsolidierung

Gegenüber der Erstkonsolidierung können sich folgende Änderungen in den Folgeperioden ergeben:

- Wenn der Kaufpreis von künftigen Ereignissen abhängt (bedingter Kaufpreis) ist eine Anpassung der Anschaffungskosten ohne zeitliche Restriktion möglich, sobald der Anpassungsbetrag wahrscheinlich und zuverlässig messbar ist; es erfolgt allerdings keine Anpassung der Anschaffungskosten, wenn eine zusätzliche Vergütung aufgrund eines Wertverfalls von hingegebenen Vermögenswerten gewährt wird.
- Eine Anpassung der Anschaffungskosten und der Kaufpreisverteilung auf die beizulegenden Zeitwerte aufgrund einer vorläufigen Verteilung ist nur innerhalb von 12 Monaten nach dem Erwerbsdatum zulässig, wenn der Wert eines fixen Kaufpreises und/oder beizulegenden Zeitwertes noch nicht endgültig ermittelt ist.
- Aktive latente Steuern, die z.B. mangels Werthaltigkeit zunächst nicht aktiviert wurden, sind nachträglich gemäß IAS 12 erfolgswirksam zu aktivieren, wenn sie die Aktivierungsvoraussetzungen erfüllen; der Goodwill ist grundsätzlich in entsprechender Höhe erfolgswirksam zu reduzieren. Eine Deckelung ergibt sich durch das Entstehen eines negativen Unterschiedsbetrags.

Angabepflichten

ED 3 sieht umfangreiche Angabepflichten für den Impairment-Test vor. Sie sind bezogen auf einzelne Segmente oder auf den Gesamtkonzern, ist der Goodwill einer Cash Generating Unit wesentlich in bezug auf den Gesamtgoodwill, so ist die Ebene der Cash Generating Unit Berichtsgegenstand. Im Einzelnen ist z.B. anzugeben

- der Goodwill-Buchwert,
- der Wertmaßstab zur Bestimmung des Recoverable Amounts (Net Selling Price bzw. Value in Use),
- der Betrag, um den die Summe der Recoverable Amounts der Cash Generating Units ihre Buchwerte übersteigt sowie
- die Schlüsselannahmen für die Bestimmung des Recoverable Amounts (Wachstumsrate künftiger Cash Flows, Zeitraum für Zahlungsprognosen etc.).

Erstanwendung

Die Neuregelungen betreffen Unternehmenszusammenschlüsse, deren Agreement Date am Tag der künftigen Veröffentlichung des endgültigen IFRS oder später ist (ED 3.77). Sie gelten auch für Goodwills, die bereits vor erstmaliger Anwendung der neuen Vorschriften entstanden sind. Die Buchwerte dieser Goodwills sind bis zum Beginn des Geschäftsjahres der Erstanwendung planmäßig abzuschreiben, danach ist die planmäßige Abschreibung zugunsten des jährlichen Impairmenttests einzustellen. Ein bestehender negativer Goodwill ist im Umstellungsjahr erfolgsneutral gegen die Gewinnrücklagen auszubuchen (ED 3.80).

IASB-Projekt Business Combinations II

Das IASB-Projekt Business Combinations II setzt die mit Phase I eingeleitete Fokussierung auf das Fair-Value-Konzept fort und behandelt einige Fragen, die in Phase I ausgeklammert waren (*Grünberger/Grünberger,* 2003, S. 218-220), z.B.

- Anwendungsfragen der Erwerbsmethode,
- Bilanzierungsfragen bei Joint Ventures nach der Fresh Start Methode und
- die Behandlung von Common Control Transactions (Unternehmenszusammenschlüsse zwischen verbundenen Unternehmen).

Das Projekt Business Combinations II beinhaltet Anwendungshinweise zur Fair Value Ermittlung (Fair Value Hierarchy). Können für Bewertungszwecke eines Vermögenswertes bzw. einer Schuld keine gleichen Vermögenswerte bzw. Schulden auf zugänglichen aktiven Märkten ermittelt werden, sind vergleichbare Vermögenswerte bzw. Schulden heranzuziehen. Ist auch dies nicht möglich, müssen geeignete Bewertungsverfahren verwendet werden. Dabei ist dem Standardsetter bewusst, dass dennoch zahlreiche Prognosen und Ermessensentscheidungen verbleiben. Weitere Regelungen zur Fair Value-Ermittlung sollen im Rahmen eines spezifischen Bewertungsprojektes und nicht im Rahmen der Vorschriften über Unternehmenszusammenschlüsse erarbeitet werden.

Abweichend von ED 3, wonach Aufwendungen, die in direktem Zusammenhang mit einem Unternehmenserwerb stehen, zu den Anschaffungskosten gehören, wird in Phase II festgelegt, dass solche Aufwendungen nicht als Teil des Fair Values des erworbenen Unternehmens anzusehen sind.

Nach den derzeit geltenden Regelungen erfolgt keine Hochrechnung des Goodwills auf Minderheitsanteile. Im Rahmen der Phase II soll unabhängig von der Beteiligungsquote des erwerbenden Unternehmens bei einem Unternehmenszusammenschluss der gesamte Goodwill angesetzt werden (Full-Goodwill-Methode). In diesem Fall ergibt sich der Goodwill als Differenz zwischen dem Fair Value des zu erwerbenden Unternehmens (betrachtet in seiner Gesamtheit) und den Fair Values der einzelnen Vermögenswerte und Schulden des zu erwerbenden Unternehmens.

Bei Unternehmenserwerben in mehreren Stufen ist vorgesehen, dass zum Zeitpunkt der Erlangung der Beherrschung sämtliche Vermögenswerte und Schulden des erworbenen Unternehmens mit ihren aktuellen Fair Values anzusetzen sind. Ein im Vergleich zu den bisherigen Ansätzen entstehender Buchgewinn oder –verlust ist erfolgswirksam in der laufenden Periode auszuweisen.

Während faktische Verpflichtungen im Erwerbszeitpunkt nicht angesetzt werden dürfen, sind bedingte Schulden (Contingent Liabilities) nach ED 3 und bedingte Vermögenswerte nach Phase II anzusetzen und mit ihrem Fair Value zu bewerten. Da die Fair Values die Wahrscheinlichkeit eines künftigen Zuflusses bzw. Abflusses eines wirtschaftlichen Nutzens beinhalten (Bilanzierung der Höhe nach als Erwartungswert), wird auf diese Wahrscheinlichkeit als Ansatzkriterium künftig verzichtet.

Beurteilung der Fair Value Bewertung nach den IASB-Projekten zu Business Combinations

In den IASB-Projekten Business Combinations I und II finden verstärkt Marktbewertungen Anwendung. Anhand einiger Beispiele soll auf die Wirkungen hingewiesen werden:

- Der Ansatz von Fair Values beinhaltet Beurteilungsspielräume sowohl bzl. der Höhe als auch der Eintrittswahrscheinlichkeit der künftigen Zahlungsvorgänge. Die Bewertung bedingter Vermögenswerte und Schulden zum Fair Value an jedem Abschlussstichtag bedeutet erhöhte Manipulationsgefahr für den jeweiligen Bilanz- als auch Ergebnisausweis. Handelt es sich um ein Passivum, so ist von einem reziproken Verhältnis von Bonität und Ergebnisausweis auszugehen: Mit sinkender Bonität reduziert sich der Fair Value der Verpflichtung und erhöht das Ergebnis und umgekehrt.

- Während IAS 38 eine Neubewertung von immateriellen Vermögenswerten zum Fair Value nur zulässt, wenn ein aktiver Markt besteht, ist nach ED 3 eine Fair Value Bewertung aller identifizierbaren (immateriellen) Vermögenswerte im Erstkonsolidierungszeitpunkt geboten, unabhängig davon, ob ein aktiver Markt besteht, denn es werden nicht die einzelnen Intangible Assets sondern ganze Unternehmen bzw. Unternehmensteile ge- und verkauft, welche ihrerseits immaterielle Vermögenswerte beinhalten.
- Die Möglichkeit, nachträglich vorläufig bestimmte Wertansätze zu ändern, ermöglicht einen Ausweis späterer Wertentwicklungen als methodische Anpassungen.
- Der Expected Cashflow Ansatz geht vom Erwartungswert mehrerer möglicher Ausprägungen der erwarteten künftigen Zahlungsstromreihe aus. Auch dies läßt erweiterte Manipulationspotenziale erkennen, welche bilanzpolitisch genutzt werden können.

4. Schlussbemerkungen

Den Vorteilen einer Fair Value Bewertung als zeitnaher und marktorientierter Wertansatz stehen Bedenken im Hinblick auf die Qualitätsanforderungen des Jahresabschlusses Entscheidungsorientierung (Relevance) und Zuverlässigkeit bzw. Nachprüfbarkeit (Reliability) gegenüber.

Der Fair Value einzelner Assets und Liabilities entspricht nicht dem Fair Value des Gesamtunternehmens bzw. seiner Segmente, welcher Entscheidungsgrundlage für Anlagedispositionen ist. Durch die Volatilität der Märkte vermischen sich im Fair Value Sachverhalte, die auf die Performancequalität des Managements schließen lassen, mit spekulativen Elementen; dies mindert die Eignung des Jahresabschlusses als Kontrollinstrument im Rahmen der Corporate Governance. Die „imparitätische“ Behandlung von Fair Value Erhöhungen (erfolgsneutral) und -verminderungen (erfolgswirksam), erschwert die Interpretation des Periodenerfolgs für Außenstehende. Bei Existenz realer Marktpreise ist die Fair Value-Ermittlung objektivierbar, die Prognoseunsicherheit bei der Abzinsung von Cashflows eröffnet allerdings große Manipulationsspielräume. Die Fair Value-Bewertung bedeutet eine Abkehr vom going concern-Grundsatz. Dies erscheint zweckmäßig bei Assets die ausschließlich zum Verkauf bestimmt sind (z.B. Trading Wertpapiere), ist dagegen problematisch bei langlebigen Gebrauchsgegenständen. Die Fair Value-Bewertung von Passiva führt bei eigener Bonitätsverschlechterung zu einer Ergebnisverbesserung.

Zusammenfassend kann die Fair Value-Bewertung gegenüber der Bewertung zu fortgeführten historischen Anschaffungskosten als Erhöhung der Entscheidungsorientierung (Relevance) bei gleichzeitiger Einschränkung der Zuverlässigkeit und Nachprüfbarkeit (Reliability) beurteilt werden.

5. Literatur

Alvarez, M. und *Biberacher, J.*, Goodwill-Bilanzierung nach US-GAAP – Anforderungen an Unternehmenssteuerung und –berichterstattung, in: Betriebs-Berater, 57. Jg. (2002), H. 7, S. 346-353.

Brücks, M. und *Wiederhold, P.*, Exposure Draft 3 „Business Combinations" des IASB, in: Kapitalmarktorientierte Rechnungslegung, 3. Jg. (2003), H. 1, S. 21-29.

Busse von Colbe, W., Ist die Bilanzierung des Firmenwerts nach dem Nonamortisation-Impairment-Ansatz des SFAS-Entwurfs von 2001 mit § 292a HGB vereinbar? In: Der Betrieb, 54. Jg. (2001), H. 17, S. 877-879.

Fladt, G. und *Feige, P.*, Der Exposure Draft 3 „Business Combinations" des IASB – Konvergenz mit den US-GAAP? In: Die Wirtschaftsprüfung, 48. Jg. (2003), H. 6, S. 249-262.

Grünberger, D. und *Grünberger, H.*, Business Combinations (Phase II): Richtungweisende Beschlüsse des IASB, in: Steuern und Bilanzen, 5. Jg. (2003), H. 13, S. 218-220.

Henselmann, K., Böcke als Gärtner? – Vorgeschichte und Stand von DRS 1a, in: Steuern und Bilanzen, 4. Jg. (2002), H. 6, S. 283-287.

Heyd, R., Internationale Rechnungslegung, Stuttgart 2003.

Kleindiek, D., Neue Goodwill-Bilanzierung nach US-GAAP und europäisches Bilanzrecht – Stellungnahme zum E-DRS 1a, in: Betriebs-Berater, 56. Jg. (2001), H. 50, S. 2572-2577.

Krawitz, N., Die Problematik des Richtlinieneinklangs bei der Goodwill-Abschreibung – Stellungnahme zum Entwurf eines Deutschen Rechnungslegungs Standards Nr. 1a, in: Deutsches Steuerrecht, 40. Jg. (2002), H. 4, S. 145-148.

Küting, K., Der Geschäfts- oder Firmenwert – ein Spielball der Bilanzpolitik in deutschen Konzernen, in: Die Aktiengesellschaft, 45, Jg. (2000), H. 2, S. 97-106.

Küting, K., *Weber, C.-P.*, *Wirth, J.*, Die neue Goodwillbilanzierung nach SFAS 142, in: Kapitalmarktorientierte Rechnungslegung 1. Jg. (2001), H. 5, S. 185-198.

Küting, K., *Weber, C.-P.* und *Wirth, J.*, Goodwill und immaterielle Vermögenswerte im Übergang auf die Anwendung des SFAS 142 in: Kapitalmarktorientierte Rechnungslegung, 2. Jg. (2002), H. 2, S. 57-66.

Lewicki, B., Die Behandlung des Firmenwerts im Konzernabschluss, in: bilanz & buchhaltung, 48. Jg. (2002), H. 1, S. 56-62.

Lüdenbach, N. und *Frohwein, N.*, Der Goodwill-Impairment-Test aus Sicht der Rechnungslegungspraxis, in: Der Betrieb, 56. Jg. (2003), H. 5, S. 217-223.

Lüdenbach, N. und *Schulz, R.*, Unternehmensbewertung für Bilanzierungszwecke – Neue Herausforderungen für den Berufsstand durch den impairment-Ansatz von FAS 142? In: Die Wirtschaftsprüfung, 47. Jg. (2002), H. 10, S. 489-499.

Mujkanovic, R., Fair Value im Financial Statement nach International Accounting Standards, Stuttgart 2002.

Pellens, B., *Basche, K.* und *Sellhorn, T.*, Full Goodwill Method, in: Kapitalmarktorientierte Rechnungslegung, 3. Jg. (2003), H. 1, S. 1-4.

Pellens, B. und *Sellhorn, T.*, Neue Goodwill-Bilanzierung nach US-GAAP, in: Der Betrieb, 54. Jg. (2001), H. 32, S. 713-720.

Pellens, B. und *Sellhorn, T.*, Minderheitenproblematik beim Goodwill Impairment Test nach geplanten IFRS und geltenden US-GAAP, in: Der Betrieb, 56. Jg. (2003), H. 8, S. 401-408.

Pfeil, O. und *Vater, H.*, „Die kleine Unternehmensbewertung" oder die neuen Vorschriften zur Goodwill- und Intangible-Bilanzierung nach SFAS No. 141 und SFAS No. 142, in: Kapitalmarktorientierte Rechnungslegung, 2. Jg. (2002), H. 1, S. 66-81.

Schildbach, T., Zeitwertbilanzierung in USA und nach IAS, in: Betriebswirtschaftliche Forschung und Praxis, 50. Jg. 1998, H. 5, S. 580–592.

Starbatty, N., Fair Value Accounting gemäß Statement of Financial Accounting Concept No. 7, in: Die Wirtschaftsprüfung, 46. Jg. (2001), H. 10, S. 543–558.

Stauber, J. und *Ketterle, T.*, Goodwill-Bilanzierung nach US GAAP, in: Der Schweizer Treuhänder, 75. Jg. (2001), H. 50, S. 955-966.

Teitler–Feinberg, E., Neue Bewertung von Goodwill und anderen immateriellen Aktiven, in: Der Schweizer Treuhänder, 75. Jg. (2001), H. 4, S. 331-338.

Theile, C. und *Pawelzik, K.*), Erfolgswirksamkeit des Anschaffungsvorgangs nach ED 3 beim Unternehmenserwerb im Konzern – Zur Bilanzierung eines excess (vormals negativer Goodwill), in: Die Wirtschaftsprüfung, 48. Jg. (2003), H. 7, S. 316-324.

Wüstemann, J. und *Duhr, A.*, Geschäftswertbilanzierung nach dem Exposure Draft ED 3 des IASB – Entobjektivierung auf den Spuren des FASB? in: Betriebs-Berater, 58. Jg. (2003), H. 47, S. 247-253.

Abbildung von F&E im Rahmen des nicht-finanziellen SEC-Reporting – Ableitung von Informationsanforderungen zum Zwecke der Entscheidungsnützlichkeit

Axel Haller[1] *und Ralph Dietrich*[2]

---

[1] Prof. Dr. *Axel Haller*, Institut für Revisions-, Treuhand- und Rechnungswesen, Johannes Kepler Universität Linz

[2] *Ralph Dietrich*, Institut für Revisions-, Treuhand- und Rechnungswesen, Johannes Kepler Universität Linz

Ziel des SEC-Reporting ist die Vermittlung von Informationen, die nützlich für Kapitalüberlassungsentscheidungen sind. Entsprechend einer ökonomischen Konvention gelten Informationen dann als entscheidungsnützlich, wenn sie der Erwartungsbildung über Höhe, zeitlichen Anfall und Risiko künftiger, auf Unternehmensebene erwirtschafteter Zahlungsmittelüberschüsse dienen, die gleichsam die Grundlage der von den Kapitalgebern zu erwartenden Ausschüttungen aus dem Unternehmen bilden. Voraussetzung für die nachhaltige Erwirtschaftung künftiger Zahlungsmittelüberschüsse sind Wettbewerbsvorteile. Für die Erhaltung bestehender Wettbewerbsvorteile bzw. für die Schaffung neuer kommt Forschung und Entwicklung eine zentrale Bedeutung zu. Vor diesem Hintergrund wird im vorliegenden Beitrag, nach einer Begründung der Bedeutung von Informationen über den F&E-Sachverhalt in einem Unternehmen für Investoren und andere Stakeholder, die vorgeschriebene Abbildung des F&E-Sachverhalts im SEC-Reporting außerhalb der Financial Statements untersucht, ein Vorschlag zur strukturierten Umsetzung der Berichtsanforderungen unterbreitet sowie die Best Practice US-amerikanischer Pharmaunternehmen expliziert.

1. SEC-Reporting

1.1 SEC-Reporting als Teilbereich der Unternehmenspublizität

Das SEC-Reporting stellt einen Teilbereich der Unternehmenspublizität dar. Unter Unternehmenspublizität sind dabei in einem weit verstandenen Sinn alle Signale und Nachrichten zu subsumieren, welche ein Unternehmen einer breiten Öffentlichkeit zugänglich macht (vgl. *Käfer,* 1960, S. 1604f.; *Moxter,* 1975, Sp. 3282). In der hier im Vordergrund stehenden Unternehmenspublizität im engeren Sinne erfolgt eine Begrenzung auf die vergangenheits-, gegenwarts- und zukunftsorientierte Selbstdarstellung der wirtschaftlichen Lage und Tätigkeit eines Unternehmens oder eines Konzerns nach außen, wobei wirtschaftliche Lage und Tätigkeit weit zu fassen ist (vgl. *Kropf,* 1993, Sp. 1669; *Pellens,* 2001, S. 1742, Sp. 2).

Hinsichtlich des Grades der Verpflichtung zur Unternehmenspublizität lässt sich eine Differenzierung zwischen freiwilliger Publizität und Pflichtpublizität vornehmen, wobei unter letzterer alle regelmäßigen und fallbezogenen Veröffentlichungen zu verstehen sind, deren Beachtung aufgrund hoheitlicher Erlasse beispielsweise durch Gesetze oder Verordnungen für bestimmte Unternehmen vorgeschrieben sind (vgl. *Pellens,* 2001, S. 1742, Sp. 2). Im Rahmen des SEC-Reporting sind an einer US-amerikanischen Börse notierte Unternehmen zu einer umfangreichen Berichterstattung gegenüber der US-Börsenaufsichtsbehörde *Securities and Exchange Commission* (SEC) verpflichtet.

1.2 Adressaten und Zweck des SEC-Reporting

Eine Konkretisierung der Publizitätspflichten setzt die Festlegung voraus, wessen Interessen die Publizität zu dienen hat. Aufgrund der Vielzahl denkbarer, heterogener Interessen ist für die Gestaltung von Vorschriften eine Komplexitätsreduktion im Sinne einer Interessentypisierung notwendig.

Als grundlegende Konvention wird deshalb die Annahme eines stets rational handelnden, nach Nutzenmaximierung strebenden homo oeconomicus getroffen. Da letztlich lediglich psychisches Einkommen, d.h. individuell erfahrenes Einkommen, Nutzen stiftet (vgl. *Fisher*, 1930, S. 10f.), dieses jedoch intersubjektiv nicht vereinbart werden kann und sich ferner auf Grund mangelnder Operationalisierbarkeit der Analyse entzieht, behilft man sich einer Konstruktion, indem man unterstellt, dass psychisches Einkommen letztlich durch Konsum (Realeinkommen) realisiert wird und Konsum wiederum durch die messbare Größe Geld(-einkommen) ermöglicht wird, so dass sich die verhaltensbezogene Konvention auf die Annahme reduzieren lässt, im Mittelpunkt des Interesses der potentiellen Berichtsadressaten stehe die Struktur des an sie fließenden Nettozahlungsstroms (vgl. *Fisher,* 1930).

Ferner gilt es, eine Eingrenzung der Adressatengruppe vorzunehmen, unterscheiden sich doch potentielle Adressatengruppen wie Mitarbeiter, Kunden, der Fiskus oder Lieferanten zwar nicht bezüglich ihres Interesses an Konsummöglichkeiten, wohl aber bezüglich der Sachverhalte, die dem an sie jeweils fließendem Nettozahlungsstrom zu Grunde liegen. Für die Beantwortung der Frage, an welche potentielle Gruppe letztlich die Publizität zu richten ist, ist nach dem vom Gesetzgeber mit der Publizitätsregulierung verfolgten Ziel zu fragen. Ziel der US-amerikanischen Pflichtpublizität ist es, die Funktionsfähigkeit des Kapitalmarktes aufgrund dessen wohlfahrtsökonomischer Bedeutung zu gewährleisten (vgl. bspw. SFAC No. 1 „*Objectives of Financial Reporting by Business Enterprises*“, Abs. 9ff.). Im Sinne des Anlegerschutzgedankens soll hierfür insbesondere betrügerisches Verhalten auf dem Kapitalmarkt verhindert werden (vgl. *Meier/Schatz,* 1989; *Merkt,* 2001).

Damit steht der Kapitalgeber, insbesondere der Anteilseigner, als Adressat der US-amerikanischen Pflichtpublizität im Mittelpunkt. Zu begründen ist dies durch die Sichtweise eines Unternehmens als nexus of contracts, wonach die meisten Stakeholder eines Unternehmens ihre finanziellen Interessen vertraglich absichern können, während die (Eigen)-Kapitalgeber als Residualempfänger das Risiko einer Interessensverletzung tragen und daher im besonderen Maße schutzwürdig sind (vgl. *Jensen/Meckling,* 1976). Während dieser Individualschutzgedanke sozialpolitisch motiviert ist, kommt ihm aus wirtschaftspolitischer Sicht kein Selbstzweck zu, dient doch vor diesem Hintergrund ein wirksamer Kapitalgeberschutz dazu, das Vertrauen in den Kapitalmarkt zu gewährleisten, was gleichsam Grundvoraussetzung dafür ist, dass der Kapitalmarkt seiner wohlfahrtsökonomisch bedeutsamen Funktion der Kapitalaufbringung und als zweitem Schritt insbesondere der Kapitallenkung in dessen produktivste Verwendung gerecht werden kann (vgl. *Ekkenga,* 1998, S. 30; *Koch/Schmidt,* (1981), S. 231ff.)

Insbesondere vor dem Hintergrund agency-theoretischer Erkenntnisse (vgl. grundlegend *Jensen/Meckling*, 1976, S. 305ff.), wonach aus der Kombination möglicher Informationsasymmetrien und Interessensdivergenzen zwischen Management und Kapitalgebern im Extremfall ein Marktversagen folgen kann, dient die auf die Verringerung von Informationsasymmetrien ausgerichtete Offenlegung von Informationen als „[...] keystone of the entire structure of federal securities legislation [...]” (*Knauss,* 1964, S. 602). Durch die von der SEC bzw. von ihr beauftragten Institutionen konkretisierten Berichtspflichten, sollen Investoren vor falschen Abbildungen wirtschaftlicher Sachverhalte, Manipulation und betrügerischen Absichten durch umfangreiche Offenlegungspflichten börsennotierter Unternehmen bewahrt werden (vgl. *Levy,* 2001, S. 1f.). So ist in der Präambel des SA 1933 als Zweck des Gesetzes auch explizit festgelegt „[...] to provide full and fair disclosure of the character of securities [...].“ Insbesondere sollen den Kapitalgebern Informationen vermittelt werden, die nützlich für deren Investitionsentscheidungen im Vorfeld eines Kapitalüberlassungsverhältnisses sowie für deren Kontrollentscheidungen nach erfolgtem Vertragsabschluss sind.

1.3 Abgrenzung des periodischen SEC-Reporting

Den Kern der Publizitätsregulierung börsennotierter Gesellschaften bilden der *Securities Act* von 1933 (SA) und der *Securities Exchange Act* von 1934 (SEA). Neben der Berichtspflicht und anderen Regelungen enthält der SEA 1934 auch die Grundlage für die Einrichtung der *Securities and Exchange Commission* (SEC) als eine Behörde, die zur Verwaltung der zur Regulierung des Kapitalmarkts verabschiedeten Gesetze mit umfangreichen sowohl exekutiven, legislativen als auch judikativen Kompetenzen ausgestattet ist. Diese Kompetenzen finden ihren Ausdruck unter anderem in den Verlautbarungen der SEC, welche der Konkretisierung der jeweiligen Gesetzesinhalte, wie insbesondere auch der Berichtspflichten dienen. Je nach Art des Unternehmens und Sachverhalts werden eine Vielzahl unterschiedlicher Berichtsanforderungen von der SEC im Rahmen so genannter Forms formuliert.

Regelungsgegenstand des auf den Primärmarkt gerichteten SA 1933 ist die Publizitätspflicht von Unternehmen, die auf dem US-amerikanischen Kapitalmarkt eine Wertpapieremission durchführen möchten. Als Voraussetzung für die Emission sieht das Gesetz die einmalige Einreichung eines Registration Statements bei der verwaltenden Behörde, der SEC, vor, welches umfangreiche Angaben sowohl über die Emission als auch das Unternehmen enthält (vgl. *Haller,* 1994, S. 22ff.).

In Abgrenzung hierzu regelt der auf den Sekundärmarkt gerichtete SEA 1934 die Publizitätsanforderungen an Unternehmen, die ein bereits emittiertes Wertpapier zum laufenden Handel an einer US-amerikanischen Börse oder im Freiverkehr zulassen möchten. Als Zulassungsvoraussetzung fordert dabei auch der SEA 1934 die einmalige Einreichung eines Registraton Statements sowie zusätzlich eine laufende Berichtspflicht, um zu gewährleisten, dass die eingereichten Angaben ständig aktualisiert werden (vgl. *Haller,* 1994, S. 24f.).

Im Rahmen dieser laufenden Publizitätspflicht gilt es dabei zwischen einer fallweisen Berichterstattung einerseits sowie einer periodischen Berichterstattung andererseits zu differenzieren. Während unter die Pflicht einer fallweisen Berichterstattung beispielsweise die Veröffentlichung von Ad-Hoc-Mitteilungen fällt, stellen die Quartalsberichterstattung sowie die Jahresberichterstattung die zentralen Pfeiler der laufenden, periodischen Publizitätspflicht dar. Diese periodische Publizitätspflicht lässt sich ferner in eine primär an die Aktionäre gerichtete Publizität einerseits sowie eine primär an die SEC gerichtete Publizität andererseits unterscheiden. Primär an die Aktionäre gerichtet sind so insbesondere der jährlich an die Anteilseigner zu versendende Annual Report sowie das Proxy Statement im Falle der Einholung von Stimmrechtsvollmachten. Primär an die SEC gerichtet und von dieser der interessierten Öffentlichkeit zugänglich gemacht, ist der Bericht gemäß Form 10-K.

Um Anlegern die Bildung eines möglichst diversifizierten Portfolios zu ermöglichen ist die Inanspruchnahme des US-amerikanischen Kapitalmarktes durch ausländische Unternehmen politisch durchaus erwünscht. Da jedoch ausländische Unternehmen auf ihren jeweilig nationalen Kapitalmärkten in der Regel eine geringere Publizitätspflicht zu beachten haben, werden für diese so genannten foreign private issuers eigene Publizitätspflichten vorgesehen, die von den entsprechenden Pflichten der US domestic issuers in bestimmten Punkten im Sinne einer Entlastung abweicht. Die periodische Jahresberichtspflicht von foreign private issuers wird in Form 20-F geregelt, das ein Pendant zur Form 10-K für US domestic issuers darstellt. Eine zusammenfassende Darstellung der SEC-Berichtspflichten bietet *Abb. 1*.

Da im Zuge der Einführung des *Integrated Disclosure Sytem* Anfang der 80er Jahre die unternehmensbezogenen Berichtspflichten nach SA 1933 sowie SEA 1934 weitestgehend vereinheitlicht wurden sowie die gemäß Form 10-K bzw. Form 20-F geforderten Publizitätspflichten auch die primär an Aktionäre gerichteten Berichte beinhalten und zusätzlich noch weiterführende Angaben fordern, beziehen sich die folgenden Ausführungen auf die einschlägigen Vorschriften der primär an die SEC gerichteten, periodischen Jahresberichterstattung, also der Publizität gemäß *Forms 10-K* bzw. *20-F*.

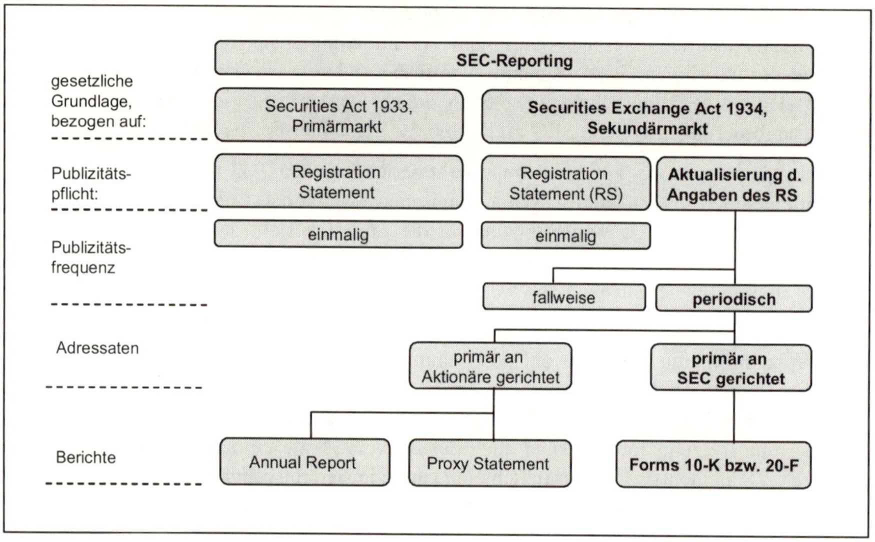

Abb. 1: Abgrenzung im SEC-Reporting

1.4 „Financial Reporting" und „Zusatzangaben" als wesentliche Säulen des periodischen SEC-Reporting

Eine strukturierte Konkretisierung der periodischen Berichtspflichten nach *Forms 10-K* bzw. *20-F* lässt sich unter Rückgriff auf die vom *Financial Accounting Standards Board* (FASB) in SFAC No. 5 „Recognition and Measurement in Financial Statements of Business Enterprises" vorgenommene Kategorisierung (vgl. *Abb. 2*) vornehmen (vgl. *FASB*, SFAC No. 5, Abs. 8).

Entsprechend dieser Systematisierung umfasst die erste Angabenkategorie die Berichte des Jahresabschlusses sowie die Notes, welche entsprechend der Definition der SEC als Financial Statements zusammengefasst werden. Eine Erweiterung der Financial Statements zum Financial Reporting i.e.S. erfolgt dabei unter Einbeziehung zusätzlicher Finanzinformationen, die aus den Daten des Rechnungswesens generiert wurden, aber nicht zu den Financial Statements zählen (vgl. auch *Brotte, 1997*, S. 13).

Wird das Financial Reporting i.e.S. schließlich erweitert um andere Mittel der Finanzberichterstattung, so ist vom Financial Reporting (i.w.S.) zu sprechen. Insbesondere sind in diesem Zusammenhang die von US domestic issuers zu erstellende Management`s Discussion & Analysis of Financial Condition and Results of Operations (MD&A) sowie die von foreign private issuers zu veröffentlichende, in Struktur und Inhalt mit dieser vergleichbare Operating and Financial Review and Prospects (OFR) zu nennen (vgl. *Bruns/Renner*, 2001, S. 7ff). Financial Reporting umfasst mithin alle Informationen, die sich direkt oder indirekt auf berichtete Daten der Finanzbuchhaltung richten. Die auf derselben Ebene angeordnete Kategorie Zusatzangaben schließlich dient gleichsam als „Auffangbecken" der Systematisierung aller anderen berichteten, in der Regel nicht-finanziellen Informationen, welche im Sinne einer Negativabgrenzung nicht dem Financial Reporting zuzuordnen, jedoch annahmegemäß von Entscheidungsnützlichkeit für den Adressaten sind.

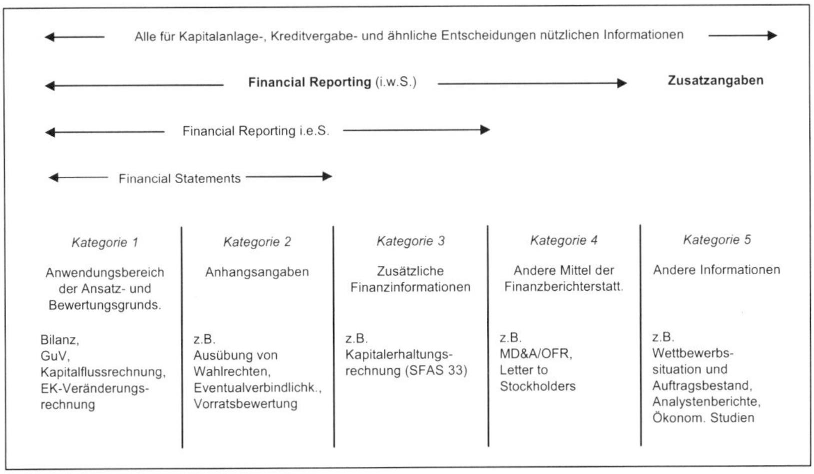

Abb. 2: Strukturierung der periodischen Berichtspflichten nach Forms 10-K bzw. 20-F (in Anlehnung an FASB, SFAC No. 5, Abs. 8.)

Von Bedeutung ist die Differenzierung zwischen *Financial Reporting* und *Zusatzangaben* insbesondere deshalb, da sich die qualitativen Anforderungen des FASB an Informationen, also insbesondere der trade-off zwischen relevance und reliability auf das Financial Reporting beziehen, während Zusatzangaben diesbezüglich höheren Freiheitsgraden unterliegen (ähnlich auch *Kuhlewind,* 1997, S. 230).

Entsprechend der Themenstellung soll im Mittelpunkt dieses Beitrages unter Ausklammerung der einschlägigen Vorschriften des Financial Reporting die Untersuchung der Abbildung des F&E-Sachverhalts im Rahmen der zu berichtenden Zusatzangaben sein.

2. Sachverhalt: „Forschung und Entwicklung"

Kennzeichnend für den Begriff „Forschung & Entwicklung" (F&E) ist dessen mangelnde einheitliche Definition in Theorie und Praxis. Zur Abgrenzung des F&E-Begriffs kann Rückgriff auf das „Frascati-Handbuch" der OECD genommen werden, wonach F&E als eine „kreative Tätigkeit, die auf einer systematischen Basis erfolgt, um den Grundstock des Wissens zu erhöhen und dieses Wissen für neue Anwendungen zu nutzen" definiert ist (*OECD*, 1994, S. 68). Eine Konkretisierung dieser allgemeinen Definition lässt sich durch eine Differenzierung von F&E in die drei Teilbereiche Grundlagenforschung und angewandte Forschung einerseits sowie Entwicklung andererseits vornehmen (vgl. *OECD*, 1994, S. 68).

Grundlagenforschung stellt dabei eine auf die Gewinnung neuer wissenschaftlicher Erkenntnisse gerichtete experimentelle oder theoretische Arbeit dar, die in der Generierung und Überprüfung von aus Gesetzmäßigkeiten abgeleiteten Hypothesen und Theorien besteht (vgl. *Bürgel* et al., 1996, S. 9; *Specht/Beckmann*, 2002, S. 17).

Aufbauend auf den Theorien der Grundlagenforschung werden unter angewandter Forschung alle auf die Realisierung konkreter Zwecke ausgerichteten Aktivitäten zur Gewinnung und Weiterentwicklung von Wissen zusammengefasst, mit dem Ziel, Kernkompetenzen aufzubauen, welche auf die Lösung praktischer Probleme gerichtet sind (vgl. *OECD*, (1994), S. 68; *Specht/Beckmann*, (1996), S. 17). Im Rahmen der angewandten Forschung werden mithin Technologien entwickelt, verstanden als ein System von anwendungsbezogenen, jedoch generalisierenden Ziel-Mittel-Aussagen (vgl. *Brockhoff*, 1994, S. 22).

Entwicklung schließlich stellt eine zweckgerichtete Auswertung und Anwendung von Forschungsergebnissen dar, um zu neuen Produkten, Verfahren oder Systemen zu gelangen bzw. um vorhandene zu verbessern (vgl. *Herzog*, 1995, S. 24). Im Rahmen der Entwicklung werden folglich Technologien in Techniken, verstanden als ein tatsächlich realisiertes, angewandtes Element einer Technologie, transformiert (vgl. *Brockhoff*, 1994, S. 22).

F&E lässt sich somit als umfassender Prozess verstehen, an dessen Anfang die Generierung von Theorien steht, die dann Grundlage der Schaffung allgemeiner Technologien sind. Technologien wiederum werden in konkrete, anwendungsbezogene Techniken umgesetzt, die gleichsam als Ergebnis des F&E-Prozesses der Verbesserung von Produkten oder Prozessen dienen (vgl. *Specht/Beckmann*, 1996, S. 18).

3. Anforderungen an ein zweckkonformes SEC-Reporting

3.1 Information in einem Entscheidungskontext

Ziel des SEC-Reporting ist die Gewährleistung des Anlegerschutzes insbesondere durch die Sicherstellung der Vermittlung entscheidungsnützlicher Informationen an die Kapitalgeber.

Eine Entscheidung ist die auf Basis rationaler Kriterien bewusst vollzogene Auswahl einer als optimal eingeschätzten Handlungsalternative (vgl. *Laux*, 1995, S. 3f.). Voraussetzung für eine rationale Entscheidung ist dabei, die möglichen Handlungsalternativen in eine präferenzabhängige Reihung der Vorziehenswürdigkeit zu bringen, wobei sich der Entscheidungsträger an dem erwarteten Ergebnis orientiert, das mit der jeweiligen Handlungsalternative verbunden ist (vgl. *Sieben/Schildbach*, 1975, S. 18).

In diesem Entscheidungskontext stellt Information entscheidungsrelevantes Tatsachenwissen dar. Die Entscheidungsrelevanz ist gleichsam konstitutiver Bestandteil der Informationsdefinition. Entscheidungsrelevanz, und damit die Qualifikation einer (Tatsachenwissen transportierenden) Nachricht als Information, wird dabei mit der Veränderung der bisher geltenden Rangfolge der Handlungsalternativen nach Verarbeitung des Nachrichtensachverhaltes umschrieben (vgl. grundlegend *Hirschleifer,* 1973, S. 31ff.). Eine Nachricht genügt mithin dann der Informationsdefinition, wenn sie geeignet ist, die Wahrscheinlichkeitsverteilung über den Eintritt künftiger Umweltzustände und der damit verbundenen Ergebniswirkung der jeweiligen Alternativen zu verändern (vgl. *Beaver,* 1998, S. 21).

Entscheidungsnützlich ist Information schließlich dann, wenn sie neben dem Kriterium der Relevanz auch das Qualitätskriterium der Verlässlichkeit erfüllt, welches auf die wirtschaftlich zutreffende Abbildung realer Verhältnisse zielt. Die Kriterien der Relevanz und Verlässlichkeit bilden mithin jeweils notwendige jedoch nicht hinreichende Bedingungen der Entscheidungsnützlichkeit (vgl. *Haller,* 2000, S. 12).

3.2 Konkretisierung des Informationsbedarfs

Für eine Konkretisierung, welche Nachrichten entscheidungsrelevant und mithin Informationen sind, gilt, dass es letztlich nicht die Nachricht per se ist, die zu einer Entscheidungsbestätigung bzw. -revision führt, sondern vielmehr der aus der Nachricht gezogene Schluss (vgl. *Krönert,* 2001, S. 39f.). Ein Schluss kann dabei erst durch die Verwendung von gesetzesartigen Hypothesen im Sinne von Wenn-Dann-Aussagen erfolgen, in welche Nachrichten gleichsam als Input Eingang finden (vgl. *Schneider,* 1983, S. 153). Erst die Kenntnis der Verwendung findenden Hypothesen macht es somit möglich, zu konkretisieren, welchen inhaltlichen Anforderungen Nachrichten genügen müssen, um als entscheidungsrelevanter Input zu dienen. Entsprechend der Annahme rational handelnder Kapitalgeber, leiten diese ihre Hypothesen aus Theorien und Modellen ab, die eine wissenschaftlich fundierte und somit begründete Ausrichtung der Entscheidung an dem Ziel der Nutzenmaximierung erlauben (vgl. *Beaver,* 1998, S. 24).

Der Nutzen aus einem Zahlungsstrom ist dabei dann maximal, wenn der Zahlungsstrom hinsichtlich ihn beschreibender Eigenschaften bestimmte Ausprägungen aufweist, die entsprechend den Präferenzen des Kapitalgebers bezüglich dieser Eigenschaften als optimal eingeschätzt werden (vgl. *Schmidt/Terrberger*, 1997, S. 47ff.). Als typische Eigenschaften eines Zahlungsstroms gelten dessen Höhe, die zeitliche Struktur sowie dessen Unsicherheit (vgl. *Moxter,* 1966, S. 38ff.). Eine Verdichtung dieser Eigenschaften zu einem Entscheidungswert erfolgt durch Modelle der Investitionstheorie, insbesondere dem Discounted Cashflow-Modell (DCF-Modell), das mithin für geeignet erachtet wird, als Grundmodell für die deduktive Ableitung entscheidungsnützlicher Informationen zu dienen (vgl. bspw. *Lange,* 1989, S.168).

Im Rahmen des DCF-Modells wird ein subjektiver Unternehmens- bzw. Anteilswert ermittelt, welcher in Ermangelung eines vollkommenen Kapitalmarktes dem Kapitalgeber dazu dient, im Vorfeld der Investitionsentscheidung eine Beurteilung der Vorteilhaftigkeit des Anteilserwerbs durch Vergleich des ermittelten Grenzpreises mit dem zu zahlenden Marktpreis vorzunehmen (Investitionsentscheidung). Nach vollzogener Investitionsentscheidung soll die Kenntnis der Wertentwicklung den Kapitalgeber in die Lage versetzen, Entscheidungen über disziplinierende Maßnahmen im Sinne einer control by voice oder einer Desinvestition im Sinne einer control by exit zu treffen, so dass Investitionsentscheidung und Kontrollentscheidung eines Kapitalgebers einer Publikumsgesellschaft letztlich auf derselben Entscheidungsgrundlage „Kenntnis der Struktur des Zahlungsstroms", also Kenntnis über Höhe, zeitlichen Zufluss sowie Unsicherheit der künftig aus dem Unternehmen fließenden Zahlungen, basieren. Entsprechend stellt auch *Moxter* fest: „Um Entscheidungen über Erwerb oder Veräußerung von Anteilen treffen zu können, braucht er [der Anteilseigner; d. Verf.] Informationen über die künftig fließenden Zahlungsströme samt deren Wahrscheinlichkeiten; [...].“ (*Moxter*, 2000, S. 2146f.).

Der Dreiteilung betriebswirtschaftlicher Ziele folgend (vgl. *Coenenberg,* 2000, S. 4ff.), bieten sich vor diesem Hintergrund grundsätzlich drei Möglichkeiten an, dem Informationsinteresse des Kapitalgebers gerecht zu werden:

- Das Management veröffentlicht unmittelbar die aus seiner Sicht prognostizierte Verteilung zukünftiger Cashflows (Zielgröße Liquidität).
- Das Management veröffentlicht zwar nicht unmittelbar zukünftig erwartete Cashflows, jedoch finanzielle Indikatoren im Rahmen der externen Rechnungslegung, die einer mittelbaren Prognose künftiger Cashflows, gleichsam einer „Prognose der Cashflow-Prognose“ dienen sollen (Zielgröße Erfolg).
- Das Management veröffentlicht weder mittelbar noch unmittelbar Cashflow bezogene Informationen, sondern vielmehr nicht in Geld ausdrückbare Angaben über die unternehmensspezifischen Stärken und Schwächen, quasi als „mittelbarste“ Grundlage einer Prognose künftiger Cashflows (Zielgröße Erfolgspotential).

Die erste Möglichkeit hätte den Vorteil, dass die gewährte Information unmittelbar vom Kapitalgeber in das DCF-Modell eingesetzt werden kann und somit von hoher Relevanz ist. Dies gilt umso mehr, als die Prognose vom Management stammt, das hierfür als Unternehmens- und Brancheninsider als am qualifiziertesten zu gelten hat. Dass diese Art von Information dennoch nicht als entscheidungsnützlich erachtet wird, liegt an der mit Prognosen unvermeidbar verbundenen Subjektivität, welche einem intendiert opportunistisch handelndem Management Raum für eine Informationsgestaltung entsprechend dessen eigenen Interessen und (potentiell) entgegen der Interessen der Kapitalgeber eröffnet (vgl. bspw. *Jensen,* 1986, S. 323ff.). Das Kriterium der Verlässlichkeit ist folglich nicht erfüllt; das Vertrauen in den Kapitalmarkt kann hierdurch nicht gefördert, sondern vielmehr beeinträchtigt werden.

Zur Gewährleistung von Verlässlichkeit wurden als Instrument zur Berichterstattung finanzieller Informationen die Financial Statements entwickelt, die mittelbar der Prognose künftiger Zahlungsüberschüsse bzw. der Ermittlung des Unternehmenswertes dienen sollen. Einer statischen Auffassung folgend, stellt in diesem Zusammenhang die Bilanz ein Instrument dar, das erlaubt, durch entsprechende Erfassung und Bewertung von Vermögenswerten und Schulden den Unternehmenswert (aus Sicht des bilanzierenden Unternehmens) zu approximieren. Im Sinne einer dynamischen Betrachtung, kann der im Rahmen der GuV ermittelte Jahresüberschuss potentiell zur Approximation des ökonomischen Gewinns herangezogen werden, welcher dann, verstanden als ewige Rente, der Abschätzung des Ertragswertes des Unternehmens dienen kann (vgl. *Moxter,* 1982, S. 142ff.).

Als dritte Möglichkeit wird in der Literatur die umfassende Abbildung des Erfolgspotentials des Unternehmens diskutiert (vgl. *Streim,* 1995, S. 703ff.). Erfolgspotential ist dabei als Bündel nachhaltig wirksamer Wettbewerbsvorteile zu verstehen (vgl. *Coenenberg,* 2003, S. 918). Der Kapitalgeber soll die berichteten Informationen, zusammen mit Informationen aus anderen Quellen, insbesondere auch über unternehmensexterne Chancen und Risiken, durch Kombination adäquater Modelle des strategischen Controlling mit dem jeweilig individuellem Erfahrungsschatz eigenständig in subjektive Cashflow-Prognosen transformieren (vgl. *Coenenberg,* 2000, S. 16ff.). Vor dem Hintergrund der vom Kapitalgeber empfundenen Verlässlichkeit der prognostizierten Cashflows kommt dieser Berichtsvariante unter isolierter Betrachtung mithin die höchste Bedeutung zu. Gänzlich verloren geht dabei jedoch das relevante Wissen des Managements über die jeweils konkrete Rolle des berichteten Erfolgspotentials für die Generierung künftiger Cashflows.

Aus der Erkenntnis, dass Erfolgspotential eine notwendige Voraussetzung für Erfolg, und Erfolg seinerseits notwendige Voraussetzung für die Liquiditätsrealisierung ist (*Abb. 3*), die drei grundsätzlichen Berichtsgrößen also in einem Vorsteuerverhältnis stehen, und Liquidtät ferner wiederum notwendig für die Schaffung von Erfolgspotential ist (vgl. *Coenenberg,* 2000, S. 7), erscheint - in Bezug auf die Entscheidungsnützlichkeit - eine kombinierte Berichterstattung anstelle einer ausschließlichen Abbildung einer dieser Steuerungsgrößen für die Kapitalgeber, aber auch für andere Stakeholder, am vorteilhaftesten.

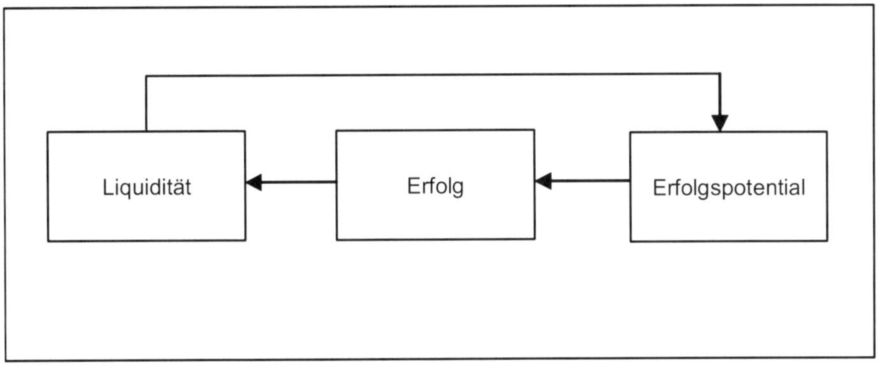

Abb. 3: Vorsteuerverhältnis zentraler Berichtsgrößen (in Anlehnung an Coenenberg, 2003, S. 918)

Durch die Kombination dieser Berichtsmöglichkeiten lässt sich insbesondere der Nachteil einer mangelnden Verlässlichkeit einer Veröffentlichung von Cashflow-Prognosen durch eine parallele Berichterstattung des Erfolgspotentials, gleichsam verstanden als die, die Prognose begründenden Annahmen des Managements, reduzieren. Der Kapitalgeber hat so die Möglichkeit, die „Richtigkeit" der Managementprognosen zu überprüfen. Während diese Vorgehensweise entscheidungsnützliche Informationen liefert, ist sie aus der Perspektive einer Cost/Benefit-Betrachtung abzulehnen. Dem potentiellen Nutzen stehen nämlich - in letzter Konsequenz von den Anteilseignern zu tragende - Kosten gegenüber, die insbesondere aus drohenden Rechtsstreitigkeiten bei falschen Prognosen sowie aus drohenden Wettbewerbsnachteilen bei einer umfassenden Veröffentlichung von Erfolgspotentialen und hieraus erwarteten Ergebnissen (Cashflows) resultieren.

Vor dem Hintergrund des Relevanz/Verlässlichkeit-Trade-off sowie des Cost/Benefit-Trade-off am geeignetsten erscheint folglich die Kombination aus Veröffentlichung von vergangenheitsbezogenen Finanzdaten, deren Erstellung konkreten Regeln folgt und die von Dritten geprüft sind, mit wettbewerbsunschädlichen jedoch die Finanzdaten plausibilisierenden (und deren Prognosewert erhöhenden) Angaben über das Erfolgspotential.

Gestützt wird diese logisch-deduktiv abgeleitete These insbesondere auch durch Ergebnisse empirisch-induktiver Arbeiten über die Konkretisierung des Informationsbedarfs (vgl. *Coenenberg/Haller,* 1993, S. 557ff.). Grundtenor der kapitalmarktorientierten Studien ist, dass Financial Statements Entscheidungsnützlichkeit zu kommt, darüber hinaus jedoch auch andere Informationen die Investitionsentscheidung der Kapitalgeber determinieren (vgl. für einen zusammenfassenden Überblick *May,* 1991, S. 313). Dass es sich dabei insbesondere um nicht-finanzielle Angaben handelt, ist eine zentrale Erkenntnis verhaltensorientierter Studien (vgl. für einen zusammenfassenden Überblick *Lange,* 1989, S. 79ff.).

Die bislang umfangreichste Verdichtung logisch-deduktiv sowie empirisch-induktiv gewonnener Erkenntnisse erfolgte Anfang der 1990er Jahre in den USA durch ein Special Committee des *American Institute of Certified Public Accountants* (AICPA). Auf Basis der aus der Studie Improving Business Reporting gewonnenen Erkenntnisse leitete das Committee einen Katalog an Vorschlägen zur Reform des Status Quo der Berichterstattung ab, mit dem Ziel, die Nützlichkeit veröffentlichter Unternehmensinformationen zu erhöhen (vgl. *AICPA*, 1994; auch *Haller/Dietrich*, 2001, S. 164ff.). Dabei handelt es sich neben den finanziellen Angaben des Jahresabschlusses und Hintergrundinformationen über das Unternehmen insbesondere um

- nicht-finanzielle Angaben, in Form von übergeordneten betrieblichen Kennzahlen und Leistungsgrößen, die das Management zur Unternehmensführung einsetzt;
- die Analyse finanzieller und nicht-finanzieller Angaben aus Sicht der Unternehmensführung, insbesondere unter Nennung von Ursachen für Veränderungen der finanziellen, operativen und leistungsbezogenen Angaben sowie unter Nennung von Eigenschaften und Auswirkungen wichtiger Entwicklungen;
- zukunftgerichtete Unternehmensinformationen, insbesondere über Chancen und Risiken, einschließlich derjenigen, die sich aus wichtigen Entwicklungen ergeben, sowie über Pläne der Unternehmensführung und kritische Erfolgsfaktoren.

Während die Bedeutung finanzieller Informationen zur mittelbaren Cashflow-Prognose vom Special Committee unterstrichen wird, erfolgt eine Konkretisierung der Informationsdimensionen des zu berichtenden Erfolgspotentials durch die Forderung nach gegenwärtigen sowie zukunftsgerichteten nicht-finanziellen Informationen und deren Analyse.

3.3 Erfolgspotential und F&E

Wie bereits oben angeführt steht die Beurteilung der erwarteten Struktur des künftig an ihn fließenden Nettozahlungsstroms, bestimmt durch den Saldo aus Einzahlungen und Auszahlungen auf Unternehmensebene, im Mittelpunkt des Kapitalgeberinteresses. Die Fähigkeit eines Unternehmens zur Erwirtschaftung eines Nettozahlungsstroms wird neben der Attraktivität der Branche insbesondere durch die relative Position des Unternehmens innerhalb seiner Branche determiniert (vgl. hierzu und zum Folgenden *Porter,* 1999). Die relative Position ist dabei gekennzeichnet durch das Erfolgspotential des Unternehmens, d.h. dessen nachhaltige Wettbewerbsvorteile.

Wettbewerbsvorteile entstehen aus dem Wert, den ein Unternehmen für seine Abnehmer schaffen kann, soweit dieser die Kosten der Wertschaffung übersteigt. Ein höherer Wert ergibt sich dabei aus dem Angebot zu einem Preis, der für gleichwertige Leistungen unter dem Preis der Konkurrenten liegt oder aus einer einzigartigen Leistung, die einen im Vergleich zur Konkurrenz höheren Preis mehr als kompensiert. Im ersten Fall besteht der Wettbewerbsvorteil in der Kostenführerschaft, die gleichbedeutend mit niedrigen Auszahlungen für die Leistungserstellung ist. Im zweiten Fall besteht der Wettbewerbsvorteil in der Leistungsdifferenzierung, die wiederum den Nettozahlungsstrom durch hohe realisierbare Einzahlungen positiv beeinflusst.

Wettbewerbsvorteile erwachsen dabei aus den vielen einzelnen Tätigkeiten des Unternehmens. Jede dieser Tätigkeiten kann einen Beitrag zur relativen Kostenposition eines Unternehmens leisten oder eine Differenzierungsbasis schaffen. Zur Verdeutlichung der unterschiedlichen potentiellen Quellen von Wettbewerbsvorteilen kann die von *Porter* entwickelte Wertkette dienen, die ein Unternehmen in strategisch relevante Tätigkeiten gliedert (vgl. *Porter,* 1999, S. 67ff.). *Rappaport* erweiterte die Wertkette konzeptionell um die Möglichkeit der Darstellung der Cashflow-Generierung (*Abb. 4*) (vgl. *Rappaport,* 1995, S. 89).

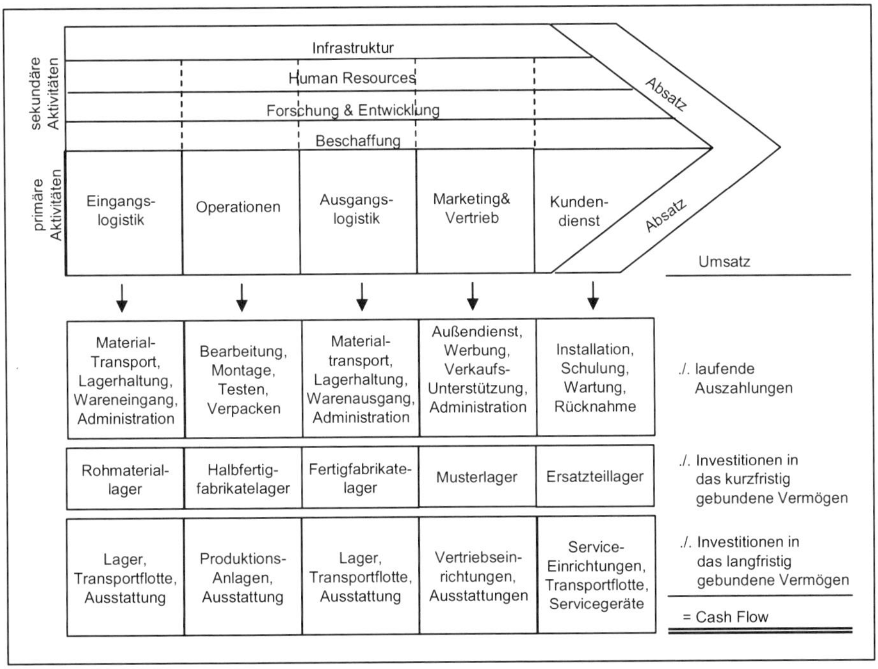

Abb. 4: Einordnung von F&E in die Verbindung zwischen Wertkette und Cashflow (in Anlehnung an Rappaport,1995, S. 89)

Führt das Unternehmen die in der Wertkette dargestellten (Wert-)Aktivitäten besser oder billiger durch als seine Konkurrenten, so verschafft es sich einen Wettbewerbsvorteil (vgl. *Porter,* 1999, S. 63ff.). Wertaktivitäten als Bausteine des Wettbewerbsvorteils lassen sich in fünf primäre Aktivitäten sowie in die, die primären Aktivitäten unterstützenden, sekundären Aktivitäten gliedern. Als eine der zentralen unterstützenden Aktivitäten gilt dabei die Technologieentwicklung als Ergebnis von Theorien und als Grundlage neuer Techniken (F&E-Prozess). *Porter* bezeichnet den technologischen Wandel als eine der wichtigsten Triebkräfte im Wettbewerb, so dass Forschung und Entwicklung (*Porter* spricht hier von Technologieentwicklung) in den meisten Branchen für den Wettbewerbsvorteil von Bedeutung ist, in einigen, wie beispielsweise der Pharmaindustrie, kommt ihr eine Schlüsselrolle zu (vgl. *Porter,* 1999, S. 73 und 223).

In den meisten Unternehmen wird dabei eine breite Palette von Techniken eingesetzt, sowohl als Grundlage von Produktmerkmalen als auch im Rahmen unternehmensinterner

Prozesse. F&E-Aktivitäten können mithin jede der in Wertaktivitäten verkörperte Technologien unterstützen und betreffen somit nicht nur die direkt mit dem Endprodukt verbundenen Techniken, sondern vielmehr auch internen Prozessen zu Grunde liegende Bereiche, wie Telekommunikation für die Auftragserfassung oder Automatisierung im Rechnungswesen. Eine grobe Differenzierung der F&E-Aktivitäten kann entsprechend in Bemühungen um Produkt- und um Verfahrensverbesserungen vorgenommen werden.

Dabei wirkt sich eine entwickelte Technik dann auf Wettbewerbsvorteile aus, wenn sie für die relative Kostenposition oder Differenzierung eine wichtige Funktion einnimmt, also wenn sie entweder im Rahmen unternehmensinterner Prozesse oder im Rahmen der unternehmerischen Leistung die Kostenantriebskräfte oder die Einflussgrößen der Einmaligkeit beeinflusst. Die hierdurch erreichte Verbesserung der relativen Position des Unternehmens innerhalb der Branche spiegelt sich dann im Unternehmenserfolg wieder, was seinen Ausdruck schließlich in einem für die Kapitalgeber vorteilhafteren Nettozahlungsstrom aus dem Unternehmen findet. Dieser Zusammenhang zwischen F&E-Aktivitäten bzw. daraus resultierenden Techniken und Unternehmenserfolg ist auch durch eine Vielzahl empirischer Studien nachgewiesen worden (für einen kritischen Überblick über diese Studien siehe *Gierl/Kotzbauer,* 1992).

Je nach Branche in der ein berichtendes Unternehmen tätig ist, ist mithin eine umfassende Abbildung des F&E-Sachverhalts als bedeutende Berichtsdimension des Erfolgspotentials sowohl durch theoretisch, deduktiv orientierte Überlegungen als auch empirische Erkenntnisse begründbar. Vor diesem Hintergrund ist die Regelung der F&E-Publizität im Rahmen der Berichtspflichten des periodischen SEC-Reporting von Interesse.

4. Umsetzung der SEC-Vorschriften über die Abbildung des F&E-Sachverhalts

4.1 Die Definition von F&E im Rahmen des Reporting

Während F&E oben allgemein definiert und abgegrenzt wurde, gilt es im Rahmen einer Untersuchung konkreter Abbildungsregeln die Frage zu beantworten, wie der F&E-Sachverhalt in den für das SEC-Reporting maßgeblichen Vorschriften konkret definiert ist. Da die SEC hinsichtlich der Abschlusserstellung im wesentlichen die Einhaltung der US GAAP voraussetzt, die ihrerseits durch das FASB (bzw. dessen Vorgängergremien) entwickelt oder konkretisiert werden, ist hierbei primär die vom FASB gewählte Definition zu betrachten. Diese ist in SFAS No. 2 „Accounting for Research and Development Costs" formuliert und lehnt sich an jene an, die von der *National Science Foundation* (NSF), deren Aufgabe in der Erhebung von Daten über Forschung und Entwicklung in den USA besteht, entwickelt wurde. Während allerdings die NSF dabei in Übereinstimmung mit der Definition der OECD zwischen Grundlagenforschung, angewandter Forschung und Verwertung der Forschungsergebnisse differenziert, erfolgt gemäß SFAS No. 2 eine Differenzierung lediglich zwischen Forschung einerseits und Entwicklung andererseits.

Entsprechend wird Forschung definiert als geplante Suche oder kritische Ermittlung mit dem Ziel, neues Wissen in der Hoffnung zu entdecken, dass dieses Wissen nützlich sein wird, um neue Produkte oder Dienstleistungen zu entwickeln bzw. um existierende Produkte oder Verfahren zu verbessern. (*FASB*, SFAS No. 2, Abs. 8(a))

Entwicklung umfasst die Umsetzung von Forschungsergebnissen oder von sonstigem Wissen in einen Plan bzw. Entwurf für ein neues Produkt, einen neuen Prozess oder für eine bedeutende Verbesserung eines existierenden Produkts oder Verfahrens, welches verkauft oder intern genutzt werden soll (*FASB*, SFAS No. 2, Abs. 8(b)).

Die SEC selbst gibt keine Definition von F&E in ihren Verlautbarungen..

4.2 Relevante SEC-Vorschriften für die Abbildung des F&E-Sachverhalts außerhalb des Financial Reporting

Im Rahmen der als Zusatzangaben zu qualifizierenden Berichtskategorien gilt für einen US domestic issuer als grundsätzlich relevante Berichtskategorie für die Abbildung des F&E-Sachverhalts insbesondere Item 1 „Business" der Form 10-K, welche bezüglich der konkreten Berichtsinhalte auf Item 101 der Regulation S-K verweist. Gegenstand der Berichtspflicht ist entsprechend eine Beschreibung der Geschäftstätigkeit. Neben einer Darstellung der allgemeinen Geschäftsentwicklung gilt es hierbei gemäß Item 101 (c) Reg. S-K insbesondere segmentbezogen, verbal das Geschäft des Unternehmens zu beschreiben, indem unter anderem auf den Entwicklungsstand angekündigter Produkte, die Wettbewerbssituation sowie auf die Bedeutung von Patenten einzugehen ist. In den Instructions to Item 101 (c) wird dabei grundsätzlich die Berichterstattung von sowohl quantifizierbaren als auch nur qualitativ beschreibbaren Sachverhalten dann gefordert, wenn das berichtende Management diesem Sachverhalt Bedeutung als Determinante der künftigen Profitabilität des Unternehmens beimisst.

Aus Sicht eines foreign private issuer leitet sich eine mögliche Berichtspflicht über F&E-bezogene Angaben insbesondere aus Item 4 sowie Item 5 der Form 20-F ab.

Item 4 „Information on the Company" verlangt grundsätzlich, vergleichbar mit den Anforderungen nach Item 1 der Form 10-K, die Beschreibung des Unternehmens und die Konkretisierung der Geschäftstätigkeit. So gilt es im Rahmen der Publizitätsanforderung des Item 4 (b) insbesondere auch über den Entwicklungsstand angekündigter Produkte, die Wettbewerbssituation sowie die Bedeutung von Patenten einzugehen.

In Abgrenzung zu Item 7 „Management`s Discussion and Analysis of Financial Condition and Results" (MD&A) der Form 10-K enthält Item 5 „Operating and Financial Review and Prospects" (OFR) der Form 20-F, als das für foreign private issuers geltende Äquivalent der MD&A, eine explizite Berichtspflicht über die im Unternehmen während der letzten drei Jahre vorgenommene F&E-Tätigkeit. Da Item 5 der Form 20-F grundsätzlich dem Financial Reporting zuzuordnen ist, steht entsprechend eine Diskussion finanzieller Daten im Mittelpunkt. Bezüglich der zu berichtenden qualitativen Angaben wird an dieser Stelle in den veröffentlichten Forms 20-F häufig auf die entsprechenden Ausführungen innerhalb des Item 4 verwiesen.

Kennzeichnend für die SEC-Vorschriften zur Veröffentlichung von Zusatzinformationen ist folglich, dass diese keine umfassenden, konkretisierten und detaillierten Bestimmungen enthalten, was und wie über den F&E-Sachverhalt zu berichten ist. Vielmehr sind F&E-bezogene Angaben insbesondere von forschungsintensiven Unternehmen zu machen, bei denen dem F&E-Sachverhalt unter der entsprechenden Berichtskategorie Bedeutung zukommt und dieser somit im Hinblick auf den Entscheidungsnutzen wichtig und damit berichtspflichtig ist. Lediglich im Rahmen des Item 5 der Form 20-F werden, innerhalb des Financial Reporting deplaziert, explizit F&E-bezogene Angaben gefordert, die Berichtspflicht wird jedoch ebenfalls nicht weiter konkretisiert. Welche Inhalte konkret zu berichten sind, und welcher Struktur ein gegebenenfalls notwendiger F&E-Bericht zu folgen hat, lässt die SEC mithin gänzlich offen und überlässt die diesbezüglich zu treffenden Entscheidungen den berichtenden Unternehmen.

4.3 Vorschlag einer zweckkonformen Umsetzung der SEC-Vorschriften

Unabhängig von ihrem finanziellen oder nicht-finanziellen Charakter sind zweckkonforme Informationen über F&E darauf gerichtet, dem Kapitalgeber eine Prognose der künftig an ihn fließenden Cashflows sowie deren Unsicherheit zu ermöglichen Im Rahmen des Financial Reporting sollten somit entsprechend Angaben über getätigte, F&E-bezogene Auszahlungen bzw. Aufwendungen, unter Beachtung des Wesentlichkeitskriteriums, für zentrale F&E-Projekte gemacht werden. Da sich der F&E-Prozess häufig über einen langen Zeitraum erstreckt, kommen dabei neben vergangenheitsorientierten Angaben insbesondere zukunftsgerichteten Angaben hohe Bedeutung zu, sofern diese mit dem Konkurrenzschutzgedanken vereinbar sind. Für die Cashflow-Prognose von Interesse ist somit eine Antwort auf die Frage, in welchem Umfang ein konkretes F&E-Projekt künftige Ausschüttungsmöglichkeiten belasten wird. Den Auszahlungen gegenüberzustellen sind dabei zukunftsorientierte Angaben, welche Einzahlungen das Unternehmen bei erfolgreichem Abschluss des F&E-Projektes erwartet (vgl. *Pohle*, 1992, S. 100). Dieser Informationsanforderung kommt beispielsweise die *SCHERING AG* nach. (vgl. *SCHERING AG* (Hrsg.), 2001, S. 4).

Werden vom Unternehmen im Rahmen des Financial Reporting entsprechende Finanzdaten berichtet, so können Zusatzangaben über F&E als Grundlage des, den Finanzdaten zu Grunde liegenden, Erfolgspotentials der Plausibilisierung bzw. Begründung der berichteten Daten dienen, um hierdurch deren empfundene Verlässlichkeit zu erhöhen. Entsprechend der grundsätzlichen Differenzierung in F&E-bezogene Auszahlungen und Einzahlungen lässt sich eine nicht-finanzielle Abbildung des F&E-Sachverhalts differenzieren in die Berichtsdimensionen Faktoreinsatz (Input) als Auszahlungsdeterminante und F&E-Ergebnis (Output) als (potentieller) Einzahlungsdeterminante (vgl. *Brockhoff*, 1982, S. 243; für einen Vorschlag zur Berichterstattung finanzieller, F&E-bezogener Daten vgl. *Brockhoff*, 2001, S. 56).

Im Rahmen der Output-bezogenen Angaben ist es das Ziel, die Kapitalgeber entsprechend der zum Berichtszeitpunkt vorherrschenden Kenntnis über den Stand sowie den bisherigen und erwarteten Verlauf des Projekts zu informieren. Im Rahmen der Input-bezogenen Angaben gilt es entsprechend, die Adressaten über den Status Quo der Ressourcen (Ist) sowie über die für die Vollendung des Projektes erwartungsgemäß benötigten Ressourcen (Soll) zu informieren. Entsprechend den grundsätzlichen Anforderungen der Kapitalgeber an Informationen gilt es dabei für beide Berichtsdimensionen (Input und Output), neben einer Beschreibung der Informationsdimension „Status Quo" insbesondere auch die Informationsdimensionen „zukunftsgerichtete Angaben" sowie „Analysen des Managements" zu gewähren.

Eine mit dem Reporting-Ziel der SEC konforme Berichterstattung von F&E-bezogenen Zusatzangaben sollte somit die in *Abb. 5* strukturiert dargestellten Berichtsinhalte haben.

Eine Konkretisierung eines entsprechenden F&E-Berichts hängt davon ab, inwieweit eine Analysierbarkeit der berichteten Angaben durch messbare Einsatzfaktoren und Forschungsergebnisse ermöglicht werden kann. Eine auf Mengenangaben beruhende quantitative Berichterstattung kann eine Rangordnung oder einen objektiven Vergleich der quantifizierten

Größen und deren Veränderungen ermöglichen. Ist eine quantitative Abbildung nicht möglich, so sollte eine verbale, nicht-quantifizierende Berichterstattung erfolgen.

Welche Sachverhalte dabei zu berichten sind, hängt letztlich von dem jeweiligen, dem Bericht zu Grunde liegenden F&E-Projekt und dessen individueller Merkmale ab. Im Rahmen der Abbildung des Input-bezogenen Status Quo wären Angaben beispielsweise über die Anzahl der Mitarbeiter im F&E-Bereich sowie deren Ausbildung oder Informationen über wesentliche Forschungseinrichtungen sinnvoll (vgl. *Kuhn,* 1992, S. 106ff.). Als Leitlinie gilt, dass alle dem Unternehmen für F&E-Aktivitäten zur Verfügung stehenden materiellen sowie immateriellen Ressourcen abzubilden sind, die gleichsam die Grundlage für die Realisierung der laufenden F&E-Projekte bilden. Dies ermöglicht dem sachverständigen Kapitalgeber bzw. Analysten (als Filter für den Kapitalgeber), im direkten Vergleich mit Konkurrenten, die Erfolgsaussichten des Unternehmens zu beurteilen, als erstes das Projekt erfolgreich abzuschließen. Zum anderen bildet der Ressourcenbestand bzw. dessen Entwicklung die Grundlage dafür, um die Plausibilität des berichteten F&E-Aufwandes zu überprüfen, sowie um erste Erwartungen über künftig notwendig werdende, F&E-bezogene Auszahlungen zu bilden.

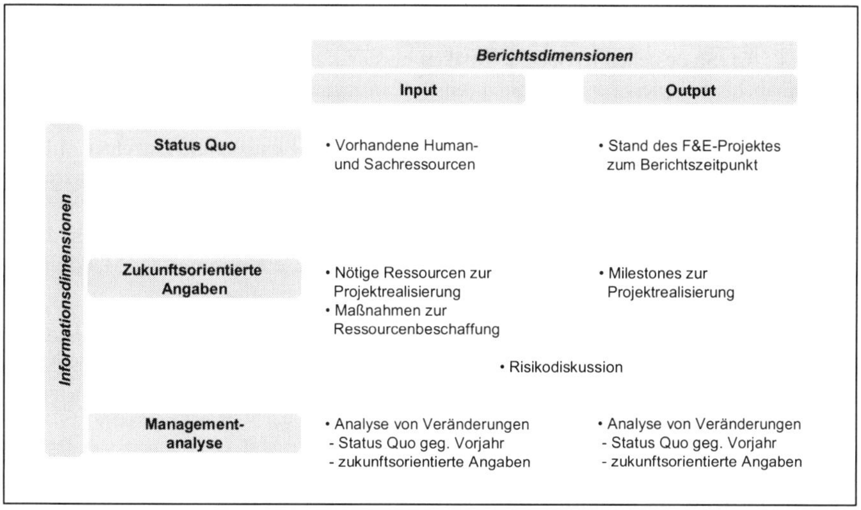

Abb. 5: Inhaltliche Strukturierung einer zweckkonformen Gestaltung F&E-bezogener Zusatzangaben im Rahmen des SEC-Reporting (mit beispielhaften Angaben)

Im Rahmen zukunftsorientierter Angaben über den künftig notwendigen Ressourceninput wären mithin alle Angaben zu subsumieren, die dem Kapitalgeber eine erweiterte Erwartungsbildung über künftige im Zusammenhang mit der Projektrealisierung zu erwartende Auszahlungen ermöglichen. Insbesondere wären hier Angaben über die Rekrutierung oder Akquisition notwendigen Know-hows oder dem Erwerb eines für den weiteren Prozess nötigen Schutzrechtes zu berichten.

Im Zuge der Managementanalyse gilt es schließlich die Entwicklung des Status Quo bzw. neue Erkenntnisse bezüglich künftig notwendiger Ressourcen durch das Management als

Insider vor dem Hintergrund der Bedeutung für den F&E-Prozess sowie, unter gegebenenfalls einer Unterbreitung von Lösungsvorschlägen zu kommentieren.

Im Rahmen der Output-bezogenen F&E-Publizität gilt es im Zuge der Informationsdimension „Status Quo“ den Stand des jeweiligen F&E-Projektes abzubilden, indem das Projekt einer der Phasen Grundlagenforschung, angewandte Forschung oder Entwicklung zu geordnet wird. Von besonderer Bedeutung ist dabei die Berichterstattung über erreichte Milestones, verstanden als Projektabschnitte, insbesondere dann, wenn die Erreichung dieser Milestones, wie beispielsweise für Pharmaunternehmen typisch, von einer Regulierungsbehörde vorgeschrieben ist. Diese Angaben sollen den Kapitalgeber in die Lage versetzen, den Fortschritt des Projekts und damit die Wahrscheinlichkeit der erfolgreichen Realisierung beurteilen zu können.

Ebenso diesem Zweck soll eine Skizzierung der noch zu realisierenden Milestones im Rahmen der zukunftsorientierten Berichterstattung dienen. Das Wissen über den noch „zu absolvierenden Weg“ ermöglicht es dem Kapitalgeber, neben einer Einschätzung der Erfolgswahrscheinlichkeit insbesondere auch die Angaben über noch benötigte Ressourcen im Rahmen der Inputpublizität zu plausibilisieren. Von besonderer Bedeutung im Rahmen der zukunftsorientierten Angaben ist auch eine Diskussion der F&E-bezogenen Risiken. Als F&E-bezogene Risiken werden dabei in der Literatur das „Technische Risiko“, das „Zeitrisiko“, das „Kostenrisiko“ sowie das „Verwertungsrisiko“ diskutiert (vgl. *Specht/Beckmann,* 1996, S. 25ff.). So bestehen technische Risiken hinsichtlich des Ergebnisses eines Entwicklungsvorhabens, beispielsweise aus Sicht eines Pharmaunternehmens in der Nichtanerkennung einer klinischen Studie als Voraussetzung der Erreichung eines Milestone durch die Regulierungsbehörde. Zeitrisiken entstehen hinsichtlich der Realisierung einzelner Milestones in der geforderten Zeit und können insbesondere bei hieraus resultierender Verspätung der Markteinführung zu Wettbewerbsnachteilen führen. Kostenrisiken betreffen Unsicherheiten bezüglich der im Rahmen des F&E-Prozesses einzusetzenden Ressourcen. Unter Verwertungsrisiko sind alle Unsicherheiten zu subsumieren, die sich mit der Markteinführung des neuen Produktes ergeben.

Vergleichbar der Inputpublizität gilt es im Rahmen der Managementanalyse schließlich kontextbezogene Hintergrundinformationen über bedeutsame Entwicklungen, deren Implikationen sowie eventuell vom Unternehmen geplante oder bereits eingeleitete gegensteuernde Maßnahmen zu geben.

4.4 Ergebnisse des *Business Reporting Research Project* des FASB über die „Best Practice“ F&E-bezogener Zusatzangaben ausgewählter Pharmaunternehmen

Im Januar 2001 veröffentlichte das FASB die Ergebnisse einer, als Business Reporting Research Project (BRRP) bezeichneten, empirischen Studie, welche unmittelbar auf der 1994 veröffentlichten Studie Improving Business Reporting des *Special Committee on Financial Reporting* des *American Institute of Certified Public Accountants* (AICPA) aufbaut (vgl. *FASB*, 2001; zusammenfassend *Haller/Dietrich,* 2001a, S. 206ff). Ziel der Studie des FASB war eine empirisch deskriptive Bestandsaufnahme, inwieweit US-amerikanische Unternehmen freiwillige Angaben bzw. eine Umsetzung unkonkretisierter Berichtspflichten der SEC vornehmen. Das Ergebnis soll Unternehmen gleichsam als „Best Practice“ zur Orientierung für ihre Publizitätsgestaltung dienen.

Aufgrund der Bedeutung von F&E für den Erfolg von Unternehmen der forschenden Pharmaindustrie wurde die F&E-Publizität ausgewählter Pharmaunternehmen speziell untersucht. In die Untersuchung einbezogen war die Unternehmenspublizität von *Abbott Laboratories*, *Allergan Inc.*, *ALZA Corp.*, *American Home Products Corp.*, *Merck & Co. Inc.*, *Pfizer Inc.*, *Schering-Plough Corp.* sowie von *Warner-Lambert Co.*

Die Ergebnisse des BRRP werden nachfolgend in *Abbildung 6* entsprechend oben vorgeschlagener Berichts- und Informationsdimensionen wiedergegeben. Zur Vermeidung von Bedeutungsverzerrungen wird dabei auf eine Übersetzung verzichtet. Da die Publizität mehrerer Unternehmen untersucht wurde, treten fallweise Redundanzen im Rahmen der identifizierten Berichtsinhalte auf

Inputpublizität	**Status Quo**	• Number of employees involved in Research & Development (R&D) within a company; • Strategic R&D alliances.
	Zukunftsorientierte Angaben	• Plans to use more external R&D and to investigate alliance opportunities; • Planned capital expenditures by product line.
	Management-analyse	• Impact of technology on R&D expenditures and product research; • Trends in the industry and their potential impact on the company's current and planned product lines.
Outputpublizität	**Status Quo**	• Detailed description of company's R&D portfolio; • Number of drugs currently in development; • Description of products in development and product development agreements with strategic alliance partners; • Development status of significant pharmaceutical products ; • R&D activities within all product lines; • Product pipeline and market size; • Listing of products, including those with expanded uses, those under regulatory review, and those in advance-stage development; • R&D activities being supplemented by in-licensing, technological collaborations, joint ventures and acquisition efforts, including the establishment of research relationships with academic institutions and individual researchers; • History of product approvals and its management of the Food and Drug Administration (FDA) regulatory process; • Discussion by a certain company that it does not disclose all of the products in development at any particular time for competitive reasons.

	Zukunftsorientierte Angaben	• Anticipated development completion dates; • Clinical trial results; • Global product approvals and clinical milestones; • FDA approval process; • FDA approvals/nonapprovals of specific pharmaceuticals; • Listing of promising compounds and the number of potential new products expected; • Potential therapeutic areas and R&D activities within those areas; • Product lines in which R&D spending will be concentrated; • Process and key steps for obtaining new drug clearance from FDA.
	Management-analyse	• Discussion of the three critical factors for the company's continued success: research "pipeline," new products, and alliances; • Discussion of difficulties in producing a product within industry manufacturing guidelines; • Discussion of trends in the industry and their potential impact on the company's current and planned product lines.

Abb. 6: Best Practice der F&E-Publizität ausgewählter US-amerikanischer Pharmaunternehmen (Quelle: FASB (Hrsg.), 2001).

5. Thesenförmige Zusammenfassung und Fazit

- In einer durch Marktunvollkommenheiten, insbesondere Informationsasymmetrien zwischen Kapitalgebern und Management gekennzeichneten realen Welt kommt der umfassenden Versorgung der Kapitalgeber mit entscheidungsnützlichen Informationen eine zentrale Bedeutung für die Gewährleistung der Kapitalaufbringungs- und Kapitallenkungsfunktion des Kapitalmarktes zu.

- Grundlegende Annahme für die Konkretisierung entscheidungsnützlicher Informationen ist, dass Kapitalgeber stets rational handelnde Individuen sind, die nach einer Maximierung ihres Nutzens streben. Als Nutzenpotentialindikator finden für analytische Zwecke Zahlungsmittel Verwendung. Entscheidungsnützliche Informationen dienen mithin der Erwartungsbildung über die Struktur des aus dem Unternehmen an den Kapitalgeber fließenden Zahlungsstroms hinsichtlich dessen Dimensionen Höhe, zeitlichem Zufluss und Unsicherheit. Eine Differenzierung diesbezüglich grundsätzlich entscheidungsrelevanter Informationen lässt sich in die Kategorien Cashflow, Erfolg und Erfolgspotential vornehmen.

- Eine Konkretisierung der zu vermittelnden Informationen wird in den USA von der Securities and Exchange Commission (SEC) vorgenommen. Die Berichtspflichten lassen sich dabei grundsätzlich differenzieren in das Financial Reporting sowie in die Gewährung von Zusatzinformationen. Während erstere Kategorie alle Informationen umfasst, die sich direkt oder indirekt auf Daten der Finanzbuchhaltung, also auf den Cashflow oder den Erfolg beziehen, dienen letztere im Sinne einer Negativabgrenzung der

Berichterstattung aller anderen annahmegemäß entscheidungsnützlichen Informationen, insbesondere der Beschreibung des Erfolgspotentials.

- Erfolgspotential ist definiert als nachhaltiger Wettbewerbsvorteil. Wettbewerbsvorteile können in Form einer Kostenführerschaft oder aufgrund eines differenzierten Leistungsangebots bestehen. Grundlage von Wettbewerbsvorteilen sind neue Techniken in Produkten oder Prozessen. Neue Techniken entstehen auf Grundlage neuer Technologien, diese wiederum auf Grundlage neuer Theorien bzw. Erkenntnisse. Dieser Zusammenhang drückt den grundsätzlich dreigeteilten Forschungs- und Entwicklungsprozess aus. Im Rahmen der Grundlagenforschung entstehen Theorien bzw. Erkenntnisse; im Rahmen der angewandten Forschung entstehen Technologien; im Rahmen der Entwicklung entstehen Techniken. Die Bedeutung von F&E für die Schaffung von Wettbewerbsvorteilen begründet die Entscheidungsnützlichkeit F&E-bezogener Informationen.
- Im Rahmen der Anforderungen des SEC-Reporting wird keine konkretisierte, explizite Berichtspflicht über F&E gefordert. Vielmehr ist eine solche Pflicht indirekt aus den Publizitätsanforderungen über die Beschreibung der Geschäftstätigkeit dann abzuleiten, wenn dem F&E-Sachverhalt im konkreten Fall für das berichtende Unternehmen für die Schaffung von Wettbewerbsvorteilen tatsächlich Bedeutung zukommt. Für die Bestimmung der zu berichtenden Informationen ist dann auf deduktiv und empirisch gewonnene Erkenntnisse über Informationsanforderungen der Kapitalgeber zurückzugreifen. Im vorliegenden Beitrag wird vorgeschlagen, eine Berichterstattung bezüglich der Berichtsdimensionen „Faktoreinsatz für F&E“ (Input) und „F&E-Ergebnis“ (Output) sowie der Informationsdimensionen „Status Quo“, „zukunftorientierte Angaben“ sowie „Managementanalyse“ zu strukturieren.
- Anregungen über die Konkretisierung einer F&E-Publizität können aus den Ergebnissen der empirischen Studie Business Reporting Research Project des FASB über die Best Practice der freiwilligen Berichterstattung ausgewählter US-amerikanischer Unternehmen generiert werden. Am Beispiel der untersuchten Pharmaunternehmen ist wenig überraschend ersichtlich, dass, entsprechend der Bedeutung von F&E für die Schaffung von Wettbewerbsvorteilen, in dieser Industrie umfangreich F&E-bezogene Angaben berichtet werden. Dies geschieht jedoch asymmetrisch. Während die Berichtsdimension „Output“ ausführlich insbesondere durch Beschreibung des Status Quo sowie schwebender Zulassungsverfahren und -prozesse berücksichtigt wird, erscheint die Berichtsdimension „Input“ vergleichsweise vernachlässigt.

6. Literatur

AICPA (Hrsg.), Improving Business Reporting – A Customer Focus, New York 1994.

Beaver, W. H., Financial Reporting: An Accounting Revolution, 3. Aufl., Upper Saddle River, New Jersey 1998.

Brockhoff, K., Forschung und Entwicklung im Lagebericht, in: WPg, 35. Jg. (1982), H. 4, S. 237-247.

Brockhoff, K., Forschung und Entwicklung, 4. Aufl., München/Wien 1994.

Brockhoff, K., Neue Herausforderungen an die Berichterstattung über Forschung und Entwicklung, in: *Boysen, K.* et al. (Hrsg.), Der Wirtschaftsprüfer und sein Umfeld zwischen Tradition und Wandel zu Beginn des 21. Jahrhunderts, Festschrift zum 75. Geburtstag von *Hans-Heinrich Otte*, Düsseldorf 2001, S. 49-65.

Brotte, J., US-amerikanische und deutsche Geschäftsberichte – Notwendigkeit, Regulierung und Praxis jahresabschlussergänzender Informationen, Wiesbaden 1997.

Bruns, H.-G. und *Renner, W.*, Finanzanalytische SEC-Berichterstattung in der Form 20-F – „Operating and Financial Review and Prospects, in: BfuP, 53. Jg. (2001), H. 1, S. 7-26.

Bürgel, H. D., *Haller, C.* und *Binder, M.*, F&E-Management, München 1996.

Coenenberg, A. G., Grundlagen der strategischen, operativen und finanzwirtschaftlichen Unternehmenssteuerung, in: *Busse von Colbe, W.* et al. (Hrsg.), Betriebswirtschaft für Führungskräfte, Stuttgart 2000, S. 3ff.

Coenenberg, A. G. Jahresabschluss und Jahresabschlussanalyse, 19. Aufl., Landsberg/Lech 2003.

Coenenberg, A. G. und *Haller, A.*, Externe Rechnungslegung, in: *Hauschildt, J.* und *Grün, O.* (Hrsg.), Ergebnisse empirischer betriebswirtschaftlicher Forschung, Festschrift für *E. Witte*, Stuttgart 1993.

Ekkenga, J., Anlegerschutz, Rechnungslegung und Kapitalmarkt, Tübingen 1998.

FASB, SFAS No. 2, Accounting for Research and Development Costs", Norwalk, USA 1974.

FASB (Hrsg.), Business Reporting Research Project, Norwalk, USA 2001.

Fisher, I., The theory of interest, New York 1930.

Gierl, H. und *Kotzbauer, N.*, Der Einfluss des F&E-Aufwandes auf den wirtschaftlichen Erfolg von Industrieunternehmen, in: ZfbF, 44. Jg. (1992), H. 11, S. 974-989.

Haller, A., Die Grundlagen der externen Rechnungslegung in den USA, Diss., 4. Aufl., Stuttgart 1994.

Haller, A., Wesentliche Ziele und Merkmale US-amerikanischer Rechnungslegung, in: *Ballwieser, W.* (Hrsg.), US-amerikanische Rechnungslegung, 4. Aufl., Stuttgart 2000, S. 1-27.

Haller, A. und *Dietrich, R.*, Kapitalmarktorientierte Gestaltung der Lageberichterstattung, in: KOR, 1. Jg. 2001, H. 2, S. 164-174.

Haller, A. und *Dietrich, R.*, Freiwillige Unternehmensberichterstattung in den USA: Ergebnisse des *Business Reporting Research Project* des FASB, in: KOR, 1. Jg. (2001a), H. 4, S. 206-211.

Herzog, R., Forschungs- und Entwicklungskulturen, in: *Herzog, R.* (Hrsg.), F&E-Management in der Pharma-Industrie, Aulendorf 1995, S. 24-26.

Hirshlifer, J., Where are we in the Theory of Information? In: AER, Vol. 63 (1973), No. 1, S. 31-39.

Jensen, M. C., Agency Costs of Free Cashflow, Corporate Finance, and Takeovers, in: AER, Vol. 76 (1986), No. 2, S. 323-329.

Jensen, M. C. und *Meckling, W. H.*, Theory of the Firm: Managerial Behavior, Agency Costs and Ownership Structure, in: JoFE, Vol. 3 (1976), No. 4, S. 305-360.

Käfer, K., Konzentration und Publizität, in: *Arndt, H.* (Hrsg.), Die Konzentration in der Wirtschaft, Berlin 1960.

Knauss, R. L., A Reappraisal of the Role of Disclosure, in: MLRev., No. 607, Feb. 1964.

Koch, H. D. und *Schmidt, R. H.*, Ziele und Instrumente des Anlegerschutzes, in: BFuP 33. Jg. (1981), H. 3, S. 231-250.

Krönert, B., Grundsätze informationsorientierter Rechnungslegung, Diss., Sternenfels 2001.

Kropff, B., Publizität, in: *Chmielewicz, K.* und *Schweitzer, M.* (Hrsg.), Handwörterbuch des Rechnungswesens, 3. Aufl., Stuttgart 1993.

Kuhlwind, A. M., Grundlagen einer Bilanzrechtstheorie in den USA, Diss., Frankfurt/Main 1997.

Kuhn, W., Forschung und Entwicklung im Lagebericht, Diss., Hamburg 1992.

Lange, C., Jahresabschlussinformationen und Unternehmensbeurteilung, Stuttgart 1989.

Laux, H., Entscheidungstheorie, 3. Auflage, Berlin u. a. 1995.

Levy, S. M., Regulation of Securities – SEC Compliance and Practice, New York 2001.

May, A., Zum Stand der empirischen Forschung über Informationsverarbeitung am Aktienmarkt – Ein Überblick, in: ZfbF, 43. Jg. (1991), H. 3, S. 313-335.

Meier-Schatz, C. H. J., Wirtschaftsrecht und Unternehmenspublizität, Zürich 1989.

Merkt, H., Unternehmenspublizität, Tübingen 2001.

Moxter, A., Grundsätze ordnungsmäßiger Bilanzierung und Stand der Bilanztheorie, in: ZfbF, 18. Jg. (1966), H. 1, S. 29-58.

Moxter, A., Publizität, in: *Grochla, E.* und *Wittmann, E.* (Hrsg.), Handwörterbuch der Betriebswirtschaftslehre, 4. Aufl., Stuttgart 1975, Sp. 3282-3288.

Moxter, A., Betriebswirtschaftliche Gewinnermittlung, Tübingen 1982.

Moxter, A., Rechnungslegungsmythen, in: BB, 55. Jg. (2000), S. 2143-2149.

OECD (Hrsg.), Frascati Manual 1993 – The Measurement of Scientific and Technological Activities – Proposed Standard Practice for Surveys of Research and Experimental Development, Paris 1994.

Pellens, B., Publizität, in: *Gerke, W.* und *Steiner, M.* (Hrsg.), Handwörterbuch des Bank- und Finanzwesens, 3. Aufl., Stuttgart 2001.

Pohle, K., Biotechnologie-Unternehmen und ihre Besonderheiten in der Rechnungslegung, in: *Boysen, K.* (Hrsg.), Der Wirtschaftsprüfer vor innovativen Herausforderungen, Festschrift für *Hans-Heinrich Otte*, Stuttgart, 1992, S. 93-107.

Porter, M. E., Wettbewerbsvorteile, Frankfurt/New York 1999.

Rappaport, A., Shareholder Value – Wertsteigerung als Maßstab für die Unternehmensführung, Stuttgart 1995.

SCHERING AG (Hrsg.), Geschäftsbericht 2001, Berlin 2001.

Schmidt, R. H. und *Terberger, E.*, Grundzüge der Investitions- und Finanzierungstheorie, 4. Aufl., Wiesbaden 1997.

Schneider, D., Rechtsfindung durch Deduktion von Grundsätzen ordnungsmäßiger Buchführung aus gesetzlichen Jahresabschlußzwecken, in: StuW, 60. Jg. (1983), H. 13, S. 141-160.

Sieben, G. und *Schildbach, D.*, Betriebswirtschaftliche Entscheidungstheorie, 4. Aufl., Düsseldorf 1994.

Specht, G./Beckmann, C., F&E-Management, Stuttgart 1996.

Streim, H., Zum Stellenwert des Lageberichts im System der handelsrechtlichen Rechnungslegung, in: *Elschen, R.* et al. (Hrsg.), Unternehmenstheorie und Besteuerung, Festschrift zum 60. Geburtstag von *D. Schneider*, Wiesbaden 1995, S. 703-721.

4. Spezielle Intangibles

Intangibles weisen eine enorme inhaltliche Spannweite auf, was die Beschäftigung mit ihnen erschwert. Vielfach findet daher eine Fokussierung auf ein spezielles Thema statt. In Diskussionen zu Intangibles findet sich nahezu immer beispielhaft das Thema des „Markenwertes“, da dieser im Zuge eines Unternehmensverkaufes häufig auch Gegenstand einer separaten Bewertung ist. Neben dem Markenwert von Verbrauchsgütern gewinnt zunehmend auch der Markenwert von Investitionsgütern an Bedeutung. Er steht in engem Zusammenhang mit dem impliziten Wissen (der Mitarbeiter) und dem expliziten Wissen (z.B. von Patenten) eines Unternehmens. Die weit verbreitete Bezeichnung `Intellectual Capital´ trägt dieser enormen Bedeutung bereits Rechnung. Diese Bedeutung zeigt sich auch in der verpflichtenden Veröffentlichung von Wissensbilanzen für Forschungsinstitute. Virtuelle Organisationen und Netzwerke basieren auf dem Grundgedanken, dass zentrale Assets einer Organisation nicht ortsgebunden sind. Die besonderen Aspekte von Intangibles in Netzwerken stellen damit einen weiteren besonders relevanten Themenbereich dar. Dies trifft auch auf die zunehmend knappen ökologischen Ressourcen zu, in deren Folge ein umweltkonformes Auftreten von zentraler Bedeutung für den Wert und den wirtschaftlichen Erfolg eines Unternehmens ist.

Wertorientiertes Markenmanagement

Ronald Fiedler [1]

---

[1] *Ronald Fiedler*, Stern Stewart & Co., München

Marken müssen aufgrund ihrer wichtigen Funktion als Verbindung zwischen dem Unternehmen und der Außenwelt in ein Konzept wertorientierter Unternehmensführung integriert werden. Eine Implementierung in der Praxis erfolgt nur zögerlich vor dem Hintergrund der Kritik an bestehenden Konzepten der Markenbewertung. Der hier vorgestellte Ansatz von Stern Stewart & Co. zeigt auf, dass weniger der absolute Wert als die Ableitung von Handlungsempfehlungen für ein wertorientiertes Markenmanagement notwendig sind.

1. Bedeutung der Marke

Nach der Beschäftigung mit wertorientierten Kennzahlen und der Optimierung von materiellen Vermögensgütern erfährt jetzt die Bedeutung der Steuerung immaterieller Vermögensgüter stärkere Beachtung. Marken nehmen hier eine besonders hohe Priorität ein, da sie als direktes Bindeglied zwischen dem Unternehmen und der Außenwelt fungieren. Die Marke ermöglicht es, Produkte zu bestimmten Preisen und bestimmten Mengen abzusetzen. Für viele Unternehmen stellt deshalb die Marke das wertvollste aller Vermögensgüter dar (vgl. *Keller*, 1998, S. 5). Die Sicherung der Abnahme der eigenen Leistungen und Erzeugnisse veranlasst zunehmend Unternehmen auch abseits der traditionellen Konsumgüterindustrie, sich mit der Thematik zu befassen. So verkündete beispielsweise *Boeing* im Jahr 2000 den Start ihrer ersten weltweiten Werbekampagne, die darauf abzielte, die Marke als Instrument zur Erreichung der Unternehmensziele bewusst einzusetzen (vgl. *Boeing*, 2000).

Im Gegensatz zu anderen immateriellen Vermögensgütern, bildet sich für Marken immer mehr ein eigener Markt heraus, der es notwendig macht, Preise zu formulieren. Doch nicht nur beim Kauf von Markenrechten, sondern auch bei Firmenakquisitionen allgemein nehmen Marken einen großen Stellenwert ein. So wurde bei der Übernahme von *Compaq* durch *Hewlett Packard* lange über die Möglichkeiten und Auswirkungen für beide starke Marken diskutiert, bevor man sich für eine Zweimarkenstrategie entschied (vgl. *Hewlett Packard*, 2002).

2. Verbindung von Marke und Wertorientierung

Das Wesen der Wertorientierung besteht darin, auf alle getätigten Investitionen eine angemessene Mindestverzinsung zu erzielen, die sich im Vergleich mit Opportunitätsanlagemöglichkeiten ergibt. Erst wenn die geforderte Mindestverzinsung erzielt und darüber hinaus ein positives Ergebnis erwirtschaftet wird, spricht man von Wertschaffung. Überrenditekonzepte, wie z.B. der Economic Value Added (EVA™ – EVA™ ist ein eingetragenes Warenzeichen von *Stern Stewart & Co.*) messen diesen Übergewinn periodisch und können zur Bewertung von Investitionen und Akquisitionen angewendet werden, indem zukünftige Übergewinne abdiskontiert betrachtet werden (vgl. *Stern/Shiley/Ross*, 2002, S. 35 ff.).

Investitionen erfolgen, außer z.B. in den Produktionsablauf oder die Vertriebsstruktur, ebenso in die Marke.

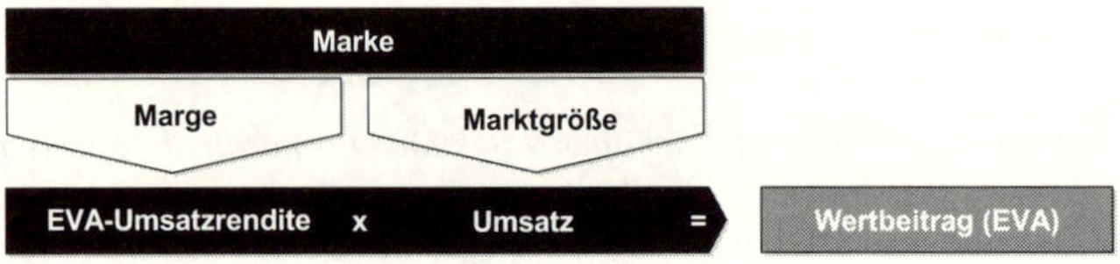

Abb. 1: Verbindung von Marke und Wertbeitrag

Strategische Markeneinführungen, Imagekampagnen, Kundenbindungsmaßnahmen etc. erstrecken sich in ihrer Wirkungsdauer auf längere Zeit und erfüllen so das wesentliche Merkmal einer Investition. Auch in diesem Bereich sollten deshalb langfristige Projekte als Investitionen angesehen und behandelt werden. Die Etablierung von Verantwortung für solche Investitionen ist der erste Schritt in Richtung Wertorientierung im Bereich der Markenführung.

Marken können prinzipiell auf zwei Arten zur Wertschaffung beitragen, welche je nach Segment unterschiedlich gewichtet sein können. Einerseits ermöglicht die Marke bei der Preisfestsetzung die Erzielung einer erhöhten Marge. Andererseits kann eine Erhöhung des Wertbeitrages über eine Ausdehnung der Menge über erhöhte Kundenbindung oder Erschließung neuer Märkte erreicht werden. Man denke hier nur an Beispiele im Automobilbau, wo innerhalb der Konzerne ähnliche Produkte unter unterschiedlichen Marken angeboten werden. So sind Baugruppen für *VW* und *Skoda* fast identisch, dennoch unterscheiden sich die Verkaufspreise und Absatzmengen der Modelle erheblich.

Beide Pfade zur Wertsteigerung lassen sich direkt in Verbindung zum Wertbeitrag bringen. Definiert man die Marge als EVA-Umsatzmarge (EVA/Umsatz), so ergibt diese in Verbindung mit dem Umsatz den gesamten Wertbeitrag der Periode (vgl. *Abb. 1*).

Die konzeptionelle Verbindung mit der Wertschaffung einer Periode alleine ist zu wenig, um Marken in ein wertorientiertes Steuerungssystem zu integrieren. Das Ergebnis der Markenführung zeigt sich erst an den Wertbeiträgen der kommenden Perioden. Viele Ansätze beschäftigen sich daher primär mit der Berechnung eines Markenwertes. Auffallend hierbei ist die Unterschiedlichkeit der Ansätze, welche sich auch in teilweise signifikanten Unterschieden der errechneten Markenwerte niederschlägt (vgl. *Nöcker*, 2002, S. 17). Für das Management entscheidend ist jedoch nicht die punktgenaue Abschätzung des Betrages, sondern vielmehr die Ableitung von Handlungsempfehlungen aus wertorientierten Gesichtspunkten. In einem wertorientierten Steuerungssystem können strategische und operative Entscheidungen der Markenführung nach Wertgesichtspunkten gefällt und überprüft werden. Ziel eines neuen Ansatzes ist es, die wesentlichen Kritikpunkte bestehender Bewertungsmodelle zu beseitigen und sich vor allem mit den praktischen Anwendungen und Handlungsempfehlungen zu befassen. Zu diesem Zweck werden zuerst Anforderungen an ein wertorientiertes Managementsystem entworfen.

3. Anforderungen an ein wertorientiertes Markenmanagementsystem

Die isolierte Ermittlung eines Markenwertes alleine stellt keine Implementierung von wertorientierten Gesichtspunkten im Markenmanagementsystem sicher. Die Anforderungen beziehen sich deshalb neben der Bewertungsmethode vor allem auf Aspekte der praktischen Anwendung.

3.1 Objektivität und Transparenz

Ein häufiger Kritikpunkt bestehender Ansätze ist deren Intransparenz. Vielfach sind Expertenmeinungen bezüglich des zukünftigen Potenzials einbezogen oder die Verbindung von einzelnen Komponenten ist nicht klar ersichtlich. Aus der Sicht des Unternehmens erfolgt die Berechnung in einer „black box". Insbesondere auf einem Gebiet, wo sich noch kein Standard herausgebildet hat, hilft Transparenz dabei, Glaubwürdigkeit zu vermitteln. Basis aller Überlegungen müssen fundierte Daten sein. Bei der Verwendung von qualitativen Markenbeurteilungen sollten diese mit Hilfe von empirischen Untersuchungen ermittelt werden. Neben der Objektivität der qualitativen Daten muss auch die Transparenz der quantitativen Parameter und der Berechnungsmethode gewährleistet sein. Nur vor diesem Hintergrund kann eine sinnvolle Interpretation der Ergebnisse erfolgen und konkrete Anhaltspunkte für das Management und ihre Maßnahmenplanung abgeleitet werden.

3.2 Wertbezug

Zur Abschätzung der quantitativen Auswirkungen von Marken greifen viele Modelle auf Daten der Buchhaltung zurück. In vielen Fällen werden Ergebnisgrößen übernommen, ohne dass eine Aussage über die Wertschaffung möglich ist. Als Basis für quantitative Überlegungen müssen jedoch Zahlen verwendet werden, die den Anspruch haben, wertorientierte Informationen zu beinhalten. So müssen insbesondere vom Ergebnis die Opportunitätskosten für das investierte Geschäftsvermögen abgezogen werden.

3.3 Vorausschau

Die Wirkung von Marken erstreckt sich auf einen langfristigen Zeitraum. Insbesondere bei neuen Marken können die derzeitigen Wertbeiträge negative Beträge ausweisen, da bestimmte Aufbaumaßnahmen hohe Kosten verursachen und Umsätze erst im Steigen begriffen sind. Nichts desto weniger kann die Basis für eine erfolgreiche Performanceentwicklung in der Zukunft bereits jetzt in der Marke verankert sein. Abschätzungen über die Auswirkungen der Marke in der Zukunft müssen daher in die Bewertung mit einfließen. Finanzkennzahlen, die den Erfolg aktueller oder historischer Perioden messen, und qualitative Faktoren, die sich auf die Kundenwahrnehmung der Marke als Basis heutiger Kaufentscheidungen beziehen, sind zu wenig, um das zukünftige Potenzial abzuschätzen. Zusätzlich müssen vorlaufende Indikatoren ermittelt werden, welche eine Aussage hinsichtlich des zukünftigen Potenzials der Marke zulassen. Dies betrifft sowohl die qualitative Seite des Markenprofils als auch die Aussagen bezüglich der quantitativen Wertbeiträge. So sind neben den derzeitigen Wertbeiträgen auch Beurteilungen über zukünftige Perioden zu integrieren.

3.4 Praktikabilität

Die operative Anwendbarkeit der Erkenntnisse aus der Markenbewertung ist dann gewährleistet, wenn ein Verantwortlicher daraus direkte Implikationen für seine Entscheidungen ableiten kann. Das sind beispielsweise für die Marketingabteilung Rahmenbedingungen und Schwerpunkte für Kommunikationskampagnen, für die Produktionsabteilung Auswertungen bezüglich Qualität und für den Vertrieb Anregungen hinsichtlich erwarteter Serviceniveaus. Aussagen zu diesen Bereichen lassen sich nur ableiten, wenn sich neben dem reinen Wert klare Verbindungen zu markenspezifischen Werttreibern aufzeigen lassen. Die Identifizierung der Werttreiber ermöglicht eine detailliertere Einbettung in ein Controllingsystem, in dem die Entwicklung einzelner Maßnahmen und Auswirkungen auf die einzelnen werttreibenden Attribute gesteuert und verfolgt werden können. Ein leistungsfähiges Controllingsystem ist notwendig, um unternehmerische Entscheidungen zur Markenführung auf eine rationale Basis zu stellen (vgl. *Meffert/Koers*, 2002, S. 404).

4. Lösungsansatz

Bei der Entwicklung des Ansatzes für ein wertorientiertes Markenmanagement von *Stern Stewart & Co.* wurden die Anforderungen an eine wertorientierte Markensteuerung berücksichtigt. Die Grundbausteine des Ansatzes sind die Markenanalyse, die Wettbewerbsanalyse und die Wertbeitragsanalyse. Durch diese werden die Bereiche der qualitativen Markenaspekte, der Quantifizierung aus Wertgesichtspunkten sowie der Informationen über die Branche abgedeckt. Diese Bausteine werden in Kombination für praktische Anwendungen der Markenbewertung, der operativen Markenführung und der strategischen Portfolioanalyse genutzt (vgl. *Abb. 2*).

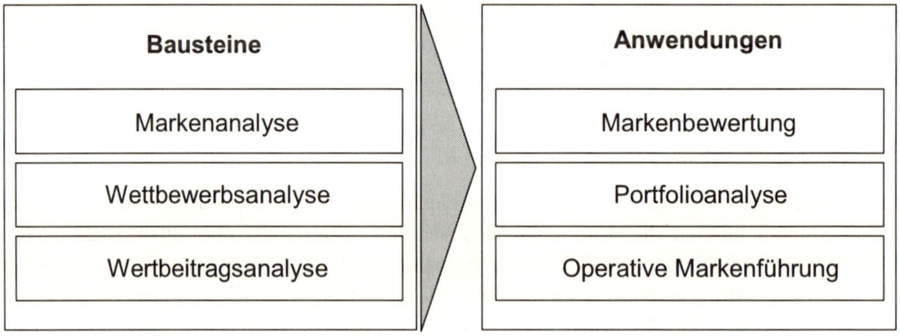

Abb. 2: Bausteine und Anwendungen des Ansatzes eines wertorientierten Markenmanagements

4.1 Markenanalyse

Als Basis für objektive Informationen über die Kundenwahrnehmung von Marken werden empirische Untersuchungen durchgeführt. Wichtig hierbei ist, dass diese Untersuchungen regelmäßig und konsistent erfolgen. Die Markenstärke kann mithilfe von vier Indikatoren charakterisiert werden (vgl. *Young & Rubicam*, 2001): Differenzierung, Relevanz, Ansehen und Bekanntheit:

- Die **Differenzierung** gibt Auskunft darüber, in welchem Maße eine Marke als besonders wahrgenommen wird und sich dadurch aus Sicht des Kunden von den Marken der Wettbewerber unterscheidet. Differenzierung ist eine wesentliche Voraussetzung dafür, höhere Preise am Markt durchzusetzen.

- Die **Relevanz** gibt an, wie angemessen und passend eine Marke von den Konsumenten wahrgenommen wird. Ist die Relevanz hoch, dann besteht ein vergleichsweise höheres Potenzial, Kunden zu gewinnen und zu halten.
- Das **Ansehen** definiert, wie beliebt die Marke ist und wie sehr sie geachtet wird. Das Ansehen steht in Beziehung zu der Art, wie eine Marke die eigenen Versprechungen und bestehende Kundenwünsche erfüllt.
- Die **Bekanntheit** sagt etwas über das Wissen über die Marke von Seiten des Kunden aus. Bekanntheit ist das Ergebnis aller Marketing- und Kommunikationsanstrengungen und der Erfahrungen, welche die Kunden bereits mit der Marke gemacht haben.

Alle vier Indikatoren ergeben die Markenstärke. Differenzierung und Relevanz stehen für die vorlaufenden Indikatoren, sie werden in Kombination auch als Markenkraft bezeichnet. Noch bevor umfangreiche Erfahrungen mit der Marke getätigt wurden, zeigen Differenzierung und Relevanz Möglichkeiten der zukünftigen Entwicklung auf. Differenzierung ist eine Grundvoraussetzung dafür, Zusatznutzen zu vermitteln und damit erhöhte Margen zu erzielen. Die Relevanz gibt darüber Auskunft, welcher potenzielle Kundenkreis prinzipiell in Betracht kommen würde, unabhängig von derzeit erzielten Verkaufsmengen. Ein Differenzierungswert über dem Relevanzwert deutet auf weiteres Wachstumspotenzial hin. Ein höherer Relevanzwert als Differenzierungswert gibt eine Tendenz zum „commodity product" an, mit ein großen Anzahl an vergleichbaren Produkten und verstärkter Wichtigkeit des Preises als Verkaufsargument. Vertreter des deutschen Lebensmitteleinzelhandels finden sich beispielsweise in dieser Kategorie.

Ansehen und Bekanntheit wiederum stellen nachlaufende Indikatoren dar und werden in Kombination auch als Markenstatur bezeichnet. Beide Indikatoren sind erheblich von historischen Erfahrungen mit der Marke abhängig. Ein hohes Ansehen und geringere Bekanntheit drücken ein Verlangen aus, die Marke näher kennen zu lernen. Luxusmarken sind häufig in dieser Kategorie zu finden. Im Fall von geringem Ansehen in Kombination mit hoher Bekanntheit ist die Marke nicht nur wenig beliebt, die Konsumenten sind sich in ihrem Urteil wegen ihrer Erfahrungen auch sicher.

In einem idealtypischen Lebenszyklusmodell erfolgt der Aufbau der Marke über starke Differenzierung, andere Indikatoren folgen schrittweise. Führende Marken bestechen durch gute Werte in allen vier Indikatoren. Absteigende Marken verlieren zuerst an Differenzierung, ein Anzeichen für eine stärkere Vergleichbarkeit mit den Konkurrenzprodukten sowie einer Entwicklung hin zur Massenware und erhöhtem Preisdruck. Werden keine Gegenmaßnahmen ergriffen, droht der schrittweise Verfall aller Indikatoren.

Neben den vier Indikatoren werden zusätzlich Markenattribute erhoben. Diese Attribute drücken Empfindungen der Kunden gegenüber der Marke aus, wie z.B. Spaß, Tradition, Innovation und so weiter. Mithilfe der Analyse der Indikatoren und der empfundenen Markenattribute lassen sich so Rückschlüsse über die Gründe für die Indikatorenposition herausfinden. Korrelationsanalysen zeigen, welche Attribute den größten Einfluss auf die vier Indikatoren haben.

4.2 Wettbewerbsanalyse

Die Wettbewerbsanalyse dient zur Abklärung des Untersuchungsobjektes durch die Bestimmung der für die Untersuchung relevanten Marktteilnehmer. Hierbei ist entscheidend, dass der Betrachtungskreis nicht zu eng gefasst wird. Dies ist aus zwei Blickwinkeln sinnvoll: Zum einen existiert in vielen Segmenten die Möglichkeit, das bestehende Geschäft auf benachbarte Segmente auszuweiten. Beispiel hierfür ist die Ausdehnung der Marke *Nivea* von der reinen Körperpflege auf Kosmetikprodukte. Etwaige Wechselwirkungen sollten daher in die Analyse integriert werden. Zum anderen können Segmente, welche ähnliche Kriterien z.B. hinsichtlich der Kaufsituation, des Wettbewerbes oder der Preissensitivität erfüllen, dienlich sein zur Gewinnung neuer Erkenntnisse bezüglich der Einflussfaktoren auf das Markenprofil. Auf diese Weise beinhaltet eine Analyse nicht nur die in einem Segment bereits heute angewendeten Markenmuster, sondern liefert gleichzeitig Ansatzpunkte für neue Ideen außerhalb des Segmentes.

Zusätzlich wird in der Wettbewerbsanalyse die Entwicklung des relevanten Marktes hinsichtlich der Wettbewerber, Chancen und Risiken, Wettbewerbsintensität, etc. untersucht. Diesbezügliche Erkenntnisse fließen ein in die Interpretation von Markenwerten sowie die Plausibilisierung von internen Planzahlen.

4.3 Wertbeitragsanalyse

In der Wertbeitragsanalyse steht die Ermittlung heutiger und zukünftiger Wertbeiträge für einzelne Marken im Vordergrund. Die Berechnung des derzeitigen Wertbeitrages pro Marke erfolgt, indem vom operativen Ergebnis die Kapitalkosten auf Basis des zurechenbaren Geschäftsvermögens abgezogen werden. Die zukünftigen Wertbeiträge spiegeln sich im immateriellen Wert des Unternehmens wider. Zur Ermittlung wird vom gesamten Marktwert das materielle Geschäftsvermögen abgezogen. Für Monobrand Unternehmen ergibt sich damit direkt der immaterielle Wert pro Marke. Bei mehreren Marken müssen Annahmen bezüglich der Aufteilung des Marktwertes getroffen werden.

4.4 Markenbewertung

Die Markenbewertung verknüpft die Ergebnisse der Bausteine in einem mathematischen Modell. Dabei steht im Vordergrund die Korrelation qualitativer Markendaten mit den entsprechenden Wertkennzahlen für die relevanten Marktteilnehmer. Auf diese Weise können mithilfe von statistischen Verfahren Implikationen für den Markenanteil innerhalb des immateriellen Wertes gewonnen werden. Dabei werden externe Faktoren außerhalb des Markenwertes herausgefiltert und ein segmentspezifisches Bewertungsmodell aufgebaut, welches die Markenstärke mit dem Markenwert in Verbindung setzt.

Auf Basis dieses Modells ergibt sich der absolute Wert der einzelnen Marken im Wettbewerb auf Basis der aktuellen Markenstärke. In Verbindung mit dem derzeit erzielten Wertbeitrag lässt sich dieser aufteilen in einen Anteil, der dem Niveau des gegenwärtigen Wertbeitrages entspricht und einen Anteil, der sich aus Erwartungen hinsichtlich zukünftiger Steigerungen dieses Wertbeitrages ergibt.

Um Möglichkeiten der Verbesserung des Markenwertes auszumachen, ist vor allem der Einfluss der einzelnen Indikatoren auf den Markenwert entscheidend. Aufgrund der Transparenz der Inputfaktoren sowie der statistischen Verfahren, lässt sich nachvollziehen, wie die verschiedenen Faktoren das Ergebnis beeinflussen.

4.5 Portfolioanalyse

Die Portfolioanalyse ist das geeignete Instrument, um die strategische Bedeutung und die notwendigen Entscheidungen hinsichtlich einzelner Marken im Portfolio sichtbar zu machen.

Zunächst wird eine Priorisierung der Marken innerhalb des Portfolios vorgenommen. Vor dem Hintergrund der Markenbewertung bezieht sich diese auf die Potenziale hinsichtlich zukünftig zu erwartender Wertbeiträge. Anhand dieser Priorisierung erfolgt die Zuteilung finanzieller und personeller Ressourcen.

In einem zweiten Schritt werden für diejenigen Marken, welche derzeit nur ein geringes Potenzial aufweisen, sowohl Repositionierungsstrategien als auch Exitstrategien geprüft. Zu evaluieren sind die Möglichkeit und Kosten einer Repositionierung sowie ihr erwarteter Nutzen durch Erhöhung der Markenstärke und damit einhergehenden Auswirkung auf den Markenwert. Übersteigt der Nutzen aus einer stärkeren Marke die damit verbundenen Kosten, ist eine Repositionierung vorzunehmen. Ist dies nicht der Fall, müssen verschiedene Exitoptionen gegeneinander abgewogen werden.

Drittens werden die aus der Bewertung abgeleiteten zukünftig erwarteten Wertbeiträge mit den internen Planzahlen verglichen. Die Bewertung lässt sich mit einigen Annahmen in erwarteten Wertbeiträgen der Marke in den zukünftigen Perioden ausdrücken. Ein Abgleich mit internen Planungsrechnungen gibt Auskunft, ob man sich selbst in der Lage sieht, entsprechende Wertbeiträge zu generieren. Ist dies nicht der Fall, müssen entweder die notwendigen Kompetenzen aufgebaut oder die Marke verkauft werden. So verlangt beispielsweise das Management von Luxusmarken andere Kompetenzen als die Führung von Marken für Massenartikel, auch wenn diesen das gleiche Ausgangsprodukt zugrunde liegt. Andererseits können aber auch positivere Erwartungen mit der Marke verknüpft werden: Sind die Pläne optimistischer, so bestehen neben dem bereits ausgewiesenen Potenzial weitere Möglichkeiten zur Steigerung des Markenwertes. Die geplanten Maßnahmen der Markenführung sind dann bezüglich ihres Effektes auf die möglichen Auswirkungen auf die Markenstärke zu überprüfen.

4.6 Operative Markensteuerung

Die operative Markensteuerung zielt darauf ab, unter Wertgesichtspunkten geeignete Maßnahmen zu identifizieren, umzusetzen und die Ergebnisse nachzuhalten. Das Kernstück hierfür ist ein wertorientiertes Controllingsystem.

Um Erkenntnisse über die Auswirkungen operativer Entscheidungen auf die Marke herauszufinden, müssen die Werttreiber der Marke analysiert werden. Diese sind jene Markenattribute, welche in Abstimmung mit der Markenstrategie die effizienteste Verbesserung der Markenindikatoren ermöglichen. Die Wechselwirkungen zwischen den einzelnen Werttreibern und der Markenstärke und damit auch dem Markenwert werden zunächst in einem Modell integriert. Auf diese Weise können zukünftige qualitative und quantitative Auswirkungen durch die Beeinflussung einzelner Werttreiber simuliert werden.

Auf Basis der zu beeinflussenden Werttreiber können nun Maßnahmen identifiziert werden. Durch Vergleich der Kosten einzelner Maßnahmen oder Maßnahmenbündel und abgeschätzter Wertentwicklung lassen sich fundierte Aussagen über die erwartete Effizienz der Maßnahmen treffen.

In der regelmäßigen Anwendung werden die einzelnen Werttreiber sowie die tatsächlich realisierten Wertbeiträge nachgehalten und mit der Planung verglichen. Auf diese Weise können die Entwicklung kontrolliert und gegebenenfalls zusätzliche Eingriffe vorgenommen werden.

Die Erkenntnisse und Anwendungen der Markensteuerung und –analyse sind nicht auf Marketingabteilungen beschränkt. Viele Abteilungen haben mit ihren Entscheidungen Einfluss auf die Wahrnehmung der Marke, seien es das Qualitätsniveau der Produktion oder der Servicegrad des Vertriebs. Ein integriertes Controllingsystem eignet sich dazu, die Auswirkungen auch anderer Bereiche auf die Marke transparent zu machen und so eine stärker markengeprägte Denkweise im gesamten Unternehmen zu etablieren. Ebenso können aggregierte Inhalte zur verbesserten Information des Top Managements genutzt werden.

Eine Möglichkeit, diese Funktionen in einem Modell zu vereinen, bietet beispielsweise der von *Stern Stewart & Co.* entwickelte Value Contribution Monitor™ (vgl. *Abb. 3* – Value Contribution Monitor™ ist ein Warenzeichen von *Stern Stewart & Co.*). In diesem umfassenden Werttreibersystem können die Entwicklung der Werttreiber sowie die Maßnahmenplanung gemeinsam gesteuert werden. Außerdem bietet er einen standardisierten Rahmen für die unternehmensinterne Kommunikation.

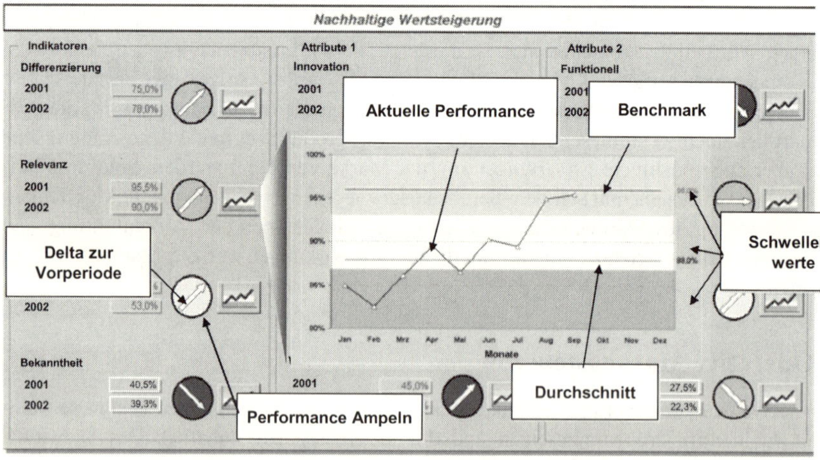

Abb. 3: Beispiel Value Contribution Monitor™

5. Zusammenfassung

Viele Markenkonzepte stellen die Berechnung eines Markenwertes in den Mittelpunkt wertorientierter Markenführung. Für die operative Anwendbarkeit ist jedoch weniger die absolute Höhe dieses Wertes gefragt als die Einbindung der Marke in ein wertorientiertes Managementsystem, welches die Ableitung von strategischen und operativen Handlungsempfehlungen erlaubt. Der Ansatz von *Stern Stewart & Co.* zeigt auf, wie die diesbezüglichen Anforderungen zu erfüllen sind.

Die drei Grundbausteine Marken-, Wettbewerbs- und Wertbeitragsanalyse erfüllen den Anspruch an Objektivität. Die Ermittlung der qualitativen Markendaten erfolgt aus empirischen Untersuchungen und die Quantifizierung geschieht auf Basis von Buchhaltungsdaten und Bewertungen des Kapitalmarktes. Transparenz entsteht durch die Dokumentation aller Inputparameter und Rechenschritte. Um die Vorausschau zu gewährleisten, werden auf der einen Seite im Hinblick auf die qualitativen Markenfaktoren die vorlaufenden Indikatoren Differenzierung und Relevanz abgefragt. Auf der anderen Seite werden mithilfe der Bestimmung des immateriellen Wertes Abschätzungen hinsichtlich zukünftig erwarteter Wertbeiträge in die Analyse aufgenommen. Durch Rücksichtnahme auf das investierte Geschäftsvermögen wird die Analyse auf eine wertorientierte Basis gesetzt.

Die Praktikabilität des Ansatzes kommt in den Anwendungsbereichen der Markenbewertung, operativen Markenführung und Portfolioanalyse zum Ausdruck. Bei der Markenbewertung werden die Bausteine zur Berechnung eines Markenwertes verknüpft. Mit der Portfolioanalyse werden Anhaltspunkte hinsichtlich des strategischen Handlungsbedarfes und strategischer Optionen identifiziert. In der operativen Markensteuerung führen Werttreiber zu einer praktischen Anwendbarkeit des Ansatzes in der Maßnahmensteuerung und Erfolgsmessung. Zusätzlich lassen sich die Ergebnisse nutzen, um den Einfluss und die Stellhebel der Marke innerhalb des Unternehmens zu kommunizieren.

Zusammengenommen ergibt sich daraus ein durchgängiger Ansatz, der die Wertorientierung in der Entscheidungsfindung, Maßnahmensteuerung und im Reporting der Markenführung etabliert.

6. Literatur

Boeing, Boeing Global Advertising Campaign Rolls out, News Release 1.05.2000, www.boeing.com/news/releases/2000/news_release_000050lc.html.

Esch, F.-R., Die Marke als Wertschöpfer, in: Frankfurter Allgemeine Zeitung Nr. 71, 25.03.2002, S. 25.

Hewlett Packard, The new HP is ready, Press release 7.05.2002, www.hp.com/hpinfo/newsroom/press/07may02a.htm.

Keller, K. L., Strategic brand management – building, measuring, and managing brand equity, Upper Saddle River 1998.

Meffert, H. und *Koers, M.*, Identidtitätsorientiertes Markencontrolling. Grundlagen und konzeptionelle Ausgestaltung, in: *Meffert, H.,Burmann, C.* und *Koers, M.* (Hrsg.), Markenmanagement. Grundfragen der identitätsorientierten Markenführung, Wiesbaden 2002, S. 403–428.

Mildenberger, M., Aktien bekannter Markenhersteller gehen gestärkt aus der Krise, in: Handelsblatt Nr. 57, 21.03.2002, S. 33.

Nöcker, R., Nur wenige Unternehmen kennen den Wert ihrer Marke, in: Frankfurter Allgemeine Zeitung Nr. 179, 5.08.2002, S. 17.

Stern, J. M., Shiely, J. S. und *Ross, I.*, Wertorientierte Unternehmensführung mit Economic Value Added. Strategie, Umsetzung, Praxisbeispiele, München 2002.

Young & Rubicam, What is the BrandAsset® Valuator? 2001, www.yr.com/knowledge/what.php.

Bedeutung, Bewertung und Steuerung von Investitionsgütermarken

Catharina Kriegbaum-Kling [1]

---

[1] Dr. *Catharina Kriegbaum-Kling*, TRUMPF GmbH + Co. KG, Ditzingen

Marken stellen häufig hohe immaterielle Werte dar. Zur Bewertung und Steuerung von Marken in der Konsumgüterindustrie wurden in den letzten Jahrzehnten zahlreiche Forschungsarbeiten und empirische Untersuchungen durchgeführt. Investitionsgütermarken erfahren erst in jüngster Zeit vermehrt Aufmerksamkeit, obwohl auch hier Wettbewerbsvorteile durch eine konsequente Markensteuerung erzielt und immaterielle Werte geschaffen werden können. In diesem Beitrag werden die Bedeutung von Investitionsgütermarken, die Unterschiede zu Konsumgütermarken sowie die Übertragbarkeit von Markenbewertungsmethoden auf Investitionsgüter untersucht und ein wichtiger Teil der Markensteuerung von Investitionsgütermarken, die Schaffung eines einheitlichen Markenauftritts, beispielhaft dargestellt.

1. Marken als immaterielle Vermögenswerte

1.1 Bedeutung von Marken für Unternehmen

Marken stellen für viele Unternehmen einen bedeutenden Werttreiber dar. In einer Befragung von deutschen Unternehmen gaben 53 % der Befragten an, dass Marken zu den wichtigsten Einflussgrößen auf den Unternehmenserfolg gehören (vgl. *PricewaterhouseCoopers/Sattler,* 1999, S. 11). Sie stellen folglich einen hohen immateriellen Wert für Unternehmen dar. Dieser äußert sich auf unterschiedliche Weise (vgl. *Srivastava/Shervani/Fahey,* 1998, S. 2ff.), unabhängig davon, für welche Güterart die Marke verwendet wird:

Viele Marken erlauben höhere Einnahmen durch Preisprämien gegenüber markenlosen Produkten und Lizenzgebühren sowie geringere Ausgaben (z. B. durch treue Kunden, die offener für eigene Marketingmaßnahmen sind). Die erhöhten Einnahmen und verringerten Ausgaben führen zu höheren Freien Cashflows.

Auch der Zeitpunkt des Eintreffens von Freien Cashflows kann durch Marken verändert werden, denn Marken können dazu führen, dass Freie Cashflows schneller erzielt werden. Dies wird beispielsweise mit der erhöhten Reaktionsbereitschaft der Kunden auf Marketingmaßnahmen sowie der leichteren Markteinführung von Neuprodukten erklärt.

Ist mit Marken eine höhere Kundenzufriedenheit und Kundentreue verbunden, kann mit höherer Sicherheit davon ausgegangen werden, dass auch in Zukunft Freie Cashflows erzielt werden, d. h. das Risiko der Freien Cashflows verringert sich. Sind die Kunden mit einer Marke zufrieden und entsprechend treu in ihrer Kaufentscheidung, führt dies dazu, dass sie für Marketingaktivitäten und Abwerbemaßnahmen der Konkurrenz weniger empfänglich sind. Dazu kommt, dass starke Marken besser in der Lage sind, Krisen zu überstehen, was ebenfalls das Risiko verringert. Beispielsweise machte im Jahr 1997 die A-Klasse der *Daimler-Benz AG* Schlagzeilen, als sie im so genannten „Elch-Test" einfach umkippte. Diese Krise konnte die starke Marke *Mercedes-Benz* aber bestens überstehen. Nach nur einem Jahr nach der Krise wurde die *Mercedes-Benz*-A-Klasse bereits wieder als „... auch von der Sicherheit her – anerkanntes Automobil in der Öffentlichkeit" bezeichnet (*Töpfer*, 1999, S. 20).

Neben dem Nachfrager- bzw. Kundenverhalten sind Marken auch angesichts anderer Interessengruppen des Unternehmens von Bedeutung. Marken können nicht nur zur Kreditsicherung verwendet werden, sondern auch generell die Kapitalbeschaffung erleichtern. Beim Rating der Banken zur Kreditvergabe können starke Marken ebenso bedeutsam sein wie bei Anlageentscheidungen von Investoren. Diese werden eher geneigt sein, in ein Unternehmen mit einer bekannten Marke zu investieren, denn wie bei potentiellen Kunden wirkt bei potentiellen Investoren die Bekanntheit von Marken als psychischer Entlastungsfaktor. Das heißt, dass angesichts der Angebotsvielfalt nicht alle Investitionsmöglichkeiten geprüft werden können und daher bekannte Dinge favorisiert werden. Marken werden dann für umso besser gehalten, je bekannter sie sind. Durch die Marke werden verdichtete Informationen vermittelt und die Wahrnehmung des gesamten Angebots beeinflusst (vgl. *Kroeber-Riel/Weinberg*, 1999, S. 291).

Eine starke Marke kann darüber hinaus die Suche nach potentiellen qualifizierten Mitarbeitern erleichtern und Suchkosten verringern. Eine Stellenanzeige wird eher wahrgenommen werden, wenn darin Bekanntes enthalten ist. Insofern können Marken auch im Wettbewerb um Arbeitskräfte von Bedeutung sein. Dies ist angesichts des Mangels an qualifizierten technischen Fachkräften gerade für die Investitionsgüterindustrie nicht zu unterschätzen. Daneben können sich auch die Mitarbeiter mit einer starken Marke identifizieren, was motivationsfördernd wirken und das Zusammengehörigkeitsgefühl steigern kann.

Die Bedeutung von Marken für Unternehmen ergibt sich also insbesondere durch ein besonderes Verhalten der Interessengruppen (Nachfrager, Kapitalgeber, Arbeitskräfte, Öffentlichkeit, Handel), das zu höheren Freien Cashflows beim Unternehmen führt. Den Barwert dieser durch die Marke verursachten Freien Cashflows nennt man Markenwert (vgl. *Abb. 1*).

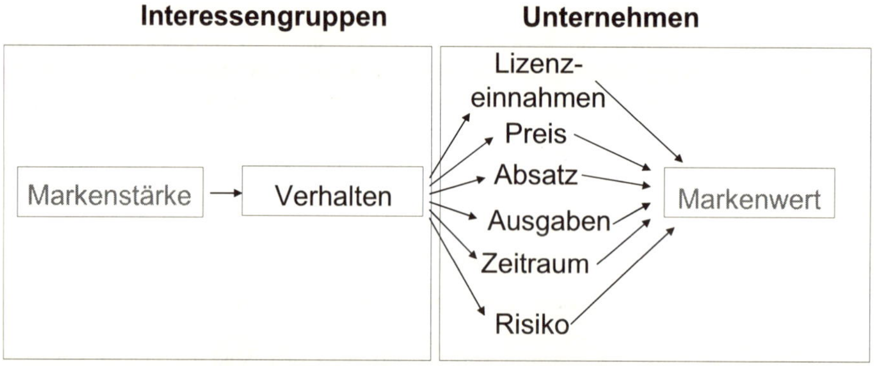

Abb. 1: Einflussfaktoren auf den Markenwert

Aufgrund der Homogenität zahlreicher Konsumgüter erfolgt die Produktdifferenzierung oft vollständig über Marken. Die Höhe des Marketingbudgets und die durchgeführten Maßnahmen zur Steigerung und zum Erhalt von Markenbekanntheit, Image und Produktqualität sind dann Hauptursachen für den Markterfolg und die Gewinnung von Marktanteilen (z. B. bei Backpulver, Vanillezucker). Entsprechend findet die Marke als immaterieller Wert im Konsumgüterbereich in Wissenschaft und Praxis bereits seit Jahrzehnten Berücksichtigung. In der Investitionsgüterindustrie wurde sie dagegen lange Zeit vernachlässigt.

Für langlebige Investitionsgüter wird, im Gegensatz zu Konsumgütern, häufig eine mindere Bedeutung von Marken eingeräumt, da deren Kauf mit einem höheren Anschaffungspreis sowie einer längeren Nutzungsdauer verbunden ist und deshalb die Anschaffung als Investition betrachtet wird, die mit einem umfassenden Informationsbeschaffungsprozess über die einzelnen Produktmerkmale einhergeht. Die Beschaffung von Investitionsgütern wird als rationaler Vorgang beschrieben, bei dem ausschließlich die Technik, die Leistung und der Preis analysiert und bewertet wird. Daher wird des öfteren die Ansicht vertreten, dass Marken nur in „irrational" oder „emotional" getroffenen Kaufentscheidungen eine Rolle spielen. Geht dagegen ein ausgedehnter Informationseinholungsprozess voraus, soll die Kenntnis der objektiven Produktmerkmale überwiegen (vgl. *Merbold*, 1995, S. 415).

Eine Studie von *PricewaterhouseCoopers/Sattler* im Jahr 1998 unterstützte diese These. Eine besonders hohe Wertschätzung von Markenwerten wurde bei Konsumgüterherstellern ermittelt. Des Weiteren wurde festgestellt, dass die befragten Unternehmen eine Zunahme des Markenwertes am stärksten im Dienstleistungsbereich, am wenigsten im Industriegütersektor erwarteten. Industriegüterhersteller hielten dabei auch eine adäquate Markenbewertung für am unbedeutendsten (vgl. *PricewaterhouseCoopers/Sattler*, 1999, S. 9). Allerdings waren Konsumgüterhersteller in der Studie überrepräsentiert (vgl. *PricewaterhouseCoopers/Sattler*, 1999, S. 17). Auch die Studie von *Günther/Kriegbaum* aus dem Jahr 1999 zeigte Unterschiede im Markenmanagement von Herstellern kurz- und langlebiger Wirtschafsgüter, beispielsweise bezüglich des Vorhandenseins eines markenverantwortlichen Bereichs (vgl. *Kriegbaum*, 2001, S. 253).

Doch immer mehr Investitionsgüterhersteller erkennen, dass auch in ihrer Branche Marken ein erhebliches Differenzierungspotential bieten. So hat sich beispielsweise der *Verband Deutscher Maschinen- und Anlagenbauer (VDMA)* in den letzten Jahren mit der Bedeutung von Marken in der Investitionsgüterindustrie beschäftigt. Eine vom *VDMA* im November 2001 durchgeführte Untersuchung bei 195 *VDMA*-Mitgliedsfirmen hat gezeigt, dass Markenpolitik bei Maschinen- und Anlagenbauern an Bedeutung gewinnt und die Markenorientierung in den letzten zehn Jahren deutlich angestiegen ist. Es wird außerdem erwartet, dass der Einfluss von Marken auf die Kaufentscheidung und damit auch auf den Unternehmenserfolg weiter steigt (vgl. *Behlke*, 2002, S. 22).

Gerade die hohe technische Komplexität von Investitionsgütern, sich ständig weiterentwickelnde Technologien und ein oftmals großes, unüberschaubares Angebot machen es für die Käufer oft schwer, objektive Alternativenvergleiche durchzuführen. In Anbetracht hoher Investitionssummen gewinnt dann gerade das Vertrauen in eine Marke und in das dahinterstehende Unternehmen an Bedeutung. Lange Nutzungsdauern verlangen außerdem hohe Ersatzteilverfügbarkeit und Servicebereitschaft, was eher von bekannten, traditionsreichen Marken erwartet wird. Schließlich dienen bekannte Marken auch dazu, dass sie überhaupt in den Informationseinholungsprozess der Kunden integriert werden. „Eine erfolgreiche Marke sorgt in jedem Geschäft für nachhaltige Sondergewinne, denn sehr viele Käufer sehen sie als Versprechen an, einen Wert von bestimmter Art und Höhe zu bekommen" (*Ward/Light/Goldstine,* 2000, S. 76).

Allerdings unterscheidet sich die Bedeutung von Marken bei Industriegütern je nach Informationseffizienz des Beschaffungsprozesses sowie nach der Bedeutung des Beschaffungsrisikos, wie eine *McKinsey*-Studie bei 560 Unternehmen unterschiedlicher Branchen gezeigt hat. Während die Bedeutung von Marken für Schaltanlagen und Werkzeugmaschinen als relativ hoch eingestuft wurde (Bewertung von 3,21 auf einer Skala von 0 bis 5), erreichten

Industriechemikalien nur einen geringen Wert der Markenbedeutung (Bewertung von 1,97) (vgl. *Caspar/Hecker/Sabel*, 2002, S. 45).

Insoweit ist es auch nicht für alle Investitionsgüter sinnvoll und möglich, einen immateriellen Wert durch eine konsequente Markensteuerung zu schaffen. In der ersten umfassenden Forschungsarbeit zur Markenpolitik im Investitionsgüterbereich von *Kemper* wurde deshalb ein Verfahren entwickelt, mit dem geprüft werden kann, ob im Einzelfall durch den Markenaufbau Vorteile für das Unternehmen erzielt werden können (vgl. hierzu *Kemper*, 2000, S. 139ff.). Die folgenden Ausführungen gehen von Gütern aus, bei denen grundsätzlich Markenbildungspotential vorhanden ist und insoweit durch Marken immaterielle Werte geschaffen werden können.

1.2 Notwendige Bestandteile

Nicht jedes markierte Produkt oder eingetragene Markenzeichen stellt für Unternehmen von vornherein einen immateriellen Wert dar. Erst wenn das Markenzeichen auch genutzt und mit ihm der Zukunftserfolgswert des Unternehmens, d. h. der Unternehmenswert, positiv beeinflusst werden kann, stellt die Marke einen Wert für das Unternehmen dar. Die Nutzung erfolgt aber immer nur in Verbindung mit einem Produkt oder einer Dienstleistung.

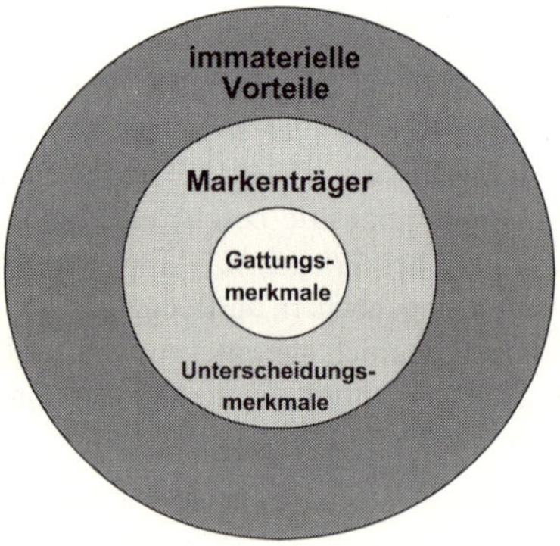

Abb. 2: Bestandteile einer Marke

Eine Marke mit immateriellem Wert hat demnach mehrere Bestandteile, die zu einer bestimmten Wahrnehmung der Marke führen und die ein bestimmtes Markenimage hervorrufen (vgl. *Abb. 2*) (vgl. *Kriegbaum*, 2001, S. 38f.):

- Den Markenträger, also die Ebene des Produktes, die die tatsächlichen, objektiv wahrnehmbaren, materialisierten Merkmale des Produktes umfasst (Produktkern). Derartige Produktmerkmale beschreiben lediglich die Produktgattung (sog. no-frills-Version, z. B. Blechschneidemaschine).
- Werden die Gattungsmerkmale durch spezifische Erkennungsmerkmale ergänzt (z. B. Herkunftsbezeichnungen, Herstellerangaben, verwendete Farben) kann der Markenträger erst zur Marke werden. Die Produktmerkmale dienen durch ihre Signalwirkung zur Unterscheidung von Wettbewerbsprodukten und stellen das Produktäußere dar. Hierzu gehört auch das Markenzeichen, das die Herkunft eines Produktes bzw. einer Dienstleistung anzeigt. Es besteht aus einem oder mehreren Wörtern, einem bestimmten Design,

Bild, Logo etc. und kann beim Patentamt registriert werden (z. B. blaue Blechschneidemaschine mit Firmen-Logo).

- Über die Unterscheidungsmerkmale hinaus kann eine Marke immaterielle Vorteile bringen, also Zusatznutzen durch subjektive Merkmale (z. B. Assoziationen mit High Tech, Zuverlässigkeit, Glaubwürdigkeit).

Demnach ist eine Leistung bzw. ein Produkt zwar die Basis einer Marke. Das Markenzeichen kann die Leistung bzw. das Produkt aber mit Assoziationen anreichern, so dass zwei identische Produkte durch unterschiedliche Markenzeichen von Grund auf verschiedene Images haben können (vgl. *Park/Jaworski/MacInnis,* 1986, S. 136).

„A product is something that offers functional benefit (e. g. ... a car). A brand name is a name, symbol, design, or mark that enhances the value of a product beyond its functional purpose. Depending on which perspective is considered, the brand can have added value to the firm..." (*Farquhar*, 1989, S. 24f.). Entsprechend dieser Sichtweise ist eine Marke immer mit einem Produkt bzw. Markenträger verbunden und kann oft auch nicht isoliert vom markenführenden Unternehmen betrachtet werden, da die Assoziationen mit der Marke häufig von anderen zum Unternehmen zugehörigen Charakteristika abhängen. Dies gilt besonders für Investitionsgüter, da diesen hohes technologisches Know-how zugrunde liegt.

1.3 Funktionen von Marken

Ähnlich zu Konsumgütermarken übernehmen Investitionsgütermarken bestimmte Funktionen, die zur Wertschaffung beitragen:

In erster Linie übernehmen Marken eine Unterscheidungsfunktion (auch Herkunfts- und Identifizierungsfunktion), die durch das Markenzeichen und die damit erreichte Herausstellung aus der anonymen Masse der Produkte erreicht wird. Dadurch kann das hinter der Marke stehende Produkt auch beworben werden (Werbefunktion). Tatsächlich ist das Markenzeichen das bedeutendste Identifikationsmerkmal eines Produktes. Durch die Werbung für die Marke kann gezielt ein Markenimage aufgebaut werden, das an sich homogene, gleichartige Produkte in der Wahrnehmung der Käufer unterschiedlich erscheinen lässt. Dadurch heben sich Marken gegenüber Konkurrenzmarken ab. Diese Funktion nennt man auch Produktprofilierungsfunktion.

Da das positive Image durch einen Markentransfer auch auf andere Produkte übertragen bzw. transferiert werden kann, lassen sich für neue Produkte leichter Kunden gewinnen und außerdem Verbundeffekte mit Produkten anderer Marken realisieren (Synergiebildungs- und Nachfragegewinnungsfunktion).

Da Marken außerdem weniger mit Nachfrageschwankungen zu kämpfen haben, wird ihnen auch eine Stabilisierungsfunktion zugesprochen. Dies ist insbesondere für Investitionsgüterbranchen zentral, die stark von konjunkturellen Zyklen beeinflusst werden.

Marken erfüllen des Weiteren eine Monopolisierungsfunktion, d. h. durch den rechtlichen Schutz der Marke müssen Wettbewerber gleichartige Produkte oder Dienstleistungen anders bezeichnen. Eine Marke besteht aber, wie oben ausgeführt, nicht nur aus dem Produktkern, sondern gerade auch aus spezifischen immateriellen Vorteilen und Assoziationen, die ein Produkt einmalig werden lassen können. So schaffen Marken zumindest ein „mentales" Patent für neuartige Produkte, falls nur der Name, nicht aber das zugrundeliegende Produkt bzw. die Technologie geschützt ist.

Wenn Produkte durch ein Markenzeichen gekennzeichnet sind, kommt dadurch auch zum Ausdruck, dass Produkte dieser Marke gleich in Qualität, Aufmachung etc. sind. Marken erfüllen daher in dieser Hinsicht auch eine Güte- und Garantiefunktion. Käufer dieser Marke können davon ausgehen, dass sie kein Risiko eingehen, sondern immer ein gleich gutes Produkt erhalten (vgl. *Kemper*, 2000, S. 98f.; *Kriegbaum*, 2001, S. 45f. und die dort angegebene Literatur).

2. Markenrelevante Besonderheiten von Investitionsgütern

Investitionsgüter sind Leistungen, die von Organisationen (Nicht-Konsumenten) beschafft werden (*Engelhardt/Günther*, 1981, S. 24). Die Käufer von Investitionsgütern sind keine privaten Einzelpersonen, die die gekauften Güter selbst verwenden, sondern Unternehmen, nicht-kommerzielle oder staatliche Organisationen. In den Beschaffungsprozess von Investitionsgütern sind häufig Beschaffungsteams aus mehreren Personen involviert, häufig aus unterschiedlichen Bereichen (z. B. Einkauf, Entwicklung, Controlling, Produktion). Dazu kommt, dass das Investitionsgut, z. B. eine Maschine, nicht nur zur Herstellung eines Produktes, sondern für die unterschiedlichsten Produkte verwendet werden kann. So kann beispielsweise eine Werkzeugmaschine zur Blechbearbeitung für die Herstellung von Designerblechkannen über Waschmaschinentrommeln bis hin zu Gehäusen von Mobiltelefonen verwendet werden. Insofern ist die Zielgruppe oft weniger leicht abgrenzbar und adressierbar als für Konsumgüter.

Oft handelt es sich außerdem nicht um Massengüter, sondern um maßgeschneiderte Konzepte, die zudem immer häufiger um Dienstleistungen (z. B. Schulungen, Beratung, Finanzierung) ergänzt werden. Die Kundenanzahl und die Komplexität der jeweiligen Investitionsgüter können sehr unterschiedlich sein, je nachdem, ob es sich um Produktgeschäft, Zulieferer-, System- oder Anlagengeschäft handelt (vgl. *Backhaus*, 2003, S. 324).

Eine weitere Besonderheit von Investitionsgütern ist deren abgeleitete Nachfrage, d. h. die Nachfrage ist abhängig von den Gütern, die auf den Investitionsgütern erstellt werden (z. B. hängt die Investition in eine Werkzeugmaschine von der wirtschaftlichen Entwicklung der Produkte ab, die auf dieser Werkzeugmaschine gefertigt werden). Insofern ist eine Bewerbung zur Nachfragegenerierung wie bei Konsumgütern wirkungslos.

Schließlich sind bei Investitionsgütern, anders als bei Konsumgütern, Käufer und Verkäufer meist in direkten Verhandlungen, in denen sie sich gegenseitig beeinflussen können. Dies bietet für den Investitionsgüterhersteller aber nicht nur Vorteile. Während anonyme Werbekampagnen von Konsumgüterherstellern zentral gesteuert und koordiniert werden können, beeinflusst bei jeder Transaktion zwischen Käufer und Verkäufer jeder einzelne Vertriebs- und Servicemitarbeiter das Image der Marke beim Kunden. Umso wichtiger ist es, dass die verfolgte Markenstrategie von allen Mitarbeitern verstanden und gelebt wird.

Auf der anderen Seite bietet dies auch andere Möglichkeiten der Unternehmens- und Angebotsdarstellung gegenüber den Kunden. Während bei Konsumgütern die zentrale Form der Kommunikation über Medienwerbung und über Zwischenhändler stattfindet, besteht bei Investitionsgütern im Entscheidungsprozess durch intensive Beratungen in der Geschäftsanbahnung, in der Verhandlungsphase bis hin zur Wartung und Gewährleistung nach dem Kauf ein intensiver Kundenkontakt. Damit gewinnt für Investitionsgüter die Darstellung des Unternehmens als Ganzes, nicht nur der Marke, an Bedeutung. Insofern werden weniger einzelne Produktmarken, sondern Firmen-, Dach- und Herstellermarken markenpolitisch eingesetzt und das Unternehmen als Ganzes als Marke positioniert.

Die bereits erwähnte *VDMA*-Studie hat gezeigt, dass 58,8 % der befragten Unternehmen eine gemischte Dach- und Produktmarkenstrategie und 26,1 % eine reine Dach- bzw. Unternehmensmarkenstrategie verfolgen (vgl. *VDMA*, 2001). Dach- und Unternehmensmarkenstrategien, die mehrere oder alle Produkte unter sich vereinen, bieten den Vorteil, dass sich alle Marketingaktivitäten auf eine Marke konzentrieren lassen, um eine Markenpersönlichkeit aufzubauen und die Bekanntheit dieser Marke insbesondere auch bei Nicht-Kunden, d. h. in der Öffentlichkeit, bei Banken, bei potentiellen Arbeitnehmern, erhöht werden kann. *Backhaus* weist jedoch darauf hin, dass die Vor- und Nachteile bei der Wahl des Markentyps im Einzelfall zu beurteilen sind (vgl. *Backhaus*, 2003, S. 419).

Wenn eine Dach- und Unternehmensmarkenstrategie verfolgt wird, steht das Unternehmen als Ganzes und damit das gesamte Firmenimage im Fokus der Markenpolitik. Das bedeutet aber im Gegensatz zu Einzelmarken, dass sich die Marke kaum vom Unternehmen trennen lässt. Die Marke hat zwar einen bedeutenden Einfluss auf den immateriellen Wert des Unternehmens, hängt aber eng mit anderen, auch immateriellen Faktoren des Unternehmens zusammen. Zielt die Positionierung einer Marke beispielsweise auf Spitzentechnologie, Spitzenqualität und -service ab, hängen damit hohe Anstrengungen in allen Unternehmensbereichen zusammen, die das Unternehmen zum Technologie- und Qualitätsführer machen und die unmittelbar auf den immateriellen Wert der Marke wirken. Bei manchen Konsumgütern reicht es dagegen unter Umständen aus, eine starke Marketingabteilung zu haben.

3. Bewertung und Steuerung von Investitionsgütermarken

3.1 Markenstärke und Markenwert als Zielgrößen der Markensteuerung

Traditionelle Performance-Maße zur Steuerung von Marken (z. B. Markenumsatz, Markengewinn) stehen im Widerspruch zum notwendigen langfristig angelegten Markenaufbau und -erhalt, da sie kurzfristig orientiert sind und den Markenverantwortlichen keinen Anreiz für Investitionen in die Marke liefern. Aktivitäten zur Stärkung von Marken kosten Geld und können kurzfristig den Gewinn mindern. Beispielsweise mindern Aktivitäten zur Qualitätssteigerung oder Forschungs- und Entwicklungsaufwendungen das Periodenergebnis und führen erst in späteren Perioden zu höheren Rückflüssen. Ebenso lässt sich der Wert von Öffentlichkeits- oder Marketingabteilungen kurzfristig nur schwer nachweisen.

Aufgrund des hohen immateriellen Potentials von Marken darf deren Steuerung aber nicht nur kurzfristig und nebenbei erfolgen. Vielmehr muss „... die langfristige Markenpflege, der Aufbau neuer Marken und kreativer Marken-Strategien in den Fokus des Tagesgeschäfts treten ..." sowie „... sichergestellt sein, daß alle Handlungen des Unternehmens im Einklang mit der Markenbotschaft stehen." (vgl. *Franzen*, 1995, S. 57) Daher ist eine konsequente Markenpolitik zu verfolgen, die bei Dach- bzw. Unternehmensmarken dazu dient, ein positives Firmenimage aufzubauen und das Unternehmen einheitlich und glaubwürdig darzustellen und bekannt zu machen.

Ein umfassendes Steuerungssystem von Marken sollte sowohl monetäre als auch nicht-monetäre Zielgrößen berücksichtigen. Nicht-monetären Indikatoren der Markenstärke (wie z. B. dem Bekanntheitsgrad oder dem Markenimage) wird ein Vorsteuerungs- und Frühwarncharakter zugesprochen, da sie Hinweise auf zukünftige Entwicklungen geben können, für die monetäre bzw. finanzielle Informationen in der Regel zu spät kommen. Die Markenstärke (auch Markenkraft, Markenvitalität, Brand Strength, Brand Power genannt) ist die Voraussetzung für den Markenwert. Sie repräsentiert die gegenwärtigen und zukünftigen Stärken einer Marke im Umfeld des Unternehmens, das die Marke nutzt (vgl. *Srivastva/Shocker*, 1991, S. 6). In der Literatur existiert eine Vielzahl an unterschiedlichen Definitionen zur Markenstärke, die sich hauptsächlich danach unterscheiden, aus welcher Sicht die Stärke der Marke beschrieben wird und wie die Markenstärke operationalisiert werden kann (vgl. *Bekmeier-Feuerhahn*, 1998, S. 36; *Irmscher*, 1996, S. 59). Die Ansätze haben aber gemeinsam, dass sie zur Messung der Markenstärke für Konsumgüter konzipiert wurden.

Monetäre Zielgrößen sind wichtig, da sonst materielle und immaterielle Vermögenswerte unterschiedlich bewertet und folglich nicht vergleichbar sind. Da der Aufbau von Marken wie jedes Investitionsprojekt mit anderen materiellen und immateriellen Investitionsprojekten um knappe Finanzmittel konkurriert, ist es hilfreich, wenn die Investitionsprojekte verglichen werden können. Auch Marken sollten daher einer Rentabilitätsbeurteilung unterzogen werden, was einer monetären Bewertung der Rückflüsse bedarf. Folglich unterstützt ein monetärer Markenwert die Einbindung der Steuerung von Marken in die Unternehmensführung, da eine gemeinsame Kommunikations- und Entscheidungsgrundlage für Marketing und Finanzwesen geschaffen wird. Der Markenwert repräsentiert das monetär quantifizierte Erfolgspotential einer Marke und ist das finanzielle Ergebnis der Markenstärke (vgl. *Srvastava/Shocker,* 1991, S. 9).

3.2 Verfahren zur Markenbewertung

Sowohl in der Theorie als auch in der Praxis herrscht bisher keine Einigkeit darüber, wie das Erfolgspotential von Marken am besten gemessen werden kann, obwohl bereits einige Forschungsanstrengungen zur Bewertung von Konsumgütermarken unternommen worden sind. Je nach gewählter Methode unterscheiden sich die ermittelten Markenwerte aber sehr, was dem Vertrauen in den Markenwert schadet und der Implementierung in Unternehmen entgegensteht.

So ist es nicht verwunderlich, dass nur 37,2 % von 129 befragten führenden deutschen Markenartikelunternehmen überhaupt eine Markenbewertung durchführen. Nur 17 % führen dabei eine monetäre Bewertung durch. 36,7 % der Unternehmen bemängeln, dass es keine geeignete Markenbewertungsmethode gäbe und weitere 20,3 % kritisieren den hohen Zeit- und Kostenaufwand (vgl. die Studie von *Kriegbaum*, 2001, S. 147ff.).

Die *VDMA*-Studie zur Bedeutung der Markenpolitik in der Investitionsgüterindustrie hat gezeigt, dass noch weniger, nämlich nur ein Bruchteil der Maschinen- und Anlagenbauer eine Markenbewertung durchführt und insgesamt nur unklare Vorstellungen zur Markenbewertung vorliegen. Von 195 teilnehmenden Unternehmen gaben nur 30 an, eine Markenbewertung durchzuführen, wobei unter Markenbewertung überwiegend „allgemeine Marktforschung" verstanden wird (vgl. *Behlke*, 2002, S. 22; *VDMA*, 2001).

Die zahlreichen in der Literatur vorherrschenden Markenbewertungsansätze (zum Überblick vgl. *Kriegbaum*, 2001, S. 89ff.) konnten sich demnach noch nicht in der Praxis, schon gar nicht in der Investitionsgüterindustrie, durchsetzen. Zudem haben die Ansätze zur Abbildung des Erfolgspotentials von Marken Schwachstellen.

Markenbewertung auf Basis von Markenkosten

In den kostenorientierten Ansätzen werden Marken anhand ihrer Anschaffungs- bzw. Herstellungskosten bewertet. Unklar ist, welche Kosten als Markenkosten zu klassifizieren und der Marke eindeutig zurechenbar sind. Während sich die Markenkosten für Konsumgüter anhand der Marketingkosten noch relativ leicht annähern lassen, rücken bei Investitionsgütern häufig produktorientierte Kosten, insbesondere für Forschung und Entwicklung, in den Vordergrund und verleihen der Marke ihren Wert. Eine objektive, allgemeingültige Abgrenzung von Markenkosten ist nicht möglich.

Zudem ist unklar, über welchen Zeitraum die Kosten erfasst werden sollen und wie Kosten für Marken, die seit Jahrzehnten bestehen, überhaupt noch ermittelt werden können (vgl. zur Kritik *Stobart,* 1989, S. 25f.; *Kapferer*, 1992, S. 300ff.).

Der mit dieser Methode ermittelte Markenwert ist außerdem vergangenheitsorientiert, während das Erfolgspotential einer Marke und damit ihr Wert zukunftsgerichtet ist. Damit wird nur der Input in die Marke betrachtet, nicht aber der Output bzw. die erzielten markenwertsteigernden Wirkungen. Dies kann falsche Steuerungsimpulse geben, da der Markenwert steigt, je mehr Kosten für die Marke anfallen, ohne dass überprüft wird, ob die Aktivitäten auch tatsächlich das zukünftige Erfolgspotential der Marke positiv beeinflussen. Das Verfahren ist nur unter der Annahme sinnvoll anwendbar, dass Marken, die hohe Kosten verursacht haben, auch einen hohen Wert darstellen. Gegen diese Annahme sprechen aber zahlreiche erfolglose Marken, die trotz hoher Investitionen keine Wertsteigerung erfuhren.

Darüber hinaus werden mit dieser Methode solche Marken unterschätzt, für die relativ wenig aufgewendet wurde, die aber trotzdem in der Lage sind, zukünftige Erfolgsbeiträge zu liefern. Aufgrund der genannten Probleme scheidet die Markenbewertung auf Basis von Markenkosten für die Markensteuerung aus.

Markenbewertung mit Hilfe von Preisprämien

Bei der Markenbewertung mit Hilfe von Preisprämien wird der Tatsache Rechnung getragen, dass mit Marken in der Regel höhere Preise als für markenlose Produkte erzielt werden. Dementsprechend basieren einige Markenbewertungsmethoden auf der durch Marken erzielbaren Preisprämien bzw. der Zusatzpreisbereitschaft für eine Marke im Vergleich zu einem markenlosen Produkt (vgl. z.B. *Aaker*, 1992, S. 39). Diese Preisprämien können entweder am Markt beobachtet oder auch durch Kundenbefragungen ermittelt werden (vgl. hierzu *Kriegbaum*, 2001, S. 187ff.). Wie bereits erwähnt, sind die Kunden von Investitionsgütern aber sehr heterogen und bestehen meist nicht aus Einzelpersonen, sondern aus Beschaffungsteams. Insofern ist eine Befragung einzelner Kunden nicht wie bei Konsumgütern durchführbar.

Wird die Preisprämie durch die Beobachtung am Markt und nicht durch die Abfrage von Zahlungsbereitschaften von potentiellen Kunden ermittelt, können sich Preisunterschiede durch unterschiedliche Kostenstrukturen oder preispolitische Zielsetzungen der Unternehmen ergeben, so dass hohe Preise nicht unbedingt auf Markenstärke schließen lassen. Abgesehen davon ist die Marktpreisbeobachtung problematisch, wenn keine vergleichbare markenlose Produktvariante existiert. Bei Investitionsgütern fehlt oft ein solcher Vergleich, wenn es sich um Auftrags- bzw. Sonderanfertigungen oder unterschiedliche Güter bezüglich technischer Daten, verwendeter Komponenten, unterschiedlicher Zusatzausstattungen (z.B. auch bezüglich ergänzender Dienstleistungsangebote), anderer Qualität usw. handelt.

Darüber hinaus werden mit der alleinigen Preisprämienbetrachtung diejenigen Marken unterbewertet, die zwar kaum teurer als markenlose Produkte sind, dafür aber einen höheren Absatz als markenlose Produkte erzielen können. Deshalb ist der markenbezogene Umsatz zu ermitteln. Allerdings lässt sich der Absatz von Investitionsgütern aufgrund der abgeleiteten Nachfrage nicht analog zur Absatzschätzung von Konsumgütern ermitteln.

Durch Abzug der markenbezogenen Kosten erhält man den Ergebnisbeitrag der Marke. Allerdings stellt sich auch hier das Problem der Abgrenzung von Markenkosten. Die Ergebnisbeitragsmethode auf Basis von Preisprämien ist zudem statisch, d.h. periodenbezogen und berücksichtigt keine zukünftig erwarteten Gewinne. Damit werden gegenwärtige Erfolge, nicht aber das Erfolgspotential der Marke abgebildet. Zur Ermittlung des zukünftigen Erfolgspotentials ist der Markenwert entsprechend kapitaltheoretischer Grundsätze und in Analogie zum Unternehmenswert als Zukunftserfolgswert aufzufassen.

Kapitalmarktorientierte Markenbewertung

Eine weitere Möglichkeit zur Ermittlung des Markenwertes stellt die Betrachtung des Unternehmenswertes am Kapitalmarkt und die Ableitung des Markenwertes aus diesem Wert dar. Unter der Annahme, dass Kapitalmärkte effizient sind, also alle verfügbaren Informationen im Aktienkurs verarbeitet wurden, müssen sich auch Veränderungen des Markenwertes durch bestimmte Marketingmaßnahmen im Aktienkurs niederschlagen (vgl. die Studie von *Simon/Sullivan,* 1992). Die Methode erlaubt es, den Markenwert von anderen Vermögensgütern zu trennen und schätzt den Wert von Marken aus Sicht des Kapitalmarktes. Allerdings ist ihre Anwendbarkeit auf börsennotierte Kapitalgesellschaften beschränkt.

Außerdem wirken nur große, bedeutende Ereignisse auf den Aktienkurs, wobei kleinere Veränderungen des Markenwertes nicht ermittelt werden können. Des Weiteren kann nur der Wert aller Marken eines Unternehmens, nicht aber der Markenwert einzelner Marken ermittelt werden. Für Dachmarken stellt dies aber kein Problem dar. Hervorzuheben ist allerdings die Abhängigkeit des Markenwertes von der allgemeinen Börsenentwicklung, die, wie die Entwicklung der letzten Jahre gezeigt hat, stark von externen Faktoren beeinflusst ist. Für die Steuerung von Marken ist die Methode daher ungeeignet.

Markenbewertung mit Markenstärkeindikatoren

Die indikatororientierten Markenbewertungsmodelle (vgl. hierzu den Überblick bei *Kriegbaum*, 2001, S. 118ff.) ermitteln in einem ersten Schritt die nicht-monetäre Markenstärke anhand von bestimmten Indikatoren, die dann im zweiten Schritt in einen monetären Wert transformiert wird. Die verwendeten Indikatoren basieren auf Erfahrungen bzw. Untersuchungen in unterschiedlichen Branchen und beinhalten keine speziell für Investitionsgüter relevanten Indikatoren. Zwar besitzen die z. B. die im Verfahren von *Interbrand* verwendeten Indikatorkategorien wie Marktführerschaft, Stabilität, Markt, Internationalität, Trend der Marke, Marketingunterstützung und rechtlicher Schutz (vgl. *Birkin*, 1994, S. 215) auch für Investitionsgütermarken Relevanz; für Investitionsgüter aber besonders wichtige Erfolgsfaktoren, z. B. Technologie- und Innovationsführerschaft, Ersatzteil-, Serviceverfügbarkeit etc. fehlen. Stattdessen werden Indikatoren verwendet, die für Investitionsgüter häufig weniger Relevanz besitzen (z. B. Handelsdurchsetzung).

Markenbewertung auf Basis von Lizenzeinnahmen

Schließlich besteht eine Möglichkeit der Markenwertbestimmung darin, den Markenwert aus den für die Nutzung der Marke erzielbaren Lizenzeinnahmen abzuleiten. Allerdings bildet die Lizenzeinnahmenmethode nur einen Teil der Einnahmenseite ab. Unberücksichtigt bleiben Einnahmen durch die eigene Nutzung der Marke sowie die komplette Ausgabenseite. Die Methode basiert zudem auf hypothetischen Annahmen zur Lizenzbestimmung, wenn nicht tatsächlich auch Lizenzen vergeben werden.

Die kurze Darstellung der Methoden mit den wichtigsten Kritikpunkten zeigt, dass zwar einige Forschungsanstrengungen zur Ermittlung von Markenwerten in den letzten Jahren unternommen wurden. Ein für Investitionsgütermarken geeignetes Verfahren zur Ermittlung monetärer Markenwerte wurde bisher aber noch nicht entwickelt. Dies bedeutet aber nicht, dass somit keine Steuerung von Marken im Investitionsgüterbereich möglich ist. Vielmehr ist der Schwerpunkt auf ein durchgängiges Marketingkonzept für Investitionsgüter zu setzen, welches in einem ersten Schritt qualitativ zu formulieren ist und auf der Markenstärke als Zielgröße der Markensteuerung aufbaut.

3.3 Bekanntheit als Ausgangspunkt der Markensteuerung von Investitionsgütermarken

Ein zentraler Faktor der Markenstärke ist grundsätzlich die Bekanntheit der Marke, da erst durch sie die Marke im Kaufentscheidungsprozess berücksichtigt wird. Außerdem können nur mit einer bekannten Marke spezifische Vorstellungen verbunden sein, die aber Bestandteil der Markenstärke sind (vgl. z.B. *Srivastava/Shocker*, 1991, S. 14f.). Die Bedeutung der Bekanntheit ergibt sich in erster Linie daraus, dass die Wahrscheinlichkeit der Aufnahme der Marke in das so genannte Consideration Set (d.h. der Marken, die überhaupt in Frage kommen) sowie in das so genannte Evoked Set (d.h. der Marken, die in die engere Wahl kommen) mit erhöhter Bekanntheit steigt. Bekanntheit führt folglich zu einem höheren Absatz und damit zu einem höheren Marktanteil. Dieser Zusammenhang wurde schon vor vielen Jahren vom *Strategic Planning Institute* mit Daten der *PIMS*-Datenbank nachgewiesen (vgl. *Burke/Newman,* 1980, S. 5).

Entsprechend kann bereits mit kommunikationspolitischen Maßnahmen zur Steigerung der Bekanntheit ein Beitrag zur Schaffung von Markenwerten geleistet werden. Gerade in technologieorientierten, technikergeprägten Unternehmen wird der Markenwert, der durch kommunikationspolitische Maßnahmen erzielt werden kann, leicht unterschätzt. Dabei geht es nicht um breit angelegte Werbeaktivitäten, sondern vielmehr um eine durchgängige Positionierung des Unternehmens als Ganzes, indem sich das Unternehmen nach außen darstellt, ein positives Firmenimage erzeugt und die Bekanntheit gesteigert wird.

Die Corporate Identity ist das individuelle und unverwechselbare Denken und Handeln eines Unternehmens als abgestimmtes Ganzes und umfasst die Gestaltung des Unternehmenserscheinungsbildes (Corporate Design), die Unternehmenskommunikation (Corporate Communication) und das Unternehmensverhalten (Corporate Behavior) (vgl. *Renner*, 1991, S. 31f.). Ein stringentes Corporate Identity-Konzept ist folglich die Basis für die Steuerung von Dachmarken zur Steigerung der Bekanntheit und zur Schaffung von Markenwerten.

3.4 Markensteuerung am Beispiel der *TRUMPF*-Gruppe

Ein Beispiel für ein durchgängiges, konsequentes Corporate Identity Konzept im Investitionsgüterbereich liefert die *TRUMPF*-Gruppe. Die *TRUMPF* Gruppe zählt mit einem Umsatz von 1,2 Mrd. EUR im Geschäftsjahr 2002/2003 und rund 5.800 Mitarbeitern zu den weltweit führenden Unternehmen der Fertigungstechnik. Unter dem Dach einer Holding sind die vier Geschäftsbereiche – Werkzeugmaschinen, Lasertechnik, Elektronik/Medizintechnik und Elektrowerkzeuge – zusammengefasst. Kerngeschäft ist die flexible Blechbearbeitung zum Stanzen und Umformen, für die Laserbearbeitung und zum Biegen. Im Bereich industrieller Laser und Lasersysteme ist das Unternehmen Technologie- und Weltmarktführer. Mit 40 Tochtergesellschaften und Niederlassungen ist die Gruppe in fast allen europäischen Ländern, in Nord- und Südamerika sowie in Asien vertreten.

Die *TRUMPF*-Gruppe verfolgt konsequent eine Dachmarkenstrategie für alle Geschäftsbereiche. Nahezu alle in- und ausländischen *TRUMPF*-Gesellschaften im Werkzeugmaschinen- und Lasertechnikbereich haben im Unternehmensnamen sowie in den Produktmarken einen direkten Bezug zur Dachmarke (z. B. *TRUMPF* Lasercell, TRUMATIC). Markensteuerung heißt für *TRUMPF*, das ganze Unternehmen als Marke zu positionieren, nicht einzelne Produkte. Zu den markenpolitischen Maßnahmen gehören daher alle Aktivitäten, welche die Bekanntheit des Unternehmens steigern, Assoziationen mit der Marke *TRUMPF* fördern und die ganze Unternehmenspersönlichkeit bei Kunden und in der Öffentlichkeit durch den Aufbau eines positiven Firmenimages formen.

Das Firmenimage wird im Rahmen des Corporate Identity Konzeptes u. a. durch das Schaffen von visueller Identität in der *TRUMPF*-Gruppe (Corporate Design) beeinflusst. Zu den Gestaltungselementen gehört besonders das Logo, mit dem *TRUMPF* wahrgenommen wird. Bis zum Jahr 1984 verwendete *TRUMPF* ein Logo, das auf die Parallelität des Begriffs „*TRUMPF*“ zum Kartenspiel hinwies. Die Assoziation von *TRUMPF* mit einem Karten- oder Glückspiel passte jedoch nicht mehr zum Image eines Unternehmens, das moderne, zuverlässige Technologie anbietet. Das Logo wurde deshalb von einem sachlichen, einprägsamen, fast flaggenähnlichem und vor allem zeitlosen Signet mit einfacher Gesamtform abgelöst. Die Logo-Farben blau, schwarz und weiß sollen klar, kühl und technisch wirken. Die Einheitlichkeit der Farben, die auch bei den *TRUMPF*-Maschinen angewendet werden, haben dabei einen hohen Wiedererkennungswert.

Auch die Firmengebäude leisten ihren Beitrag zum Firmenimage, wenn sie mehr als eine Hülle für die Produktion sind und auf ihre Weise Assoziationen mit der Marke entstehen lassen. Gruppenweit wird außerdem auf einen dem Corporate Design entsprechenden Messeauftritt geachtet.

Auch das Produktdesign trägt durch wiederkehrende Formelemente und einheitliche Farbgestaltung zur Wiedererkennung der Marke bei. Bei *TRUMPF* ist das Produktdesign fester Bestandteil der Produktentwicklung. Neben der Vermittlung von Gebrauchsqualität dient das Produktdesign vor allem auch zur Differenzierung von technisch ähnlichen Produkten der Wettbewerber. Besonderer Wert wird auf ein modernes, aber nicht modisches Design gelegt. Das Design steht in direktem Zusammenhang mit der Innovationskraft von *TRUMPF*. Die *TRUMPF*-Gruppe steht weltweit für Innovation in der Fertigungstechnik. Dahinter stehen ständige Forschungs- und Entwicklungsanstrengungen, die die Assoziation der Marke *TRUMPF* mit Innovation und High Tech fördern. Insofern sind Forschungs- und Entwicklungs-Aufwendungen auch Aufwendungen für die Marke.

Als Investitionsgüterhersteller ist *TRUMPF* ist bei Nicht-Anwendern häufig weniger bekannt, da sich diese unter den Maschinen weniger vorstellen können. Eine Möglichkeit zur Steigerung der Bekanntheit ist es, eine Brücke zwischen den Maschinen und den Produkten zu schaffen, die auf den Maschinen gefertigt werden und bei einer breiteren Öffentlichkeit bekannt sind. *TRUMPF* hat hierfür in firmeneigenen Räumen, die Ausstellung „HorizonTe“ eingerichtet, mit der die Besucher in die Welt der Blechbearbeitung eingeführt werden und erleben, wo ihnen im täglichen Leben *TRUMPF* begegnet. Dadurch werden Assoziationen mit Produkten des Alltags und teilweise auch zu anderen bekannten Marken geschaffen und die Markenbekanntheit gesteigert.

4. Fazit

Auch in der Investitionsgüterindustrie können durch gezielte Markenpolitik immaterielle Markenwerte geschaffen werden. Der monetäre Markenwert als Zielgröße der Markenpolitik erfordert aber speziell für Investitionsgüter noch einige Forschungsanstrengungen, da sich vorhandene Markenbewertungsmethoden für Konsumgütermarken nicht ohne weiteres auf Investitionsgüter übertragen lassen. Der Fokus der Markensteuerung von Investitionsgütern muss daher im ersten Schritt auf qualitativen Zielgrößen liegen. Der Anfang kann, aufgrund der Besonderheiten von Investitionsgütern und der häufigen Verfolgung von Dachmarkenstrategien, mit einem durchgängigen Corporate Identity Konzept gemacht werden. Ein einheitlicher, durchgängiger Markenauftritt erhöht die Wiedererkennung, steigert die Bekanntheit und schafft damit, oft schon mit wenigen Mitteln, einen immateriellen Wert und damit einen Beitrag zum Unternehmenserfolg.

5. Literatur

Aaker, D. A., Management des Markenwertes, Frankfurt a. M./New York 1992.

Backhaus, K., Industriegütermarketing, 7. Auflage, München 2003.

Behlke, J., Die Bedeutung der Marke in der Investitionsgüterindustrie, in: VDMA Nachrichten, o. Jg. (2002), H. 9, S. 21-22.

Bekmeier-Feuerhahn, S., Marktorientierte Markenbewertung. Eine konsumenten- und unternehmensbezogene Betrachtung, Wiesbaden 1998.

Birkin, M., Assessing Brand Value, in: *Stobart, P.* (Hrsg.): Brand Power, London 1994.

Burke, W. L. and *Newman, R. G.*, Brand Awareness and Profitability, in: *The Strategic Planning Institute* (Hrsg.): PIMSLetter, Nr. 24, Cambridge, Mass. 1980.

Caspar, M. und *Hecker, A.* und *Sabel, T.*, Markenrelevanz in der Unternehmensführung – Messung, Erklärung und empirische Befunde für B2B-Märkte, in: *Backhaus, K., Meffert, H.*, u. a. (Hrsg.), MCM/McKinsey-Reihe zur Markenpolitik, Arbeitspapier Nr. 4, Münster 2002.

Engelhardt, W. H. und *Günther, B.*, Investitionsgütermarketing; Anlagen; Einzelaggregate; Teile; Roh- und Einsatzstoffe; Energieträger, Stuttgart 1981.

Farquhar, P. H., Managing Brand Equity, in: Marketing Research, 1. Jg. (1989), H. 3, S. 24-33.

Franzen, O., Marken-Controlling effizient gestalten, in: Markenartikel, 57. Jg. (1995), H. 2, S. 57-61.

Franzen, O., *Trommsdorff, V.* und *Riedel, F.*, Ansätze der Markenbewertung und Markenbilanz, in: *Bruhn, M.* (Hrsg.): Handbuch Markenartikel, Band II: Markentechnik, Markenintegration, Markenkontrolle, Stuttgart 1994, S. 1373-1401.

Irmscher, M., Markenwertbegriffe, in: Markenartikel, 58. Jg. (1996), H. 2, S. 58-61.

Kapferer, J.-N., Die Marke - Kapital des Unternehmens, Landsberg/Lech/Köln 1992.

Kemper, A. C., Strategische Markenpolitik im Investitionsgüterbereich, Lohmar/Köln 2000.

Kriegbaum, C., Markencontrolling, München 2001.

Kroeber-Riel, W. und *Weinberg, P.*, Konsumentenverhalten, 7. Auflage, München 1999.

Merbold, C., Die Investitionsgütermarke, in: Markenartikel, 57. Jg. (1995), H. 9, S. 414-417.

Park, C. W., Jaworski, B. J. and *McInnis, D. J.*, Strategic brand concept-image management, in: Journal of Marketing, 50. Jg. (1986), H. 5, S. 135-145.

PricewaterhouseCoopers und *Sattler, H.*, Praxis von Markenbewertung und Markenmanagement in deutschen Unternehmen, Frankfurt a. M. 1999.

Renner, S. G., Corporate Identity, Gestaltungsformen, Informationslogistik, Unternehmensauftritt, Würzburg, 1991.

Srivastava, R. K., Shervani, T. A. and *Fahey, L.*, Market-Based Assets and Shareholder Value: A Framework for Analysis, in: Journal of Marketing, 62. Jg. (1998), H.1, S. 2-18.

Srivastava, R. K. and *Shocker, A. D.*, Brand Equity: A Perspective on Its Meaning and Measurement, Technical Working Paper, Marketing Science Institute, Report Nr. 91 – 124, Cambridge/Massachusetts 1991.

Stobart, P., Alternative Methods of Brand Valuation, in: *Murphy, J.* (Hrsg.): Brand Valuation – Establishing a True and Fair View, London 1989, S. 23-31.

Töpfer, A., Die A-Klasse, Neuwied 1999.

VDMA (Hrsg.), Bedeutung der Markenpolitik in der Investitionsgüterindustrie, Ergebnisse einer Befragung des VDMA, Oktober/November 2001.

Ward, S., Light, L. und *Goldstine, J.*, Warum Hightech-Firmen ihre Produkte zu Marken entwickeln sollten, in: Harvard Business Manager, 22. Jg. (2000), H. 1, S. 75-85.

Management von intellektuellem Vermögen in kooperativen Unternehmensnetzwerken

Klaus Valentin[1], *Bernhard Schmiedinger*[2], *Andreas Markom*[3], *Susanne Schunder*[4], *Martin Nohava*[5], *Curt Stoll*[6], *Ulrich Stolz*[7] *und Andreas Nilsson*[8]

---

[1] *Klaus Valentin*, Profactor Produktionsforschungs GmbH, Steyr, Österreich

[2] *Bernhard Schmiedinger*, Profactor Produktionsforschungs GmbH Steyr, Österreich

[3] Dr. *Andreas Markom*, Österreichische Akademie für Arbeitsmedizin, Klosterneuburg, Österreich

[4] Dr. *Susanne Schunder*, Österreichische Akademie für Arbeitsmedizin, Klosterneuburg, Österreich

[5] *Martin Nohava*, Österreichische Zertifizierungs- und Begutachtungs-GmbH, Wien, Österreich

[6] *Curt Stoll,* KACU systemic GmbH, Auenwald, Deutschland

[7] Dr. *Ulrich Stoll*, KACU systemic GmbH, Auenwald, Deutschland

[8] *Andreas Nilsson*, Institut für Informatik, Universität Göteborg, Schweden

Die geänderten wirtschaftlichen Rahmenbedingungen führen immer häufiger zu einem Zusammenschluss von Organisationen in Form von kooperativen Unternehmensnetzwerken. Für den nachhaltigen Erfolg dieser Netzwerke ist die gezielte Bündelung der intellektuellen Vermögensteile von zentraler Bedeutung. Aus diesem Grund ist es erforderlich, dass ein Managementsystem im Netzwerk implementiert wird, welches eine kontinuierliche Identifikation, Planung, Einsatz, Kontrolle und Weiterentwicklung der intellektuellen Vermögensteile sicherstellt. Der Beitrag zeigt, basierend auf den Ergebnissen des EU Forschungsprojektes Plexus Möglichkeiten für das gezielte Management des intellektuellen Vermögens in den einzelnen Phasen des Lebenszyklus von kooperativen Unternehmensnetzwerken auf.

1. Unternehmensnetzwerke als Antwort auf die geänderten wirtschaftlichen Rahmenbedingungen

Netzwerke, Unternehmenskooperationen, modulare Organisationen und virtuelle Unternehmen sind heute keine Schlagworte mehr, sondern Realität und haben inzwischen einen festen Platz in der globalen Wirtschaftwelt eingenommen (vgl. *Yoshiono*, 1995; *Faulkner*, 1995 und *Arino*, 1998). Als Folge der geänderten wirtschaftlichen Rahmenbindungen werden die klassischen Grenzen von Unternehmen aufgelöst. Anstelle von mehrstufigen Unternehmenshierarchien treten zunehmend dezentrale modulare Strukturen, welche durch Autonomie, Kooperation und indirekter Führung charakterisiert sind (vgl. *Picot*, 1998).

Als wirtschaftliche und gesellschaftliche Treiber dieser organisationalen Veränderungen können folgende Aspekte genannt werden (vgl. *Picot*, 1998 und *Brütsch*, 1999):

- Gegenwärtige globale konjunkturelle Situation.
- Globalisierung der Weltwirtschaft, nicht zuletzt hervorgerufen durch den Abbau von Handelsbarrieren.
- Ressource Wissen als zentrale wertschöpfende Komponente bei immer mehr Produkten und Dienstleistungen.
- Übergang vom Käufer- zum Verkäufermarkt in einer Vielzahl von Märkten.
- Notwendigkeit zur kontinuierlichen Verbesserung und Neuentwicklung von Produkten, Dienstleistungen und Unternehmensprozessen.
- Optimierung von Beschaffungsabläufen und -kosten durch die gezielte Auswahl von Systemlieferanten.
- Weiterentwicklung der Informations- und Kommunikationstechnologien, welche die Grenzen von Raum und Zeit relativieren.
- Wertewandel in der Arbeitwelt und Gesellschaft.

Theoretisch kann die Entstehung von kooperativen Unternehmensnetzwerken durch Ansätze der Netzwerktheorie und -ökonmie, der Principal Agent Theorie, der evolutionären und situativen Organisationstheorien erklärt werden (vgl. z.B. *Fleisch*, 2001; *Ries*, 2001; *Mintzberg*, 1991 und *Gomez*, 1992).

Die vermehrte Bildung von kooperativen Unternehmensnetzwerken ist eine Konsequenz der genannten Entwicklungen. Unter einem kooperativen Unternehmensnetzwerk wird basierend auf wissenschaftlicher Literatur (vgl. u. a. *Picot*, 1998; *Brütsch*, 1999 und *Ries*, 2000) und den in *Plexus* basierend auf empirischen Studien identifizierten Charakteristika der Zusammenschluss von mindestens zwei rechtlich selbständigen, aber durchaus wirtschaftlich horizontal oder vertikal verbundenen Unternehmen bzw. Institutionen zur Durchführung gemeinsamer wirtschaftlicher Aktivitäten verstanden. Ein kooperatives Unternehmensnetzwerk bildet sich aufgrund eines bestehenden oder latenten Marktbedürfnisses, welches von einer Organisation aufgrund der individuellen Kompetenzen, physischen Ressourcen oder unternehmenspolitischen Grundsatzentscheidungen (z.B. zu hohes Risiko bei einer alleinigen Neuentwicklung) nicht alleine befriedigt werden kann. Zentraler Beitrag der einzelnen Partner sind die spezifischen Kompetenzen, welche in diesem Beitrag als Kombination von integrierten und durch organisationale Lernprozesse koordinierte Gesamtheiten von intellektuellen Vermögensbestandteilen wie Technologien, Know-how, Prozessen und Einstellungen (vgl. *Hinterhuber*, 1996) verstanden werden. Durch diese gezielte Bündelung von Ressourcen und Kompetenzen können Synergiepotentiale lukriert werden, welche die Leistungsfähigkeit der einzelnen Unternehmen bei weitem übersteigen. Ein weiteres Charakteristikum eines kooperativen Unternehmensnetzwerks ist der Marktauftritt der Partnerunternehmen unter einem gemeinsamen Namen. Darüber hinaus zeichnet sich ein kooperatives Unternehmensnetzwerk durch ein Minimum an Managementfunktionen aus. Dadurch soll auch längerfristig die Flexibilität und Effektivität dieser Organisationsform aufrechterhalten werden. Der Einsatz von Informations- und Kommunikationstechnologien ist aufgrund der räumlichen Trennung der am Netzwerk teilnehmenden Organisationen von zentraler Bedeutung für die Integration und Koordination der Partnerleistungen.

Übergeordnetes Ziel von Organisationen für die Beteiligung an einem Netzwerk ist die Sicherstellung der nachhaltigen wirtschaftlichen Viabilität. Darüber hinaus wurden folgende, daraus abgeleitete Ziele im Rahmen der in *Plexus* durchgeführten empirischen Studie als zentrale Motivatoren für die Teilnahme an einem Netzwerk identifiziert:

- Zugang zu neuen Märkten,
- Zugang zu komplementären materiellen und immateriellen Ressourcen,
- Kostenreduktion durch Economies of Scale and Scope,
- Aufbau und Weiterentwicklung der eigenen Kompetenzen,
- Optimierung der Auslastung der vorhandenen Kapazitäten,
- Reduktion von Risiko, Kosten und zeitlichen Aufwänden bei der Neu- bzw. Weiterentwicklung von Produkten, Dienstleistungen und Prozessen.

Für den erfolgreichen Aufbau und laufenden Betrieb von kooperativen Unternehmensnetzwerken können folgende kritische Erfolgsfaktoren angeführt werden (vgl. *Ries*, 2001):

- Verträglichkeit der Organisationskulturen,
- Komplementarität von Strategien,
- Adaptivität von Prozessen und Qualität von organisationsübergreifenden Prozessschnittstellen,
- Gemeinsamen Identität und Ausrichtung,
- Vertrauensbasis zwischen den teilnehmenden Unternehmen,
- Qualität der Kompetenzen und die Leistungsfähigkeit der einzelnen Partner.

Im Zuge des 5. EU Forschungsrahmenprogramms wird im Bereich Information Society Technology (IST) unter der wissenschaftlichen Leitung von *Profactor* das internationale Forschungsprojekt *Plexus* durchgeführt, in welchem Methoden und Tools für ein systematisches und zielorientiertes Management von intellektuellem Vermögen im Kontext von Unternehmensnetzwerken entwickelt werden. Weitere Teilnehmer dieses Projektes sind die *Österreichische Akademie für Arbeitsmedizin*, die *Österreichische Zertifizierungs- und Begutachtungs-GmbH*, die *KACU systmic GmbH*, die Universität Göteborg, die Universität Madrid, *Interconsult Bulgarien* und das spanische IT-Unternehmen *Indra*. Die nachstehend präsentierten Ansätze und Methoden sind Ergebnisse dieses Forschungsprojektes.

2. Intellektuelles Vermögen als zentrale Ressource in kooperativen Unternehmensnetzwerken

Der Produktionsfaktor Wissen und die damit in unmittelbaren Zusammenhang stehenden intellektuellen Vermögensteile werden in der Epoche der Wissensgesellschaft zu den zentralen Ressourcen für Organisationen aller Art. Erst durch die gezielte Identifikation, Bewertung, Nutzung und Weiterentwicklung dieser intellektuellen Ressourcen kann eine Differenzierung vom Wettbewerb erfolgen und somit die Viabilität des organisationalen Systems nachhaltig gesichert werden. Der Einfluss der drei klassischen Produktionsfaktoren Kapital, Boden und Arbeit auf den Erfolg von Unternehmen wird durch die neuen wirtschaftlichen Rahmenbedingungen, in welchen sich kooperative Unternehmensnetzwerke bewegen relativiert. Die Zunahme der Bedeutung des intellektuellen Vermögens als zentrale wettbewerbs beeinflussende Größe betrifft dabei aber nicht nur Organisationen des tertiären Wirtschaftssektors. Auch für Organisationen des ersten und zweiten Sektors nimmt die Bedeutung des intellektuellen Vermögens aufgrund der steigenden Wissensintensität bei der Produktion immer mehr zu. Darüber hinaus ist das intellektuelle Vermögen die Grundlage für Innovation und Effizienz- und Effektivitätssteigerung ist (vgl. *Wilke*, 2001 und *Stewart*, 1998).

In der vorliegenden Arbeit wird bewusst von intellektuellem Vermögen und nicht von dem in der wissenschaftlichen Literatur am häufigsten verwendeten Begriff des intellektuellen Kapitals (vgl. u. a. *Edvinsson*, 1998; *Teece*, 2002 und *Lev*, 2000) gesprochen, da die Aufmerksamkeit des Netzwerkmanagements auf jene intellektuellen Ressourcen gerichtet werden soll, die bewusst für die Wertschöpfung im Netzwerk eingesetzt werden und somit eine wesentliche Grundlage für die zukünftige Ertragsfähigkeit des kooperativen Netzwerkes bilden.

Diese Zunahme der Bedeutung des intellektuellen Vermögens spiegelt sich beispielsweise in der Marktwert-Buchwert-Differenz von vielen börsennotierten Unternehmen wieder. Diese Differenz, welche als Goodwill bezeichnet wird, kann zu einem beträchtlichen Teil den intellektuellen Vermögensteilen zugeordnet werden. „Studien zeigen, dass ca. 80% des Markt- bzw. Unternehmenswertes auf intangible Vermögensteile zurückzuführen sind. Lag der geschätzte Anteil der intangiblen Vermögensteile Mitte der 80er-Jahre noch durchschnittlich bei 30% bis 40%, so wird er heute insbesondere bei wissensintensiven Unternehmen auf bis zu 90% geschätzt. Dies bedeutet, dass nur ca. 10% des Unternehmenswertes durch das Rechnungswesen erfasst werden.“ (vgl. *Schnabel*, 2001) Diese Ergebnisse zeigen deutlich, dass die intellektuellen Vermögensteile für die Shareholder eine zentrale Rolle bei der Bewertung von Unternehmen spielen. Darüber hinaus untermauern diese Erkenntnisse die Tatsache, dass intellektuellen Vermögensteile Aufschluss über die zu erwartende zukünftige Ertragskraft einer Organisation geben.

Die Ergebnisse der in *Plexus* durchgeführten empirischen Studie belegen, dass sich der Großteil der befragten Organisationen über die Bedeutung der intellektuellen Vermögensgegenstände für die Ertragskraft bzw. die nachhaltige Viabilität einer Organisation im Klaren ist. Ein weiteres Ergebnis ist allerdings die Tatsache, dass es nur ein kleiner Prozentsatz von Unternehmen und Netzwerken ist, der bewusst und systematisch diese intellektuellen Vermögensgegenstände managt.

3. Die Elemente des intellektuellen Vermögens

In den nachstehenden Absätzen wird eine Definition von intellektuellem Vermögen und dessen Struktur vorgestellt. Sowohl Definition, als auch Struktur wurden im Rahmen des EU Forschungsprojektes *Plexus* basierend auf bestehenden theoretischen und empirischen Ansätzen und neuen empirischen Erkenntnissen entwickelt. Oberstes Ziel bei der Entwicklung der Struktur war die Schaffung einer Möglichkeit zur vollständigen und systematischen Identifikation, Beschreibung und Darstellung des intellektuellen Vermögens eines kooperativen Unternehmensnetzwerkes bzw. der am kooperativen Netzwerk teilnehmenden Partnerorganisationen.

Unter intellektuellem Vermögen werden in diesem Artikel all jene intangiblen, im wesentlichen wissensbezogenen Ressourcen einer Organisation verstanden, die identifiziert, bewertet und gemessen werden können. Die intellektuellen Vermögensteile tragen wesentlich zur Wertschöpfung des Netzwerks bei und müssen deshalb gezielt in Hinblick auf die Erreichung der mittel- und langfristigen Organisationsziele entwickelt werden (vgl. *Valentin*, 2003).

Nach der in Plexus entwickelten Struktur kann zwischen drei Kategorien von intellektuellem Vermögensteilen unterschieden werden.

- **Humanvermögen**: Das Humanvermögen beinhaltet die Fähigkeiten und Fertigkeiten von Mitarbeitern, welche für die Leistungserstellung und Administration des kooperativen Netzwerkes eingesetzt werden. Darüber hinaus spielen physische und psychologische Aspekte, wie Arbeitszufriedenheit, intrinsische Motivation oder Commitment aufgrund deren Bedeutung für die Produktivität der Mitarbeiter eine wesentliche Rolle (vgl. *Valentin*, 2003). Das Unternehmen bzw. das kooperative Unternehmensnetzwerk, welches diese humanbezogenen Ressourcen bewusst für die Wertschöpfung einsetzt, ist selbstverständlich nicht Eigentümer der Individuen an sich, sondern besitzt nur „Verfügungsrechte" in Bezug auf die geforderten und vereinbarten Leistungen.
- **Organisationales Vermögen**: Unter organisationalem Vermögen werden all jene organisationsbezogenen intellektuellen Vermögensbestandteile verstanden, welche die zielgerichtete Kooperation und Koordination der Organisationsmitglieder ermöglichen (vgl. *Valentin*, 2003). Dabei kann, abgeleitet von dem Managementansatz der *Hochschule St. Gallen* (vgl. *Bleicher*, 1995), zwischen intellektuellen Vermögensteilen auf drei Ebenen unterschieden werden.

  1. Die Normative Ebene, deren Aufgabe die Entwicklung und Definition der grundsätzlichen Zielsetzungen, Prinzipien, Normen und Regeln ist, enthält die Organisationskultur, Organisationspolitik und Organisationsverfassung als Elemente.

2. Die Strategische Ebene ist Konsequenz des normativen Überbaus, positioniert die Organisation in ihrer Umwelt und soll die Rahmenbedingungen für die Erreichung der mittel- und langfristigen Organisationsziele sicherstellen. Dieser Bereich beinhaltet die Organisationsstruktur und die Organisationsstrategie als Elemente.
3. Die operative Ebene beinhaltet die organisationsspezifischen Prozesse. Diese werden als die logische Aneinanderreihung von Aktivitäten verstanden, die zur Befriedigung des Bedürfnisses eines Prozesskunden erforderlich sind. Prozesse zeichnen sich durch einen identifizierbaren In- und Output aus. Neben organisationsinternen Prozessen entlang der Wertschöpfungskette zählen auch Prozesse, welche durch einen Kontakt mit externen Stakeholdern charakterisiert sind, wie z.B. Beschaffungs-, Informations- bzw. Distributionsprozesse zu dieser Vermögenskategorie.

- **Kodifizierte intellektuelle Vermögensteile**: Diese Vermögenskategorie beinhaltet all jene kodifizierten intangiblen Vermögensteile, die keine physische Substanz haben, die der innehabenden Organisation Rechte und juristischen Schutz verleihen und die mit der Organisation untrennbar verbunden sind. Diese Vermögensteile sind das explizite Ergebnis der intellektuellen Aktivitäten von Individuen und können zumeist Einzelpersonen zugeordnet werden (vgl. *Smith*, 2000). Das kodifizierte intellektuelle Vermögen lässt sich in vier Subkategorien unterteilen:

1. Die erste Kategorie subsumiert alle geistigen Schutz- und Urheberrechte über die eine Organisation verfügt. Dazu zählen unter anderem Marken, Logos, Patente, Lizenzen, Technische Beschreibungen, Copyrights und Software (vgl. *Smith,* 2000).
2. Unter der zweiten Kategorie, den Stakeholder bezogenen intellektuellen Vermögensteile, werden all jene Verträge, Datenbanken und Listen verstanden, welche die Zusammenarbeit mit Stakeholdern regeln (z.B. Kunden- oder Lieferantenverträge) oder Angaben zu ihnen enthalten.
3. Die dritte Kategorie wird als prozessbezogene kodifizierte intellektuelle Vermögensteile bezeichnet. Vermögensteile dieser Subkategorie beinhalten Datenbanken, Listen und Beschreibungen über die Prozesse entlang der Wertschöpfungskette.
4. Abschließend zählen produktbezogene kodifizierte intellektuelle Vermögensteile zu dieser Kategorie. Produktbeschreibungen, Produktdatenbanken gehören dieser Kategorie an.

Neben der Identifikation der intellektuellen Vermögensteile ist auch deren Dokumentation für ein effektives Management unerlässlich. Folgende Aspekte sollten dabei bei einer umfassenden Dokumentation berücksichtigt werden:

- Bezeichnung des intellektuellen Vermögensgegenstandes,
- Zuordnung zu einer Kategorie des intellektuellen Vermögens,
- Kurze verbale Beschreibung,
- „Eigentümer" des intellektuellen Vermögens,
- Indikatoren, welche eine Darstellung des Wertes und dessen Entwicklung ermöglichen,
- Indikatorenwerte,
- Periodizität bzw. Anlässe einer Bewertung,
- Dokumente, welche nähere Informationen zu den intellektuellen Vermögensteil beinhalten,

- Dokumentation der Maßnahmen, welche zur Weiterentwicklung des intellektuellen Vermögensbestandteiles umgesetzt wurden.

Die Zuordnung der intellektuellen Vermögensbestandteile zu den einzelnen Komponenten der Struktur (vgl. *Abb. 1*) und deren Beschreibung bildet die Grundlage für ein zielführendes Management des intellektuellen Vermögens im kooperativen Unternehmensnetzwerk.

4. Ansätze zur Bewertung des intellektuellen Vermögens in Unternehmensnetzwerken

Neben der Identifikation und Darstellung von intellektuellen Vermögensgegenständen ist die periodische bzw. anlassbezogene Messung und Bewertung unerlässlicher Bestandteil für ein systematisches Management dieser Vermögensgegenstände. Im Rahmen der in Plexus durchgeführten Studie, wurden auch die Anforderungen an etwaige Bewertungsmethoden für das intellektuelle Vermögen im Kontext von Unternehmensnetzwerken erhoben. Die rasche und einfache Anwendbarkeit der Bewertungsmethoden und die Nachvollziehbarkeit der Ergebnisse wurden dabei seitens der befragten Unternehmen als zentrale Anforderungen identifiziert.

Folgende Zielsetzungen sind mit der Messung und Bewertung von intellektuellen Vermögensgegenständen im Kontext von Unternehmensnetzwerken verbunden:

- Identifikation von geeigneten Netzwerkpartnern;
- Unterstützung des Netzwerkmanagements bei der gezielten Identifikation, Bewertung, Anwendung und Weiterentwicklung des intellektuellen Vermögens des kooperativen Unternehmensnetzwerkes;
- Schaffung einer Grundlage für die Verteilung von Gewinnen in Netzwerken;
- Verbesserung des Images durch gezielte Kommunikation des intellektuellen Vermögens und dessen Wert.

Grundsätzlich kann zwischen zwei unterschiedlichen Formen der Bewertung von intellektuellem Vermögen unterschieden werden:

- Monetäre Bewertung der intellektuellen Vermögensgegenstände;
- Indikatorenbasierte qualitative Bewertung der einzelnen Vermögensgegenstände,

Bei der monetären Bewertung der intellektuellen Vermögensgegenstände kann zwischen Ansätzen zur summarischen Bewertung (d.h. Gesamtbewertung des intellektuellen Vermögens) und jenen der Bewertung von einzelnen Vermögensgegenständen unterschieden werden. Grundsätzlich ist anzumerken, dass eine Vielzahl von Methoden (z.B. die klassischen Methoden aus dem Bereich der Unternehmensbewertung) theoretisch für das intellektuelle Vermögen an sich und die einzelnen Vermögensteile angewandt werden könnten. Bei der Auswahl der geeigneten Methode ist jedoch der Aufwand, welcher mit der Erhebung der für die Bewertung notwendigen Informationen verbunden ist und die Qualität des verfügbaren Datenmaterials, dem Informationswert gegenüber zu stellen.

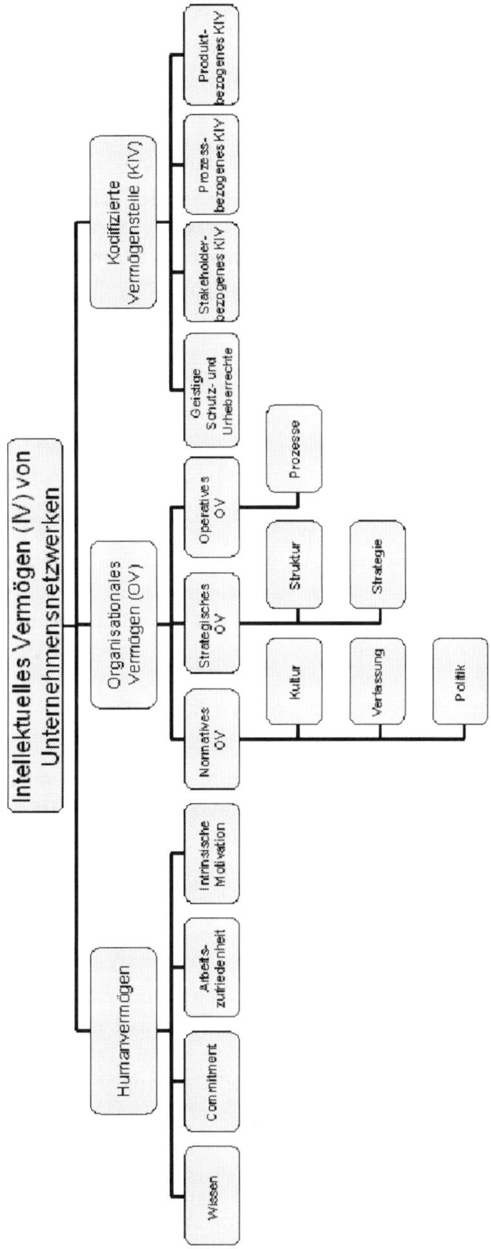

Abb. 1: Struktur des Intellektuellen Vermögens

Als Beispiel für eine summarische Bewertung des intellektuellen Vermögens kann der vom kooperativen Unternehmensnetzwerk erzielte ökonomische Mehrwert herangezogen werden. Der Mehrwert berechnet sich aus der Differenz zwischen den Erlösen, die aus der betrieblichen Tätigkeit des Netzwerkes resultieren und all jenen Kosten die sich auf physische Ressourcen (z.B. Materialaufwand, Maschinenaufwände etc.) bzw. das eingesetzte finanzielle Kapital beziehen. Die Personalkosten werden dabei als Teil des Mehrwerts gesehen, welcher durch den gezielten Einsatz von Fähigkeiten und Fertigkeiten der Netzwerkmitarbeiter erwirtschaftet wird. Dieser Mehrwert ist eine einfach zu erhebende Maßzahl für die Effektivität und die Effizienz des Einsatzes der intellektuellen Vermögensteile im kooperativen Unternehmensnetzwerk. Um sich dem tatsächlichen Wert des im Unternehmensnetzwerk vorhandenen intellektuellen Vermögens weiter anzunähern, wird der errechnete Wert mit der Fluktuationsrate und der Abwesenheitsrate multipliziert. Diese beiden Indikatoren führen zu einer Berücksichtigung der Verfügbarkeit des intellektuellen Vermögens.

Des Weiteren kann die schon angesprochene Martkwert-Buchwert Differenz als Indikator für den Gesamtwert des intellektuellen Vermögens herangezogen werden (vgl. *Sveiby*, 1997). Dabei ist allerdings anzumerken, dass diese Verhältniskennzahl gerade bei Unternehmensnetzwerken nur schwierig zu erheben ist, da ein unmittelbarer Marktwert zumeist nicht vorhanden ist. Sollte das kooperative Netzwerk über Informationen bezüglich des Marktwertes verfügen, so muss dieser im Kontext des Bewertungsanlasses, des Bewertungsziels und in Hinblick auf die gegebenen externen Einflussfaktoren, die nicht vom Netzwerk beeinflussbar sind, interpretiert werden. Als weiterer möglicher Indikator für den Wert des intellektuellen Vermögens eines Unternehmens kann Tobin's q herangezogen werden. (vgl. *Bontis*, 1999) Dieser Indikator stellt dem Marktwert einer Organisation, den Reproduktionswert bzw. dem Buchwert gegenüber. Die genannten Kennzahlen eignen sich vor allem für einen Branchenvergleich und für die Darstellung der Effektivität und Effizienz des Einsatzes des intellektuellen Vermögens über mehrere Perioden hinweg.

Für die Bewertung der „anderen kodifizierten Vermögensteile" und der „intellektuellen Beiträge eines Partners" schlägt das *Plexus* Konsortium die Anwendung der klassischen Bewertungsverfahren Kosten-, Einkommens- und Marktansatz vor. Die Wahl der geeigneten Bewertungsmethode ist dabei vom Bewertungsanlass (z.B. Offerierung von Personalleistungen, Verkauf von Lizenzen etc.) und der Verfügbarkeit der für die Bewertung notwendigen Informationen abhängig (vgl. *Smith*, 2001).

Aufgrund der Tatsache, dass eine monetäre Bewertung von intellektuellen Vermögensgegenständen oftmals aufgrund deren Spezifität nicht möglich ist, bzw. aufgrund des Aufwandes und des aus einer monetären Bewertung erzielbaren Informationsgehaltes nicht sinnvoll erscheint, wurden basierend auf Erkenntnissen aus empirischen Studien und theoretischen Erklärungsmodellen Dimensionen für die einzelnen Vermögensteile abgeleitet. Unter Dimensionen werden dabei Attribute von intellektuellen Vermögensbestandteilen verstanden, die diese beschreiben und die einen wesentlichen Einfluss auf die durch den Einsatz erzielbare Wertschöpfung haben. Basierend auf diesen Dimensionen wurden Indikatoren abgeleitet, die Ausdruck einer Dimension sind und die relativ einfach zu erheben sind.

5. Zielorientiertes Management von intellektuellem Vermögen in Unternehmensnetzwerken

Die nachfolgenden Subkapitel geben einen kurzen Überblick über die Managementaktivitäten in den einzelnen Phasen des Lebenszyklus von Unternehmensnetzwerken (vgl. *Ries*,

2001). Dabei wird im Besonderen auf die Aktivitäten des Netzwerkmanagements in Bezug auf das zielorientierte Management von intellektuellem Vermögen eingegangen.

5.1 Phase 1: Strategische Grundsatzentscheidung bzw. Initiierung

Ausgangspunkt für die Gründung eines Unternehmensnetzwerkes ist die Erkenntnis einer Organisation, dass die Befriedigung eines vorhandenen bzw. latenten Markbedürfnisses aufgrund der im Unternehmen vorhandenen Informationen wahrscheinlich nicht durch die eigenen Ressourcen erfolgen kann. Aus diesem Grund ist in dieser ersten Phase über die Sinnhaftigkeit einer Initiierung eines kooperativen Unternehmensnetzwerkes zu entscheiden. Dabei werden in einem ersten Schritt alle möglichen Kooperationsformen gesammelt und in Hinblick auf ihre Brauchbarkeit (rechtliche und organisatorische Rahmenbedingungen der Kooperation) für die Erreichung der Kooperationszielsetzung untersucht. In einem weiteren Schritt werden die potentiell geeigneten Kooperationsmodelle in Hinblick auf den Fit zu den unternehmensspezifischen Strategien (z.B. lang- vs. kurzfristige Bindung an andere Organisationen) geprüft. Im Anschluss an diese grundlegende Entscheidung werden verschiedene, auf mögliche Kooperationen bezogene, Rahmenbedingungen (z.B. Zielsetzungen der Kooperation, mögliche Kooperationsbereiche, Ressourcenanforderungen an potentielle Netzwerkpartner oder Anzahl der Partner) ausformuliert und für die nächste Phase, der Anbahnung und Vorbereitung auf die Netzwerkgründung, aufbereitet.

In Bezug auf das intellektuelle Vermögen ist in dieser Phase die Identifikation, Darstellung und Bewertung der eigenen intellektuellen Vermögensteile durchzuführen. Basierend auf den daraus resultierenden Informationen werden Anforderungsprofile an potentielle Netzwerkpartner erstellt, welche die benötigten intellektuellen Vermögensteile, deren Dimensionen und Indikatoren bzw. Soll-Indikatorenwerte beinhalten. Diese Anforderungsprofile unterstützen das Management der Organisation bei der Identifikation von potentiell passenden Organisationen.

5.2 Phase 2: Anbahnung, konzeptionelle Vorbereitung bzw. Formierung

Basierend auf den in Phase 1 erhobenen Informationen wird nun das Netzwerk aufgebaut. Die Identifikation und Selektion von geeigneten Partnern ist zentraler Aspekt dieser Lebenszyklusphase. Dabei ist die Urteilsbildung über die Fähigkeiten und Ressourcen eines Unternehmens/einer Organisation in einer bestimmten Form eines Netzwerkes zu kooperieren von zentraler Bedeutung. Die in Phase 1 entwickelten Anforderungsprofile für intellektuelle Vermögensteile sind dabei dem Netzwerkmanagement bei der Selektion von potentiellen Netzwerkpartnern behilflich.

Die Identifikation und Selektion von geeigneten Netzwerkpartnern verläuft in der Praxis in den meisten Fällen innerhalb des gegenwärtigen Umfeldes eines Unternehmens. Aufgrund gemeinsam abgewickelter Aufträge und Projekte entscheiden sich viele Unternehmen aufgrund der bereits vorliegenden für einen Netzwerkpartner. Dadurch wird allerdings eine Vielzahl von Kooperationsmöglichkeiten ausgeschlossen, die vor allem in Hinblick auf die erfolgreiche Zielerreichung effektiver wären. Je weniger persönliche Erfahrungen mit einem potentiellen Partner vorhanden sind, desto wichtiger werden quantifizierbare, objektive und nachvollziehbare Indikatoren über intellektuelle Vermögensteile (z.B. Ausbildungsniveau, Regelmäßigkeit der Zertifizierungen, Alter und Kapazität der physischen Ressourcen, etc.) und Referenzprojekte. Durch die sorgfältige und umfassende Ausarbeitung von den genannten Anforderungsprofilen kann somit das bestehende Vertrauensdefizit reduziert werden und der Pool von möglichen Netzwerkpartnern vergrößert werden.

Sind die „optimalen“ Netzwerkpartner identifiziert, werden die Konditionen der Zusammenarbeit gemeinsam festgelegt. Dabei sind unter anderem die nachstehenden Punkte zu verhandeln und schriftlich in Form z.B. eines Vertrages oder eines „Memorandum of Understandings“ festzuhalten:

- Vision, Mission, strategische Ziele des kooperativen Unternehmensnetzwerkes;
- Beiträge der Netzwerkpartner in Bezug auf physische Ressourcen und Intellektuelle Vermögensteile;
- Regelung der Gewinn- und Verlustverteilung zwischen den Netzwerkpartnern;
- Definition der im Netzwerk wahrzunehmenden organisationalen Funktionen;
- Definition der Rollen der einzelnen Netzwerkpartner;
- Kompetenzen der einzelnen Partner im Netzwerk;
- Beschreibung der Kernprozesse im Netzwerk (z.B. Akquisitions-, Produktions-, Distributionsprozesse);
- Definition der Spielregeln in Bezug auf die Aufnahme von neuen und Sanktionierung bzw. Ausschluss von bestehenden Partnern;
- Auftritt des kooperativen Unternehmensnetzwerkes am Markt;
- Rechtsform des kooperativen Unternehmensnetzwerkes;
- Dokumentationsrichtlinien für die Aktivitäten des Netzwerkes;
- Gemeinsamer Ressourcen- und Kompetenzaufbau bzw. –entwicklung.

Ziel dieser Regelungen ist neben der formalen Sicherstellung des nachhaltigen Netzwerkbetriebes die Transparenz über die Beiträge, Rechte und Verpflichtungen der einzelnen Netzwerkpartner. Diese Vereinbarung wird somit zu einem wichtigen Instrument für den Aufbau von Vertrauen zwischen den einzelnen Netzwerkpartnern.

Für den Erfolg des Netzwerkes ist es weiters unerlässlich, dass diese Entscheidungen von allen Partnern getragen werden und die einzelnen Regelungen regelmäßig in Hinblick auf ihre Brauchbarkeit für die einzelnen Partnerorganisationen und das Umfeld des Netzwerkes überprüft werden. Nur wenn für die einzelnen Partner die individuelle Zielerreichung sichergestellt ist, werden diese ihre vereinbarten Leistungen in bester Qualität und innerhalb der vereinbarten Zeit zur Verfügung stellen.

Um die nachhaltige Überlebensfähigkeit des Netzwerkers sicherzustellen ist es, wie in der vorangegangenen Aufzählung bereits kurz dargestellt wurde auch notwendig, eine Aufbau- und Ablauforganisation für das Netzwerk zu definieren. Dadurch wird eine Organisation geschaffen, welche ein zielorientiertes und umfassendes Management des Netzwerkes erlaubt. Eine besondere Rolle nimmt dabei das Management des intellektuellen Vermögens ein. Es sind daher Funktionen für das kooperative Unternehmensnetzwerk und entsprechende Rollen zu definieren, welche die laufende Erhebung, Dokumentation, Integration und zielorientierte Weiterentwicklung des intellektuellen Vermögens koordiniert und sicherstellt.

5.3 Phase 3: Betriebsphase

Unternehmensnetzwerke werden durch den freiwilligen Zusammenschluss von zwei oder mehreren Partnern gegründet, um gemeinsame Ziele zu verfolgen. In dieser Phase wird das Netzwerk nun operativ aktiv. Aufträge werden akquiriert und abgewickelt. Damit beginnt auch ein permanenter Lern- und Austauschprozess zwischen den Netzwerkpartnern. Aufgrund der gesammelten Erfahrungen kommt es zu einer kontinuierlichen organisatorischen und inhaltlichen Weiterentwicklung des Netzwerkes an sich und aller Partnerorganisationen.

Wie jede andere Organisationsform benötigt, wenn auch häufig in eingeschränktem Umfang, jedes Unternehmensnetzwerk ein Managementsystem, welches Planungs-, Steuerungs-, Kontroll- und Verbesserungsprozesse beinhaltet.

Im Rahmen des EU Forschungsprojektes *Plexus* wurde basierend auf ISO 9004 ein Managementsystem für Unternehmensnetzwerke unter besonderer Berücksichtigung des Managements von intellektuellem Vermögensteile entwickelt (vgl. *Abb. 2*). Die nachfolgenden Absätze sollen einen Überblick über die Kernelemente dieses Managementsystems geben (vgl. *Nohava*, 2003).

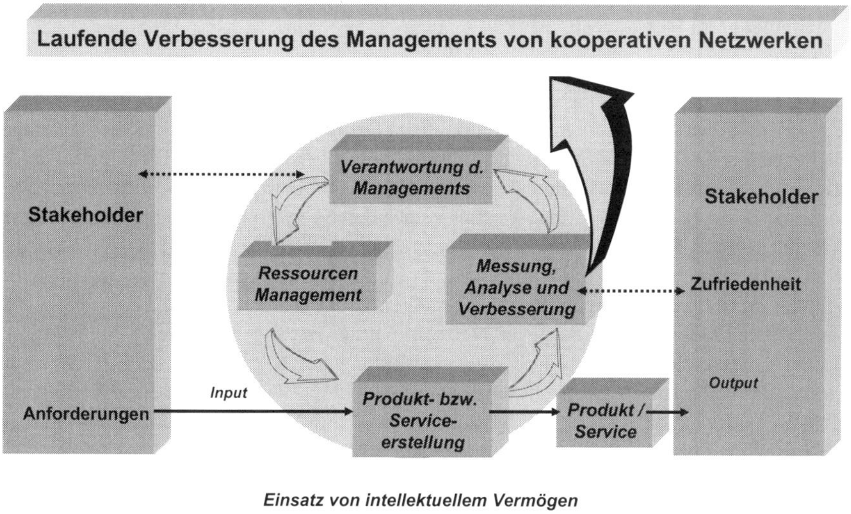

Abb. 2: Das Plexus Managementsystem von kooperativen Unternehmensnetzwerken

Verantwortung des Managements

Grundlage eines effektiven Netzwerkmanagements ist das zielorientierte und engagierte Verhalten der Mitglieder des Netzwerkmanagements, welches sich im Idealfall aus Vertretern aller Partnerorganisationen zusammensetzt. Es ist Aufgabe dieses Managementteams, basierend auf den Anforderungen und Bedürfnissen der Stakeholder, wirtschaftlichen Rahmenbedingungen und individuellen Partnerinteressen, übergeordnete Netzwerkziele, eine entsprechende Netzwerkpolitik und daraus abgeleitete mittel- und langfristige Netzwerkstrategien zu entwickeln, zu dokumentieren und netzwerkintern und –extern zu kommunizieren. Diese werden kontinuierlich einer Revision in Hinblick auf die Gültigkeit im geänderten

Organisationskontext unterzogen. Des Weiteren sind Umsetzungspläne zur Realisierung der Netzwerkziele zu erarbeiten und deren Umsetzung anhand definierter Indikatoren laufend zu kontrollieren und zu steuern. Neben der Entwicklung der genannten Maßnahmenpläne ist die Schaffung einer gemeinsamen Organisationskultur erforderlich, welche die Realisierung der übergeordneten Ziele erlaubt.

Darüber hinaus müssen in Hinblick auf die intellektuellen Vermögensteile, die für die Wertschöpfung eingesetzt werden sollen, abgeleitet aus der Netzwerkpolitik und den damit verbundenen Netzwerkstrategien Ziele und Maßnahmenpakete definiert werden, welche die Qualität (z.B. Entwicklung geeigneter Aus- und Weiterbildungsprogramme) und den zielorientierten Einsatz dieser (z.B. durch Instrumentarien des Wissensmanagements wie z.B. Dokumentenmanagementsysteme, Expertensysteme oder Lessons-Learnt-Datenbanken) laufend verbessern und an die Anforderungen der Stakeholder anpassen.

Eine weitere wesentliche Aufgabe des Netzwerkmanagements ist die interne und externe Kommunikation der intellektuellen Vermögensteile des Unternehmensnetzwerkes. Die Veröffentlichung von Berichten über die intellektuellen Vermögensbestandteile des Netzwerkes und deren Wert bzw. Wertentwicklung leistet einen wesentlichen Beitrag zur Vertrauensbildung seitens aller Stakeholder.

Management von Ressourcen

Es ist Aufgabe des Netzwerkmanagements sicherzustellen, dass die zur Erreichung der Netzwerkziele und zur Abwicklung eines Auftrages erforderlichen physischen und intellektuellen Ressourcen zur richtigen Zeit, am richtigen Ort und in zur Abwicklung des Auftrages entsprechender Qualität und Menge zur Verfügung stehen. Aus diesem Grund ist es erforderlich schon vorab den Ressourcenbedarf zu planen und bei etwaigen identifizierten Ressourcendefiziten (z.B. gegenwärtig verfügt keiner der Netzwerkpartner über Kompetenzen im Bereich Finite Elemente Simulation) entsprechende Alternativstrategien zu deren Ausgleich auszuarbeiten. Bei kooperativen Unternehmensnetzwerken bedeutet dies, dass bei Defiziten im Bereich des intellektuellen Vermögens zwischen den Möglichkeiten Akquisition über den Markt, Aufnahme neuer Netzwerkpartner mit entsprechenden intellektuellen Vermögensbestandteilen oder einer Eigenentwicklung gewählt werden kann. Entscheidet sich das Unternehmensnetzwerk für eine Eigenentwicklung von intellektuellen Vermögensteilen, so sind z.B. für die spezifische Weiterentwicklung des Humanvermögens spezielle Aus- und Weiterbildungspläne zu erarbeiten und umzusetzen.

Management der Produkt- und Dienstleistungserstellung

Es ist Aufgabe des Netzwerkmanagements durch die Definition und Dokumentation von Prozessen und den darin enthaltenen wertschöpfenden und administrativen Aktivitäten eine effektive und effiziente Produkt- und Dienstleistungserstellung, welche eine Befriedigung der unternehmensinternen und -externen Prozesskunden erlauben, sicher zu stellen. Die geplanten Ressourcen sind den einzelnen Aktivitäten zuzuteilen. Um die Leistungsfähigkeit der einzelnen Prozesse laufend zu verbessern sind Input-, Throughput und Outputgrößen zu prognostizieren und mit den tatsächlichen Werten zu vergleichen.

In Bezug auf das Management des intellektuellen Vermögens bedeutet dies, dass die Mitarbeiter in den für die Abwicklung der Aktivitäten notwendigen Kenntnissen geschult werden, dass ihnen Prozess- und Aktivitätenbeschreibungen und –anleitungen zur Verfügung gestellt werden und ihre Tätigkeiten durch auf die Aufgaben abgestimmte Informations- und Kommunikationssysteme unterstützt werden.

Messung, Analyse und Verbesserung

Es ist Aufgabe des Top Managements effektive und effiziente Methoden im Managementsystem des kooperativen Netzwerkes zu implementieren, welche eine systematische und umfassende Bewertung aller intellektuellen Vermögensteile erlauben. Bei Implementierung der in diesem Beitrag vorgeschlagenen Bewertungsmethoden bedeutet dies, dass relevante Dimensionen für die einzelnen Vermögensteile identifiziert werden, geeignete Indikatoren bestimmt und erhoben werden. Wichtig in diesem Zusammenhang ist die intensive Auseinandersetzung mit den erhobenen Werten im Kontext des gesamten Unternehmensnetzwerkes. Wie bereits dargestellt wurde, beeinflussen sich die Indikatoren für die einzelnen Vermögensteile zum Teil gegenseitig. Für ein effektives Management bedeutet dies, dass versucht werden muss signifikante Zusammenhänge zu identifizieren und bei einer späteren Maßnahmenableitung zu berücksichtigen. Neben diesen wechselseitigen Abhängigkeiten sind auch laufend jene Vermögensteile zu identifizieren, die für die Wertschöpfung und damit für auch für die Schaffung von Kundenwerten von zentraler Bedeutung sind. Werte für das Unternehmen können nur dann geschaffen werden, wenn alle Stakeholder nachhaltig zufrieden gestellt werden können. Aus diesem Grund ist es wichtig in periodischen Abständen Stakeholderbefragungen durchzuführen, welche Aufschluss über die wahrgenommenen Leistung und Differenzierungsmerkmale zur Konkurrenz geben. Die Berücksichtigung dieser Informationen erlaubt in Hinblick auf die Weiterentwicklung der intellektuellen Vermögensgegenstände eine gezieltere, auf den Stakeholderwert fokussierte Ableitung von Maßnahmen.

5.4 Phase 4: Auflösung

Die Entscheidung über die Auflösung der Kooperation kann auf unterschiedlichen Gründen beruhen. Beispielhaft sind nachstehend einige davon genannt:

- Strategische Neuausrichtung einzelner Partner;
- Veränderte Marktsituation;
- Konflikte zwischen den Partnern;
- Diskrepanz zwischen Erwartungen (Ziele) und Wirklichkeit (Realisierung);
- Auflösung aufgrund Zielerreichung oder zeitlichem Ende.

Die Phase der Auflösung beinhaltet die Abwicklung der letzten gemeinsamen Aufträge, die Aufteilung der gemeinsam aufgebauten Ressourcen und die Rückgabe bzw. Rückvergütung der tangiblen und intangiblen Vermögensteile, die in das Netzwerk eingebracht wurden. Zur Bewertung der intellektuellen Vermögensteile und die Aufteilung etwaiger Verkaufserlöse können die in Abschnitt 3 dargestellten monetären Bewertungsmethoden angewandt werden.

6. Fazit

Die in diesem Beitrag präsentierten intellektuellen Vermögensteile spielen eine wesentliche Rolle für den nachhaltigen Erfolg von kooperativen Unternehmensnetzwerken. Aufgrund ihrer Bedeutung sollte das zielorientierte Management dieser bislang zum Teil vernachlässigten Vermögensart zu einem fixen Bestandteil eines umfassenden Managementsystems auch im Kontext von kooperativen Unternehmensnetzwerken werden. In jeder Lebenszyklusphase eines kooperativen Netzwerkes sind, wie in diesem Beitrag dargestellt wurde,

Aktivitäten zu deren Management unerlässlich. Identifikation, laufende Bewertung, zielorientierte Planung und Einsatz und eine kontinuierliche Weiterentwicklung sind zentrale Bestandteile eines umfassenden Managements von intellektuellem Vermögen für jede Form von Organisation. Um solch ein Managementsystem erfolgreich implementieren zu können und den Wert der intellektuellen Vermögensteile stetig zu erhöhen sind Methoden zu identifizieren, welche leicht handhabbar, leicht umsetzbar und ziel- bzw. ergebnisorientiert sind. Die in diesem Beitrag dargestellten und im Rahmen des EU Forschungsprojektes *Plexus* entwickelten Methoden sollen ein zielorientiertes Management von intellektuellem Vermögen ermöglichen. Damit soll auch sichergestellt werden, dass die durch die Kooperation in Unternehmensnetzwerken erzielbaren Synergien auch tatsächlich realisiert werden können. Die Kommunikation der intellektuellen Vermögensteile und deren Wertentwicklung nach innen und nach außen schafft Vertrauen in die Organisation und erlaubt eine klare Differenzierung vom Wettbewerb.

7. Literatur

Ariño, A. und *de la Torre, J.*, Learning from failure: towards an evolutionary model of collaborative ventures, in: Organisation Science, Vol. 9 (1998), Nr. 3.

Bleicher, K., Integriertes Management, Frankfurt/Main 1996.

Bontis, N. et al., The Knowledge Toolbox: A Review of the Tools Available To Measure and Manage Intangible Resources, in: European Management Journal, Vol. 17 (1999), Nr. 4.

Brütsch, D., Virtuelle Unternehmen, Zürich 1999.

Faulkner, D., International strategic alliances: co-operating to compete, Berkshire 1995.

Fleisch, E., Das Netzwerkunternehmen, Berlin 2001.

Gomez, P., Neue Trends in der Konzernorganisation, in: ZfO, 61. Jg. (1992), H. 3, S. 166 – 172.

Hinterhuber, H. H., Kundenzufriedenheit durch Kernkompetenzen, München 1997.

Hinterhuber, H. H., Strategische Unternehmensführung, I. Strategisches Denken: Vision, Unternehmenspolitik, Strategie, 6. Auflage, Berlin/New York 1996.

Lev, B., Intangibles, Washington 2001.

Mintzberg, H., The effective Organisation: Forces and Forms, in: *De Wit, B.* und *Meyer, R.* (Hrsg.), Strategy – Process, Content, Context, New York 1994.

Next Generation Research Group, Knowledge Measurement, Arthur Andersen, October 1998.

Nohava, M. et al.: *Plexus* Workpackage 4 – Network Management Methodology; Easy to apply manual comprising Management Techniques to build up, manage, organize and monitor Cooperative Network Organizations, Wien 2003.

Picot, A., Reichwald, R. und *Wigand, R.T.*, Die grenzenlose Unternehmung: Information, Organisation und Management; 3. überarbeitete Auflage, Wiesbaden 1998.

Ries, A., Controlling in virtuellen Netzwerken, Wiesbaden 2001.

Schnabel, U., Wertorientiertes Management von Intangible Assets, in: Praxis der Wirtschaftsinformatik, Performance Measurement, HMD (2001), S. 227 – 236.

Smith, G., Valuation of intellectual property and intangible assets, New York 2000.

Stewart, T., Der vierte Produktionsfaktor, München 1998.

Sveiby, K., The new organizational wealth, San Francisco 1997.

Sydow, J., Management von Netzwerkorganisationen – zum Stand der Forschung, in: *Sydow, J.* (Hrsg.), Management von Netzwerkorganisationen, 2001.

Teece, D., Managing Intellectual Capital, New York 2002.

Valentin, K. et al., *Plexus* Workpackage 3 – Intellectual Asset Structure and Assessment Model; An easy to handle and easy to use methodology for the identification, measurement and management of Intellectual Assets in Business Cooperative Networks, Steyr 2003.

Wilke, H., Systemisches Wissensmanagement, Stuttgart 2001.

Yoshiono, M. Y. and *Rangan, U. S.*, Strategic Alliances, Boston, Mass. 1995.

Immaterielle und ökologische Ressourcen im Rechnungswesen

Edeltraud Günther[1] *und Thomas Günther*[2]

---

[1] Prof. Dr. *Edeltraud Günther*, Lehrstuhl für Betriebliche Umweltökonomie, Technische Universität Dresden

[2] Prof. Dr. *Thomas Günther*, Lehrstuhl für Betriebliches Rechnungswesen/Controlling, Technische Universität Dresden

Neben den im Rechnungswesen traditionell abgebildeten materiellen und finanziellen Ressourcen nutzen Unternehmen nicht abgebildete immaterielle und ökologische Ressourcen. Die Nutzung und Entwicklung dieser Ressourcen beeinflusst unternehmerische Entscheidungen und die zukünftige Unternehmensentwicklung. Die Autoren gehen der Frage nach, ob und, wenn ja, in welcher Form immaterielle und ökologische Ressourcen in der Unternehmenssteuerung und im Rechnungswesen berücksichtigt werden können. Dabei wird zunächst in einem Idealmodell eine Integration immaterieller und ökologischer Ressourcen in das vorhandene Rechnungswesen durch ein erweitertes Bilanzverständnis vorgeschlagen. Aufgrund der absehbaren Beschränkungen in der Umsetzung dieses Ansatzes wird nachfolgend ein dreistufiges Realmodell zur Berücksichtigung immaterieller und ökologischer Ressourcen im Rahmen des vorhandenen Rechnungswesens dargestellt.

1. Relevanz von immateriellen und ökologischen Ressourcen

Die Bedeutung immaterieller und ökologischer Ressourcen wird in jüngster Zeit unter dem Schlagwort „Corporate Citizenship" im Zusammenhang mit der Verantwortung der Unternehmen für die Gesellschaft diskutiert. Doch auch bei einer rein auf das Unternehmen selbst ausgerichteten Sichtweise spielen immaterielle und ökologische Ressourcen insofern eine Rolle, als eine Leistungserstellung ohne sie nicht möglich ist. Die unmittelbare Betroffenheit des Unternehmens ergibt sich dabei aus dem direkten Einfluss auf die obersten Zielgrößen eines Unternehmens, die Liquidität, den Erfolg (Erlöse und Kosten) sowie das Erfolgspotential (die langfristige Wettbewerbsfähigkeit). Das Erfolgspotential kann monetär ausgedrückt als Unternehmenswert gemessen werden, wobei der periodische Wertbeitrag (Value Added) die Wertsteigerung in einer betrachteten Periode misst (z. B. als EVA™ oder CVA, vgl. *Baum/Coenenberg/Günther,* 1999, S. 5ff.). Als oberste Zielgröße in diesem Beitrag wird daher der Wertbeitrag herangezogen. Dabei lassen sich auch die Kosten- und Leistungsrechnung sowie der Jahresabschluss, die direkt die Zielgröße „Erfolg" messen, dem Wertbeitrag indirekt zuordnen. Ein positiver Wertbeitrag liegt – sehr allgemein formuliert vor –, wenn der „Erfolg" die Kapitalkosten übersteigt.

Immaterielle und ökologische Ressourcen werden hierbei als „Mittel zum Zweck" betrachtet, d. h. im Hinblick auf ihre Wirkung auf Liquidität, Erlöse und Kosten sowie die langfristige Wettbewerbsfähigkeit der Unternehmen.

Analysiert man die Wirkung immaterieller und ökologischer Ressourcen auf diese obersten Zielgrößen, so kann man für alle drei Ziele eine nennenswerte Relevanz erkennen, die nachfolgend erläutert werden.

1.1 Wirkung auf die Zielgröße „Liquidität"

Bedingt durch die Nutzung immaterieller und ökologischer Ressourcen, können Anforderungen an das Unternehmen herangetragen werden, die unmittelbar liquiditätswirksam sind:

Weltweite Anmeldungen von Lizenzen stellen für mittelständische Unternehmen einen Aufwand und eine erhebliche Mittelbindung dar, die bei kurzen Innovationszyklen zu Druck auf die Verwertung des Know-hows führt. Kunden können bei einem wahrgenommenen Fehlverhalten des Unternehmens gegenüber Mitarbeitern (z. B. bei Einsatz von Kinderarbeit in Entwicklungsländern) einen Konsumboykott auslösen.

Die Berücksichtigung von Umweltauflagen in Form eines Gebotes (z. B. Grenzwerteinhaltung) erfordert die Entscheidung zwischen einer zusätzlichen Beschaffung von „unproduktivem" Kapital in Form von nachgeschalteten Prozessen (z. B. Filter) oder einer Beschaffung integrierter Technologien. Ein Vergleich der Cashflows beider Alternativen zeigt deren Liquiditätswirksamkeit. Ebenfalls von Seiten des Staates sind Stilllegungsverfügungen kurzfristig liquiditätswirksam, da Umsatzeinbrüche folgen oder kurzfristige Nachrüstungen erforderlich werden.

Die Bedeutung immaterieller und ökologischer Ressourcen zeigt sich in allen diesen Fällen in ihrer möglichen Wirkung als „K.O.-Kriterium" für die Unternehmensführung. D. h. die an das Unternehmen durch die Anspruchsgruppen - in den oben aufgeführten Beispielen Staat und Kunden - herangetragenen Anforderungen beeinflussen direkt die Liquidität und somit auch die kurzfristige Überlebensfähigkeit des Unternehmens.

1.2 Wirkung auf die Zielgröße „Erfolg"

Marken mit hoher Markenstärke erlauben Preisprämien und/oder erhöhte Absatzmengen im Vergleich zu Produkten mit geringer Markenstärke und ermöglichen daher überdurchschnittliche Renditen. Die systematische Nutzung vorhandener Kundenbeziehungen und deren Entwicklung führt nicht nur aufgrund von Kundenbindung und Loyalität zur Erhaltung der Umsätze, sondern auch zu Kosteneinsparungen durch kundeninduzierte Lernprozesse, zu Neukunden aufgrund Kundenempfehlungen und zu zusätzlichen Umsätzen durch Cross-Selling von Produkten.

Die Verknappung von ökologischen Ressourcen führt auch zu einer größeren ökonomischen Knappheit und somit zu höheren Kosten für die Unternehmen (z. B. steigende Entsorgungskosten aufgrund von Deponieraumverknappung, höhere Grundstückskosten aufgrund zunehmender Versiegelung). Darüber hinaus stellen rechtliche Regelungen, wie z. B. die Altautoverordnung, Anforderungen an die Unternehmen, die erfolgsbeeinflussend wirken. Können Unternehmen den Umgang mit ökologischen Ressourcen als Strategie für eine Nische nutzen und für ihre ökologieorientierten Produkte einen höheren Preis erzielen, ergibt sich ein erfolgssteigernder Einfluss. Beispiele hierfür sind der ökologische Landbau oder Textilien mit dem Gütesiegel „Öko-Tex Standard 100". Der durch die Entstehung einer EU-Richtlinie vorangetriebene Handel mit Treibhausgasemissionslizenzen setzt ebenfalls am klassischen ökonomischen Kalkül der Unternehmen an (vgl. *EG-Kommission*, 2001).

1.3 Wirkung auf die Zielgröße „Erfolgspotenzial"

Schließlich ist der Umgang mit immateriellen und ökologischen Ressourcen auch als langfristiger Wettbewerbsvorteil nutzbar. Dass Anleger einen verantwortungsvollen Umgang mit immateriellen und ökologischen Ressourcen honorieren, zeigt sich in der Bewertung auf den Finanzmärkten. Die Kapitalmarktstudien von *Lev* u. a. zeigen eine hohe Wertrelevanz von Indikatoren zu immateriellen Ressourcen wie z. B. Patentanmeldungen, Pattentezitate, Markenwerten, F&E-Aufwand etc. (vgl. *Lev*, 2001). Öko-Fonds oder der *Dow Jones Sustainabi-*

lity Group Index bewerten das Engagement von Unternehmen im Bereich ökologischer Ressourcen. Darüber hinaus wirkt sich ein verantwortungsvoller Umgang mit den Ressourcen auch positiv auf das Unternehmensimage aus und gewährleistet auch auf diese Weise die langfristige Überlebensfähigkeit eines Unternehmens. Allerdings bleibt im Bereich des Erfolgspotentials die Frage zu beantworten, ob Unternehmen durch ihr Engagement bezüglich immaterieller und ökologischer Ressourcen alleine ihren Unternehmenswert zu steigern vermögen oder ob dieses ein Baustein in einem umfassenden strategisch ausgerichteten Management ist („conditio sine qua non“).

Nun stellt sich die Frage, wie diese Bedeutung von den Unternehmen wahrgenommen und umgesetzt wird.

1.4 Bedeutung immaterieller Ressourcen

Die zunehmend wahrgenommene Bedeutung immaterieller Ressourcen für das betriebliche Wirtschaften ist auf einen Trend in den westlichen Volkswirtschaften zurückzuführen, der bereits seit mehreren Jahrzehnten Strukturveränderungen hervorruft: den Rückgang des primären (Landwirtschaft und Rohstoffgewinnung) und des sekundären (weiterverarbeitende Industrie) Sektors. Hingegen gewinnt der tertiäre Sektor der Dienstleistungen eine zunehmende Bedeutung. Neben grundlegenden Dienstleistungsbereichen wie z. B. dem Gesundheitsbereich nutzen auch der primäre und sekundäre Sektor neben den klassischen materiellen (Anlage- und Umlaufvermögen) und finanziellen Ressourcen (Ausstattung mit liquiden Mitteln und Verschuldungspotential) zunehmend immaterielle Ressourcen. Unternehmen nutzen immaterielle Ressourcen in Form von Kundenkapital (z. B. aufgebaute Markenwerte oder Kundenbeziehungen), Humankapital (z. B. Know-how der Mitarbeiter oder Produkt- Know-how), Innovationskapital (z. B. Patente oder die Innovationskraft) und Prozess- und Strukturkapital (z. B. Informationssysteme, Standortfaktoren).

Wie eine aktuelle Studie von *PriceWaterhouseCoopers* und der *TU Dresden* zeigt, kommt in Industrien der „New Economy“, also in eher dienstleistungsnahen Bereichen (Stichprobe: CDAX-Indizes Media, Pharma/Health, Technology, Software und Telecommunication) nach Aussage des Managements dem Human Capital mit Abstand eine herausragende Bedeutung für den Unternehmenserfolg zu (vgl. *Abb. 1*). Materiellen und finanziellen Ressourcen wird dagegen nur eine durchschnittliche Bedeutung zugewiesen (vgl. *PriceWaterhouse-Coopers/Günther/Beyer,* 2003).

Betrachtet man die Abbildung der als relevant wahrgenommenen Ressourcen in der Unternehmenssteuerung, ergibt sich ein sehr ernüchterndes Bild. Nur knapp 20 % der befragten Unternehmen, die immateriellen Werte eine hohe Bedeutung beimessen, erfassen ihr Human-, Innovations- oder Kundenkapital in quantitativer Form, knapp die Hälfte der Unternehmen „nur qualitativ“ und bis zu 25 % überhaupt nicht. Weitaus besser schneidet die Erfassung der direkten und indirekten Unternehmensprozesse ab. Hier ist sicherlich durch die in den 80-er Jahren diskutierten und eingeführten Prozessmanagement-Methoden eine bessere Ausgangssituation gegeben (vgl. *Abb. 2*). Diese Ergebnisse bestätigen frühere Ergebnisse zur praktischen Ausgestaltung von Performance Measurement-Systemen wie z. B. der Balanced Scorecard in Unternehmen (vgl. *Günther/Grüning*, 2002, S. 5ff.).

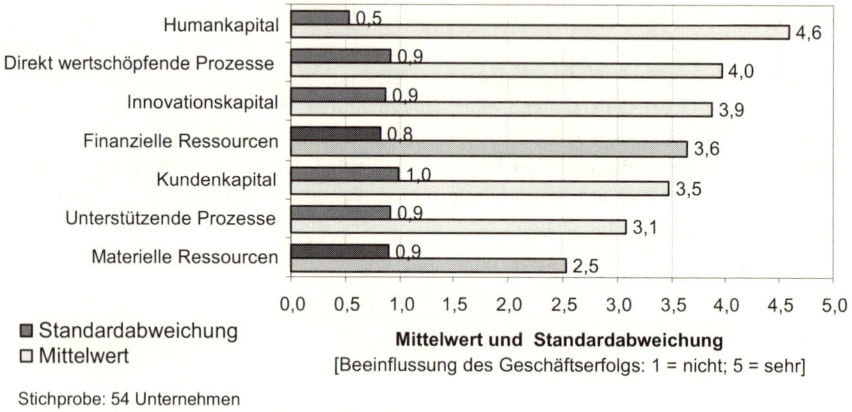

Abb. 1: Bedeutung betrieblicher Ressourcen für den Unternehmenserfolg (vgl. PriceWaterhouseCoopers/Günther/Beyer, 2003, S. 15)

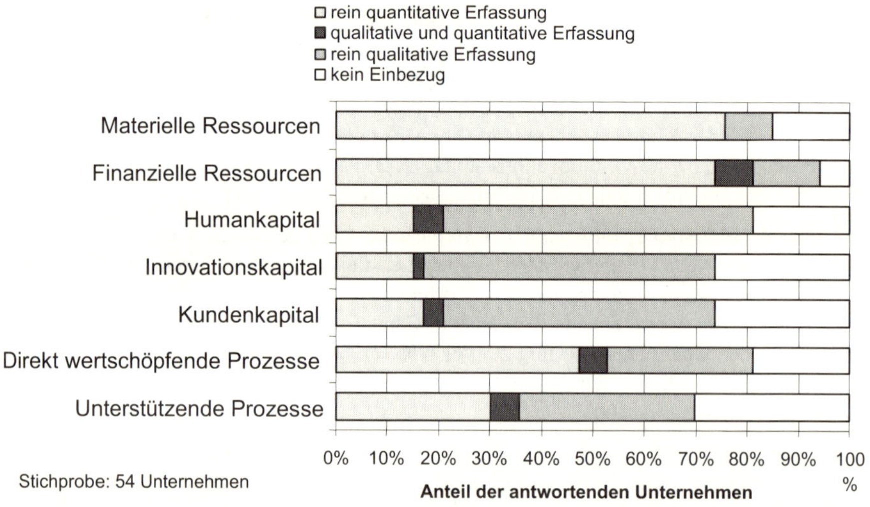

Abb. 2: Art der Messung betrieblicher Ressourcen in der Unternehmenssteuerung (PriceWaterhouseCoopers/Günther/Beyer, 2003, S. 17)

Für die Kommunikation nach außen in Form einer freiwilligen, den traditionellen Jahresabschluss und Lagebericht ergänzenden Berichterstattung ergibt sich ein vergleichbares Bild. Auch hier wird über „klassische" Ressourcen (finanzielle Ressourcen und materielle Ressourcen; letztere wurden wegen der ohnehin guten Abbildung im Jahresabschluss nicht mehr explizit betrachtet) in der Regel in quantitativer Form berichterstattet, für immaterielle Ressourcen ist die Art der angebotenen Daten (i. d. R. keine oder qualitative Daten) und deren Berichtsebene (i. d. R. nur auf Gesamtunternehmensebene) jedoch nicht deren offensichtlicher Relevanz entsprechend (vgl. *PriceWaterhouseCoopers/ Günther/Beyer,* 2003).

1.5 Bedeutung ökologischer Ressourcen

Wie historische Betrachtungen zeigen (vgl. *Weeber*, 1993) wurden ökologische Ressourcen bereits im Altertum lokal übernutzt. Allerdings hat sich die Wahrnehmung im Laufe der Zeit geändert, indem die Bedeutung anders bewertet wurde und vor allem die unmittelbare Betroffenheit und somit der Leidensdruck zunahm. Diese veränderte Wahrnehmung bei unmittelbarer Betroffenheit wird auch als NIMBY-Effekt (Not in my backyard) bezeichnet.

Unternehmen nutzen ökologische Ressourcen zur Versorgung (z. B. mit Rohstoffen), zur Entsorgung (z. B. Boden zur Aufnahme von Abfällen) sowie zur Regelung (z. B. Abbau des Phospatgehaltes im Abwasser durch natürliche Prozesse). Dabei können ökologische Ressourcen entweder ratenknapp sein (z. B. wird eine bestimmte Menge an Abwasserbelastung in einem bestimmten Zeitraum natürlich abgebaut, werden diese Grenzwerte überschritten, droht ein sog. Umkippen des Gewässers). Von Kumulativknappheit spricht man, wenn eine Ressource nach einer bestimmten Zahl von Nutzungen nicht mehr zur Verfügung steht (z. B. Erdöl zur Versorgung oder Deponieraum zur Entsorgung). Nun stellt sich auch hier die Frage nach der Bedeutung dieser Ressourcen für die Unternehmenssteuerung, nach ihrer Abbildung und nach der Kommunikation nach außen.

Eine aktuelle Studie des *ifo-Instituts* ordnet 43,1 % der Unternehmen einem umweltmanagementorientierten (25,2 %) oder nachhaltigkeitsorientierten (17,9 %) Unternehmenstypus zu. Dies verdeutlicht, dass bei den Unternehmen die Bedeutung ökologischer Ressourcen verankert ist (*ifo-Institut*, 2002).

Fragt man allerdings nach der Abbildung in unternehmerischen Entscheidungsinstrumenten, so zeigt sich auch hier ein ernüchterndes Bild. Gängige Praxis ist, dass die durch die Nutzung ökologischer Ressourcen entstehenden Kosten immer dann externalisiert werden, wenn sie nicht auch ökonomisch knapp sind. Seit die Versorgung mit Ressourcen, z. B. Wasser oder die Entsorgung von Abfällen bepreist wird, werden diese Ressourcen der Ver- bzw. Entsorgung auch in unternehmerischen Entscheidungskalkülen abgebildet.

Betrachtet man die Kommunikation der Unternehmen über den Umgang mit ökologischen Ressourcen, so zeigt sich, dass der Schwerpunkt der Berichterstattung in gesonderten Umweltberichten liegt. Eine Integration in die klassische Finanzberichterstattung erfolgt weniger. In Deutschland informieren nur ca. 5 % der Unternehmen über ihren Umgang mit ökologischen Ressourcen in der klassischen Berichterstattung. Weltweit liegt der Durchschnitt bei ca. 20 % (vgl. *KPMG*, 1999).

Ähnlich wie zum Ende der 80-er Jahre, als *Johnson* und *Kaplan* mit ihrem Buch „Relevance Lost" für die Kosten- und Leistungsrechnung die fruchttragende Frage aufwarfen, ob der state of the art noch ausreiche, die aktuellen Fragestellungen zu lösen, so scheint auch zu Beginn des 21. Jahrhunderts die Frage berechtigt, ob das praktizierte Rechnungswesen nicht falsche Schwerpunkte in seiner Ausrichtung setzt: Relevance Lost, Teil II.

2. Aktuelle Entwicklungen für das Rechnungswesen

Dass Handlungsbedarf im internen und auch im externen Rechnungswesen besteht, zeigt eine Vielzahl von Initiativen und Entwicklungen in den letzten Jahren, die eine Differenzierung oder Erweiterung des Rechnungswesens fordern:

Auf immaterielle Ressourcen bezogene Initiativen:

- Das Business Reporting Research Project des US-amerikanischen *Financial Accounting Standards Boards* (FASB) fordert eine Erweiterung der freiwilligen Berichterstattung der Unternehmen auch um Informationen über bilanziell nicht ansetzbare, immaterielle Vermögenswerte (vgl. FASB, 2001).
- Die Neuregelung der Behandlung von Goodwill in den US-GAAP durch SFAS 141 und 142 führt zu einem permanenten Impairment Test (Niederstwerttest) für einzelne immaterielle Vermögensgegenstände und wirft daher die Frage nach ihrem Wert neu auf.
- Der *Arbeitskreis „Immaterielle Werte im Rechnungswesen"* der *Schmalenbach-Gesellschaft* fordert eine Abschaffung des Aktivierungsverbotes gemäß § 248 II HGB für selbst geschaffene immaterielle Vermögensgegenstände des Anlagevermögens und eine Aktivierung innerhalb enger Grenzen (vgl. *AK Immaterielle Werte*, 2001). Ergänzend wurde jüngst ein Vorschlag für ein freiwilliges, den Jahresabschluss ergänzendes Reporting für immaterielle Werte veröffentlicht (vgl. *AK Immaterielle Werte*, 2003).
- Von Praktikern und Forschern wurden Controlling- und Berichtskonzepte für immaterielle Ressourcen entwickelt. Beispiele hierfür sind der *Skandia* Navigator (vgl. *Edvinsson/ Malone*, 1997), der Intangible Assets Monitor (vgl. *Sveiby*, 1997), der Intellectual Capital Navigator (vgl. *Stewart*, 1997), die Value Chain Scoreboard (vgl. *Lev*, 2001) oder der Intellectual Capital Report (vgl. *Austrian Research Center*, 2000 und *Maul*, 2000).
- Insbesondere im skandinavischen Raum ergänzen Unternehmen wie *Skandia*, *Celemi International*, *WM-data AB*, *KREAB*, *Jacobson & Widmark*, *Carl Bro a/s*, *Coloplast a/s* oder *Deutsche Bank AG* ihre externe Berichterstattung durch freiwillige Angaben über ihr Intellectual Capital.
- Aus Theorie und Praxis werden Forderungen nach einer kapitalmarkt- oder wertorientierten Berichterstattung (Value Reporting) erhoben (vgl. z. B. *Labhart*, 1999 und *Ruhwedel/Schulze*, 2002). Der *Arbeitskreis Externe Unternehmensrechnung der Schmalenbach-Gesellschaft* entwickelte Grundsätze für ein so genanntes Value Reporting (vgl. *AK Externe Unternehmensrechnung*, 2002). Das Reporting über immaterielle Werte stellt, wie auch die Forderungen aus der FASB-Studie und des Arbeitskreises Externe Unternehmensrechnung zeigen, einen wesentlichen Baustein hierzu dar.

Auf ökologische Ressourcen bezogene Initiativen:

- Die Empfehlung der EU-Kommission zur Berücksichtigung von Umweltaspekten in Jahresabschluss und Lagebericht von Unternehmen: Ausweis, Bewertung und Offenlegung thematisiert die Berücksichtigung von umweltschutzbedingten Verbindlichkeiten und Aufwendungen und bezieht sich ausdrücklich auf geltendes EU-Recht sowie International Accounting Standards. Stellungnahmen hierzu z. B. des Deutschen Standardisierungsrates sowie des Instituts der Wirtschaftsprüfer liegen bereits vor.
- Die Sustainability Reporting Guidelines der *Global Reporting Initiative* (vgl. *GRI*, 2002) zielen auf die Steuerung der ökonomischen, ökologischen und sozialen Leistung eines Unternehmens. Zur Steuerung werden Indikatoren vorgeschlagen. Dabei sollen die Hauptprobleme identifiziert, Anspruchsgruppen einbezogen und deren Betroffenheit bestimmt sowie der Einfluss auf die Zielfindung und die Maßnahmen des Unternehmens betrachtet werden.

- Die internationale, auch in Deutschland geltende Norm DIN EN ISO 14031 zielt auf die Messung der Umweltleistung von Unternehmen. Dabei wird zwischen operativen und managementbezogenen Leistungsindikatoren unterschieden. Darüber hinaus wird eine Messung des Umweltzustandes vorgeschlagen (vgl. *Nagus*, 2000).
- Die EG-Öko-Audit-Verordnung sieht eine freiwillige Validierung des Managementsystems von Organisationen vor. In der 2000 überarbeiteten Fassung ist die Bestimmung der Umweltleistung der zu validierenden Organisationen berücksichtigt.
- Stellvertretend für verschiedene Richtlinien, die zurzeit auf EU-Ebene erarbeitet werden, sei die Richtlinie über ein System für den Handel mit Treibhausgasemissionsberechtigungen genannt. Die Handelbarkeit von Verschmutzungsrechten überführt ökologische Ressourcen in das ökonomische Bewertungskalkül, indem die Rechte zugewiesen werden und anschließend handelbar sind.
- Das *International Financial Reporting Interpretations Committee* (IFRIC) des IASB veröffentlichte 2003 das Exposure Draft D 1 Emission Rights, in dem Verschmutzungsrechte als immaterieller Vermögensgegenstand interpretiert werden.
- Der Dow Jones Sustainability Group Index und Öko-Fonds spiegeln eine Bewertung der Nachhaltigkeit durch die Finanzmärkte wieder (*Dow Jones Sustainability Index*, 2002 und *Oekom research*, 2002).

Diese sicherlich nicht abschließenden Listen von aktuellen Aktivitäten im Sinne einer Erweiterung des Rechnungswesens veranschaulichen, dass vor allem das externe, aber auch das interne Rechnungswesen vor neuen Herausforderungen steht. Ein Vergleich der Initiativen zeigt, dass immaterielle Ressourcen deutlich stärker im „Rechnungswesen-Lager" diskutiert werden als ökologische Ressourcen. Bei letzteren überwiegen Initiativen aus dem „Umwelt-Lager". Vielleicht ist es gerade diese hauseigene Lobby, die ein zusätzliches Hemmnis für eine Weiterentwicklung des Rechnungswesens aufbaut.

3. Gemeinsamkeiten von ökologischen und immateriellen Ressourcen

Man mag sich nun vielleicht immer noch fragen, warum ökologische und immaterielle Ressourcen in diesem Beitrag zusammen betrachtet werden. Unseres Erachtens weisen beide, auf den ersten Blick verschieden erscheinende Fragestellungen, erhebliche Gemeinsamkeiten auf, die eine gemeinsame Lösung des Problems der Abbildung im Rechnungswesen erfordern:

- Wie unsere und andere Untersuchungen zeigen, sind ökologische und immaterielle Ressourcen notwendige Voraussetzungen für die Erstellung der Gesamtleistung und für den Unternehmenserfolg.
- Beide Ressourcenbündel entstehen außerhalb der „Systemgrenze" des Unternehmens. Dies hat z. B. die Konsequenz, dass Unternehmen Verschmutzungsmöglichkeiten oder Humankapital sehr wohl nutzen ohne selbst Eigentümer dieser Ressourcen zu sein. Die Rechte an diesen Ressourcen gehören Dritten bzw. der Allgemeinheit, sind also nicht explizit zugewiesen. Das behindert ihre Aktivierungsfähigkeit in der Bilanz, wenngleich diese Ressourcen quasi „geleast" werden.

- Die Verantwortung für beide Ressourcenbündel liegt in der Regel organisatorisch in nicht direkt wertschöpfenden Querschnittsbereichen (z. B. F&E-, Umweltschutz- oder Personalabteilung), sofern sich diese selbst überhaupt als Verantwortliche verstehen.
- Beide Ressourcentypen stellen Komponenten der Nachhaltigkeit (sustainable development), d. h. eine an den in Zukunft benötigten Ressourcen ausgerichteten Unternehmenssteuerung dar. Wird diese mit den drei Ausprägungen ökonomisch, ökologisch und sozial gemessen, so fließen sowohl immaterielle als auch ökologische Ressourcen ein.
- Eine Preisbildung an Märkten ist für beide Ressourcen aufgrund der oben beschriebenen Eigenheiten nur partiell möglich. Wenn eine Preisbildung möglich ist, werden die Ressourcen in der Regel bereits in der Bilanz, wie z. B. Patente und Rückstellungen für Altlasten, oder in der Gewinn- und Verlustrechnung, wie z. B. Markenkosten oder Entsorgungskosten, erfasst.
- Eine vollständige Erfassung aller Wirkungen von und auf ökologische bzw. immaterielle Ressourcen scheint nicht möglich. Eine vollständige (körperliche) Bestandsaufnahme im Sinne einer Inventur wird nicht möglich sein.
- Für die immateriellen und ökologischen Ressourcen, die erfasst werden sollen, gibt es - wie noch zu erläutern sein wird - Ansatz- und/oder Bewertungsprobleme ähnlicher Art.
- Für die Ressourcen selbst gibt es keine trennscharfen Definitionen; ihre Messung ist häufig nur qualitativ möglich, wodurch die Objektivität eingeschränkt wird.
- Grundsätze ordnungsmäßiger Buchhaltung (GoB) für den Ansatz, die Bewertung und den Ausweis sind aufgrund ihres Neuigkeitsgrad im Gegensatz zu den jahrhundertlangen Erfahrungen mit der Doppik für finanzielle und materielle Ressourcen noch nicht ausgebildet.

Bei Betrachtung dieser Gemeinsamkeiten wird ersichtlich, dass die Grenzen zwischen ökologischen und immateriellen Ressourcen fließend werden, wenngleich inhaltlich unterschiedliche Aspekte betrachtet werden. Nach dem ressourcen-orientierten Ansatz des strategischen Managements stellen beide essentielle Ressourcen für das Unternehmen dar, die vom Unternehmen genutzt werden, um Wettbewerbsvorteile auf seinen Absatzmärkten zu erringen (marktorientierter Ansatz), um letztlich Gewinne zu erzielen und positive Wertbeiträge zu erwirtschaften. Daher werden nachfolgend beide Ressourcentypen gemeinsam betrachtet, um so ihre Abbildbarkeit im Rechnungswesen zu analysieren.

4. Idealmodell: Integrierter Ansatz

Aufgabe des Rechnungswesens ist es, leistungswirtschaftliche Vorgänge monetär abzubilden. Sofern immaterielle und ökologische Ressourcen steuerungsrelevant sind, stellt sich die Frage, wie diese Ressourcen in das Rechnungswesen integriert werden können.

4.1 Integration in das vorhandene Rechnungswesen

Das vorhandene Rechnungswesen berücksichtigt Ressourcen als einzelne Vermögensgegenstände in den Aktiva des Unternehmens. Sowohl nach nationalem Handels- und Steuerrecht als auch weitgehendst nach internationalem Recht (IAS bzw. US-GAAP) werden Ressourcen als einzelne Vermögensgegenstände bzw. Wirtschaftsgüter aktiviert, sofern sie einen wirtschaftlichen Wert darstellen, einzeln veräußerbar sind und selbstständig bewertet werden können. Immaterielle und ökologische Ressourcen stellen in der Regel – wie obige Ausführungen zur Auswirkung auf die drei Zielgrößen Liquidität, Erfolg und Erfolgspotential deutlich machen – einen wirtschaftlichen Wert dar. Eine einzelne Veräußerbarkeit und separate Bewertbarkeit ist jedoch nur in Einzelfällen wie z. B. bei Patenten, Verschmutzungsrechten, Markenrechten oder selbsterstellter Software gegeben. Zudem stehen z. B. beim Humankapital oder bei ökologischen Ressourcen die Eigentums- und Verfügungsrechte nicht dem Unternehmen, sondern einzelnen Individuen oder der Allgemeinheit (sog. Allmenderessourcen) zu. Sollen jedoch immaterielle und ökologische Ressourcen umfassend und möglichst vollständig betrachtet werden, scheidet eine Integration in das vorhandene Rechnungswesen aus.

4.2 Integration in ein weiterentwickeltes Rechnungswesen

Blickt man historisch auf die Entwicklung des Rechnungswesens, so ist das von *Luca Pacioli* bereits vor über 500 Jahren entwickelte Rechnungswesen, wie wir es noch heute als Finanzbuchhaltung nutzen, stets aufgrund gesellschaftlicher und wirtschaftlicher Entwicklungssprünge überarbeitet und ergänzt worden. Die Finanzbuchhaltung folgt dem Bilanzkonzept der Nominalkapitalerhaltung, daher entsteht Gewinn (= Jahresüberschuss) dann, wenn der monetäre, nominale Wert des Netto-Vermögens (= Eigenkapitals) steigt. Die industrielle Entwicklung im 18. und 19. Jahrhundert führte durch die damalige Knappheit der Produktionsmittel zu einem engen (materiellen) Ressourcenbezug, zum Rückzug auf den Produktionsbereich und zur Substanzwertorientierung. Die damit entstandene Kosten- und Leistungsrechnung ergänzt durch Substanzerhaltungsrechnungen in späteren Jahrzehnten ist Ausdruck der Realkapitalerhaltung. Gewinn (= Betriebsergebnis) entsteht dann, wenn das (reale) betriebsbedingte Netto-Vermögen erhalten wird. Scheingewinne werden bereinigt und Substanzerhaltungsrücklagen zum Ausgleich von Preissteigerungen auf der Passivseite gebildet. Je nachdem, ob die Fremdfinanzierung berücksichtigt wird, kann dann noch zusätzlich zwischen Brutto- und Nettosubstanzerhaltung unterschieden werden.

Bereits beginnend mit bilanztheoretischen Überlegungen in den 1960er Jahren, aber vor allem angetrieben durch die zunehmende Ausbreitung der Ausrichtung am Unternehmenswert (Shareholder Value-Konzept) erfolgte eine Ausrichtung am Grundsatz der Erfolgskapitalerhaltung. Gewinn, verstanden als ökonomischer Gewinn, ergibt sich als Differenz der Zukunftserfolgswerte (= Unternehmenswerte) zu Periodenende bzw. Periodenanfang. Erfolg, verstanden als Wertschaffung, findet erst statt, wenn der Gewinn über den Kapitalkosten auf das investierte Kapital liegt.

Durch die Fair Value-Bewertung nach internationalem Handelsrecht ist ein Teil des Bilanzverständnisses auch in das externe Rechnungswesen übergeschlagen. So können z. B. immaterielle Vermögensgegenstände, die sich durch die Aufteilung des Goodwills im Fall des Erwerbs ergeben, nach SFAS 141 und SFAS 142 sowohl bei der Erst- als auch bei der jährlichen Folgebewertung (impairment test) nach der income method bewertet werden. Die income method bewertet den Vermögensgegenstand mit dem Barwert zukünftiger Nettozuflüsse, quasi als Discounted Cashflow. Ziffer 12-18 der EU-Empfehlung zur Berücksichtigung von Umweltaspekten im Jahresabschluss und Lagebericht sehen eine derartige Kapitalisierung von Umweltaufwendungen bei Vorliegen der Voraussetzungen als Vermögensgegenstand vor.

Zur Berücksichtigung immaterieller und ökologischer Ressourcen könnte nun das Bilanzverständnis wiederum in einer vierten Stufe zu einer nachhaltigen Kapitalerhaltungsrechnung erweitert werden. Bei der nachhaltigen Kapitalerhaltungsrechnung werden neben den materiellen (d. h. bereits berücksichtigten ökologischen) und finanziellen Ressourcen zusätzlich auch immaterielle und ökologische Ressourcen betrachtet, wenngleich sie dem Unternehmen nicht als Eigentümer gehören, jedoch quasi von ihm mitgenutzt werden. Nachhaltige Kapitalerhaltung liegt, der Idee des substainable developments folgend, vor, wenn alle vier Ressourcentypen (materielle, finanzielle, immaterielle und ökologische) durch das unternehmerische Wirtschaften in ihrem Kapitalstock für zukünftiges Wirtschaften erhalten bleiben. Gewinn, ausgedrückt als Wertschaffung (Integrated Value Added), entsteht erst dann, wenn die Kapitalkosten der vier Kapitalstöcke bedient sind.

Greift man die Grundidee von *Sveiby* zur unsichtbaren Bilanz auf (vgl. *Sveiby*, 1997, S. 11), kann die traditionelle in der Finanzbuchhaltung betrachtete Bilanz zusätzlich um immaterielle Ressourcen (hier analog zu *Sveiby* gegliedert in die externe Struktur, die interne Struktur und die Kompetenz der Mitarbeiter) und in Erweiterung von *Sveiby* um genutzte ökologische Ressourcen erweitert werden (Erweiterter Bilanzansatz). Auch auf der Passivseite können zusätzlich Passiva berücksichtigt werden, die durch die erweiterte Sichtweise entstehen. Dies können z. B. ungeschriebene Verpflichtungen des Unternehmens gegenüber seinen Mitarbeitern zur Erhaltung und zum Ausbau der Unternehmenskultur, Rückstellungen für geplante Maßnahmen zur Erhaltung des Markenwertes sowie bisher nicht berücksichtigte Verpflichtungen zur Erhaltung von ökologischen Ressourcen sein, für die bisher keine rechtliche Verpflichtung besteht (vgl. *Abb. 3*). Der bisher sichtbare Teil der Bilanz wird um einen bisher unsichtbaren Teil erweitert. Damit ist auch die Hoffnung verbunden, die bei vielen Unternehmen sich auseinander entwickelten Buchwerte (= Eigenkapital) und Marktwerte wieder enger zusammenzuführen, indem mit dem immateriellen Kapital und dem ökologischen Kapital Komponenten zusätzlich berücksichtigt werden, die bisher zwar für den Marktwert relevant waren, bilanzrechtlich jedoch nicht berücksichtigt werden konnten.

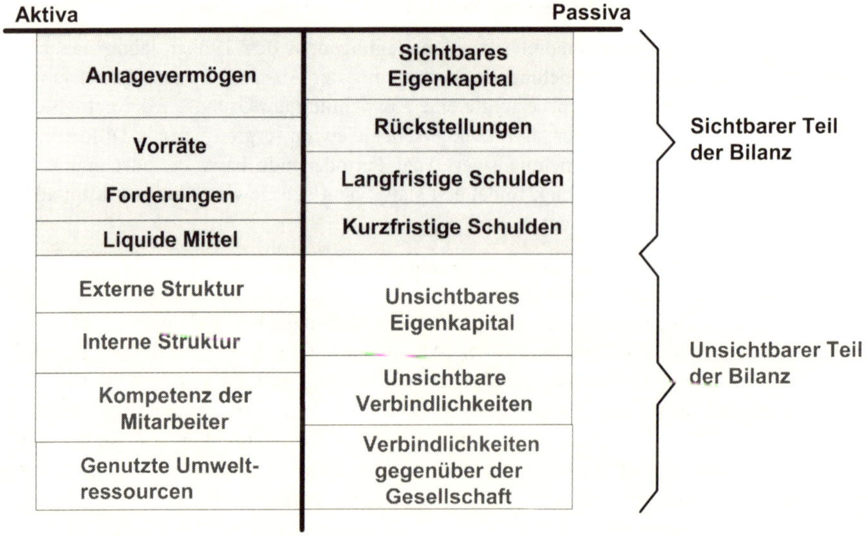

Abb. 3: Erweiterte unsichtbare Bilanz (in Erweiterung von Sveiby, 1997, S. 11)

Das jüngst veröffentlichte Exposure Draft D 1 Emission Rights des *International Financial Reporting Interpretations Committee (IFRIC)* des *IASB* stellt eine erste Entwicklung in diese Richtung dar. Es schlägt vor, kostenlos oder gegen Entgelt zugewiesene Emissionsrechte als immaterielle Vermögensgegenstand im Sinne von IAS 38 zu behandeln. Bei Zuweisung unter dem Marktpreis der entstehenden Märkte für Emissionsrechte ist ein passivischer Rechnungsabgrenzungsposten zu bilden, der über die Laufzeit des Emissionsrechtes aufgelöst wird. Die Verpflichtung der Unternehmen zur Deckung ihrer Emissionen mit Emissionsberechtigungen wird auf der Passivseite als Verbindlichkeit dargestellt. Damit erfolgt ein Bruttoausweis im Sinne der in *Abb. 3* dargestellten erweiterten unsichtbaren Bilanz. Zudem werden externe Effekte auch für den Fall kostenloser Nutzung bilanziell dargestellt (vgl. zu Details *Günther*, 2003). Eine vergleichbare Handhabung schlägt *IFRIC* auch für so genannte Renewable Energy Certificates (REC) vor.

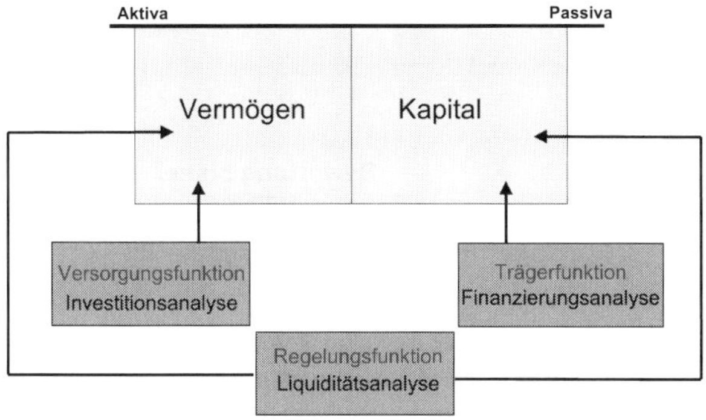

Abb. 4: Generalisierter Bilanzansatz

Die nachhaltige Kapitalerhaltungsrechnung wäre damit bezüglich ihres Grundkonzeptes in der Lage, bilanzähnliche Inventurlisten wie Ökobilanzen (für ökologische Ressourcen) und Wissensbilanzen oder das Intellectual Property Statement (vgl. z. B. *Maul*, 2000; *Austrian Research Center*, 2000) (für immaterielle Ressourcen) zu inkorporieren. Das traditionelle Bilanzverständnis mit dem Vermögen bzw. der Analyse der Investitionen auf der Aktivseite, dem Kapital bzw. der Analyse der Finanzierung auf der Passivseite und dem Abgleich beider Seiten durch die Liquiditätsanalyse würde generalisiert werden durch die Darstellung der Versorgung mit Ressourcen (Aktiva), dem Träger der Ressourcen (Passiva) und der Regelung als Abgleich von Aktiv- und Passivstruktur (vgl. *Abb. 4*).

4.3 Umsetzungsprobleme

Zur konkreten Ausgestaltung der Bilanz kann der traditionellen Vorgehensweise folgend in Ansatz (Darf/Soll/Muss eine Ressource angesetzt werden?), Bewertung (Mit welchem Wert soll die Ressource angesetzt werden?) und Ausweis (Unter welcher Position soll die Ressource angesetzt werden?) gefolgt werden. Die Frage des Ausweises soll in diesem Beitrag nicht näher betrachtet werden.

Zur Berücksichtigung immaterieller und ökologischer Ressourcen ist eine Erweiterung sowohl bezüglich des Ansatzes als auch bezüglich der Bewertung notwendig (vgl. *Abb. 5*).

So fordert z. B. der *Arbeitskreis „Immaterielle Werte im Rechnungswesen"* der *Schmalenbach-Gesellschaft* eine Aufhebung des Aktivierungsverbotes für selbst erstellte immaterielle Vermögensgegenstände des Anlagevermögens gemäß § 248 II HGB. Hierdurch könnten diejenigen immateriellen Ressourcen in der Bilanz angesetzt werden, die die abstraktive Aktivierungsfähigkeit z. B. als Vermögensgegenstand erfüllen.

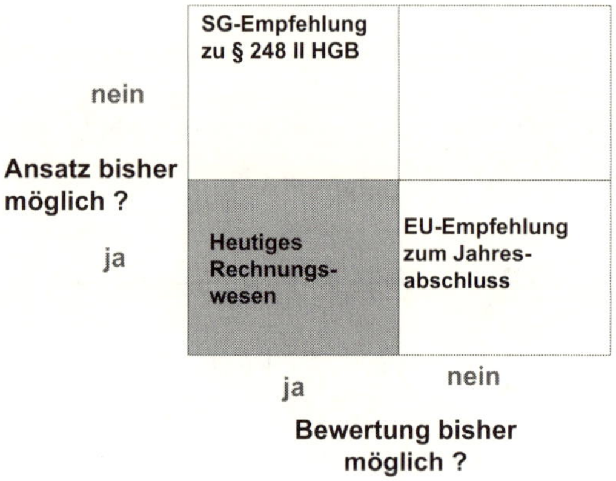

Abb. 5: Ansatz- und Bewertungsproblematik

Dennoch treten in der konkreten Umsetzung eine Reihe von Problemen auf, die die Integration von immateriellen und ökologischen Ressourcen in eine nachhaltige Kapitalerhaltungsrechnung zum heutigen Zeitpunkt als wenig erfolgversprechenden Weg erkennen lassen:

Der Ansatz immaterieller und ökologischer Ressourcen erfordert analog zur Berücksichtigung von materiellen und finanziellen Vermögensgegenständen eine eindeutige Identifizierbarkeit und Separierbarkeit einzelner Ressourcen. Immaterielle und ökologische Ressourcen können jedoch häufig nicht isoliert werden, bzw. von anderen Ressourcen getrennt werden. Sie treten quasi vernetzt auf. So ist es fast unmöglich, das Humankapital eines Unternehmens von seinen Innovationen, seinen Kundenbeziehungen oder seinem Strukturkapital zu trennen. Ebenso ergeben sich vernetzte, komplexe, kaum mehr einzeln isolierbare Auswirkungen auf die Umwelt z. B. durch die Erweiterung eines Werkes wie des *Airbus*-Werkes in Hamburg. Distanz-, Langzeit-, Synergie-, Kumulationswirkungen verstärken die Komplexität dieser Zusammenhänge noch.

Eine vollständige Darstellung aller immateriellen und ökologischen Ressourcen wird im Gegensatz zum traditionellen Verständnis einer Inventur als körperliche Bestandsaufnahme nicht möglich sein, da einerseits kein abgeschlossener Katalog von eindeutigen Definitionen vorliegt und zum anderen viele Wirkungen noch unbekannt bzw. unerforscht sind.

Durch den spezifischen Charakter immaterieller und ökologischer Ressourcen (z. B. teilweise fehlende Eigentumsrechte, Vernetzung, teilweise öffentliche Güter) ist eine Bewertung auf Güter- und Dienstleistungsmärkten nicht möglich. Dadurch fehlen Anschaffungskosten sowie Marktpreise. Daher sind allenfalls Herstellungs-, Wiederbeschaffungs- oder Reproduktionskosten ermittelbar, die nur eine beschränkte Bewertung erlauben.

Viele Aspekte, wie z. B. der Wert einer Schulungsmaßnahme oder die Erhaltung von Artenvielfalt, sind monetär derzeit nicht einfach bewertbar. Daher kann als Substitut nur auf quantitative oder qualitative Indikatoren zurückgegriffen werden, die jedoch den Einbezug in eine erweitere Bilanz mangels Aggregation über Geldwerte verhindern.

Traditionelle Grundsätze ordnungsmäßiger Buchführung (GoB) wie Richtigkeit, Willkürfreiheit, Klarheit, Vollständigkeit, Stetigkeit und Wirtschaftlichkeit sind auf immaterielle und ökologische Ressourcen nicht oder nur eingeschränkt übertragbar. Eine Bewertung des Humankapitals und des Umweltkapitals wird voraussichtlich nie willkürfrei und nie vollständig möglich sein. Dennoch zeigen unsere empirische Studien, dass die Unternehmen mit den gleichen Standards an die Berichterstattung über immaterielle Werte herangehen wie an das traditionelle externe Rechnungswesen (vgl. *PriceWaterhouseCoopers/Günther/Beyer*, 2003, S. 30). Die Berücksichtigung ökologischer Ressourcen wird bisher eher für eine getrennte Berichterstattung diskutiert. Neue GoB scheinen notwendig, um eine Berichterstattung über immaterielle und ökologische Ressourcen im Rahmen des integrierten Ansatzes überhaupt zu ermöglichen. Thesenartig können folgende ersten Grundsätze vorgeschlagen werden:

- Die Ressourcen sollen nicht vollständig, sondern alle wesentlichen Ressourcen sollen erfasst werden.
- Die Erfassung kann nicht willkürfrei, sondern soll nachvollziehbar sein.
- Die Erfassung soll richtig bezüglich des Ansatzes, kann jedoch nur plausibel bezüglich der Bewertung sein.
- Der Grundsatz der Klarheit ist durch die logische Struktur der Erfassung und Bewertung zu gewährleisten.
- Es ist besser nicht zu bewerten, als eine lähmende Diskussion über das richtige Bewertungsverfahren auszulösen.

Auch wenn die methodischen Überlegungen zur nachhaltigen Kapitalerhaltungsrechnung noch nicht abgeschlossen sind, erscheint eine Integration in das Rechnungswesen auch mit einem erweiterten Bilanzverständnis zum derzeitigen Zeitpunkt nicht möglich bzw. würde nur für einen Teil der betrachteten Ressourcen möglich sein.

5. Realmodell: Stufen-Konzept

Scheint aufgrund von praktischen Überlegungen eine zufriedenstellend vollständige Umsetzung des integrierten Ansatzes zum heutigen Zeitpunkt nicht möglich, bleibt zu fragen, wie trotzdem immaterielle und ökologische Ressourcen im Rechnungswesen so berücksichtigt werden können, dass eine vernünftige Steuerung möglich wird und die bisherige Berücksichtigung verbessert wird. Hierbei erscheint eine stufenweise Vorgehensweise vorteilhaft.

5.1 Stufe 1: Differentiated Value Added

Im traditionellen internen und externen Rechnungswesen kann in einer ersten Stufe versucht werden, immaterielle Ressourcen und ökologische Ressourcen so auszuweisen, dass diesbezügliche Entscheidungen adäquat unterstützt werden können.

So ist z. B. eine Bewertung und Steuerung von Marken nicht möglich, solange markenspezifische Kosten (z. B. spezielle Werbe- und Kommunikationsmaßnahmen, markenspezifische Produktentwicklungen oder Qualitätsmaßnahmen) und markenspezifische Erlöse nicht separat erfasst werden. Genauso kann im Bereich ökologischer Ressourcen eine aussagekräftige Umweltkostenrechnung entwickelt werden. In der Regel wird dabei im Gegensatz zum integrierten Ansatz die Struktur des Rechnungswesens nicht in Frage gestellt, sondern es folgt nur eine stärkere Differenzierung z. B. in der Kostenarten-, Kostenstellen- oder Kostenträgerrechnung. So kann auf bereits erarbeitete Strukturierungen von Umweltkosten wie z. B. nach VDI 3800 oder von Investitionen gemäß der Struktur des § 15 UStatG zurückgegriffen werden. Die Ausgestaltung der Differenzierung ist jedoch den in den jeweiligen Unternehmen gegebenen Fragestellungen im Sinne der Kosten- und Leistungsrechnung als Zweckrechnung anzupassen.

Im Bereich der immateriellen Ressourcen ist z. B. eine stärkere Zerlegung des bezahlten Goodwills aus Übernahmen im Rahmen der Purchase Price Reallocation gemäß SFAS 141 oder 142 sinnvoll, die eine pauschale Sammelposition Goodwill durch den expliziten Ausweis einzelner, wenn auch weniger immaterieller Vermögensgegenstände ersetzt. Ebenso würde die vom *Schmalenbach-Arbeitskreis* vorgeschlagene Aufhebung des Aktivierungsverbotes nach § 248 II HGB Spielraum für einen differenzierten Ausweis selbstgeschaffener immaterieller Vermögensgegenstände schaffen.

Zu berücksichtigen ist jedoch, dass der Rahmen der traditionellen Rechnungslegung bei dieser Vorgehensweise nicht angezweifelt wird und daher immaterielle und ökologische Ressourcen auch nur in diesem Rahmen erfasst werden können. Die Messung beschränkt sich daher auf diejenigen Ressourcen, die monetär gemessen werden können.

5.2 Stufe 2: Adjusted Value Added

Aufgrund des teilweise öffentlichen Charakters von immateriellen und ökologischen Ressourcen und der hierdurch bedingten beschränkten Exklusivität und Bewertbarkeit nutzen Unternehmen derartige Ressourcen, ohne der Allgemeinheit die entstehenden Kosten zu ersetzen bzw. Nutzen in Rechnung zu stellen. Als Erweiterung zur ersten Stufe wäre es daher denkbar, fiktive Kosten und Erlöse in Entscheidungsrechnungen zu integrieren. Eine systematische Integration in das traditionelle Rechnungswesen ist gar nicht direkt notwendig und würde dieses zudem erheblich verzerren. Diese fiktiven Kosten und Erlöse stellen quasi Raten für ungeschriebene Leasing- und Mietverträge dar, indem z. B. die Bildungsleistung von Universitäten oder die Trägerfunktion der Luft und Gewässer zur Aufnahme von Emissionen genutzt wird.

Für Entscheidungsrechnungen könnten nun diese kostenlosen Nutzungen bepreist werden, wodurch auch eventuell anstehende Änderungen im Unternehmensumfeld wie z. B. die Einführung und der Handel mit Verschmutzungsrechten simuliert werden können. Ebenso können z. B. Kapitalkosten auf vorhandene, jedoch nicht aktivierte Markenwerte angesetzt werden, um der Tatsache Ausdruck zu verleihen, dass unsichtbares Vermögen als Investment genutzt und in seiner Substanz erhalten werden muss.

Für derartige Entscheidungen ist der Netto-Effekt der drei entscheidungsrelevanten Kostenkategorien zu berücksichtigen (vgl. *Abb. 6*). Die Kosten der Handlungsalternative (Aktionskosten), die eventuell zum Teil auf Dritte übergewälzt werden können (überwälzbare Kosten), sind mit den Alternativen, z. B. Konsequenzen einer Reaktion Dritter zu vergleichen (Sanktionskosten). Für immaterielle Werte bedeutet dies z. B., dass die Marketingkosten durch den Transfer einer vorhandenen Marke auf einen neuen Produktbereich (Line extension) im Idealfall durch erhöhte Kundenumsätze ausgeglichen wird, ohne dass die vorhandene Marke in ihrer Markenstärke und damit ihrem Markenwert beschnitten wird. Im Umweltbereich können die Mehrkosten durch die Einführung eines systematischen Abfallmanagements durch Vereinbarungen mit Lieferanten und Kunden zu Mehrwegbehältern oder zur Verpackungsgestaltung (Überwälzung auf Lieferanten und Kunden) gemindert werden. Die sich vielleicht immer noch ergebenden Mehrkosten sind jedoch mit eventuellen Auswirkungen potentieller Sanktionen z. B. durch Nichtauswahl als Systemlieferant abzugleichen.

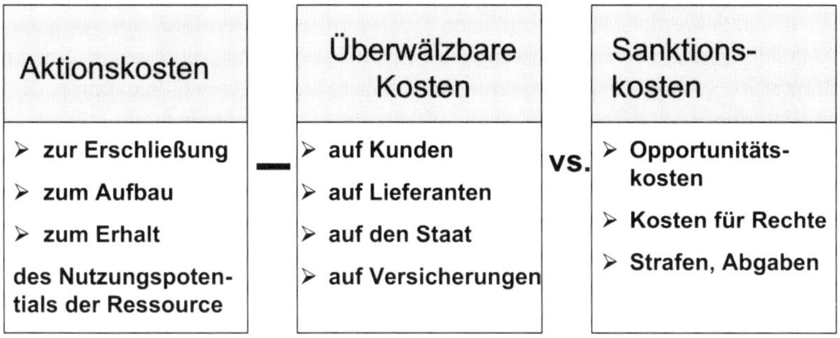

Abb. 6: Analyse des Netto-Effektes bei Ressourcenentscheidungen

Durch eine hiermit angepasste Wertbeitragsrechnung (Adjusted Value Added), die auch Opportunitätskosten und externe Effekte berücksichtigt, könnten z. B. Fehlentscheidungen im Bereich immaterieller und ökologischer Ressourcen vermieden werden.

5.3 Stufe 3: Extended Value Added

Die fiktive Berücksichtigung als Adjusted Value Added setzt jedoch voraus, dass immaterielle und ökologische Ressourcen monetär bewertet werden können. Ist eine Bewertung nicht möglich, können jedoch die versteckten Ressourcen zumindest identifiziert werden (d. h. der Ansatz ist möglich). So bietet sich als dritte und letzte Möglichkeit an, die Auswirkungen der Ressource nicht-monetär, aber quantitativ über Indikatoren zu erfassen.

Für die Auswahl und Gestaltung der Indikatoren sind hierbei einige Anforderungen zu erfüllen:

- Primäre Intention ist es, Indikatoren zu finden, die in einer Ursache-Wirkungs-Relation mit den obersten Zielgrößen des Unternehmens (wie eingangs erläutert Liquidität, Erfolg und Erfolgspotential) stehen. Dadurch werden Top-down- (Was erklärt den ökonomischen Erfolg?) und Bottom-Up-Analysen (Wie wirken sich Veränderungen der immateriellen und ökologischen Ressourcen auf die Unternehmensziele aus?) möglich. Lev fordert, dass die Indikatoren empirisch als signifikante Einflussgrößen auf den Unternehmenswert belegt sind (vgl. *Lev*, 2001, S. 164). Daher können hier Ansätze des Performance Measurement für die Messung von immateriellen und ökologischen Ressourcen genutzt werden.
- Für die immateriellen Ressourcen soll eine Messung der Wirkung der untersuchten Ressourcen (z. B. der Umsatz mit neuen Produkten als so genannte Outcome-Kennzahl für das Innovationskapital) gegenüber einer Erhebung von Outputgrössen (z. B. im Wirt-

schaftsjahr neu angemeldete Patente) präferiert werden. Outputgrößen sollen wiederum Prozessgrössen (z. B. Anzahl neuer Produkte) und diese wiederum reinen Inputgrößen (z. B. F&E-Aufwand in Relation zum Umsatz) vorgezogen werden.

- Für die ökologischen Ressourcen kann zwischen operativen Indikatoren, die das Ergebnis des Umgangs mit den Ressourcen messen (z. B. spezifische Verbrauche oder Entsorgungskosten), strategischen Indikatoren, die die Steuerung der Ressourcen abbilden (z. B. Anzahl der Standorte mit Umweltmanagementsystem oder Schulungen) und Zustandsindikatoren, die den Zustand der Umwelt außerhalb des Unternehmens messen, unterschieden werden. Letztere können für ein Unternehmen oder eine Branche dann entscheidungsrelevant werden, wenn sie einen maßgeblichen Anteil an der Nutzung bestimmter Ressourcen (z. B. Wasserqualität) haben. Gemessen werden können der absolute Wert, die Veränderung im Zeitablauf oder der Zielerreichungsgrad.
- Analog zu den Verfahren des Performance Measurement sollen Kennzahlen die Grundsätze der Validität (Wird das gemessen, was gemessen werden soll?), Reliabilität (Ist die Messung zuverlässig oder mit Zufälligkeiten behaftet?), Objektivität (Kommt ein Dritter zum selben Messergebnis?) und Effizienz (Lohnt sich der Aufwand der Messung?) berücksichtigen (vgl. z.B. *Gleich*, 2001 und *Grüning*, 2002).
- Wie bereits erläutert, lassen sich dabei sowohl Doppelzählungen (aufgrund der Verknüpfung zwischen verschiedenen immateriellen und ökologischen Ressourcen) als auch Unvollständigkeiten in der Praxis nicht vermeiden. Vielmehr soll durch die Mehrzahl und den systematischen Zusammenhang der Kennzahlen ein möglichst zutreffendes Bild des Unternehmens gezeichnet werden.

Sowohl im immateriellen als auch im ökologischen Bereich sind daher eine Vielzahl methodisch ähnlicher Ansätze entstanden, die sich auf Indikatoren stützen. Beispielhaft seien folgende Ansätze genannt:

Aggregierende Verfahren

Aggregierende Verfahren versuchen die Leistungsfähigkeit des Unternehmens in einer Kennzahl zu messen, die durch Aggregationen aus mehreren bewerteten und untereinander gewichteten Kennzahlen hervorgeht.

Beispiele für immaterielle Ressourcen:

- TechFactor™ Method von *A.D. Little* zur Bewertung von Technologien (*Hastabacka/ Khoury*, o. J.)
- IC-Index™ als ganzheitlicher Intellectual Capital-Index aus verschiedenen Einzelindikatoren (*Roos* u. a., 1997)

Beispiele für ökologische Ressourcen:

- Eco-Indicator (*PRé Consultants B.V.*, 2000)
- Umweltbelastungspunkte (*BUWAL*, 1998)
- Materialintensität pro Service-Einheit (MIPS) (*Schmidt-Bleek*, 1998)

Mehrfaktoren-Modelle

Mehrfaktoren-Modelle stellen mehrere Kennzahlen nebeneinander dar, ohne eine Aggregation vornehmen zu müssen. Zwischen den einzelnen Kennzahlen besteht in der Regel ein logischer Zusammenhang z. B. als Rechensysteme oder als Ordnungssystem z. B. in Anlehnung an die Balanced Scorecard.

Beispiele für immaterielle Ressourcen:

- *Skandia* Navigator (*Edvinsson/Malone*, 1997)
- Liahona-Homes-Modell (*Bornemann* u. a., 1999)
- Value Chain Scoreboard (*Lev*, 2001)
- Intellectual Capital Navigator (*Stewart*, 1997)
- Intangible Asset Monitor (*Sveiby*, 1997)
- Intellectual Capital Statement (*AK Immaterielle Werte*, 2003)

Beispiele für ökologische Ressourcen:

- Wirkungsindikatoren (*Umweltbundesamt*, 1999)
- Relativ-abstufende Bewertung (*Bundesumweltministerium/Umweltbundesamt*, 2001)
- Kritische Volumina (*Etterlein* u. a., 1992)

Wie die kurze Darstellung zeigt, existiert eine Vielzahl möglicher Indikatoren als auch möglicher Indikatorensysteme. In den letzten Jahren sind Bemühungen zu erkennen, die verschiedenen Vorschläge zu vereinheitlichen und Berichtsstandards zu entwickeln. Aufgrund der Vielschichtigkeit der Probleme und unternehmens- und branchenindividueller Besonderheiten bestehen jedoch Zweifel, ob eine Standardisierung zielführend ist.

6. Fazit

Die Bedeutung, die immateriellen und ökologischen Ressourcen für die Unternehmenssteuerung zukommt, stellt das Rechnungswesen vor neue Herausforderungen. Die Steuerung von immateriellen und ökologischen Ressourcen wird bisher noch in der Regel zentralen Stabsabteilungen (Personal, F&E, Marketing, Umweltschutz) überlassen. Im Controlling, dem eigentlich die Aufgabe der Koordination von Teilsystemen des Unternehmens zukommt, ist bisher das Bewusstsein hierfür noch zu gering.

Eine Integration dieser bisher versteckten und unsichtbaren Ressourcen in das Rechnungswesen erscheint aufgrund der besonderen Eigenschaften der beiden betrachteten Ressourcentypen in vollem Umfange nicht möglich. Daher wird ein Stufenmodell vorgeschlagen, dass eine monetäre Abbildung immaterieller und ökologischer Ressourcen soweit als möglich realisiert, ohne hierdurch für nur schwerfällig monetarisierbare Effekte Bewertungsprobleme einzugehen. Letztendlich ist zu gewährleisten, dass die ureigensten Ziele des Unternehmens, Liquidität, Erfolg und Erfolgspotential, auch für und mit immateriellen und ökologischen Ressourcen zieladäquat gesteuert werden können. Promotor dieser Entwicklung müssen vor allem Controllingabteilungen sein, die das fragmentiert vorhandene Know-how bündeln und eine Abbildung der Ressourcen vorantreiben.

7. Literatur

Arbeitskreis „Externe Unternehmensrechnung" der Schmalenbach-Gesellschaft für Betriebswirtschaft e.V., Grundsätze für das Value Reporting, in: Der Betrieb, 55. Jg. (2002), H. 45, S. 2337-2340.

Arbeitskreis „Immaterielle Werte im Rechnungswesen" der Schmalenbach-Gesellschaft für Betriebswirtschaft e.V., Kategorisierung und bilanzielle Erfassung immaterieller Werte, in: Der Betrieb, 54. Jg. (2001), H. 19, S. 989-995.

Arbeitskreis „Immaterielle Werte im Rechnungswesen" der Schmalenbach-Gesellschaft für Betriebswirtschaft e.V., Freiwillige externe Berichterstattung über immaterielle Werte, in: Der Betrieb, 56. Jg. (2003), H. 23, S. 1233-1237.

Austrian Research Center (Hrsg.), Intellectual Capital Report 1999, Seibersdorf 2000.

Baum, H.-G., Coenenberg, A. G. und *Günther, T.*, Strategisches Controlling, 2. Aufl., Stuttgart 1999.

Bornemann, M., Knapp, A., Schneider, U. und *Sixl, K. I.*, Holistic Measurement of Intellectual Capital, http://www.oecd.org//dsti/sti/industry/indcomp/act/ams-conf/ symposium-old.htm#holistic, Amsterdam 1999.

Bundesumweltministerium und *Umweltbundesamt* (Hrsg.), Handbuch Umweltcontrolling, München 2001.

BUWAL (Hrsg.), Bewertung in Ökobilanzen mit der Methode der ökologischen Knappheit. Ökofaktoren 1997, SRU-297-D, Bern 1998.

Dow Jones Sustainability Indexes (Hrsg.), Corporate Sustainability Assessment, http://www.sustainability-index.com/ assessment.

Edvinsson, L. and *Malone, M. S.*, Intellectual Capital: Realizing Your Company's True Value by Finding Its Hidden Brainpower, New York 1997.

Etterlein, K., Hürsch, P. und *Topf, M.*, Ökobilanzen – Ein Leitfaden für die Praxis, Mannheim 1992.

Financial Accounting Standards Board (FASB), Business Reporting Research Project, FASB, 2001.

Gleich, R., Das System des Performance Measurement, München 2001.

Global Reporting Initiative (GRI) (Hrsg.), Sustainability Reporting Guidelines, Amsterdam 2002.

Grüning, M., Performance Measurement Systeme als Managementwerkzeug, Wiesbaden 2002.

Günther, T. und *Grüning, M.*, Performance Measurement-Systeme im praktischen Einsatz. In: Controlling, 14. Jg. (2002), H. 1, S. 5-13.

Günther, E., Rechnungslegung von Emissionsrechten, in: Kapitalmarktorientierte Rechnungslegung, 3. Jg. (2003), H. 10.

Hastsbacka, M. A. and *Khoury, I. A.*, Technology Valuation – A Milestone on the Path from Art to Science, o. J., http:www.adltechnology.com/servicedocs/value3.pdf.

Ifo-Institut für Wirtschaftsforschung (Hrsg.), Industrieumfrage Nachhaltiges Wirtschaften in Deutschland, München 2002.

Kommission der Europäischen Gemeinschaften (EG-Kommission) (Hrsg.), Vorschlag für eine Richtlinie des Europäischen Parlamentes und des Rates über ein System für den Handel mit Treibhausgasemissionsberechtigungen in der Gemeinschaft und zur Änderung der Richtlinie 96/61/EG des Rates, KOM (2001) 581, 23.10.2001, Brüssel 2001.

KPMG (Hrsg.), International Survey of Environmental Reporting 1999, Amsterdam 1999.

Labhart, P.A., Value Reporting - Informationsbedürfnisse des Kapitalmarktes und Wertsteigerung durch Reporting, Zürich, 2001.

Lev, B., Intangibles: Management, Measurement, and Reporting, Washington D.C. 2001.

Maul, K.-H., Die Wissensbilanz als Teil des handelsrechtlichen Jahresabschlusses – Wissensbilanzen dargestellt am Beispiel des Jahresabschlusses von Hochschulen, in: Deutsches Steuerrecht, 47. Jg. (2000), H. 38, S. 2009-2016.

Maul, K.-H. und *Menninger, J.*, Das „Intellectual Property Statement" – eine notwendige Ergänzung des Jahresabschlusses ? In: Der Betrieb, 53. Jg. (2000), H. 11, S. 529-533.

Nagus (Hrsg.), DIN EN ISO 14031:1999: Umweltleistungsbewertung, Berlin 2000.

Oekom research (Hrsg.), Environmental Rating, http://oekom.de/ag/german/ rating.htm, 15.12.2002.

PRé Consultants B.V. (Hrsg.), The Eco-indicator 99 – A damage oriented method for Life Cycle Impact Assessment, 2. Ed., Amersfoort 2000.

PriceWaterhouseCoopers, Günther, T. und *Beyer, D.*, „Value Reporting" – Entwicklungsperspektiven der externen Unternehmensberichterstattung, Frankfurt a.M., 2003.

Roos, J., Roos, G., Dragonetti, N. C. and *Edvinsson, L.*, Intellectual Capital: Navigating in the New Business Landscape, New York 1997.

Ruhwedel, F. und *Schulze, W.*, Value Reporting: Theoretische Konzeption und Umsetzung bei den DAX 100-Unternehmen, in: Zeitschrift für betriebswirtschaftliche Forschung, 54. Jg. (2002), H. 11, S. 602-632.

Schmidt-Bleek, F., Das MIPS-Konzept, München 1998.

Stewart, T. A., Intellectual Capital: The New Wealth of Organizations, London 1997.

Sveiby, K. E., The New Organizational Wealth: Managing and Measuring Knowledge-Based Assets, San Francisco 1997.

Umweltbundesamt (Hrsg.), Bewertung in Ökobilanzen, Methode des Umweltbundesamtes zur Normierung von Wirkungsindikatoren, Ordnung (Rangbildung) von Wirkungskategorien und zur Auswertung nach ISO 14042 und 14043, Nr. 92, 1999.

Weeber, K.-W., Smog über Attika – Umweltverhalten im Altertum, Rembek 1993.

Intangible Ressourcen, Zielsystem und Management interner Wissenspotenziale

Roland Helm [1] *und Rudolf C. Meiler* [2]

---

[1] Prof. Dr. *Roland Helm*, Unilever-Stiftungslehrstuhls für Allgemeine Betriebswirtschaftslehre, Friedrich-Schiller-Universität Jena

[2] Dr. *Rudolf C. Meiler*, ThyssenKrupp Stahl AG, Duisburg

So genannte intangible Ressourcen sind für die Wettbewerbsfähigkeit eines Unternehmens von großer Bedeutung. Weithin bekannt ist auch die Tatsache, dass Wissen in Unternehmen in vielfältiger Weise vorliegt und unbedingt genutzt werden sollte, um dem Wettbewerbsdruck mit Innovationen zu begegnen. Das Wissen sollte jedoch zielorientiert „zu Tage gefördert werden", damit Ineffizienzen vermieden werden. Eine Voraussetzung dafür liegt in der Formulierung und Diffusion der Unternehmensvision und des Zielsystems mittels einer integrierten internen Kommunikation. Nach einer kurzen Einführung in die zugrunde liegenden theoretischen Grundlagen des Resource-based View of Strategy fließen die skizzierten Überlegungen in der Konzeption einer integrierten Unternehmenskommunikation zusammen.

1. Disposition

Bei einer hohen Veränderungsgeschwindigkeit des Unternehmensumfelds ist die Wettbewerbsfähigkeit langfristig vor allem dadurch zu sichern, dass alle Ressourcen in adäquater Weise genutzt werden (können). Neben den Möglichkeiten der Rationalisierung, Produktionsverlagerung und Internationalisierung kommt vor allem der Innovationstätigkeit bzw. der Innovativität eines Unternehmens eine herausragende Bedeutung zu (vgl. *Helm,* 2001, S. 30 f.). Es ist unstrittig, dass Wissen um Prozesse und Produkte in den Unternehmen in vielfacher Weise vorliegt, um Prozess- und Produktinnovationen zu generieren, es aber in den meisten Fällen zu einem großen Teil ungenutzt bleibt. Eine zentrale Aufgabe des Managements ist es daher, geeignete Rahmenbedingungen für die effiziente Wissensgenerierung und -offenlegung zu schaffen.

Im Folgenden sollen nach einem Einblick in die theoretische Bedeutung von intangiblen Ressourcen im Allgemeinen und insbesondere hinsichtlich der intangiblen Ressource Wissen für die Wettbewerbsfähigkeit von Unternehmen einige unverzichtbare Voraussetzungen für ein effizientes Management des unternehmensinhärenten Wissens aufgezeigt werden. Es ist zu zeigen, dass diese in der Formulierung einer Unternehmensvision, der Ableitung einer Zielhierarchie und einer funktionierenden integrierten internen Kommunikation dieser Aspekte zu finden sind.

2. Resource-based View of Strategy, intangible Ressourcen und Wettbewerbsfähigkeit

Folgt man dem Paradigma des Resource-based View of Strategy, so resultieren dauerhafte (finanzielle) Erfolge aus der besonderen Kombination von unternehmenseigenen, physischen und nicht-physischen Ressourcen und damit verbundenen strategischen Ressourcenvorteilen eines Unternehmens (vgl. *Barney*, 1991, S. 101; *Rühli*, 1994, S. 42). Diese Beziehungen sind wechselseitig und mit einer gewissen Dynamik versehen. Zum einen bildet die situationsadäquate Ressourcenkombination den Gegenstand von Strategien, zum anderen hat wiederum die Implementation eben dieser Strategien langfristige Auswirkungen auf die Ausstattung an Ressourcen in den Unternehmen, so dass dadurch auch neue „bedeutende" Geschäftsfelder entstehen können. Die strategische Führung von Unternehmen beinhaltet damit die effiziente Allokation von Ressourcen zur Erzielung eines maximalen Output (vgl. *Bamberger/Wrona*, 1996, S. 148).

Aus dieser Perspektive heraus ergibt sich weiterhin ein Erklärungsansatz für die Heterogenität verschiedener Unternehmen in einer Branche. Selbst wenn diese über die gleiche oder eine vergleichbare Ressourcenausstattung verfügen, so können die Unternehmen aufgrund divergierender Fähigkeiten, diese Ressourcen in geeigneter Weise zu kombinieren, auch verschiedene Erfolgspotenziale aufweisen (vgl. *Bamberger/Wrona*, 1996, S. 135).

Die Heterogenität der unternehmensindividuellen Ressourcenausstattung basiert außer auf der erfolgreichen Implementation von Strategien auf zwei weiteren Punkten.

Zum einen ist die Unvollkommenheit der Faktormärkte zu nennen. Diese wirkt aufgrund der Unterschiedlichkeit der Erwartungen der Marktteilnehmer dahingehend, dass die Preise von Ressourcen nicht den zukünftig erwarteten Erträgen aus diesen Ressourcen entsprechen, und sich diese Preise damit aus der ex ante Unsicherheit ergeben. Diese Aussage gilt für beide Seiten, also sowohl für den Verkäufer als auch für den Käufer der betreffenden Ressource. Einem Unternehmen ist es damit möglich, teilweise „bessere" Ressourcen auszuwählen oder zu beschaffen als dies einigen Wettbewerbern möglich ist. Wäre dies nicht gegeben, könnte jedes Unternehmen lediglich durchschnittliche Gewinne erwirtschaften. Dies bedeutet in der Folge einen eingeschränkten ex ante Wettbewerb um diese Ressourcen (vgl. *Peteraf*, 1993, S. 185; *Rühli*, 1994, S. 43 f.).

Zum anderen ist eine Immobilität von Ressourcen festzustellen. Die Folge ist, dass prinzipiell nicht alle Ressourcen auf Faktormärkten gehandelt werden können. Dies gilt vor allem für sogenannte „intangible resources" (vgl. *Wernerfelt*, 1984, S. 172), die sich mit organisations- oder mitarbeiterbedingten Fähigkeiten beschreiben lassen (vgl. *Grant*, 1991, S. 122). Besondere Fähigkeiten eines Unternehmens sind häufig nicht physisch vorhanden, sondern in dem komplexen Gebilde der Unternehmensorganisation oder -kultur eingebunden, also intangibel. Dementsprechend können sie nur über einen gewissen Zeitraum im Unternehmen entwickelt werden und sind damit schwer handelbar oder mit anderen Ressourcen kombinierbar. Die Übersetzung des Begriffs „intangible" als „nicht greifbar, unbestimmt, vage" trifft die Bedeutung nur unzureichend. Aus diesem Grund wird weiterhin der Begriff „(in)tangibel" verwendet. Am ehesten würde der bilanzielle Ausdruck „immaterielle Vermögensgegenstände" derartige Ressourcen beschreiben, wobei dieser jedoch im ursprünglichen Sinne nicht umfassend genug erscheint (vgl. *Helm*, 1997).

Da der Aufbau dieser zweiten Kategorie von Ressourcen mit dem Faktor Zeit verbunden ist (vgl. *Grant*, 1991, S. 123), kann davon ausgegangen werden, dass dieser Zeitverbrauch mit Investitionen in diese Fähigkeiten verbunden ist. Die Entwicklung von organisatorischen Grundlagen und nicht-physischen Potenzialen kann damit mit direkten finanziellen Aufwendungen zum Beispiel in Form von Schulungskosten oder bei einem direkten Engagement von Mitarbeitern mit Gehaltskosten etc. verbunden sein. Fallen keine direkten finanziellen Aufwendungen an, können Opportunitätskosten entstehen, da diese Mitarbeiter sozusagen nebenbei diese Fähigkeiten entwickeln und bei einem begrenzten Zeitbudget ihre eigentlichen Tätigkeiten hinten anstellen.

Barney (1991, S. 101) unterscheidet in Erweiterung der oben erwähnten binären Einteilung in tangible und intangible Ressourcen drei Kategorien von potenziellen, strategisch relevanten Ressourcen, wobei bei den beiden letztgenannten erhebliche Zusammenhänge erkennbar sind.

- Physikalische Ressourcen, wie zum Beispiel die verwendete Technologie, Anlagenausstattung, Standorte oder der Zugang zu Rohstoffen.
- Humanressourcen, wobei unter anderem die Ausbildung der Mitarbeiter, Erfahrungen oder Beziehungen genannt werden können.
- Organisatorische Ressourcen, wie zum Beispiel Innovationsfähigkeit, Kontroll- und Koordinationsfähigkeiten etc.

Wie bereits ausgeführt, kann die Bedeutung von Strategischen Geschäftseinheiten in einem Unternehmen auf eine aktuelle oder zukünftige Ressourcenheterogenität zwischen den Konkurrenten und letztlich auf unvollkommene Faktormärkte sowie auf die geschäftseinheitsindividuellen Fähigkeiten nicht frei handelbare, einzigartige Ressourcenkombinationen in der Geschäftseinheit zu entwickeln, zurückgeführt werden (vgl. *Rühli*, 1994, S. 43 f.). Dies ist gleichbedeutend mit einer Divergenz hinsichtlich der Erfolgspotenziale zwischen konkurrierenden Unternehmen.

Investitionen in die Entwicklung einzigartiger Ressourcenbündel zu langfristigen Erfolgspotenzialen resultieren in so genannten Kernkompetenzen des Unternehmens (vgl. *Prahalad/Hamel*, 1990, S. 85). Sie zeichnen sich durch technische, technologische, vertriebliche oder organisatorische Fähigkeiten des Unternehmens aus, die durch Akkumulation und Fusion verschiedener Ressourcen entstehen (vgl. *Prahalad/Hamel*, 1990, S. 82; *Rühli*, 1994, S. 44). Die Interaktion zwischen den Prozessen des Akkumulierens und des Fusionierens bedingt Lernprozesse in der Organisation, die als Katalysatoren verbesserter Leistungsfähigkeit bezeichnet werden können. Einzelne Fähigkeiten oder Ressourcen werden dabei in kürzeren, Kernkompetenzen als Ressourcenbündel in längeren Zeitabständen verbessert.

Zusammenfassend kann damit folgendes festgehalten werden. Die strategische Erfolgsbasis eines Unternehmens besteht aus der aktuellen und zukünftigen Fähigkeit, Ressourcen zu akquirieren und zu Kernkompetenzen zu fusionieren. Diese Vorgänge sind mit Investitionen im weiteren Sinne verbunden. Diese Kernkompetenzen sind aus mehreren Gründen schwer auf andere Unternehmen übertragbar und außerdem durch die Bemühungen der Konkurrenz, Rückstände hinsichtlich der Ressourcenausstattung aufzuholen, in ihrer strategischen Relevanz gefährdet (vgl. *Argyres*, 1996, S. 130). Kernkompetenzen sind die Voraussetzung für die Generierung produktspezifischer Wettbewerbsvorteile, wobei sie eine notwendige, aber nicht hinreichende Bedingung sind. Die grundlegenden Zusammenhänge des Resource-based View of Strategy sind in *Abb. 1* (in Anlehnung an *Rühli*, 1994, S. 43) zusammengefasst.

Auch Wettbewerbsvorteile und damit auch die darauf basierenden Erfolge unterliegen langfristig einer gewissen Abnutzung durch den ökonomischen Wertverlust der zugrundeliegenden Ressourcen und durch die Imitationsanstrengungen seitens der Konkurrenz (vgl. *Grant*, 1991, S. 123). Da dies bei intangiblen Ressourcen in erheblichem Umfang eingeschränkt bzw. erschwert ist, sind Wettbewerbsvorteile auf Basis intangibler Ressourcen oder auch in Kombination mit tangiblen Ressourcen besonders nachhaltig und damit wertvoll sind. Weiterhin ist evident, dass die Erhaltung der Heterogenität der Ressourcenausstattung der Unternehmen eine conditio sine qua non der Generierung nachhaltiger, strategischer Wettbewerbsvorteile darstellt. Betrachtet man dies unter einer dynamischen Perspektive, so ist offensichtlich, dass sowohl beim Aufbau von Kernkompetenzen als Basis von nachhaltigen Wettbewerbsvorteilen als auch bei ihrer späteren Verwertung Wettbewerbsbehinderungen vorliegen müssen (vgl. *Peteraf*, 1993, S. 182). Damit ist die Strategieentwicklung dahingehend zu erweitern (vgl. *Rühli*, 1994, S. 47), dass antizipativ Maßnahmen getroffen werden müssen, die günstige Voraussetzungen schaffen

- zum Aufbau von Kernkompetenzen,
- zur späteren Verteidigung der daraus resultierenden Wettbewerbsvorteile gegenüber der Konkurrenz sowie
- zur Weiterentwicklung und Anpassung aufgrund der Wandelbarkeit ihrer Bedeutung im Zeitablauf.

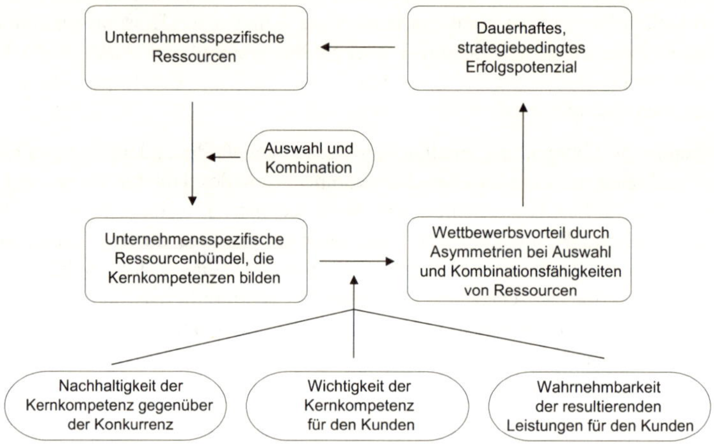

Abb. 1: Die grundlegenden Zusammenhänge des Resource-based View of Strategy

Aus den Ausführungen wird deutlich, dass der Entwicklung von intangiblen Ressourcen eine besondere strategische Bedeutung zukommt und diese die Wettbewerbsfähigkeit des Unternehmens hochgradig beeinflussen (vgl. *Hall,* 1992, S. 141).

Ohne Zweifel ist das im Unternehmen latent vorhandene Wissen ein wertvoller Vermögensgegenstand, mithin eine „intangible" Ressource, die entwickelt werden muss. Damit es in einer Organisation dazu kommen kann, sind bestimmte Voraussetzungen und geeignete Instrumente zur Offenlegung und Weiterentwicklung vorhandenen Wissens notwendig, die nachfolgend skizziert werden sollen.

3. Zweck und Notwendigkeit interner Kommunikation des Zielsystems

3.1 Notwendigkeit der Zielformulierung

In der strategischen Unternehmensplanung wird vielfach die Notwendigkeit der Formulierung adäquater Zielsetzungen zum Zweck der Überprüfung des Zielerreichungsgrades betrieblicher Aktivitäten betont (vgl. exemplarisch *Kreikebaum*, 1997).

Ziele dienen im unternehmensinternen Planungsprozess als Bewertungskriterien („Messlatten") für durchzuführende Aktionen mit dem Ergebnis, die Konsequenzen der Aktionen und damit auch die zur Auswahl stehenden Aktionen in eine eindeutige Rangfolge der Vorziehenswürdigkeit zu bringen. Sie sind demnach angestrebte zukünftige Zustände, die durch die Analyse der internen und externen Rahmenbedingungen erreichbar erscheinen. Nach der Auswahl einer Aktion dienen sie der Ergebniskontrolle und unterstützen damit den Lernprozess. Nur wenn entsprechende Ziele vorliegen, kann die Effektivität, d.h. die Prüfung der korrekten Auswahl einer Aktion, und die Effizienz, d.h. die Prüfung der korrekten Durchführung der gewählten Aktion, beleuchtet und damit steuernd in die Prozessverbesserung eingegriffen werden (vgl. *Böcker/Helm*, 2003, S. 46).

Ziele sind jedoch nicht zu Beginn eines Planungsprozesses – unabhängig von der Planungsebene – vorgegeben, sie entwickeln sich im Planungs- bzw. im Lernprozess (vgl. *Kreikebaum*, 1997, S. 65 f.). Im Prinzip bedeutet dies, dass für alle Aktivitäten im Unternehmen entsprechende Ziele in einem Zielsystem zu formulieren sind, mithin sich eine Vielzahl an Zielen eines Unternehmens ergibt, deren Inhalte im Entwicklungsprozess gegenseitig abzustimmen sind.

Üblicherweise wird hier von einer sogenannten Zielhierarchie im Unternehmen gesprochen. *Abb. 2* gibt exemplarisch den Aufbau einer derartigen Hierarchie wieder. Ausgehend von der obersten Zielebene werden die in der Pyramide darunter liegenden Ziele immer konkreter und immer mehr auf einen bestimmten Unternehmensbereich hin bezogen. Damit verbunden ist auch eine zunehmende Anzahl an Zielen auf einer Hierarchieebene. Diese sollen in vertikaler und horizontaler Hinsicht koordiniert sein, d.h. sich nicht widersprechen, demnach in Bezug auf andere Ziele konsistent sein (z.B. perfekte Produktionstechnologie vs. maximale Rentabilität). Dieser Abstimmungsbedarf bedingt eine interne Unternehmenskommunikation.

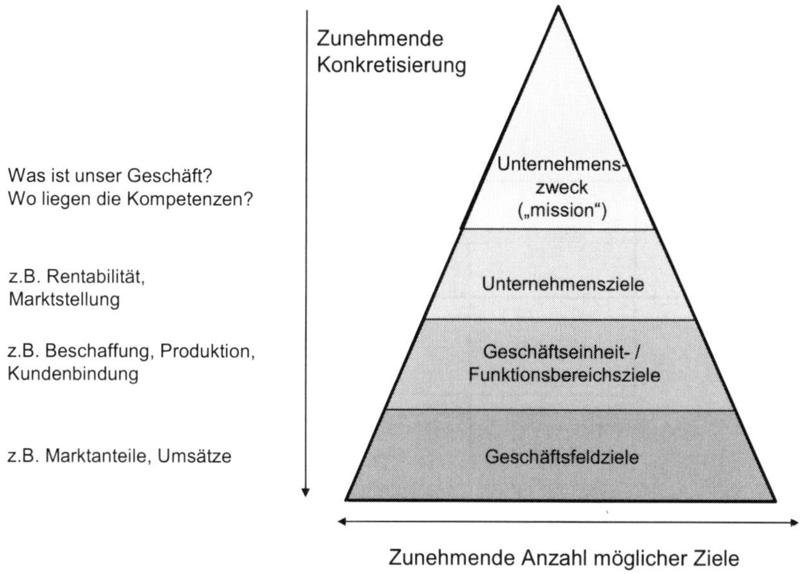

Abb. 2: Beispiel einer Zielhierarchie

Ziele wirken auf die betriebliche Entscheidungsfindung dadurch ein, dass sie zum einen die Bewertung der Alternativen von Entscheidungen einer bestimmten Hierarchieebene ermöglichen und zum anderen die Zielbildung auf nachgelagerten Hierarchieebenen wesentlich beeinflussen. Entscheidungen einer bestimmten Hierarchiestufe beeinflussen die Entscheidungen auf der nachgelagerten Entscheidungsebene vor allem dadurch, dass sie auf die Ausprägung des Zielsystems der nachgelagerten Hierarchieebene einwirken.

3.2 Unternehmensvision und -leitbild als kommunikativer Rahmen

Die übergeordneten Ziele der Zielsystemhierarchie sind an einer so genannten Unternehmensvision (oder Unternehmensphilosophie) und an einem Unternehmensleitbild festzumachen. Diese sollten einfach und für jeden verständlich formuliert sein, sich jedoch von Konkurrenten deutlich abgrenzen. Die Notwendigkeit dieser Größen für die langfristige Unternehmenssicherung ist unbestritten, wobei allerdings die Verbreitung in den Unternehmen selbst oft ungenügend ist (vgl. *Zahn/Dillerup,* 1995, S. 59).

Die Vision muss weit genug in die Zukunft reichen und grundsätzliche Inhalte kommunizieren, d.h. der Horizont für die Vision muss zeitlich und qualitativ über den Tag hinausreichen. Sie muss so zwischen Wirklichkeit und Utopie angesiedelt sein, dass sie als grundsätzliche Herausforderung verstanden werden kann, aber gleichzeitig den Glauben zulässt, gerade noch machbar zu sein. Zusammenfassend bedeutet dies (*Abb. 3*):

- Die Vision darf nicht zu nahe an der Realität, am Alltagsgeschäft sein.
- Die Vision darf nicht irreal und zeitlich nicht zu weit entfernt sein.
- Eine Vision beinhaltet das gerade noch Machbare.

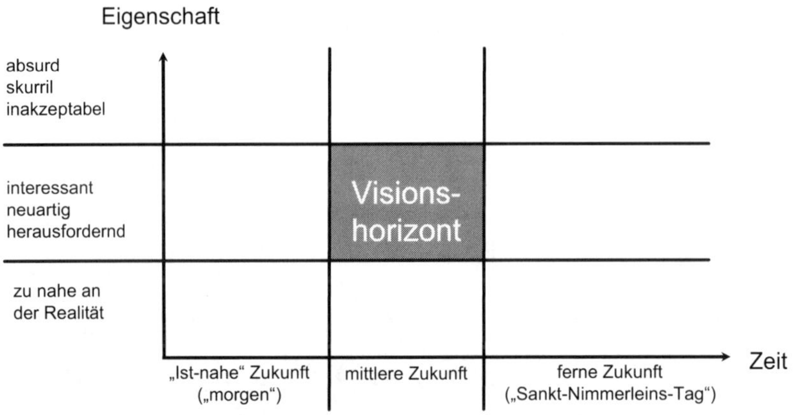

Abb. 3: Zeitliche und inhaltliche Abgrenzung einer Unternehmensvision

Simon (1990) sieht beispielsweise „die verschiedenen Aktivitäten des Unternehmens bzw. der Mitarbeiter wie Metallspäne, und die Vision als Magnet, der sie ordnet. Wenn alle im Unternehmen wissen, wo sie stehen und wo sie hin wollen, fällt es ihnen leichter, ihre Aktivitäten zu koordinieren. Erst so bewegt sich das Unternehmen auf ein gemeinsames Ziel zu".

Einige Beispiele aus der Unternehmenspraxis verdeutlichen diesen Anspruch:

- „Wir wollen, dass die Menschen auch auf kleinstem Raum und mit minimalem Energieverbrauch die besten Klang- und Seherlebnisse genießen können." Sony.
- „Unsere Aufgabe ist Kommunikation. Wir wollen Wissen demokratisieren, Information für das breite Publikum bereitstellen und Bildung und Unterhaltung produzieren." Bertelsmann.
- „Unternehmen müssen Entscheidungen fällen, Entscheidungen basieren auf Wissen. Unsere Business Information Services liefern das grundlegende Wissen, das Industrie, Handel, Dienstleistungsunternehmen und Medien der "Alten" und der "Neuen" Ökonomie benötigen, um ihre Marktentscheidungen zu treffen." *GfK AG.*

Die Unternehmensvision wird jedoch nur dann den Zweck, den Mitarbeitern im Unternehmen Ziel und Richtung zu geben, erfüllen können, wenn sie vor allem von den Führungskräften nicht nur wahrgenommen und gespeichert wird, sondern im Unternehmen kommuniziert und vorgelebt wird. Eine Vielzahl von Forschungsergebnissen aus der Sozialpsychologie zum Thema Selbstverstärkung in Lernprozessen von Individuen belegt dies in eindrucksvoller Weise (vgl. *Herkner*, 2001, S. 80 f.).

Im Gegensatz zur Vision ist ein Leitbild weniger abstrakt, dagegen konkreter, ausführlicher und mehr am Ist-Zustand als am Soll-Ziel orientiert.

Das Unternehmensleitbild ist die schriftlich formulierte Bündelung von Gedanken zu betriebsspezifischen Handlungs- und Entscheidungsprinzipien. Es setzt den Rahmen für die Zusammenarbeit mit Mitarbeitern, Kunden, Lieferanten, Wettbewerbern und Öffentlichkeit. Demnach kann das Leitbild auch als Verhaltensrichtlinie oder als ausformulierte Führungsgrundsätze angesehen werden. Ein Leitbild ist quasi die „Verfassung" des Unternehmens. Sie gewinnt aufgrund der abnehmenden Prognostizierbarkeit der Zukunft im Allgemeinen an Bedeutung (vgl. *Bea/Haas,* 1997, S. 65).

Durch die Formulierung eines Unternehmensleitbildes soll folgendes erreicht werden:

- Wertvorstellungen und Entscheidungsgrundlagen des Unternehmens werden auf einen Nenner gebracht.
- Gemeinsamkeiten werden konkretisiert.
- Der unternehmerische Kurs für die Zukunft wird konzeptionell festgelegt, eine Schematisierung wird vermieden.
- Wechselseitige Abhängigkeiten zwischen Personen, Funktionen und Unternehmensbereichen werden verdeutlicht.
- Unternehmerische Ziele werden nach innen und außen deutlicher und verständlicher.
- Die Verantwortung des Unternehmens in Wirtschaft und Gesellschaft wird aufgezeigt.

Ein sehr anschauliches Beispiel aus einem Bereich der Wirtschaft, in dem Derartiges kaum vermutet wird, entstammt der Unternehmensgruppe (mehrere Restaurants der Top-Gastronomie, Bistros, Auberges, Kochschule etc.) des französischen Meisterkochs *Alain Ducasse* (vgl. *Zipprick*, 2003). Jeder Mitarbeiter wird auf folgendes Leitbild verpflichtet:

- Wir verpflichten uns, mit Sorgfalt und Disziplin zu arbeiten und auf Details zu achten.
- Wir vermitteln intellektuelle und handwerkliche Fähigkeiten ebenso wie Sinn für Kommunikation.

- Wir sind bekannt dafür, Herausforderungen zu mögen.
- Der Schlüssel zu unserem Erfolg ist die „Evolution“ unserer Mitarbeiter.
- Wir erkennen die essenziellen Talente jedes Individuums.
- Die Überschrift zu diesen Leitsätzen lautet: „Exzellenz ernten“.

Vision und Leitbild können auch als Unternehmenspersönlichkeit in Anlehnung an die Beschreibung natürlicher Personen mittels verschiedener zu beobachtender Eigenschaften beschrieben werden.

Nach außen und innen erkennbar wird die Unternehmenspersönlichkeit durch das Auftreten des Unternehmens in Form des tatsächlichen Verhaltens – vor allem auch der Personen, die mit der „relevanten Öffentlichkeit“ und Mitarbeitern in qualitativ und quantitativ überdurchschnittlichem Kontakt stehen, des Unternehmenserscheinungsbildes und der internen und externen Unternehmenskommunikation. Als Orientierungskonzept hierzu werden vielfach Richtlinien zur so genannten Corporate Identity verwendet, die darauf abzielt, dem Auftreten des Unternehmens eine erkennbare Linie zu geben (vgl. *Dichtl/Helm,* 2002, S. 273). *Abb. 4* verdeutlicht diese Überlegungen.

Die Unternehmensvision und das Unternehmensleitbild dienen demnach dazu, alle nachgelagerten Zielbildungsprozesse als auch die Unternehmensstrategien (bzw. deren simultane Entwicklung) in einen Rahmen einzubinden.

Die Entwicklung dieser Planungsgrößen bedingt einen interaktiven Lernprozess, bei dem alle Beteiligten (auf allen hierarchischen Ebenen des Unternehmens) eben diesen Rahmen kennen und verinnerlicht haben müssen. Es wurde bereits gezeigt, dass nur dann die Unternehmensvision bzw. das Unternehmensleitbild ihren Zweck erfüllen können. Dies bedingt jedoch wiederum eine effiziente interne Unternehmenskommunikation.

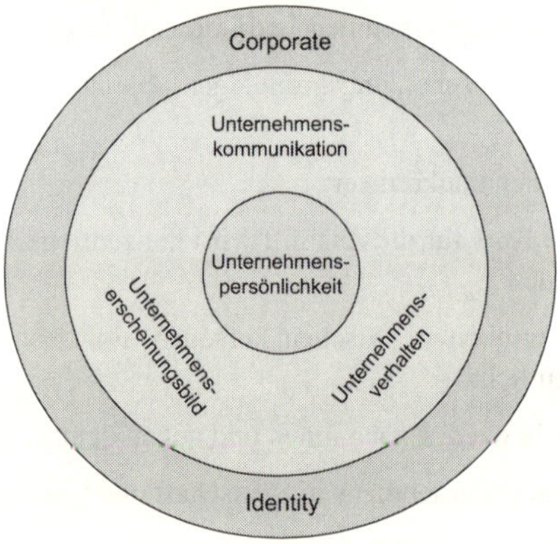

Abb. 4: Unternehmenspersönlichkeit und Unternehmensauftritt

Schließlich wird dadurch die Stringenz der entwickelten Prozesse gefördert, es wird aber auch die nachgelagerte (top-down-) Kommunikation erleichtert, um eine effiziente Umsetzung im Unternehmen zu gewährleisten. Dies wird dadurch erreicht, dass die Nachvollziehbarkeit der Planungen und damit die Delegation auf die Mitarbeiter durch den im Unternehmen bekannten Rahmen (die unternehmensinternen Rahmenbedingungen im Leitbild) unterstützt wird. Es sollte so zusagen zu „Wiedererkennungseffekten" kommen.

3.3 Resümee

Intangible Ressourcen und deren (Weiter-) Entwicklung sowie die effektive und effiziente Kombination sind unabdingbare Voraussetzungen für den langfristigen Unternehmenserfolg. Dazu müssen einerseits die Beteiligten – soweit möglich – in den Entwicklungsprozess eingebunden, andererseits die Rahmenbedingungen, d.h. Ziele und Strategien, bei der Delegation und Umsetzung bekannt und verinnerlicht sein. Beides hat eine im Unternehmen verbreitete Unternehmensvision bzw. ein präziseres Unternehmensleitbild, eine bekannte Zielhierarchie sowie eine funktionierende interne Unternehmenskommunikation als Voraussetzung.

Die Effizienz beider vorausgesetzter Bestandteile bedingt sich gegenseitig. Dies bedeutet, dass das Funktionieren einer internen Kommunikation das Vorhandensein einer Vision (eines allgemein bekannten Rahmens) voraussetzt, aber auch für die Bekanntheit und Akzeptanz der Unternehmensgrundsätze eine funktionierende interne Kommunikation notwendig ist.

4. Integrierte Unternehmenskommunikation als Voraussetzung der Entwicklung intangibler Ressourcen

Im vorangegangenen Abschnitt wurde deutlich, dass die interne Unternehmenskommunikation für die Effektivität und Effizienz der Entwicklung intangibler Ressourcen als auch deren Umsetzung im Unternehmen von hoher Bedeutung ist.

Das Konzept der Integrierten Unternehmenskommunikation (vgl. dazu auch *Bruhn,* 1997, S. 96) ist darauf ausgerichtet, aus den verschiedensten Quellen der internen und externen Kommunikation eines Unternehmens eine Einheit herzustellen, um für alle „Stakeholder" des Unternehmens, d.h. Mitarbeiter, Kunden, Verwaltungseinrichtungen, Fremdkapitalgeber etc., konsistente Informationen mittels eines ebenso konsistenten Erscheinungsbildes zu vermitteln (vgl. *Böcker/Helm,* 2003, S. 417 ff.).

Dazu ist ein vorgegebener Rahmen in Form einer Unternehmensvision bzw. eines –leitbildes notwendig. Der skizzierte Zusammenhang sowie mögliche Ausprägungen der internen und externen Kommunikation sind in *Abb. 5* dargestellt.

Im Prinzip stellt dieser Ansatz damit eine Vereinigung der externen Kommunikation – vor allem mit dem Kunden – und der internen Kommunikation – vor allem mit Mitarbeitern – dar, um Wiedersprüche in den Aussagen zu vermeiden und alle Kommunikatoren und Adressaten auf den selben Wissensstand zu bringen.

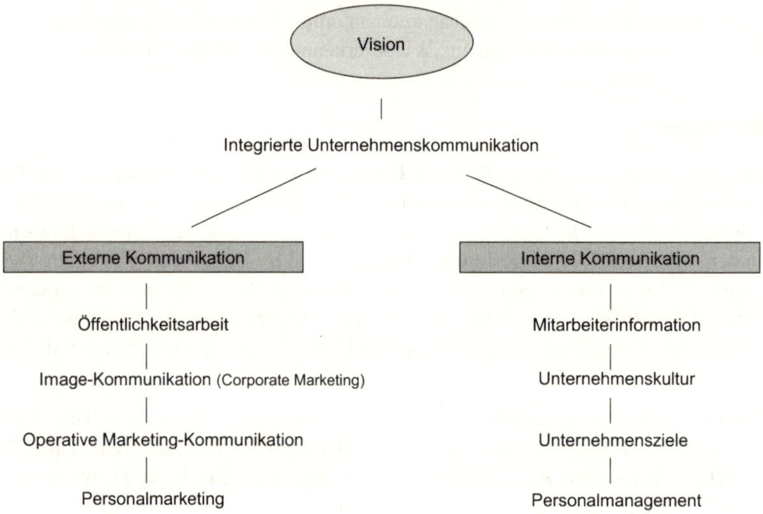

Abb. 5: Integrierte Unternehmenskommunikation

Die Umsetzung der internen Kommunikation bzw. die Integration in das – hoffentlich – bestehende (externe) Kommunikationskonzept ist nicht nur eine konzeptionelle, sondern auch eine organisatorische Aufgabe. Sie muss einerseits von Personen (federführend) vollbracht werden, die selbst in hohem Ausmaß auf der Mitarbeiterseite integrierend wirken (können), aber andererseits auch die hierarchische Kompetenz zur Integration von interner und externer Kommunikation haben. Vielfach sind damit auch strukturelle Veränderungen im Unternehmen verbunden. Sinnvoll ist es, wenn Mitarbeiter unterschiedlicher Abteilungen und somit auch unterschiedlicher Subkulturen daran arbeiten (vgl. *Bruhn,* 1997, S. 146 ff.).

5. Interne Kommunikation des Zielsystems und Management interner Wissenspotenziale

Zu Beginn dieses Beitrags wurde aufgezeigt, dass effiziente Lernprozesse im Unternehmen einen vorgegebenen Rahmen in Form eines kommunizierten und akzeptierten Unternehmensleitbildes und ein entsprechendes strategisches Zielsystem als Basis benötigen. Eine Prozessverbesserung kann nur unter dieser Voraussetzung in sinnvoller Weise erfolgen, da nur dann alle Beteiligten wissen, „wohin der Weg führen soll". Anders formuliert, alle Beteiligten sollen wissen, was mit den Aktivitäten erreicht werden soll.

Lernprozesse im Unternehmen bzw. Abläufe im Unternehmen ganz generell basieren auf Informationen, die ermittelt, verarbeitet und gespeichert werden müssen. Informationen werden vielfach als die wertvollste – intangible – Ressource eines Unternehmens bezeichnet (vgl. *Zahn,* 2001, S. 380). Diese sind im Allgemeinen an Mitarbeiter im Unternehmen oder an Personen, die in irgendeiner Art und Weise mit dem Unternehmen in Beziehung stehen, gebunden. Sollen diese beispielsweise in rechnergestützten Informationsmanagementsystemen abgelegt, verarbeitet und anderen Beteiligten zur Verfügung gestellt werden, so müssen Anreize geschaffen werden, dass die Personen ihre Informationen offen legen und der „Allgemeinheit" (oder einem berechtigten Teil der Unternehmensmitarbeiter) zur Weiterverarbeitung zur Verfügung stellen, diese aber auch ständig aktualisieren. Durch diese Vernetzung vieler Informationen – in vielen Fällen wiederum durch interpersonalen Informationsaustausch – entsteht schließlich Wissen (vgl. *Zahn,* 2001, S. 378 ff.).

Wissensmanagement beinhaltet demnach eine intelligente Sammlung von Informationen durch die Schaffung von anreizkompatiblen Designs für den Einzelnen sowie gleichzeitig die intelligente zieladäquate Bündelung dieser Informationen und die Schaffung einer geeigneten Ausgangsbasis für die Weiterverarbeitung dieser intangiblen Ressourcen, sozusagen die „Veredelung des Rohmaterials Information".

Das Management des im Unternehmen vorhandenen Wissens umfasst damit nicht nur die Organisation der technischen Möglichkeiten, sondern auch den dieser Problematik vorgelagerten Aspekt der den individuellen Informations- bzw. Wissensträger betreffenden Bereich der Motivation zur Weitergabe und -entwicklung von Wissen (vgl. *Kubitschek/Meckl,* 2000).

Abgesehen von der Notwendigkeit geeignete – monetäre, aber auch nicht-monetäre (wie zum Beispiel Belobigungen und entsprechende interne Kommunikation, entsprechende persönliche Weiterbildungschancen) – Anreizstrukturen zu diesem Zweck zu entwickeln, ist es aber auch notwendig, die in jeder Organisation vorhandenen Informationen, die im Allgemeinen zu facettenreich sein dürften, dahingehend zu kanalisieren, dass vor allem die für die Unternehmensentwicklung relevanten Informationen offen gelegt bzw. artikuliert werden. Für den unwahrscheinlichen Fall, dass alle vorhandenen, mehr oder minder relevanten Informationen tatsächlich freigegeben werden, würde dies mit Sicherheit die Informationsverarbeitungskapazitäten im Unternehmen sprengen. Eine ungenügende Verarbeitung und Würdigung führt anschließend zur Demotivation der Mitarbeiter und damit einhergehend zur Verminderung der Informationsfreigabe unter das Ausgangsniveau (*Abb. 6*). Dies ist zu vermeiden.

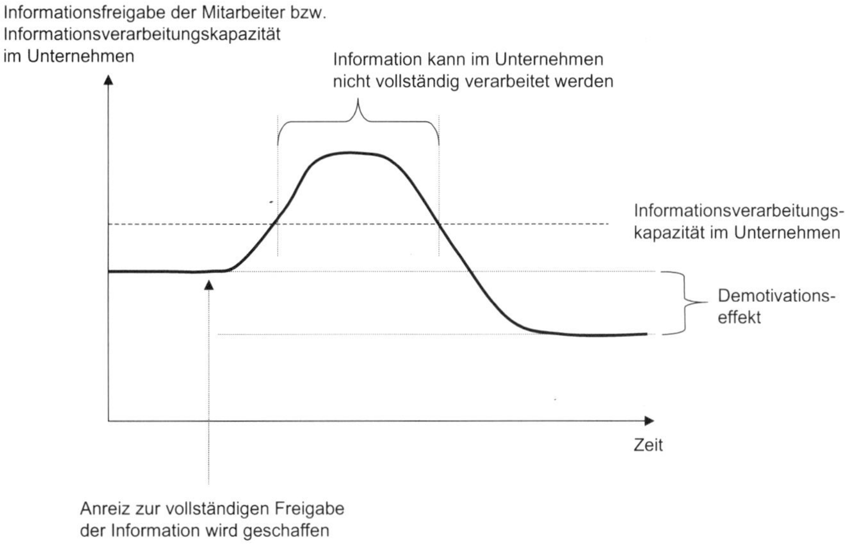

Abb. 6: Auswirkung mangelnder Informationsverarbeitungskapazität im Unternehmen

Die Aufgabe der zieladäquaten Informationsselektion kann in erster Linie nur dem einzelnen Mitarbeiter obliegen. Zu diesem Zweck muss ihm aber bekannt sein, was im Unternehmen erreicht werden soll, mithin „wohin das Schiff steuern soll". Im Groben wird dies durch die Kommunikation der Unternehmensvision als auch des Zielsystems erreicht. Sind diese bei den Mitarbeitern bekannt, können diese relevante und irrelevante Informationen unterscheiden und erkennen zumindest im Ansatz einen nicht-monetären Anreiz in der Unternehmensweiterentwicklung und im Unternehmensfortbestand.

Auch die Informationsverknüpfung zu Wissen erfordert die Kenntnis, was mit der Bündelung erreicht werden soll. Dies bedeutet, auch die Installation eines Wissensmanagements im Unternehmen hat eine adäquate Kenntnis der Unternehmensvision als Voraussetzung. Nur unter dieser Bedingung ist eine Weiterentwicklung von Informationen zu „brauchbarem" Wissen, d.h. eine sinnvolle Weiterverarbeitung dieser besonders wertvollen, im Unternehmen vorliegenden intangiblen Ressource, möglich.

Die Umsetzung von Wissen in verbesserte Prozesse und Leistungen für den Markt wiederum muss in einer Art Rückkoppelung durch die Kanäle der Internen Kommunikation zu den Mitarbeitern gelangen. Dies sollte die Motivation verbessern. Die neuen diesbezüglichen Informationen werden anschließend wiederum aufbereitet, um von neuem in den Prozess der Wissensveredelung einzugehen. Die vorangegangenen Ausführungen sind in *Abb. 7* illustriert. Die Gestaltung und Verbesserung des Leistungserstellungsprozesses im Unternehmen beinhaltet dabei sowohl die Produkt- als auch die Prozessgestaltung bzw. –optimierung bzw. auch –variation im Sinne einer Diversifikation in neue produktbezogene Geschäftsfelder und/oder Märkte.

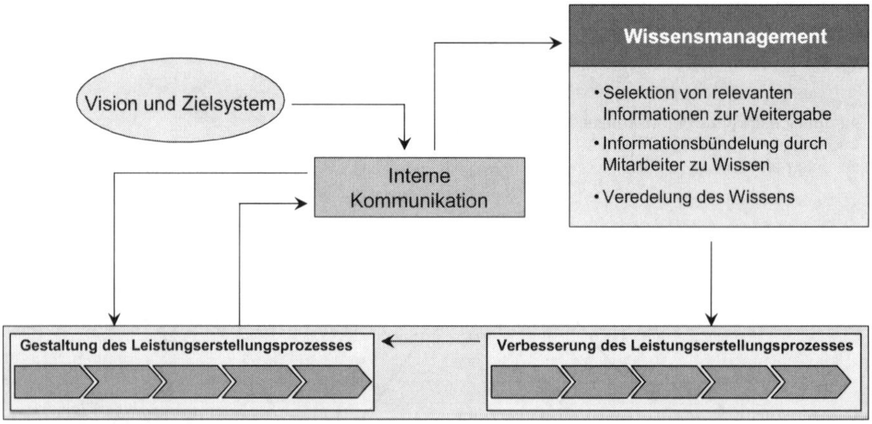

Abb. 7: Einbettung der Internen Kommunikation in das Wissensmanagement

In diesem Zusammenhang wird als Instrument zur Vermeidung von Entwicklungs- und Austauschbarrieren von Informationen bzw. Wissen der Aufbau eines Menschen- bzw. Organisationsbildes angeführt, das opportunistisches Verhalten soweit möglich unterbindet. Dazu stehen im Wesentlichen folgende Bausteine zur Verfügung (vgl. *Kubitschek/Meckl*, 2000, S. 749 f.):

- Schaffung einer Unternehmenskultur und dazu als Voraussetzung eines Unternehmensleitbildes, in welcher die Entwicklung und Weitergabe von Informationen und Wissen als positiver Wert gesehen wird.
- Personalauswahlverfahren, bei dem zukünftige Mitarbeiter gewonnen werden, die eben diese Werte bereits in sich tragen.
- Personal- und Organisationsentwicklungsverfahren zur Vermittlung dieser Werte und zur Schaffung der Voraussetzungen eines individuellen Wissensmanagements.
- Bildung von Teamstrukturen und Förderung entsprechender persönlicher Kontakte zur Förderung der Bereitschaft, Wissen weiterzugeben. Aufgrund dieser Maßnahme soll das gegenseitige Vertrauen gesteigert werden und damit einhergehend die Zusammenarbeit, d.h. auch die zielorientierte Kombination von Informationen. Weiterhin wird dadurch sozialer Druck gegenüber unkooperativen Kollegen aufgebaut. Hier ist nochmals die Bedeutung des Unternehmensverhaltens zu erwähnen. Führungskräfte sollten ein Verhalten gegenüber Mitarbeitern praktizieren, das der Abschottung von Wissen im Sinne eines „Wissen ist Macht“ - Symptoms entgegenwirkt.
- Vorgelebter kooperativer Führungsstil und entsprechende Karrierebilder sowie adäquate Führungskräfteauswahlverfahren.

Dies alles sollte jedoch nicht darüber hinwegtäuschen, dass vor allem in größeren Unternehmen der Aufbau von Vertrauen, die Förderung „altruistischen Verhaltens“ sowie die glaubhafte Kommunikation dieser Bausteine durch die größere Anzahl an hierarchischen Ebenen und die Anonymität der Mitarbeiter in erheblichem Umfang erschwert wird. Gerade vor diesem Hintergrund ist die Ausgestaltung der Instrumente der internen Kommunikation äußerst bedeutsam.

6. Zusammenfassung

Wissen in Organisationen ist gemäß den Überlegungen des Resource-based View of Strategy eine der wertvollsten intangiblen Ressourcen eines Unternehmens. Diese Ressource ist aber in den meisten Fällen nur zu einem Bruchteil genutzt, da es aus vielfältigen Gründen nicht offen gelegt bzw. adäquat verarbeitet wird bzw. werden kann. Um eine effiziente Verarbeitung von offen gelegtem Wissen zu gewährleisten ist es notwendig, den Mitarbeitern einen geeigneten Rahmen aufzuzeigen.

Die bisherigen Ausführungen haben ergeben, dass zur Ziel- und Strategieentwicklung eine Unternehmensvision bzw. ein Unternehmensleitbild vorgegeben sein muss. Weiterhin ist auch bei der Umsetzung die dementsprechende Kenntnis bei den Beteiligten notwendig, da dadurch die Nachvollziehbarkeit und damit zusammenhängend auch die Delegation erleichtert wird. Sowohl die Unternehmensvision, das Leitbild als auch die strategischen und operativen Ziele sowie die dazugehörigen Strategien sind in geeigneter Weise zu kommunizieren. Dabei ist darauf zu achten, dass die Mitarbeiter einer Hierarchieebene auf einen vergleichbaren Wissensstand kommen, somit alle relevanten Informationen über das Unternehmen diffundieren.

Die interne Unternehmenskommunikation ist schließlich auch ein unabdingbarer Bestandteil eines funktionierenden Wissensmanagements im Unternehmen. Nur bei der Kenntnis der groben Zielsetzungen in Form von Vision und Leitbild können vorhandene Informationen zielgerecht weitergegeben, verarbeitet als auch weiterentwickelt werden. Last but not least ist in diesem Zusammenhang die Motivationskomponente dieser Verbindung der beiden wichtigen Bestandteile einer Unternehmensentwicklung (Unternehmensvision, siehe Abschnitt 3 und Informationen, siehe Abschnitt 5) für die Mitarbeiter zu nennen.

7. Literatur

Argyres, N., Evidence on the Role of Firm Capabilities in Vertical Integration Decisions, in: Strategic Management Journal, Vol. 17 (1996), No. 4, S. 129-150.

Bamberger, I. und *Wrona, T.*, Der Ressourcenansatz und seine Bedeutung für die Strategische Unternehmensführung, in: ZfbF – Schmalenbachs Zeitschrift für betriebswirtschaftliche Forschung, 47. Jg. (1996), H. 2, S. 130–153.

Barney, J., Firm Resources and Sustained Competitive Advantage, in: Journal of Management, Vol. 2 (1991), No. 1, S. 99–120.

Böcker, F. und *Helm, R.*, Marketing, 7. Aufl., Stuttgart 2003.

Bea, F. X. und *Haas, J.*, Strategisches Management, 2. Aufl., Stuttgart 1997.

Bruhn, M., Kommunikationspolitik, München 1997.

Dichtl, E. und *Helm, R.*, Marketing, in: *Bea, F. X., Dichtl, E.* und *Schweitzer, M.* (Hrsg.), Allgemeine Betriebswirtschaftslehre: Leistungsprozess, 8. Aufl., Stuttgart 2002, S. 213–286.

Grant, R. M., The Resource-based Theory of Competitive Advantage: Implications for Strategy Formulation, in: California Management Review, Vol. 33 (1991), No. 3, S. 114–135.

Hall, R., The Strategy Analysis of Intangible Resources, in: Strategic Management Journal, Vol. 13 (1992), No. 5, S. 135–144.

Helm, R., Internationale Markteintrittsstrategien, Köln 1997.

Helm, R., Planung und Vermarktung von Innovationen, Stuttgart 2001.

Herkner, W., Sozialpsychologie, 2. Aufl., Bern 2001.

Kreikebaum, H., Strategische Unternehmensplanung, 5. Aufl., Stuttgart 1997.

Kubitschek, C. und *Meckl, R.*, Die ökonomischen Aspekte des Wissensmanagements – Anreize und Instrumente zur Entwicklung und Offenlegung von Wissen, in: ZfbF – Schmalenbachs Zeitschrift für betriebswirtschaftliche Forschung, 51. Jg. (2000), H. 7, S. 742–761.

Peteraf, M. A., The Cornerstones of Competitive Advantage: A Resource-based View, in: Strategic Management Journal, Vol. 14 (1993), No. 6, S. 179-191.

Prahalad, C. K. and *Hamel, G.*, The Core Competence of the Corporation, in: Harvard Business Review, May-June, 1990, S. 79-91.

Rühli, E., Die Resource-based View of Strategy: Ein Impuls für einen Wandel im unternehmungspolitischen Denken und Handeln? In: *Gomez, P., Hahn, D., Müller-Stewens, G.* und *Wunderer, R.* (Hrsg.), Unternehmerischer Wandel: Konzepte zur organisatorischen Fundierung, Wiesbaden 1994, S. 31-57.

Simon, H., Herausforderung Unternehmenskultur, Stuttgart 1990.

Wernerfelt, B., A Resource-based View of the Firm, in: Strategic Management Journal, Vol. 5 (1984), No. 5, S. 171-180.

Zahn, E., Informationstechnologie und Informationsmanagement, in: *Bea, F. X., Dichtl, E.* und *Schweitzer, M.* (Hrsg.), Allgemeine Betriebswirtschaftslehre: Führung, 8. Aufl., Stuttgart 2001, S. 376–428.

Zahn, E. und *Dillerup, R.*, Beherrschung des Wandels durch Erneuerung, in: *Reichwald, R.* und *Wildemann, H.* (Hrsg.), Kreative Unternehmen. Spitzenleistungen durch Produkt- und Prozessinnovationen, Stuttgart 1995, S. 35–76.

Zipprick, J., Wie wird man zur Legende? In: McK Wissen 05 (2003), S. 30-37.

Wissensbilanzen zur internen Steuerung und externen Berichterstattung von Forschungseinrichtungen

Jürgen Blum [1] *und Robert Borrmann* [2]

---

[1] Prof. Dr. *Jürgen Blum*, Zentrum für Wissenschaftsmanagement Speyer e.V.

[2] *Robert Borrmann*, Horváth & Partner GmbH, Stuttgart

Das Thema Messung und Bewertung immaterieller Vermögensgegenstände gewinnt gegenwärtig in der Betriebswirtschaftslehre zunehmend an Bedeutung. Im Unterschied zu klassischen Industrieunternehmen nimmt die Bedeutung von Wissen - „wenn man wüsste, was man und was Andere wissen" - und damit verbunden die Notwendigkeit sich mit dieser Thematik zu beschäftigen zu. In Forschungsorganisationen, deren eigentliche Funktion es ist, Wissen zu produzieren und dieses auch in Form von Innovationen an den Markt zu transferieren, muss dies umso mehr gelten. Der vorliegende Beitrag zeigt den aktuellen Stand der Praxis und Forschung sowie die Weiterentwicklungsbedarfe dieses Instrumentes im Rahmen des Wissenschaftsmanagements auf.

1. Einleitung

Die wissenschaftliche Forschung wird als einer der entscheidenden Wettbewerbsträger der Zukunftsfähigkeit gehandelt, um der stagnierenden Weltwirtschaft neue Wachstumsimpulse zu geben. In der Regierungserklärung von Bundeskanzler Schröder vom 14. März 2003 heißt es, dass unser Wohlstandsniveau nur gehalten werden kann, „ [...] wenn wir in dieser schwierigen wirtschaftlichen Situation verstärkt in Bildung und Forschung investieren" (*Schröder*, 2003). Angesichts leerer öffentlicher Kassen und des damit verbundenen Wettbewerbs um diese Gelder, muss sich die öffentliche Förderung von Forschung einer vermehrten Rechtfertigung unterziehen, da diese Art der Mittelverwendung verstärkt in Konkurrenz zu alternativen Formen im öffentlichen oder privaten Sektor gesehen wird (vgl. *Brockhoff/Gerwin*, 2001, S. 9).

Im Spannungsfeld der steigenden Bedeutung der Wissenschaft einerseits und den leeren öffentlichen Kassen andererseits ergibt sich für die Wissenschaft vermehrt die Notwendigkeit, den Einsatz ihrer Mittel effektiv und effizient zu gestalten. Darüber hinaus muss auch dem Zuwendungsgeber die Möglichkeit gegeben werden, öffentliche Mittel auf Basis aussagekräftiger Informationen zielgerichtet einzusetzen und ausgewählte Schwerpunkte zu fördern, um mit ihnen den größtmöglichen Gesamtnutzen zu erzielen:

„In der staatlichen Förderung wissenschaftlicher Einrichtungen soll ein Paradigmenwechsel von der Bestandssicherung (Input) zur Ergebnisorientierung (Output) und von der Detail- zur Globalsteuerung stattfinden. Gegenstand der Förderung und Finanzierung werden künftig nicht mehr die Einrichtungen an sich, sondern ihre wissenschaftlichen Ergebnisse bzw. Serviceleistungen sein." (*Bund-Länder-Kommission für Bildungsplanung und Forschungsförderung*, 2003, S. 3)

Wie aber kann ein derart gefordertes Instrument aussehen, welches zum einen die Beurteilung wissenschaftlicher Leistungen durch Externe erlaubt und zum anderen den Wissenschaftlern bzw. ihren Organisationen ermöglicht, die Fördermittel möglichst effektiv und effizient im Sinne einer maximalen Zielerreichung bei gegebenem Mitteleinsatz zu allokieren?

Wissensbilanzen werden zunehmend als adäquates Instrument gesehen, welches es erlaubt, die beiden Forderungen zu unterstützen. Der vorliegende Beitrag zeigt ausgehend von der Vorstellung eines Wissensbilanzmodells und seiner Steuerungs- und Berichtsfunktion das

Potenzial auf, welches dieses Instrument zur Lösung dieser Problemstellung bietet. Hierbei werden die Grenzen der zur Zeit bestehenden Ansätze kritisch reflektiert und an den Anforderungen, die an ein Koordinationsinstrument gestellt werden, gespiegelt, um anschließend Weiterentwicklungspotenziale abzuleiten.

Im Juli des Jahres 2003 fand in Graz ein Expertengespräch zum Thema Wissensbilanzen statt, welches vom *Zentrum für Wissenschaftsmanagement*, dem *Zentrum für Hochschulentwicklung*, den *Austrian Research Centers Seibersdorf* sowie der technischen Universität Graz veranstaltet wurde. Dieser Artikel nimmt die aktuellen Ausführungen der Kongressteilnehmer auf und lässt diese Ergebnisse in die Ausführungen einfließen.

2. Grundverständnis von Wissensbilanzen

Nach *Koch*, der als wissenschaftlich-technischer Geschäftsführer für die Erstellung der ersten Wissensbilanz der *Austrian Research Centers Seibersdorf* (vgl. *ARCS*, 1999) verantwortlich war, gründet sich die Entstehung des Terminus Wissensbilanz auf sprachliche Differenzen zwischen Experten aus den administrativen Bereichen der *Austrian Research Centers Seibersdorf* und den Forschern der wissenschaftlichen Abteilungen (Aussage von *Koch* beim Expertengespräch in Graz 2003). Die Funktion und Interpretation von Wissensbilanzen in der Literatur und in der Praxis findet sich in erster Linie nicht als reine Gegenüberstellung von Mittelherkunft und Mittelverwendung wieder, sondern als betriebswirtschaftliches Instrument, welches einer neuen Form der Berichterstattung dient, indem es über die Entwicklung der Intangible Assets Auskunft gibt (vgl. *Bornemann/Leitner*, 2002, S. 342). *Bornemann* und *Leitner* machen darauf aufmerksam, dass Intangible Assets weder synonym mit dem Wissensbegriff an sich oder dem gesamten Wissensbestand der Organisation zu sehen sind (vgl. *Bornemann/Leitner*, 2002, S. 342). „An intangible asset is a claim to future benefits that does not have a physical or financial [...] embodiment" (*Lev*, 2001, S. 5). Somit stellen sich Wissensbilanzen als Tool zur Berichterstattung über die verwendeten Ressourcen und ihre erzielte Wertschöpfung dar.

Praktische Ansätze der Visualisierung von Intellectual Capital finden sich bereits 1993, als die Frage nach der Bilanzierung organisationalen Wissens bei Industrieunternehmen aufkam. Der schwedische Finanzdienstleister *SKANDIA* war eines der ersten Unternehmen, welches die Öffentlichkeit mit einer solchen neuartigen Form der Bilanzierung konfrontierte und als Beilage zu den traditionellen Bilanzdaten veröffentlichte: der erste Versuch einer Art von Wissensbilanz, die die bisher eher pauschal als „goodwill" bezeichneten Aktivposten systematischer darzustellen versuchte (vgl. *Probst/Raub/Romhardt*, 1999, S.20 f. und zu den Aktivierungskriterien von Intangible Assets nach US-Gaap vgl. *Schreiber*, 2002).

Nonaka postuliert, dass „Analyzing the organization in terms of its design and capability to process information imposed by the environment no doubt constitutes an important approach to interpreting certain aspects of organizational activities." (*Nonaka*, 2002, S.437) und zeigt damit die Notwendigkeit nach einem Steuerungsinstrument zur betriebswirtschaftlichen Bewertung der verwendeten Ressourcen und ihrer erzielten Wertschöpfung auf. Folglich ist eine Wissensbilanz nicht nur ein notwendiges Tool für privatwirtschaftliche Unternehmen, sondern auch ein Instrument für Organisationen aus dem Umfeld der Öffentlichen Verwaltung, dem New Public Management, wobei der Begriff „New-Public-Management" als Oberbegriff für die weltweit relativ einheitliche „Gesamt-Bewegung" von Verwaltungsreformen fungiert, die sich im Wesentlichen mit der Modernisierung öffentlicher Einrichtungen und neuen Formen der Gestaltung von Prozessen und Strukturen staatstragender Leistungen befassen (vgl. *Budäus*, 1994, S. 30ff.).

3. Ziele der Wissensbilanzierung

Die Funktion von Wissensbilanzen leitet sich aus zwei Hauptaufgabenbereichen ab (vgl. hierzu *Leitner*, 2003, S. 24; *Bornemann/Leitner*, 2002, S. 346 und *Biedermann/Graggober/Sammer*, 2002, S. 54).

Zum einen sollen sie, wie die inhaltliche Begriffsbestimmung nahe legt, der Berichterstattung dienen. Die Berichterstattung in Form der Rechenschaftslegung, d.h. der Zusammenstellung der relevanten Informationen und ihre Weitergabe in geeigneter Verdichtung und Form, beinhaltet, dass die berichtenden Einrichtungen der Öffentlichkeit und dem Zuwendungsgeber aufzeigen, welche Forschungstätigkeiten bzw. -ergebnisse mit den Fördergeldern geleistet wurden (vgl. *Brockhoff/Gerwin*, 2001, S. 8).

Zum anderen soll durch Wissensbilanzen die interne Steuerung ermöglicht bzw. verbessert werden. Es besteht die Notwendigkeit, eine richtungweisende Strategie innerhalb der Forschungsorganisationen zu entwickeln und umzusetzen, was die Formulierung strategischer Ziele, Maßnahmen und Indikatoren umfasst.

Aus der Berichterstattungsfunktion leitet sich ein weiteres Ziel, das mit Wissensbilanzen zu erfüllen gesucht wird, ab. Es gilt ein staatliches Finanzierungsverfahren zu entwickeln, welches die die richtige Balance zwischen Autonomie und Steuerung schafft: Wie eingangs erwähnt, findet in der Forschungsförderung ein Umbruch von der Input- zur Outputförderung statt. Der Wissenschaftsrat, der die Bundesregierung und die Regierungen der Länder berät und die Aufgabe hat, Empfehlungen zur inhaltlichen und strukturellen Entwicklung der Hochschulen, der Wissenschaft und der Forschung sowie des Hochschulbaus zu erarbeiten, fordert dahingehend, dass der Staat (bei entsprechender Gegenleistung der Forschungsorganisationen) keine Detailsteuerung leisten kann und sollte, sondern sich auf übergreifende forschungspolitische Prioritäten beschränkt, die es in enger Interaktion mit sämtlichen beteiligten Akteuren zu entwickeln gilt (vgl. *Wissenschaftsrat*, 2003, S. 5). Das bestehende System der Forschungsförderung unter dem Gesichtspunkt der Effizienz zu betrachten und zu optimieren genügt demnach nicht. Es erfordert vielmehr eine Verständigung der Beteiligten über ihre Ziele und Vertrauen in eine konstruktive Zusammenarbeit.

In Österreich ist die Erstellung von Wissensbilanzen seit 2002 gesetzlich verankert: §13 Leistungsvereinbarung in den „Allgemeinen Bestimmungen zum Organisationsrecht des Universitätsgesetztes“ verlangt beginnend mit dem Jahr 2005: „Jede Universität hat der Bundesministerin oder dem Bundesminister im Wege des Universitätsrats jeweils bis 30. April eine Wissensbilanz über das abgelaufene Kalenderjahr vorzulegen. Gesondert darzustellen sind zumindest

1. der Wirkungsbereich, gesellschaftliche Zielsetzungen sowie selbst definierte Ziele und Strategien;
2. das Intellektuelle Vermögen, differenziert in Human-, Struktur- und Beziehungskapital;
3. die in der Leistungsvereinbarung definierten Leistungsprozesse mit ihren Outputgrößen und Wirkungen.“ (*Nationalrat Österreich*, 2002, S. 1277)

Die Ziele der Wissensbilanzierung können wie folgt zusammengefasst werden (vgl. *Abb. 1*).

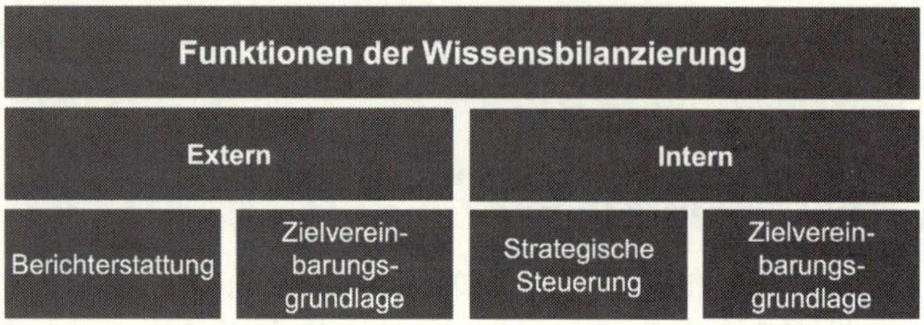

Abb. 1: Funktionen der Wissensbilanzierung

Sämtliche Funktionen der Wissensbilanz sind nicht isoliert zu betrachten, sondern interdependent. „Forschungsförderung hat immer Auswirkungen auf die Dynamik der Wissenschaft, selbst wenn mit ihr keine expliziten Steuerungsabsichten verbunden sind. Die strategische Gestaltung der Forschungsförderung soll dazu dienen, diese Auswirkungen so zu moderieren, dass die Leistungsfähigkeit des Wissenschaftssystems optimiert wird" (*Wissenschaftsrat*, 2003, S. 5).

Trotz des engen Zusammenhangs zwischen Forschungsförderung und interner Steuerung von Wissenschaftseinrichtungen darf nicht übersehen werden, dass Ziele, die aus der Funktion von Wissensbilanzen abgeleitet werden, höchst unterschiedlich sein können. Während die externe Nutzung die bessere Positionierung im Wettbewerb um das Ansehen in der Öffentlichkeit und um Fördermittel zum Ziel hat, ist die interne Nutzung auf die Verbesserung der internen Steuerung ausgerichtet. Es erscheint somit erforderlich, sich entweder auf die interne oder externe Funktion zu beschränken, oder das Instrument so auszugestalten, dass es - unter Umständen in verschiedenen Ausprägungen - allen Anforderungen der Adressaten (Öffentlichkeit, Zuwendungsgeber, Management) gerecht wird. Der Weiterentwicklungsbedarf der Wissensbilanz wird unter anderem auch in der Wissensbilanz des DLR gesehen: „Gleichwohl ist sicher, dass wir nicht am Ende ihrer Entwicklung sind" (*Blum*, 2001, S. 3). Dies ist eine Problematik, die zurzeit noch nicht gelöst ist und weiterer Forschung bedarf.

4. Aufbau und Anwendung von Wissensbilanzen

4.1 Der Aufbau von Wissensbilanzen

Die bestehenden Modelle des Intellectual Capital Reporting (zu den einzelnen Modellen vergleiche beispielsweise die Zusammenfassungen von *Bontis*, 2001 und *Edvinsson/Brünig*, 2000, S. 20ff.) unterteilen sich in deduktiv-summarische und induktiv analytische Ansätze zur Bewertung Intellektuellen Kapitals. Deduktiv-summarische Ansätze, wie beispielsweise der Economic Value Added (EVA), der Market Value Added (MVA), die Markt-/ Buchwert Relation, *Tobin's* Q oder der Calculated Intangible Value (CIV) lassen aus der Differenz von Markt- zu Buchwert eines Unternehmens auf dessen Intellektuelles Kapital schließen. Induktiv-analytische Ansätze ermitteln das Intellektuelle Kapital eines Unternehmens mittels spezifischer Indikatoren. Diese Indikatoren eröffnen zwei signifikant unterschiedliche Perspektiven. Aus einer logistisch orientierten Prozesssicht in so genannte Prozessmodelle, wobei die Prozesssicht einen detaillierten Blick auf die In- und Outputflüsse entlang der gesamten Wissensproduktion ermöglicht und aus Potenzialsicht in Strukturmodelle, welche die organisationsinternen Veränderungen der Aktivposten Human-, Struktur- und Beziehungskapital analysieren.

Bezüglich ihres Aufbaus zeigen sich die bisher veröffentlichten Wissensbilanzen der *Austrian Research Centers* in Seibersdorf (vgl. *ARCS*, 1999 und Folgejahre), die des *Deutschen Zentrums für Luft- und Raumfahrt* in Köln (vgl. *Blum*, 2000 und Folgejahr) sowie die Wissensbilanzen des Schwerpunktbereiches Betriebswirtschaft der *Montanuniversität Leoben* (vgl. *WBW*, 2001 und Folgejahr) einheitlich, indem sie Strukturmodelle und Prozessmodelle verbinden. Ausgehend vom Selbstverständnis, der Rolle und der Vision der Wissenschaftseinrichtung im gesellschaftlichen Kontext werden zunächst die Wissensziele definiert. Darauf aufbauend erfolgt eine differenzierte Modellbeschreibung und eine Charakteristik der aufzunehmenden Potenziale anhand der Dreiteilung in Human-, Struktur- und Beziehungskapital, wobei der Begriff des Intellektuellen Kapitals im bilanziellen Sinne eher dem Begriff des Intellektuellen Vermögens entspricht (zum Terminus Kapital in Abgrenzung zum Vermögen vgl. *Maul*, 2000, S. 2012). Aus dieser Strukturierung lässt sich die Transformation der Potenziale in reale Ergebnisse in den Kernprozessen identifizieren. Aufgrund der Natur öffentlich geförderter Forschung können diese Ergebnisse sowohl finanzieller als auch immaterieller Art und forschungs-, wirtschafts- und gesellschaftsbezogen sein.

Abb. 2 zeigt den Aufbau des Wissensbilanz Modells des *Deutschen Zentrums für Luft- und Raumfahrt*, welches auf dem Modell der *ARCS*, 1999, S. 13 aufbaut und in Zusammenarbeit von DLR und ARCS weiterentwickelt wurde:

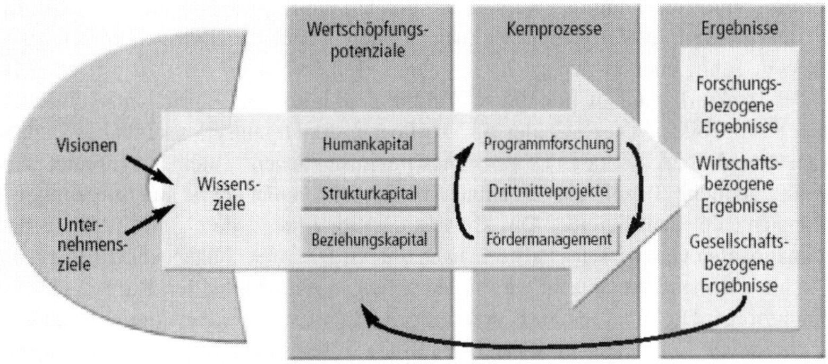

Abb. 2: Wissensbilanz Modell des DLR (Quelle: Blum, 2000, S. 7)

Einen offenen Aspekt stellt die Möglichkeit der Quantifizierbarkeit von Wissenschaftsleistungen dar. Der Input wissenschaftlicher Leistungen ist über die Säulen der Kostenrechnung (Kostenartenrechnung, Kostenstellenrechnung und Kostenträgerrechnung) mittels Ressourcenzuordnung, wie beispielsweise über Stundenaufschreibungen auf Forschungsprogramme, möglich.

Im Gegensatz zur Darstellung des Inputs von Wissenschaftseinrichtungen führt die Beschreibung des Outputs gerade in Wissenschaftseinrichtungen zu beträchtlichen Problemen. Da das wirtschaftliche Ergebnis öffentlich geförderter Forschungseinrichtungen per Definition ausgeglichen ist, besteht zwar der Zwang zum wirtschaftlichen Handeln, einen Indikator für einen messbaren „Erfolg" einer Wissenschaftseinrichtung erhält man daraus jedoch nicht. Pragmatische Plausibilitätsbetrachtungen führen zu der nahe liegenden These, dass die Preisbildung, die eine Rentabilitätsberechnung ermöglichen würde, für wissenschaftliche Leistungen bzw. ihre Indikatoren zumindest zurzeit nicht sinnvoll durchgeführt werden kann. Zu groß sind die Unsicherheiten, die sich z.B. aus der Burteilung des monetären Wertes einzelner Publikationen oder wissenschaftlicher Auszeichnungen ergeben können.

Die Frage nach dem Mehrwert, den Wissensbilanzen hierzu leisten können, steht noch offen. Denkbar ist hier eine Lösung, die kurzfristig eine qualitative Beschreibung und mittelfristig eine quantitative nicht monetäre Darstellung erlaubt. Überlegungen zur nicht monetären Darstellung auf einer einheitlichen Skala unter Einsatz von Fuzzy Logic finden sich bei *Haase*, 1999. Für eine monetäre Darstellung, wenn sie überhaupt möglich ist, fehlt derzeit eine konzeptionelle Vorstellung.

4.2 Strategische Steuerung von Wissenschaftseinrichtungen mit Wissensbilanzen

Die interne Steuerung leitet sich aus der Strategie der Forschungseinrichtung ab, aus der die Wissensziele generiert werden, die von der Institution entlang der Kernprozesse erreicht werden sollen. Die konsequente Quantifizierung der Wissensziele in Form geeigneter Indikatoren, die in Kennzahlen beschrieben werden, ist hierbei die Ausgangsbasis der konkreten Zieldefinition.

Ein Beispiel soll diese Wirkung illustrieren:
Das *Deutsche Zentrum für Luft- und Raumfahrt* (DLR) hat sich im Jahr 2000 als Kernprozess der Programmforschung das Wissensziel gesetzt, die Kapazitäten auf diejenigen Bereiche zu konzentrieren, in denen es nachweisbar zu den hochqualifiziertesten Forschungszentren der Welt zählt, oder in den nächsten Jahren eine solche Position zu erreichen sucht. Seitens des Strukturkapitals wurde dieses Ziel mit dem Indikator „Einrichtung von Center of Excellence" abgebildet. Diese Center of Excellence, die Organisationseinheiten darstellen, die im DLR nach Entscheidung des Vorstands mit zusätzlichen Mitteln ausgestattet werden, sollen die fachliche Exzellenz, mögliche Marktchancen der Forschungsergebnisse und Transferpotenziale identifizieren. Das Beziehungskapital wird durch die Zusammenarbeit mit internationalen Gastwissenschaftlern ausgedrückt. Die forschungsbezogenen Ergebnisse lassen sich u.a. anhand der Anzahl der Publikationen je Wissenschaftler, Rufe an Hochschulen, Lehraufträge, Diplomarbeiten, Dissertationen, Habilitationen, Beteiligungen an Projekten der *Deutschen Forschungsgemeinschaft* (DFG), Erfolgsquote bei EU-Projekten und im Verhältnis der Prime-contractor Verträge zu den Gesamt-EU-Projekten beschreiben. Für diese Indikatoren werden nun Sollgrößen definiert. Sie können dabei höchst unterschiedliche Bemessungsparameter darstellen: bei einigen Indikatoren kann „nur" eine quantitative Verbesserung gegenüber dem aktuellen Stand angestrebt werden, wie z.B. Steigerung der Publikationstätigkeit, bei anderen Indikatoren ist es möglich, einen absoluten Wert anzusetzen (z.B. die Erhöhung der Anzahl der Center of Excellence von vier auf sechs bis zehn).

Der Vergleich mit ausgewählten Partnern im Sinne eines Benchmarking hilft hierbei, die eigene Position zu beurteilen und in Form einer Stärken- und Schwächenbilanz gegenüberzustellen. So können Felder mit akutem Handlungsbedarf identifiziert und priorisiert werden. Die Steuerungswirkung wird durch die Gegenüberstellung des Ziels im Sinne eines Planwertes mit dem am Jahresende tatsächlich erreichten Wert und daraus resultierenden Handlungsempfehlungen generiert.

4.3 Externe Berichterstattung mit Wissensbilanzen

Die interne Steuerung im Sinne der Erreichung aus der individuellen Strategie abgeleiteter Wissensziele kann sich konfliktär zur externen Berichterstattung und zur Leistungsvereinbarung mit Zuwendungsgebern auswirken. Während die interne Steuerung ihr Augenmerk auf Felder richtet, in denen Handlungsbedarf besteht, sollte die Außendarstellung für die Öffentlichkeit und die Zuwendungsgeber eher auf erreichten Zielen oder aber auf ausgewogenen Berichten erreichter und nicht erreichter Ziele beruhen.

Würde die Zielgruppe, an die berichtet wird, nicht oder nur unzureichend differenziert, besteht die Gefahr, durch eine übertriebene Darstellung von Schwächen im Wettbewerb um Fördermittel und öffentliche Anerkennung eine schlechte Position zu erreichen. Während diese Gefahr einer zu detaillierten Darstellung von noch nicht erreichten Zielen im Rahmen der Berichterstattung durch hohe Bedeutung der Finanzierung durch die öffentliche Hand und die eigene Außendarstellung eher gering erscheint, gewinnt die zweite Gefahr großes Gewicht. Werden sämtliche Ziele in der Wissensbilanz berücksichtigt, sind auch Ziele enthalten, die bereits erreicht wurden oder bei denen kein dringender Handlungsbedarf besteht. Dies würde die interne Steuerung durch zu viele Indikatoren lähmen. Damit liegt der Schluss nahe, dass die interne Steuerung und die externe Berichterstattung auf zumindest teilweise unterschiedlichen Indikatoren basieren sollten.

Eine Möglichkeit besteht darin, einen allgemeinen Teil zu veröffentlichen, der ein mit anderen Forschungseinrichtungen vergleichbares Set an Indikatoren beinhaltet und nur einen - für die interne Steuerung in manchen Teilen darüber hinaus gehenden und in anderen Teilen weniger differenziert betrachteten - spezifischen Teil einbezieht. Diese Alternative würde die zur Berichterstattung und gegebenenfalls Zuwendungsbemessung herangezogenen Indikatoren als eine Art Zielsystem begreifen, welches die Definition des erhofften Outputs einer Wissenschaftsreinrichtung enthält. Es würde nur denjenigen Aspekten „Management-Attention" geliefert, die kritisch sind oder zumindest einen priorisierten Handlungsbedarf im Sinne der Gesamtstrategie aufweisen.

4.4 Zielvereinbarungen auf der Grundlage von Wissensbilanzen

Zielvereinbarungen mit dem Zuwendungsgeber dienen einer Organisation zur Fundierung von Entscheidungen über die angestrebten Ziele und erlauben eine Definition des Tätigkeitsspektrums. Sie sind das Ergebnis von Verhandlungen zwischen Staat und Forschungseinrichtung und stellen den finalen Kontrakt und damit die konsensuale Lösung im Hinblick auf die einzuleitenden Entwicklungen und die festgelegten Ziele dar. In diesem Sinne werden die zugeteilten Finanzen an Aussagen über die Ziele selbst sowie an die vereinbarten Sollgrößen, auf die die Ergebnisse des Handelns Bezug nehmen, gekoppelt (vgl. *Ziegele*, 2000, S.335).

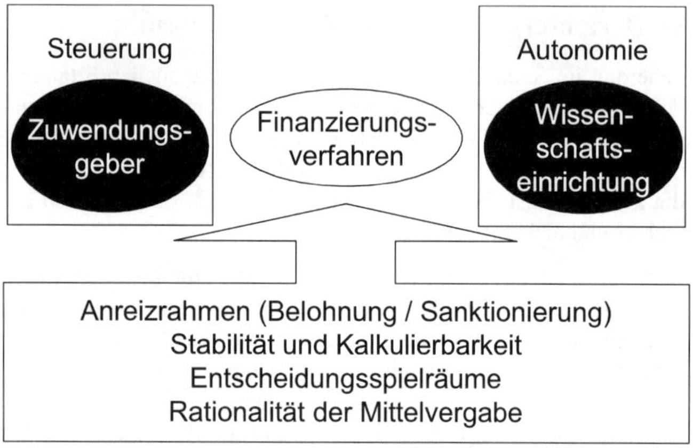

Abb. 3: Anforderungen an ein Finanzierungsverfahren (in Anlehnung an Ziegele, 2000, S. 332f.)

Wie aus *Abb. 3* ersichtlich, erfordert diese Sichtweise nach *Ziegele*, 2000, S.332f. von einem geeigneten Finanzierungsverfahren, dass es:

- einen wettbewerblichen Anreizrahmen schafft, der Aufgabenerfüllung, Leistung und Innovationsfähigkeit von Forschungseinrichtungen finanziell belohnt bzw. geringen Erfolg in diesen Bereichen finanziell sanktioniert;
- eine stabile und kalkulierbare Finanzierung der per staatlichem Auftrag übernommenen Aufgaben sicherstellt;
- tatsächliche Entscheidungsspielräume belässt;
- den Anspruch auf die staatliche Globalzuweisung – als Folge der Rationalität des Verfahrens – der Höhe nach plausibel begründet und legitimiert.

Wissenschaftliche Einrichtungen werden traditionell vom Staat über Instrumente der Kameralistik und des Inkrementalismus (d.h. der Fortschreibung von Haushalten mit prozentualen Zu- oder Abschlägen je nach Tarifentwicklung bzw. Haushaltslage) gesteuert. Anhand von vier ausgewählten Aspekten wird deutlich, warum die herkömmlichen Instrumente die genannten Anforderungen nicht erfüllen und in welche Richtung die Veränderung des alten zu einem neuen Steuerungsmodell gehen sollte (vgl. *Abb. 4*).

Während nach dem alten Steuerungsmodell die Inputorientierung im Vordergrund steht und den propagierten Paradigmenwechsel von der Bestandssicherung (Input) zur Ergebnisorientierung (Output) bzw. von der Detail- zur Globalsteuerung nur mangelhaft berücksichtigt, basiert die zielgerichtete Steuerung des neuen Modells auf einer Outputorientierung, die wettbewerbsorientiert ausgestaltet ist. Diese Umorientierung steht in engem Zusammenhang mit Beschränkungen der steuerungspolitischen Autonomie der Wissenschaftseinrichtungen. Dezentrale Entscheidungen werden forciert und lassen durch grobe Steuerungsvorgaben die wenigen Lenkungsziele von hoher Priorität in den Vordergrund treten. Dies führt auch zu einer Erhöhung der Flexibilität. Diese Flexibilität wird zusätzlich untermauert durch das Vermeiden einer Ex-ante-Steuerung, die in erster Linie zur Zielbeitragsmessung vorab festgelegter Ergebnisse verwendet wird. Eine ordnungspolitische Rahmensetzung schafft finanzielle Bedingungen und damit Anreize und Sanktionen von autonomen Entscheidungen, statt durch Eingriffe regulativ Ergebnisse feinzusteuern.

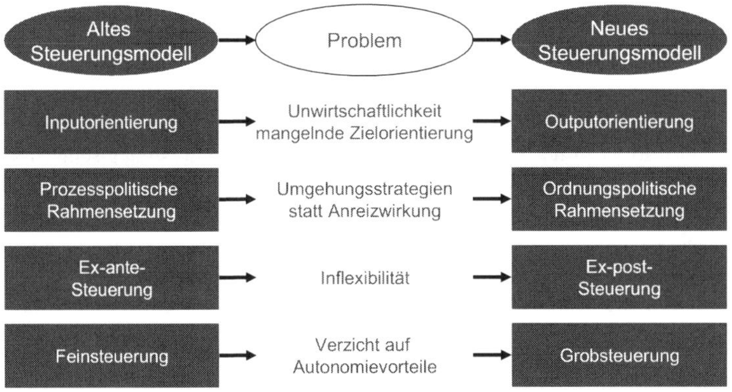

Abb. 4: Veränderungen des Steuerungsmodells (Quelle: Ziegele, 2000, S. 333)

Die dargestellten Anforderungen an ein Finanzierungsverfahren und an Veränderungen des Steuerungsmodells zeigen, dass Wissensbilanzen durchaus das Potenzial haben, zur Unterstützung der Zielvereinbarung zwischen Zuwendungsgeber und Wissenschaftseinrichtung zu werden. Die Realisierung dieses Potenzials setzt jedoch voraus, dass diese externe Zielvereinbarung innerhalb der Organisation - beispielsweise durch interne Zielvereinbarungen - konsequent umgesetzt wird.

5. Fazit und Ausblick

Obwohl der Aufwand, der mit der Wissensbilanzierung verbunden ist, von Organisation zu Organisation in Abhängigkeit von der Ausgangssituation variiert, kann er doch mit rund 200 TEuro Initialaufwand und im laufenden Betrieb mit rund 150 TEuro per anno beziffert werden. Die Darstellung des Nutzens ist jedoch ungleich schwieriger. Je nach Funktion, die mit der Wissensbilanzierung verfolgt wird, können für die Nutzenbewertung meist nur qualitative Aussagen zu Rate gezogen werden. Die Entwicklungszeit einer Wissensbilanz kann, sofern die Unterstützung in der Organisation für dieses Projekt vorhanden ist, unter einem Jahr betragen.

Auch wenn der Gedanke, die Leistung einer Universität oder eines Forschungszentrums mittels weniger oder einer einzigen - welcher Dimension auch immer entstammenden - Kennzahl zu beurteilen nicht plausibel erscheint, erfolgt dies heute bereits beispielsweise in Rankings von Universitäten.

Die monetäre Quantifizierung der Leistung bzw. ihrer Indikatoren, die es (theoretisch) ermöglichen würde, die Forschungsförderung mit anderen öffentlich Aufgaben, wie beispielsweise der Landesverteidigung, zu vergleichen, beruht häufig auf spekulativen Hypothesen. Die Messung des Inputs, insbesondere jedoch die Gewichtung und die Messung der Indikatoren des Outputs, können betriebswirtschaftlich nur dann sinnvoll erfolgen, wenn es einen einheitlichen Wertekonsens der Gesellschaft gibt, der dies ermöglicht. Die summarischen Kosten, die eine Organisation für den Erhalt eines Nobelpreises aufgewendet hat, sind trotz der extremen Langfristigkeit der zugrunde liegenden Forschungstätigkeiten vergleichsweise einfacher zu ermitteln, als der Wert, den die Einrichtung und eine Volkswirtschaft bereit sind für das gestiegene Ansehen und die gestiegene Wettbewerbsfähigkeit aufzuwenden.

Wissensbilanzen wollen und können Peer Reviews als klassisches Instrument der Beurteilung von Forschungsleistungen nicht ersetzen. Sie stellen vielmehr ein Instrument dar, welches über Indikatoren in der einen Ausprägung die konsequente strategische Steuerung in Zeiten gestiegener Autonomie ermöglicht und die in einer anderen Ausprägung einen Mehrwert für die Transparenz in der Wissenschaftslandschaft zu Zeiten vermehrten Rechtfertigungsdruckes liefert. Der aktuelle Stand der Forschung und Praxis liefert gute Ansätze, aber zumindest in Deutschland, zurzeit noch keine in den Zielen und in der Methodik abgesicherte Basis für die Mittelzuweisung durch öffentliche Zuwendungsgeber. Eine derartige Einschränkung zeigt die Wissensbilanz des DLR, in der darauf hingewiesen wird, dass Neuland betreten wird und, dass „die Zahlen der Wissensbilanz [...] auf Ihre Aussagekraft zu analysieren und prognostisch zu bewerten sein [werden]" (*Blum*, 2000, S. 3).

Die Frage, ob Wissensbilanzen überhaupt eingeführt werden sollen, wurde für Österreich durch den Gesetzgeber beantwortet. Auch in Deutschland werden Bewertungssysteme für die Großforschung und in deren Sog auch für Universitäten kommen. Das Wissenschaftsmanagement sollte die Chance nutzen, den nationalen wie auch internationalen Bürokratien, die nach aussagekräftigen Bewertungsmethoden suchen, mit rationalen wissenschaftsadäquaten Lösungsvorschlägen entgegen und zuvor zu kommen. Nicht Zurückhaltung der „autonomen Wissenschaft“ ist geboten, sondern Gestaltung ist gefragt, die die Wissenschaftlerinnen und Wissenschaftler vom Nutzen von Wissensbilanzen für sie überzeugt.

6. Literatur

ARCS, Wissensbilanz des *Austrian Research Centers Seibersdorf*, Seibersdorf 1999.

Biedermann, H., Graggober, M. und *Sammer, M.*, Die Wissensbilanz als Instrument der Steuerung von Schwerpunktbereichen am Beispiel eines Universitätsinstitutes, in: *Bornemann, M.* und *Sammer, M.* (Hrsg.), Anwendungsorientiertes Wissensmanagement, Ansätze aus der betrieblichen und universitären Praxis, Wiesbaden 2002, S. 53-72.

Blum, J., Wissensbilanz des *Deutschen Zentrums für Luft- und Raumfahrt*, Köln 2000.

Blum, J., Wissensbilanz des *Deutschen Zentrums für Luft- und Raumfahrt*, Köln 2001.

Bontis, N., Assessing Knowledge Assets: A Review of the Models Used to Measure Intellectual Capital, in: International Journal of Management Reviews, Vol. 3 (2001), Nr. 3, S. 41-60.

Bornemann, M. und *Leitner, K.-H.*, Entwicklung und Realisierung einer Wissensbilanz für eine Forschungsorganisation, in: *Pawlowsky, P.* und *Reinhard, R.* (Hrsg.), Wissensmanagement für die Praxis, Methoden und Instrumente zur erfolgreichen Umsetzung, Neuwied 2002, S. 335-365.

Bornemann, M. und *Sammer, M.* (Hrsg.), Anwendungsorientiertes Wissensmanagement, Ansätze und Fallstudien aus der betrieblichen und universitären Praxis, Wiesbaden 2002.

Brockhoff, K. und *Gerwin, J.*, Nicht länger im Verborgenen - Jahresabschlusspublizität von Forschungszentren, in: Wissenschaftsmanagement, 7. Jg. (2001), H. 1, S. 8-13.

Budäus, D., Public Management. Konzept und Verfahren zur Modernisierung öffentlicher Verwaltungen, Berlin 1994.

Bund-Länder-Kommission für Bildungsplanung und Forschungsförderung, Mindestanforderungen an Programmbudgets und Handreichung für die Erstellung von Programmbudgets in Einrichtungen der *Wissenschaftsgemeinschaft Gottfried Wilhelm Leibniz* (WGL), Bonn 2003

Choo, C. W. und *Bontis, N.* (Hrsg.), The Strategic Management of Intellectual Capital and Organizational Knowledge, Oxford 2002.

Haase, V., Active Scorecard, in: *Lausen, G., Oberweis, A.* und *Schlagpeter, G.* (Hrsg.), Informatik und formale Beschreibungsverfahren, Stuttgart 1999, S.78-92.

Lausen, G., Oberweis, A. und *Schlagpeter, G.* (Hrsg.), Informatik und formale Beschreibungsverfahren, Stuttgart 1999.

Leitner, K. H., Wissensbilanzierung. Ein neues Berichtswesen für Forschungsorganisationen und Hochschulen, in: Wissenschaftsmanagement, 9. Jg. (2003), H. 2, S. 20-24.

Lev, B., Intangibles. Management, Measurement, and Reporting, Washington 2001.

Maul, K. H., Wissensbilanzen als Teil des handelsrechtlichen Jahresabschlusses – Wissensbilanzen dargestellt am Beispiel des Jahresabschlusses von Hochschulen, in: DStR, 38. Jg. (2000), H. 47, S. 2009-2016.

Nationalrat Österreich, Bundesgesetzblatt, 9. August 2002.

Nonaka, I., A dynamic Theoriy of Organizational Knowledge Creation, in: *Choo, C. W.* and *Bontis, N.* (Hrsg.), The Strategic Management of Intellectual Capital and Organizational Knowledge, Oxford 2002, S. 437-462.

Pawlowsky, P. und *Reinhard, R.* (Hrsg.), Wissensmanagement für die Praxis, Methoden und Instrumente zur erfolgreichen Umsetzung, Neuwied 2002.

Probst, G., Raub, S. und *Romhardt, K.*, Wissen managen - Wie Unternehmen ihre wertvollste Ressource optimal nutzen, Frankfurt 1999.

Schreiber, S., Der Ansatz von Intangible Assets nach US-GAAP. Zentrale Aktivierungskriterien – relevante Verlautbarungen – Systembildung, Wiesbaden 2002.

Schröder, G., Regierungserklärung von Bundeskanzler Schröder vor dem Deutschen Bundestag, Berlin 14. März 2003.

Titscher, S., Winckler, G., Biedermann, H. u. a. (Hrsg.), Universitäten im Wettbewerb: zur Neustrukturierung österreichischer Universitäten, München u.a. 2000.

WBW - Institut für Wirtschafts- und Betriebswissenschaften der Montanunivsersität Leoben, Wissensbilanz, Leoben 2001.

Wissenschaftsrat, Strategische Forschungsförderung. Empfehlungen zu Kommunikation, Kooperation und Wettbewerb im Wissenschaftssystem, Drs. 5654/03, Essen 2003.

Ziegele, F.: Mittelvergabe und Zielvereinbarungen - Finanzierungsinstrumente eines Neuen Steuerungsmodells im Verhältnis Staat - Hochschule. Überlegungen und Erfahrungen in den deutschen Bundesländern, in: *Titscher, S., Winckler, G., Biedermann, H.* u. a. (Hrsg.), Universitäten im Wettbewerb: zur Neustrukturierung österreichischer Universitäten, München u.a. 2000, S.331-386.

Methoden der Patentbewertung

Martin Kloyer [1]

---

[1] Dr. *Martin Kloyer,* Lehrstuhl für Allgemeine Betriebswirtschaftslehre, Produktion und Industriebetriebslehre, Friedrich-Schiller-Universität Jena

Der Patentwert ist der Barwert der erwarteten Einzahlungsüberschüsse aus seiner Nutzung. Er hat eine subjektive und eine objektive Dimension und wird durch eine Reihe endogener und exogener Determinanten bestimmt. „Klassische", gesetzlich geregelte Bewertungsverfahren berücksichtigen weder die Determinanten noch sind sie wegen ihrer Vergangenheitsorientierung in der Lage, einen Erwartungswert zu erfassen. Neuere Ansätze, die sich auf empirisch geprüfte Indikatoren und Expertenbefragungen stützen, versuchen diese Mängel zu beheben.

1. Patentbewertung als Gegenstand der Betriebswirtschaftslehre

Die Bewertung von Patenten als „intangible assets" spielt in mehreren Bereichen der Betriebswirtschaftslehre eine Rolle. In einigen Bereichen ist sie rechtlich eindeutig geregelt. Dies gilt für die Bilanzierung und für gerichtliche Auseinandersetzungen bei Patentverletzungen, in denen Schadensersatzleistungen bestimmt werden. In anderen Bereichen aber existieren keine rechtlichen Regelungen. Gemeint sind insbesondere der Patenthandel, die Bewertung von Patenten bzw. Patentportfolios durch den Kapitalmarkt und die Patentbewertung als Grundlage für Entscheidungen über Herstellung und Vermarktung von materiellen Endprodukten. In diesen Zusammenhängen kann ein Patent nur als Erwartungswert der Einzahlungsüberschüsse aus der Verwertung definiert werden. Die rechtlich vorgesehenen und praktizierten Bewertungsverfahren werden dieser Perspektive nicht gerecht. Umso relevanter ist die Frage nach der Operationalisierung des Patentwertes im Sinne eines Erwartungswertes.

Der Beitrag stellt zunächst die für die Thematik bedeutenden rechtlichen Grundlagen dar und definiert den Patentwert. Im nächsten Schritt werden Determinanten des Patentwerts vorgestellt und erläutert. Daraufhin werden Möglichkeiten der Operationalisierung der Determinanten und damit der Patentbewertung eingeführt und kritisch betrachtet. Der Beitrag schließt mit einer knappen Zusammenfassung.

2. Rechtliche Grundlagen und Definition des Patentwertes

In diesem Abschnitt seien jene rechtlichen Grundlagen der Patentierung dargestellt, deren Kenntnis für die nachfolgende Diskussion der betriebswirtschaftlichen Bewertungsmethoden erforderlich ist. In diesem Zusammenhang ist eine differenzierende Betrachtung der deutschen (Deutsches Patentgesetz (DPatG)), europäischen (Europäisches Patentübereinkommen (EPÜ)) und internationalen (Patent Cooperation Treaty (PCT)) Regelungen weitgehend verzichtbar, weil sich keine prinzipiellen Unterschiede ergeben würden.

Eine zum Patent anmeldbare Erfindung im Sinne des DPatG und des EPÜ ist „eine Lehre zum praktischen Handeln, deren beanspruchter Gegenstand oder deren beanspruchte Tätigkeit technischer Natur, realisierbar und wiederholbar ist und die Lösung einer Aufgabe durch technische Überlegungen darstellt" (*Schulte,* 2001, § 1 Rdn. 16).

„Patente werden für Erfindungen erteilt, die neu sind, auf erfinderischer Tätigkeit beruhen und gewerblich anwendbar sind" (§ 1, Abs. 1 DPatG; Art. 52, Abs. 1 EPÜ).

Neu ist eine Erfindung, wenn sie zum Zeitpunkt der Anmeldung weder zum tatsächlichen noch zum fiktiven Stand der Technik gehört. Ersterer umfasst alles, was der Öffentlichkeit irgendwann, irgendwo und irgendwie zugänglich gemacht wurde (vgl. *van Venrooy,* 1996, S. 54; *Rogge,* 1998, S. 187; *Schulte,* 2001, § 3, Rdn. 10 und 17). Letzterer umfasst alles, was zwar noch nicht der Öffentlichkeit zugänglich gemacht, aber bereits zum Patent angemeldet wurde (vgl. *Bernhardt/Krasser,* 1986; *van Venrooy,* 1996, S. 59; *Rogge,* 1998, S. 187). Im europäischen Verfahren zählen dabei ausschließlich europäische Anmeldungen zum fiktiven Stand, im deutschen allerdings alle weltweit erfolgten Anmeldungen, die Schutz für Deutschland beanspruchen (vgl. *Schulte,* 2001, § 3, Rdn. 26). Die Hürde der Neuheit ist damit im deutschen Verfahren höher als im europäischen.

Das Kriterium der erfinderischen Tätigkeit bzw. der Erfindungshöhe ist erfüllt, wenn die Erfindung für den Fachmann, ausgehend vom Stand der Technik, nicht nahe liegt, wobei hier nicht am fiktiven, sondern nur am tatsächlichen Stand der Technik gemessen wird (vgl. § 4, Satz 2 PatG bzw. Art. 56, Satz 2 EPÜ; *Schulte,* 2001, § 4, Rdn. 14). Es kann also das Kriterium der erfinderischen Tätigkeit erfüllt sein, ohne dass eine Neuheit vorliegt. Das Ausmaß der Erfindungshöhe hängt von der Komplexität der technischen Problemlösung ab.

Am Kriterium der gewerblichen Anwendbarkeit scheitern die meisten Ergebnisse der Grundlagenforschung, da dort in der Regel keine konkreten Anwendungsmöglichkeiten genannt werden können (vgl. *Brockhoff,* 1999, S. 103).

Nach der Anmeldung erstellt das europäische Patentamt automatisch - und das deutsche auf Antrag - einen Rechercheberičht, der dem Anmelder den Stand der Technik mitteilt, aber noch keine Prüfung der genannten Kriterien beinhaltet (vgl. *Schulte,* 2001, § 43, Rdn. 6). Diese Prüfung erfolgt auf Antrag und kann mit der Erteilung oder Ablehnung des Patents enden oder aber auch zu Entgegenhaltungen und der Auflage von Nachbesserungen führen.

Ein erteiltes Patent kann, wenn es nicht an durch den Wettbewerb angestrengten Einspruchs-, Beschwerde- oder Nichtigkeitsverfahren (Die verschiedenen Verfahrenstypen sind unterschiedlichen Instanzen zuzuordnen - im Folgenden wird für alle drei Verfahrensarten der Begriff des Einspruchsverfahrens verwendet) scheitert, 20 Jahre aufrechterhalten werden.

Eine Patentschrift gliedert sich grob in folgende Abschnitte (vgl. z. B. *Hubmann/Götting,* 1998, S. 148ff.): Auf die Beschreibung des Standes der Technik folgt die Darstellung des technischen Problems bzw. der Aufgabe, hierauf die Beschreibung der Lösung und schließlich die Formulierung der unabhängigen und der von diesen abhängigen, weil auf sie Bezug nehmenden, Ansprüche (Anspruchsumfang).

Ein Patent ist das Recht bzw. – betriebswirtschaftlich gesehen - die Realoption (vgl. *Pakes,* 1986) auf die exklusive Nutzung einer Erfindung. Seine Nutzung ist als Investition zu betrachten, die mit Einzahlungen und Auszahlungen verbunden ist. Dementsprechend ist der Patentwert der Barwert der erwarteten Einzahlungsüberschüsse aus der Nutzung des Patents. Die Bestimmung des Patentwerts setzt Wahrscheinlichkeitsaussagen zum Eintritt von alternativen Umweltzuständen und zu in diesen Umweltzuständen erzielbaren Einzahlungsüberschüssen voraus (vgl. zu den Voraussetzungen der Bestimmung des Werts von Optionen z.B. *Laux,* 1993). Mit anderen Worten: Die für den Fall der Patentdurchsetzung zu erwartenden Einzahlungsüberschüsse müssen mit der Durchsetzungswahrscheinlichkeit multipliziert werden (vgl. *Harhoff/Reitzig,* 2001).

3. Determinanten des Patentwerts

Der Patentwert wird von einer Reihe von Determinanten bestimmt. Diese sind danach zu unterscheiden, inwieweit sie vom Anmelder beeinflussbar sind (endogen vs. exogen) und inwiefern sie den subjektiven, vom Inhaber wahrgenommenen Wert oder aber den objektiven Wert beeinflussen (vgl. *Abb. 1*).

Determinante	**Beeinflussbarkeit durch den Inhaber: endogen/exogen**	**Wertdimension: subjektiv/objektiv**
Prozessinterne Unsicherheit	Exogen	Objektiv
Marktunsicherheit	Exogen	Objektiv
Rechtliche Unsicherheit	Exogen/endogen	Objektiv
Stand der Technik	Exogen	Objektiv
Erfindungshöhe	Exogen	Objektiv
Umgehungsmöglichkeiten	Exogen	Objektiv
Massierung	Exogen	Objektiv
Kontrolle über die komplementären Ressourcen	Exogen/endogen	Subjektiv
Restnutzungsdauer	Exogen/endogen	Objektiv
Anspruchsumfang	Endogen	Objektiv
Anmeldungszeitpunkt	Endogen	Objektiv
Patentfunktion „Ausschlussrecht"	Exogen/endogen	Objektiv
Patentfunktion „Lizenzpatent"	Exogen/endogen	Objektiv
Patentfunktion „Sperrpatent"	Exogen/endogen	Subjektiv
Patentfunktion „Verhandlungsmassepatent"	Exogen/endogen	Objektiv/subjektiv

Abb. 1: Determinanten des Patentwerts

3.1 Exogene Determinanten

Zunächst einmal wird der (objektive) Patentwert durch das Ausmaß der Unsicherheit zum Zeitpunkt der Erteilung beeinflusst. Dabei treten drei Dimensionen der Unsicherheit auf. Wenn Erfindungen aus frühen Forschungsphasen angemeldet werden, sobald die Bedingung der gewerblichen Anwendbarkeit erfüllt ist, besteht – erstens - die technologisch bedingte, innovationsprozessinterne Unsicherheit, ob auf der Basis der Erfindung jemals ein Endprodukt wird vermarktet werden können (vgl. *Gilbert/Newberry*, 1982). Diese Unsicherheit ist insbesondere in F&E- und wettbewerbsintensiven Branchen anzutreffen, in denen ein Wettlauf um die Neuheit besteht. Daneben steht – zweitens - die Unsicherheit, ob am Ende der eigenen Produktentwicklung noch ein Markt existiert (vgl. *Gilbert/Newberry*, 1982; *Pakes*, 1986). Schließlich kann sich bis dahin ein „dominant design“ (*Afuah/Utterback,* 1997, S. 185). durchgesetzt haben, das von dem, welches der eigenen Entwicklung zugrunde gelegt wurde, abweicht. Schließlich besteht – drittens - insofern rechtliche Unsicherheit, als das Verhalten des Patentamts im Prüfverfahren und das der Wettbewerber in Hinblick auf Einsprüche nicht vorhersehbar sind. Teilweise ist diese Unsicherheit allerdings endogen, da mit der Dimensionierung des Anspruchsumfangs die Größe der Angriffsfläche bestimmt wird (siehe unten).

Der (objektive) Patentwert hängt darüber hinaus von Merkmalen der Erfindungsqualität ab. Dabei handelt es sich nicht um die Neuheit und die gewerbliche Anwendbarkeit. Diese sind dichotom ausgeprägte Voraussetzungen der Patentierbarkeit und damit keine Determinanten, die entsprechend ihrer Ausprägung den Patentwert beeinflussen. Determinanten des Patentwerts sind aber die metrisch skalierbaren Größen des Standes der Technik, der Erfindungshöhe (Komplexität der technischen Problemlösung) und der Umgehungsmöglichkeiten (*Gallini*, 1993). Unter Umgehung ist die Substitution der geschützten durch eine alternative Technologie zu verstehen. Umgehungsmöglichkeiten ergeben sich insbesondere bei einer kumulativen Technologieentwicklung (vgl. z.B. *Scotchmer*, 1991; *Chang*, 1995), die dadurch gekennzeichnet ist, dass Produkte auf mehreren, von verschiedenen Inhabern gehaltenen Patenten beruhen. Grundlegend ist hier, dass ein hoher Stand der Technik mit einem entwickelten und damit potentiell ertragreichen Markt einhergeht. Außerdem ist festzustellen, dass mit zunehmender Erfindungshöhe die Umgehungsmöglichkeiten abnehmen und dass es umso schwieriger ist, eine große Erfindungshöhe zu realisieren, je höher der Stand der Technik ist. In Hinblick auf die zu erwartenden Einzahlungsüberschüsse ergeben sich vier realtypische Ausprägungskombinationen von Erfindungshöhe und Stand der Technik. Intuitiv einsichtig ist, dass ein geringer Stand der Technik in Verbindung mit geringer Erfindungshöhe einen niedrigen Einzahlungsüberschuss verspricht. Bei hohem Stand der Technik und großem Ausmaß erfinderischer Tätigkeit ist ein hoher Barwert der Einzahlungsüberschüsse zu erwarten, da hohes Marktpotential und große Umgehungsschwierigkeiten zusammentreffen. Nicht eindeutig ist die Situation bei niedrigem Stand der Technik verknüpft mit großer Erfindungshöhe. Hier ist anzunehmen, dass sich Einzahlungsströme verzögert einstellen, weil der Markt für die Erfindung erst entwickelt werden muss. Sind umgekehrt der Stand der Technik hoch und die Erfindungshöhe niedrig, wirken sich die hohen Wahrscheinlichkeiten für Entgegenhaltungen, Einsprüche und Umgehungserfindungen negativ auf den Erwartungswert aus.

Die Determinante der „Massierung“ beschreibt den Umstand, dass der (objektive) Wert des einzelnen Patents umso geringer ist, je mehr Patente einem Produkt zugrunde liegen bzw. dass der Wert eines Patents umso höher ist, je mehr Produkte auf ihm beruhen (*Cohen* et al., 2000). Der Grad der Massierung schwankt branchenabhängig stark (*Acs/Audretsch*, 1988). Typischerweise ist er bei einer kumulativen Technologieentwicklung groß.

Intuitiv einsichtig ist, dass der (subjektive) Wert eines Patents aus der Perspektive des Inhaberunternehmens auch davon abhängt, inwieweit es die komplementären Ressourcen zur Herstellung und Vermarktung materieller Endprodukte kontrolliert (*Teece*, 1986). Investitionen in die komplementären Ressourcen, die vor der Patentverwertung getätigt werden müssen, mindern den Barwert der erwarteten Einzahlungsüberschüsse. Diese Determinante ist allerdings längerfristig endogen, da es dem Unternehmen prinzipiell freisteht, komplementäre Ressourcen zu erwerben oder zu veräußern.

Schließlich sei die – ebenfalls teilweise endogene – Determinante der Restnutzungsdauer genannt. Zwar wirkt sich eine längere Restnutzungsdauer positiv auf den (objektiven) Patentwert aus. Dem wirkt aber die Tatsache entgegen, dass insbesondere in den ersten Jahren der Patentlaufzeit Unsicherheit bezüglich Einsprüchen und Verletzungen des Wettbewerbs besteht.

3.2 Endogene Determinanten

Hinsichtlich der endogenen Determinanten ist an erster Stelle der Anspruchsumfang zu nennen (vgl. *Gilbert/Shapiro*, 1990). Die Grenzen werden hier durch das Minimum der Patentierbarkeit und das Maximum der Erfindungshöhe gezogen. Die Wirkung auf den (objektiven) Patentwert ist ambivalent. Eine Zunahme des Anspruchsumfangs führt zwar zu einem Anstieg des anvisierten Marktpotentials, aber auch zu einer höheren Wahrscheinlichkeit von Entgegenhaltungen und Einsprüchen. So können sich die Wirkungen auf den Erwartungswert neutralisieren. Die Wahrscheinlichkeit der Erteilung und Aufrechterhaltung des Patents (Durchsetzungswahrscheinlichkeit) hängt dabei entscheidend davon ab, welche Rückzugsoptionen (Unteransprüche) ein Anmelder in Hinblick auf mögliche Entgegenhaltungen im Prüfverfahren und Einsprüche durch Wettbewerber vorsieht.

Einfluss auf den (objektiven) Patentwert hat schließlich auch der Zeitpunkt der Anmeldung. Bei einer Verzögerung der Anmeldung ist zwar häufig durch fortgesetzte Forschung und Entwicklung eine größere Erfindungshöhe erzielbar. Es steigt aber auch die Wahrscheinlichkeit, durch einen Konkurrenten überholt zu werden (vgl. *Hall/Ham,* 1999, S. 13).

3.3 Interdependenz der Determinanten, Patentfunktionen und Durchsetzungswahrscheinlichkeit

Es bestehen zahlreiche Wechselwirkungen zwischen den beschriebenen Determinanten. In der Realität finden sich typische Kombinationen von Ausprägungen der exogenen Determinanten, die jeweils eine typische Form der Patentnutzung in Verbindung mit bestimmten endogenen Entscheidungsspielräumen zulassen.

Im Wesentlichen sind vier Formen der Patentnutzung bzw. Patentfunktionen zu unterscheiden (vgl. z. B. *Hermans,* 1991, S. 86f.; *Schewe,* 1993, S. 348f.; *Mordhorst,* 1994, S. 127; *Rahn,* 1994, S. 378f.; *Faix,* 1998, S. 50ff. und 350ff; *Haupt,* 2000, S. 66f.; *Harhoff/Reitzig,* 2001; *Knight,* 2001, S. 52f.; *Rebel*, 2001):

- Das Patent dient zum Ausschluss des Wettbewerbs bei aktiver eigener Nutzung der Erfindung (Ausschlusspatent).
- Das Patent wird als Verhandlungsmasse für cross-licensing eingesetzt (Verhandlungsmassepatent).
- Das Patent wird auslizenziert (Lizenzpatent).
- Das Patent dient zur Wettbewerbssteuerung, indem es Forschungsfelder besetzt (Sperrpatent).

Die Patentfunktion kann nicht gänzlich frei gewählt werden. Sie ist daher eine der gemischt exogenen-endogenen Determinanten des Patentwerts. Der Spielraum bei der Wahl der Patentfunktion wird bestimmt durch die Erfindungshöhe und die Umgehungsmöglichkeiten, den Grad der Massierung sowie das Ausmaß der Kontrolle über die komplementären Ressourcen. Voraussetzung für eine erfolgreiche Nutzung als Ausschlusspatent ist eine große

Erfindungshöhe in Verbindung mit geringen Umgehungsmöglichkeiten, die eine längerfristige Aufrechterhaltung des Patents gegenüber Einsprüchen des Wettbewerbs ermöglichen. Diese Bedingungen sind typischerweise bei einer nicht-kumulativen Technologieentwicklung in Verbindung mit einer geringen Massierung gegeben. Hinzukommen muss die Kontrolle des Patentinhabers über die komplementären Ressourcen, die erst die exklusive Nutzung für eigenständige Herstellung und Vermarktung von materiellen Endprodukten ermöglicht. Diese Bedingungen schaffen üblicherweise die Grundlage für einen großen Anspruchsumfang und damit für ein Ausschlusspatent. Wenn von den vier Bedingungen die Kontrolle über die komplementären Ressourcen fehlt, hat der Inhaber die Wahl zwischen Lizenz- und Sperrpatent. Dabei erfordert das Sperrpatent stärker eingeschränkte Umgehungsmöglichkeiten als das Lizenzpatent. Wenn im Vergleich mit dem Ausschlusspatent nicht nur die komplementären Ressourcen fehlen, sondern darüber hinaus auch eine geringe Erfindungshöhe, viele Umgehungsmöglichkeiten sowie eine hohe Massierung zu verzeichnen sind, dann bietet sich die Verwendung des Patents als Verhandlungsmasse für crosslicensing an.

Es ist offensichtlich, dass sich die Patentfunktion auf die Art des Patentwerts auswirkt. So besitzen ein Ausschluss- und ein Lizenzpatent einen objektiven Wert. Der Wert eines Verhandlungsmasse- und eines Sperrpatents hingegen hat sowohl eine objektive als auch eine subjektive Komponente.

Nach der oben eingeführten Definition ist der Patentwert das Produkt aus den für den Fall der Patentdurchsetzung zu erwartenden Einzahlungsüberschüssen und der Durchsetzungswahrscheinlichkeit. Es wurde bereits erwähnt, dass die Durchsetzungswahrscheinlichkeit durch die Bemessung des Anspruchsumfangs beeinflusst werden kann. Letztlich aber erschwert die Determinante der Unsicherheit über das Verhalten der Prüfer und Wettbewerber die Patentwertbestimmung erheblich. Die Unsicherheit besteht darüber, ob und wie Prüfer und Wettbewerber behindernd agieren werden (vgl. *Reitzig*, 2002, S. 163). Diese Unsicherheit wird noch dadurch verstärkt, dass eigene Reaktionen auf Prüfer- und Wettbewerberverhalten und deren Erfolgsaussichten kaum antizipiert werden können.

4. Operationalisierung der Determinanten des Patentwerts

Ansätze zur Bestimmung des Patentwerts können vor dem Hintergrund der bisherigen Ausführungen nur dann akzeptiert werden, wenn sie den Einfluss der Determinanten berücksichtigen und durch ihre Zukunftsorientierung in der Lage sind, den Erwartungswert der Einzahlungsüberschüsse zu ermitteln.

4.1 Rechtliche Regelungen

Es existiert eine Reihe von rechtlich geregelten Ansätzen zur Patentbewertung. §139 DPatG nennt drei alternative Verfahren zur Berechnung von Schadensersatz im Falle einer Patentverletzung:

- Der entgangene Gewinn entspricht der Differenz zwischen hypothetischem Gewinn des Patentinhabers bei Nicht-Verletzung und dem Gewinn bei eingetretener Verletzung.
- Der Verletzergewinn ist der Gewinn, den der Patentverletzer aus der Vermarktung des auf dem verletzten Patent beruhenden Produkts erwirtschaftet.

- Die hypothetische Lizenzgebühr ist die Gebühr, die zum Zeitpunkt der Verletzung vereinbart worden wäre.

Abgesehen davon, dass keiner der drei Ansätze die genannten Determinanten des Patentwerts berücksichtigt, ist allen der Mangel gemeinsam, dass sie ausschließlich vergangenheitsorientiert sind, mithin keinen Erwartungswert ermitteln können.

Eine weitere Regelung zur Patentbewertung findet sich in §§9 und 10 ArbeitnehmererfindungsG. Dort wird insbesondere der sogenannte Schätzwert eingeführt, der jenem Betrag entspricht, den der Arbeitgeber einem externen Erfinder zahlen müsste. Diese Methode ist zwar insofern kompatibel mit der hier eingeführten Definition des Patentwerts, als der hypothetische Kaufpreis dem Erwartungswert entspricht. Allerdings findet sich kein weiterer Hinweis zur Operationalisierung der Determinanten des Patentwerts.

Die Patentbewertung im Rahmen der Bilanzierung ist ebenfalls gesetzlich geregelt. § 248, Abs. 2 HGB untersagt die bilanzielle Aktivierung immaterieller Güter im Anlagevermögen. Allerdings ist mittlerweile auf der Grundlage des Gesetzes zur Kontrolle und Transparenz im Unternehmensbereich (KonTraG) sowie des Kapitalaufnahmeerleichterungsgesetzes (KapAEG) die Bilanzierung nach US-GAAP oder IAS möglich (*Schildbach*, 2000). Danach sind erworbene immaterielle Vermögensgegenstände mit den Anschaffungskosten, selbst erstellte mit den Herstellungskosten anzusetzen (*KPMG*, 1999, S. 66f.). Die Anschaffungskosten anzusetzen, deckt sich ebenso wie der hypothetische Kaufpreis mit der hier vertretenen Definition des Patentwerts. Dies kann nicht für die Herstellungskosten gelten, die naturgemäß keine Zahlungserwartungen abbilden.

4.2 Lizenzgebühr

Zukunftsorientiert und hinsichtlich der Berücksichtigung der Determinanten weiter fortgeschritten sind Ansätze aus der Literatur zur Lizenzierung. Die dort angestellten Überlegungen zur Bewertung eines auszulizenzierenden Objekts schließen den Spezialfall der Patentlizenz ein (vgl. z. B. *Stumpf/Groß*, 1998). Danach haben nicht nur die Herstellungskosten der Erfindung, sondern eben auch – zukunftsorientiert – die Nutzungsdauer und Marktmerkmale in die Bewertung einzufließen. Insbesondere ist zu prüfen, welche zusätzliche Absatzmenge zu welchem zusätzlichen Preis und damit welcher Zusatzgewinn durch den Erwerb der Patentlizenz realisiert werden kann. Dabei ist die zu erwartende Wettbewerbsintensität bzw. der Grad der zu erwartenden Monopolstellung des Lizenznehmers zugrundezulegen. Die Ansätze zur Bestimmung der Lizenzgebühr erfüllen zwar das Kriterium der Zukunftsorientierung, lassen aber eine Reihe der genannten Determinanten des Patentwerts unberücksichtigt.

4.3 Empirisch überprüfte Indikatoren

Empirische Untersuchungen haben für eine Reihe von objektiv messbaren Indikatoren einen positiven Zusammenhang mit dem Patentwert ermittelt. Im Folgenden sind die bestätigten Indikatoren aus solchen Untersuchungen beschrieben, die die abhängige Variable des Patentwerts valide operationalisiert haben. Zu den validen Operationalisierungen zählen (in Klammern die nachfolgend verwendeten Abkürzungen):

- Erstens die erzielten, auf ein Patent zurückzuführenden Einzahlungsüberschüsse (AV Einz),

- zweitens die Einspruchswahrscheinlichkeit (die auf einen hohen Patentwert schließen lässt)(AV Einspr),
- drittens die Patentqualität (AV Qual) und
- viertens die Patenterteilung (AV Ert).

Operationalisiert wird jeweils der objektive Patentwert.

Bestätigt wurde der positive Einfluss des (exogenen) Indikators der Vorwärtszitate (vgl. *Albert* et al., 1991 (AV Einz); *Harhoff* et al., 1999 (AV Einz); *Harhoff/Reitzig, 2000* (AV Einspr); *Lanjouw/Schankermann,* 1999 (AV Qual)). Dies sind jene Zitate, die das Patent selbst erhält, wenn es späteren Anmeldungen als Stand der Technik durch die Prüfer entgegengehalten wird.

Bestätigt wurde auch der (endogene) Indikator der Familiengröße (vgl. *Guellec/van Pottelsberghe de la Potterie,* 2000 (AV Ert); *Harhoff* et al., 1999; *Harhoff/Reitzig*, 2000; *Lanjouw/Schankermann*, 1999). Darunter ist die Zahl der Länder zu verstehen, in denen Patentschutz für die Erfindung besteht.

Auch der (endogene) Indikator der Eigentumsverhältnisse wurde bestätigt (vgl. *Guellec/van Pottelsberghe de la Potterie*, 2000); *Harhoff/Reitzig*, 2000; *Lanjouw/Schankermann*, 2000 (AV Einspr)). Danach sind von Unternehmen gehaltene Patente wertvoller als von Einzelpersonen gehaltene.

Schließlich wurde auch der (teilweise endogene) Indikator des Anspruchsumfangs bestätigt (vgl. *Lanjouw/Schankermann*, 1999; *Lanjouw/Schankermann* 2000).

Prinzipiell sind die Indikatoren zwar geeignet, den Patentwert als Erwartungswert zu operationalisieren. Allerdings wird von den oben beschriebenen Determinanten allein der Anspruchsumfang und damit indirekt und tendenziell auch die Erfindungshöhe operationalisiert.

4.4 Expertenbewertung

Bezüglich der anderen Determinanten scheint es gegenwärtig keinen anderen Ansatz der Operationalisierung als den der subjektiven Bewertung durch Experten zu geben. Dabei kann ein Unternehmen die in seinem spezifischen Umfeld relevanten Determinanten ermitteln, indem es für die einzelnen Patente seines Portfolios die Ausprägungen der Determinanten bewerten lässt (vgl. zur Operationalisierung *Reitzig,* 2002, S. 188ff.) und die Korrelationen mit dem Patentwert bestimmt. Letzterer kann – in Anlehnung an die oben erwähnte Schätzwertmethode – entsprechend der hier eingeführten Definition des Patentwertes als Erwartungswert der Einzahlungsüberschüsse ermittelt werden, indem Experten nach ihrer Zahlungsbereitschaft für das hypothetisch von einem Konkurrenten gehaltene Patent befragt werden. Ein neu zu bewertendes Patent des Unternehmens müsste dann lediglich hinsichtlich der als relevant ermittelten Determinanten eingeschätzt werden.

5. Zusammenfassung

Die Ermittlung des Wertes des Intangible Assets „Patent" ist nicht nur im Rahmen der Bilanzierung und in gerichtlichen Auseinandersetzungen von Bedeutung. Sie ist auch Voraussetzung für den Handel mit Patenten, für die Unternehmensbewertung am Kapitalmarkt und für Entscheidungen über Herstellung und Vermarktung von materiellen Endprodukten. In diesen Zusammenhängen ist der Patentwert als Erwartungswert der Einzahlungsüberschüsse aus der exklusiven Nutzung des Patents zu verstehen. Zu seiner Bestimmung ist der Einfluss einer Reihe von Determinanten zu berücksichtigen. Probleme bei der Bestimmung liegen in der Ermittlung der für ein konkretes Unternehmen spezifischen Determinanten und in der Bemessung der Wirkung der Determinanten auf den Patentwert. Die gesetzlich vorgesehenen Methoden der Patentbewertung berücksichtigen weder die Determinanten noch sind sie zukunftsorientiert und damit in der Lage, den Erwartungswert zu bestimmen. In der Literatur wird zwar für eine Reihe von objektiv messbaren und prinzipiell zukunftsorientierten Indikatoren ein Zusammenhang mit dem Patentwert empirisch nachgewiesen. Von den Determinanten werden dabei aber nur der Anspruchsumfang und die Erfindungshöhe operationalisiert. Daher empfiehlt es sich für ein Unternehmen, über eine Expertenabschätzung der Determinanten und des Werts des eigenen Patentportfolios die in seiner Situation relevanten Determinanten und ihre Wirkungsrichtung und -intensität zu ermitteln.

6. Literatur

Afuah, A. N. and *Utterback, J. M.*, Responding to structural industry changes: A technological evolution perspective, in: Industrial and Corporate Change, Vol. 6 (1997), No. 1, S. 183-202.

Albert, M. B., *Avery, D.*, *Narin, F.* and *McAllister, P.*, Direct validation of citation counts as indicators of industrially important patents, in: Research Policy, Vol. 20 (1991), No. 3, S. 251-259.

Acs, Z. J and *Audretsch, D. B.*, Innovation in large and small firms: An empirical analysis, in: American Economic Review, Vol. 78 (1998), No. 2, S. 678-690.

Bernhardt, W. und *Krasser, R.*, Lehrbuch des Patentrechts – Recht der Bundesrepublik Deutschland, Europäisches und Internationales Patentrecht, 4. Auflage, München 1986.

Brockhoff, K., Forschung und Entwicklung. Planung und Kontrolle, 5.Auflage, München, Wien 1999.

Chang, H. F., Patent scope, antitrust policy, and cumulative innovation, in: RAND Journal of Economics, Vol. 26 (1995), No. 1, S. 34-57.

Cohen, W. M., *Nelson, R. R.* and *Walsh, J. P.*, Protecting their intellectual assets: Appropriability conditions and why U. S. manufacturing firms patent (or not), Cambridge 2000.

Faix, A., Patente im strategischen Marketing – Sicherung der Wettbewerbsfähigkeit durch systematische Patentanalyse und Patentnutzung, Berlin 1998.

Gallini, N. T., Patent policy and costly imitation, in: RAND Journal of Economics, Vol. 23 (1992), No. 1, S. 52-63

Gilbert, R. and *Newberry, D.*, Preemptive patenting and the persistence of monopoly, in: American Economic Review, Vol. 72 (1982), No. 3, S. 514-526.

Gilbert, R. and *Shapiro, C.*, Optimal patent length and breadth, in: RAND Journal of Economics, Vol. 21 (1990), No. 1, S. 106-112.

Guellec, D. and *van Pottelsberghe de la Potterie, B.*, Analysing patent grants, Brüssel 2000.

Hall, B. H. and *Ham, R.M.*, The patent paradox revisited: Determinants of patenting in the US semiconductor industry, in: NBER Working Paper, 1999, Nr. 7062.

Harhoff, D., Scherer, F. and *Vopel, K.*, Citations, family size, opposition and the value of patent rights, LMU München/Harvard University/ZEW Mannheim 1999.

Harhoff, D. and *Reitzig, M.*, Determinants of opposition against EPO patent grants: The case of pharmaceuticals and biotechnology, LMU München 2000.

Harhoff, D. und *Reitzig, M.*, Strategien zur Gewinnmaximierung bei der Anmeldung von Patenten. Wirtschaftliche und rechtliche Aspekte als Entscheidungsgrößen beim Schutz von Erfindungen, in: Zeitschrift für Betriebswirtschaft, 71. Jg. (2001), H. 5, S. 509-529.

Haupt, R., Industriebetriebslehre. Einführung. Management im Lebenszyklus industrieller Geschäftsfelder, Wiesbaden 2000.

Hermans, J.-P., Strategies for protection of innovations. The case of a research-intensive multinational corporation, in: *Täger, U.* and *Witzleben, A. von* (Hrsg.), Strategies for the protection of innovation. Proceedings of the first European Congress on industrial property rights and innovation. Patinnova'90, Luxembourg, 1991, S. 83-88.

Hubmann, H. und *Götting, H.-P.*, Gewerblicher Rechtsschutz, 6. Aufl., München 1998.

Knight, H. J., Patent Strategy for Researchers and Research Managers, 2. Aufl., Chichester et al. 2001.

KPMG, International accounting standards, Stuttgart 1999.

Lanjouw, J. O. and *Schankerman, M.*, The quality of ideas: Measuring innovation with multiple indicators, NBER 1999.

Lanjouw, J. O. and *Schankerman, M.*, Characteristics of patent litigation: A window on competition, CEPR 2000.

Laux, C., Handlungsspielräume im Leistungsbereich des Unternehmens: Eine Anwendung der Optionstheorie, in: Zeitschrift für betriebswirtschaftliche Forschung, 45. Jg. (1993), H. 11, S. 933-958.

Mordhorst, C. F., Ziele und Erfolg unternehmerischer Lizenzstrategien, Wiesbaden 1994.

Pakes, A., Patents as options: Some estimates of the value of holding European patent stocks, in: Econometrica, Vol. 54 (1986), No. 478, S. 755-784.

Rahn, G., Patentstrategien japanischer Unternehmen, in: GRUR Int., 1994, S. 377-382.

Rebel, D., Gewerbliche Schutzrechte. Anmeldung – Strategie – Verwertung. Ein Praxishandbuch, 3. Auflage, Köln et al. 2001.

Reitzig, M., Die Bewertung von Patentrechten – Eine theoretische und empirische Analyse aus Unternehmenssicht, Wiesbaden 2002.

Rogge, R., Der Neuheitsbegriff unter besonderer Berücksichtigung kollidierender Patentanmeldungen, in GRUR Int., 1998, S. 186-189.

Schewe, G., Kein Schutz vor Imitation, in: Zeitschrift für betriebswirtschaftliche Forschung, 45. Jg. (1993), H. 4, S. 344-360.

Schildbach, T., Ansatz und Bewertung immaterieller Anlagewerte, in: *Ballwieser, W.* (Hrsg.): US-amerikanische Rechnungslegung, Stuttgart, 2000, S. 99-138.

Schulte, R., Patentgesetz mit Europäischem Patentübereinkommen. Kommentar auf der Grundlage der deutschen und europäischen Rechtsprechung, 6. Auflage, Köln et al. 2001.

Stumpf, H. und *Groß, M.*, Der Lizenzvertrag, 7. Aufl., Heidelberg 1998.

Scotchmer, S., Standing on the shoulders of giants: Cumulative research and the patent law, in: Journal of Economics Perspectives, Band 5 (1991), No. 1, S.29-41.

Teece, D. J., Profiting from technological innovation. Implications for integration, collaboration, licensing and public policy, in: Research Policy, Vol. 15 (1986), No. 6, S. 285-305.

Venrooy, G.J. van, Patentrecht einschließlich Arbeitnehmererfinderrecht und Vertragsrecht, Düsseldorf 1996.

5. Empirie und Praxisbeispiele

Einen erheblichen Mangel in der Diskussion um Intangibles stellt die Dominanz konzeptioneller Ansätze gegenüber praktischen Arbeiten dar. Insbesondere gesicherte und verallgemeinerbare empirische Erkenntnisse liegen nur in relativ geringem Umfang vor, wie dies allerdings auch typisch für ein neues Themengebiet ist. Symptomatisch für ein neues Themengebiet ist auch die Dominanz explorativer Arbeiten mit Fallstudiencharakter. Von besonderem Interesse auch für zukünftige Forschungen sind dabei Längsschnittanalysen, die die Entwicklung über die Zeit dokumentieren. Da die Beschäftigung mit Intangibles in den skandinavischen Ländern bereits seit längerer Zeit erfolgt, können zwei wegweisende Pioniere der Intangibles-Forschung derartige Untersuchungsergebnisse präsentieren. Besondere Beachtung verdient die Querschnittsanalyse auf Basis vorhandenem Datenmaterials, die anhand der systematischen Auswertung der Geschäftsberichte führender Unternehmen Beispiele und Lösungsansätze für ein geeignetes Reporting über immaterielle Vermögenswerte ableitet.

Externes Reporting über immaterielle Vermögenswerte

Gerhard Speckbacher[1], *Stefan Güldenberg* [2] *und Raoul Ruthner* [3]

---

[1] Prof. Dr. *Gerhard Speckbacher*, Institut für Unternehmensführung, Wirtschaftsuniversität Wien, Österreich

[2] Dr. *Stefan Güldenberg*, Institut für Unternehmensführung, Wirtschaftsuniversität Wien, Österreich

[3] *Raoul Ruthner*, Institut für Unternehmensführung, Wirtschaftsuniversität Wien, Österreich

9. Fazit

10. Literatur

In modernen wissensintensiven Unternehmen basiert die Wertschaffung in zunehmendem Maße auf dem Aufbau und dem Management immaterieller Vermögenswerte. Die enormen Schwankungen der Markt-/Buchwertrelationen in vielen Unternehmen sind ein Zeichen zunehmender Unsicherheit des Anlegerpublikums über die Wertrelevanz von Investitionen in immaterielles Vermögen. Ein geeignetes Reporting, das „Überraschungen" und damit Volatilitäten und Kapitalkosten reduziert, kann aufgrund der Besonderheiten immaterieller Vermögenswerte nicht analog zum traditionellen Rechnungswesen physischer Vermögenswerte erfolgen. Die systematische Auswertung der Geschäftsberichte führender Unternehmen zeigt diesbezügliche Defizite auf. Daneben wird über Beispiele und Lösungsansätze eines geeigneten Reportings über immaterielle Vermögenswerte berichtet.

1. Zunehmende Bedeutung immaterieller Vermögenswerte

Die vergangenen Jahrzehnte waren durch eine fundamentale Veränderung der Ressourcenstruktur von Unternehmen gekennzeichnet. Während die Wertschöpfung traditioneller Industrieunternehmen primär auf physische Vermögensgegenstände, wie Grundstücke, Rohstoffe oder Produktionsanlagen gegründet war (vgl. dazu etwa *Chandler,* 1990), werden in den wissensbasierten Unternehmen des Informationszeitalters Überrenditen und damit Wertsteigerungen hauptsächlich durch Strategien zur Schaffung und zum Einsatz immaterieller Vermögenswerte erzielt. Nicht mehr Größen- oder Bereichsvorteile bei der Nutzung physischen Vermögens sichern in erster Linie die Wettbewerbsfähigkeit, sondern Innovationen, Kundenbeziehungen, Mitarbeiterwissen, Brand Capital oder nachhaltige Kooperationen mit Zulieferern. Als Beleg hierfür zitieren *Kaplan/Norton* (2001, S. 88) verschiedene Studien, nach denen in den letzten zwanzig Jahren der Anteil der Buchwerte physischer Vermögensgegenstände am Marktwert von Unternehmen kontinuierlich von 62 auf 20 Prozent abgenommen hat. Demnach wird inzwischen nur mehr ein kleiner Teil der tatsächlichen Werte und der damit verbundenen Wertschaffung in Unternehmen durch das traditionelle Rechnungswesen abgebildet. Tatsächlich legt das traditionelle Rechnungswesen einen Erfolgsbegriff zugrunde, der im Kern den Zuwachs des physischen (Netto-)Vermögens misst. Immaterielle Vermögenswerte bleiben weitgehend unberücksichtigt und Investitionen in immaterielles Vermögen vermindern sogar als Kosten den Periodenerfolg nach der traditionellen Sichtweise (vgl. dazu im Einzelnen auch *Ruthner/Speckbacher,* 2002). Damit scheint offensichtlich, dass das traditionelle Rechnungswesen als Steuerungs- aber auch als Kommunikationsinstrument ausgedient hat.

2. Immaterielles Vermögen als Differenz zwischen Markt- und Buchwerten?

Interessanterweise wird als wesentlicher Indikator für die zunehmende Bedeutung immateriellen Vermögens und für den damit einhergehenden Bedeutungsverlust des traditionellen Rechnungswesens zumeist das in den vergangenen Jahrzehnten beobachtbare voranschreitende Auseinanderklaffen von Buchwerten und Marktwerten, die als „eigentlicher Gesamtwert“ von Unternehmen interpretiert werden, herangezogen. Betrachtet man jedoch die Entwicklung der Markt-Buchwert-Relation ausgewählter Unternehmen in den vergangenen Jahren, so zeigt sich ein differenzierteres Bild (vgl. *Abb. 1*). Insbesondere stellt sich die Frage, ob immaterielles Vermögen tatsächlich einfach als Differenz zwischen dem Marktwert und dem Buchwert von Unternehmen ermittelt werden kann. Eine derartige Ermittlung würde offensichtlich das Problem negativen immateriellen Vermögens aufwerfen.

Marktwert und Buchwert ausgewählter Unternehmen im Jahr 2000 und 2002

	2000			2002		
	Marktwert*)	Buchwert*)	M/B-Ratio	Marktwert*)	Buchwert*)	M/B-Ratio
ThyssenKrupp	8.489	9.196	0,92	5.973	9.149	0,65
DaimlerChrysler	45.821	38.000	1,21	38.199	39.421	0,97
Deutsche Telekom	97.250	59.441	1,64	41.028	66.019	0,62
Commerzbank	16.426	11.558	1,43	3.998	13.104	0,31
Allianz	98.472	31.302	3,15	24.328	49.013	0,50
Bayer	40.972	16.256	2,52	14.380	17.020	0,84

*) in Mio. EUR

Gibt es negatives immaterielles Vermögen?

Abb. 1: Marktwert/Buchwert-Ratio für ausgewählte Unternehmen im Jahr 2000 und 2002 (vgl. Pellens, 2002)

Zwar beinhaltet der Marktwert immaterielle Vermögenswerte, da der Marktwert einen zukunftsbezogenen Gesamtwert des Unternehmens abbildet, während der Buchwert dem Prinzip der vergangenheitsorientierten, summativen Einzelbewertung vorwiegend physischer Vermögensgegenstände folgt. Allerdings ist auch der Marktwert von Unternehmen vorsichtig zu beurteilen, da dieser auf zukünftigen Erwartungen der Anleger, gelegentlich auch auf deren spekulativen Überlegungen beruht. Gerade im Hinblick auf die langfristige und oft schwer durchschaubare sowie vermittelbare Wirkungsweise von Investitionen in immaterielles Vermögen ist nicht ohne weiteres davon auszugehen, dass das Anlegerpublikum stets über die notwendigen Informationen für eine realistische Bepreisung immateriellen Vermögens verfügt. Die scheinbar einfache Gleichung:

Marktwert = Buchwert des physischen Vermögens + Wert des immateriellen Vermögens

und die damit verbundene Negativabgrenzung immateriellen Vermögens als „derjenige Teil des Marktwertes, der nicht im Buchwert abgebildet ist“, ist höchst fragwürdig, da hier eine Wertadditivität suggeriert wird, die es nicht gibt. Ähnliches gilt für Ansätze, die immaterielles Vermögen als „unsichtbaren“ Bestandteil der Bilanz interpretieren (vgl. z.B. *Sveiby*, 1997, S. 11).

Physisches Vermögen kann für sich genommen von wesentlich geringerem Wert sein, als in der strategisch sinnvollen Kombination mit entsprechendem Humankapital, Markenpotential und Kundenbeziehungen. Andererseits existiert im Hinblick auf immaterielle Vermögensbestandteile sogar (abgesehen von Ausnahmen, wie Markenrechten) häufig überhaupt kein „stand alone value". Vielmehr ist von stark synergetischen Beziehungen zwischen immateriellem und physischem Vermögen auszugehen. Eine additive Separation in physisches und immaterielles Vermögen wird dem nicht gerecht.

Buchwerte von Unternehmen basieren im Kern auf der summativen Bewertung einzelner, hauptsächlich physischer Vermögensgegenstände zu historischen Kosten (Anschaffungs- bzw. Herstellungskosten). Sie orientieren sich also an historischen Kosten (Vergangenheitsorientierung), vernachlässigen Synergien aus der kombinierten Verwendung von Vermögensgegenständen (summative Einzelbewertung) und beschränken sich im Wesentlichen auf physische Vermögensgegenstände.

Marktwerte weichen schon deshalb von Buchwerten in der Regel ab, weil der Wert von Vermögensgegenständen höher aber auch geringer sein kann, als deren historische Kosten. Ein positiver/negativer Kapitalwert besagt ja gerade, dass der Zukunftswert eines Vermögensgegenstandes höher/geringer ist als dessen Anschaffungskosten (vgl. dazu auch *Speckbacher*, 2002). Die Kosten der Erstellung eines physischen Vermögensgegenstandes sind zumeist nur eine grobe Näherung für deren Wert, viel mehr gilt dies allerdings für immaterielle Werte. Die Kosten der Erstellung von Know-how, Reputation, Brand Capital und anderen immateriellen Vermögenswerten sind tatsächlich, wie *Kaplan* und *Norton* (2001, S. 89) es ausdrücken, „poor approximations" für deren Wert. Beziehungen zu Mitarbeitern oder Kunden können in einem zukunftsträchtigen Unternehmen mit einer erfolgreichen Strategie wesentliches immaterielles Vermögen darstellen. Sie können aber andererseits ebenso eingegangene Leistungsverpflichtungen des Unternehmens repräsentieren, denen keine entsprechende „Gegenleistung" mehr gegenübersteht und hierdurch wertmindernd wirken.

Ähnlich problematisch ist die Bewertung von Synergien. Der Wert mehrerer Vermögensgegenstände übersteigt im Falle einer sinnvollen kombinierten Verwendungsmöglichkeit die Summe der Einzelwerte. Fehlt jedoch eine Strategie zur sinnvollen kombinierten Verwendung von Vermögensgegenständen, so kann deren Wert leicht unter ihre Anschaffungs-/Herstellungskosten sinken.

Insofern kann es keineswegs verwundern, wenn der Gesamtwert von Unternehmen (auch von einem rationalen und gut informierten Anlegerpublikum) höher, gelegentlich aber auch geringer als deren Buchwert eingeschätzt wird.

3. Besonderheiten des externen Reportings über immaterielle Vermögenswerte

Auf das eigentliche Problem weisen die zu beobachtenden großen Schwankungen in den Markt-/Buchwert-Relationen hin: Neu auftretende Informationen können die Einschätzung des Wertes materiellen, insbesondere aber immateriellen Vermögens stark verändern. Während die Berichterstattung über physische Vermögensgegenstände allerdings quantitativ, systematisiert, standardisiert und relativ umfassend im Geschäftsbericht bzw. im Rahmen der Zwischenberichterstattung erfolgt, existiert eine Berichterstattung über immaterielles Vermögen allenfalls in Ansätzen. Wie unsere nachfolgend vorgestellte Erhebung belegt, werden Informationen hierüber in Geschäftsberichten – wenn überhaupt – zumeist unsystematisch, überwiegend qualitativ, einzelfallweise und unzusammenhängend nach außen kommuniziert. Die dadurch bestehende Ungewissheit führt tendenziell zu höheren Schwankungen der Bewertung durch das Anlegerpublikum und damit insbesondere zu höheren Kapitalkosten.

Besondere Schwierigkeiten für die Berichterstattung zum immateriellen Vermögen erwachsen, wie die obigen Überlegungen verdeutlicht haben, vor allem aus der Tatsache, dass eine isolierte Bewertung immaterieller Vermögenswerte auf der Basis historischer Kosten kaum Aussagegehalt besitzt. Immaterielles Vermögen besitzt vielmehr in aller Regel lediglich indirekten Wert. Mitarbeiterzufriedenheit, technologisches Know-how oder gute Beziehungen zu Zulieferern wirken sich nicht unmittelbar auf finanzielle Ergebnisse aus und ihr Wert im Unternehmen ist stark von Kontextfaktoren, wie der Unternehmensstrategie oder organisationalen Gegebenheiten, abhängig. Insofern kann immateriellen Vermögenswerten nur in Ausnahmefällen ein sinnvoller isolierter Wert zugewiesen werden, über den nach außen berichtet werden kann. Vielmehr ist es wichtig, zu kommunizieren, über welche Kombinationsprozesse immaterielle Werte im Rahmen einer geeigneten Strategie letztlich in finanzielle Ergebnisse transformiert werden sollen.

Im Folgenden werden, nach grundlegenden Ausführungen zum Begriff und zur Systematisierung immaterieller Vermögenswerte, Ergebnisse einer Erhebung zur Publikation von Informationen über immaterielle Vermögenswerte in Geschäftsberichten von Unternehmen des deutschsprachigen Raumes vorgestellt.

4. Merkmale immaterieller Vermögenswerte

Die grundsätzliche Gemeinsamkeit immaterieller Vermögenswerte mit physischem Vermögen besteht darin, dass beide Vermögensarten das Ergebnis vergangener Transaktionen bzw. Ereignisse sind und zukünftige Erfolgspotentiale für Unternehmen repräsentieren (vgl. *Canibano/García-Ayuso/Sánchez*, 2000, S. 104). Innerhalb der vorherrschenden Rechnungslegungssysteme (IAS, US-GAAP, HGB) werden immaterielle Vermögenswerte allerdings sehr restriktiv definiert. So definieren die International Accounting Standard (IAS) immaterielle Vermögenswerte als „Nonmonetary assets, without physical substance, held for use in production or supply of goods or services or for rental to others, or for administrative purposes, which are identifiable and are controlled by the enterprise as a result of past events, and from which future economic benefits are expected to flow." (*Epstein/Mirza*, 2001, S. 329).

Trotz der Uneinheitlichkeit der in der Fachliteratur vorgeschlagenen Definitionen für immaterielle Vermögenswerte gibt es einige wichtige Kriterien, die in fast allen gängigen Definitionen eine Rolle spielen (vgl. *Canibano/García-Ayuso/Sánchez*, 2000, S. 106): Immaterielle Vermögenswerte sind demnach:

- identifizierbare (und separierbare),
- nicht-monetäre bzw. nicht-physische
- Quellen zukünftiger Unternehmenserfolge,
- die entweder erworben oder intern zu ermittelbaren Kosten erstellt wurden,
- durch eine endliche Lebensdauer gekennzeichnet sind,

- einen eigenständigen Marktwert, unabhängig vom Unternehmen aufweisen und
- sich im Eigentum oder unter Kontrolle des Unternehmens, als Ergebnis vergangener Transaktionen oder Ereignisse befinden.

Die beiden letzten Punkte wirken dabei besonders restriktiv. Investitionen in immaterielle Erfolgsquellen zeichnen sich im Unterschied zu Investitionen in physische Vermögenswerte gerade typischerweise durch relativ unklare Verfügungsrechte bezüglich der dadurch entstandenen Potentiale aus, was letztlich insbesondere deren isolierte Handelbarkeit und damit deren isolierte Bewertbarkeit durch Marktpreise stark beeinträchtigt. Beispielsweise kann das Know-how in den Köpfen der Mitarbeiter nur bedingt als durch das Unternehmen kontrollierbar angesehen werden. Mitarbeiter mit spezifischen Kenntnissen und Fähigkeiten können das Unternehmen verlassen und diese Kenntnisse und Fähigkeiten beispielsweise in einem anderen Unternehmen gewinnbringend umsetzen. Ebenso sind Kundenbeziehungen oft an bestimmte Mitarbeiter gebunden und damit nicht uneingeschränkt in der Verfügungsmacht der Unternehmenseigentümer. Unklare bzw. uneinheitliche Verfügungsrechte behindern aber die isolierte Handelbarkeit auf Märkten und damit eine Bewertung zu Marktpreisen.

Im Folgenden verwenden wir eine in dieser oder ähnlicher Form in der Literatur häufig zu findende Einteilung immaterieller Vermögenswerte (vgl. z.B. *Sveiby*, 1997, S. 8. ff.). Diese beginnt auf der Ebene der Mitarbeiter des Unternehmens als wesentliche Quelle insbesondere immaterieller Vermögenswerte (Humankapital), erstreckt sich weiter auf die von den Mitarbeitern geschaffenen internen Strukturen (Strukturkapital) und schließlich auch auf externe Strukturen im Sinne der Beziehungen zu externen Stakeholdern, wie Kunden und Zulieferern (Beziehungskapital). Die nachfolgende *Abb. 2* kombiniert diese Einteilung mit einer auf der Ordinatenachse veranschaulichten Betrachtung des unterschiedlichen Grades der Absicherung von Verfügungsrechten seitens der Unternehmenseigentümer.

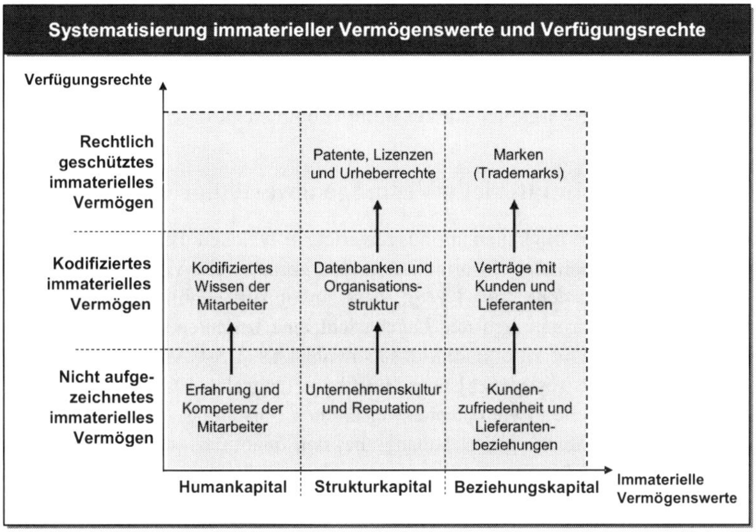

Abb. 2: Systematisierung immateriellen Vermögens und Verfügungsrechte

Hierbei wird deutlich erkennbar, dass lediglich Teile des immateriellen Vermögens in der Verfügungsmacht der Eigentümer stehen und eine Überführung in Richtung besser abgesicherter Verfügungsrechte nur sehr bedingt möglich ist. Gerade im Hinblick auf die Zwecke des externen Reportings ist der Grad der Verfügungsmacht seitens der Eigentümer von entscheidender Bedeutung für die Einschätzung der Wertrelevanz der entsprechenden Position.

5. Erhebung zum externen Reporting immaterieller Vermögenswerte

Im Zuge einer Studie am Institut für Unternehmensführung der Wirtschaftsuniversität Wien wurden die Geschäftsberichte der ATX 20 und DAX 30 Unternehmen des Geschäftsjahres 2001 auf deren Informationen zu immateriellen Vermögenswerten hin untersucht. Insbesondere wurde dabei systematisch erhoben, in welchem Umfang und in welcher Weise (qualitative/quantitativ, kontextbezogen/isoliert) entsprechende Informationen in welchen Teilen des Geschäftsberichtes kommuniziert werden. Im Folgenden werden die wichtigsten Ergebnisse der Studie zusammenfassend dargestellt, wodurch ein Überblick über den derzeitigen Stand des Reportings über immaterielle Vermögenswerte in den Geschäftsberichten der großen börsennotierten Unternehmen gegeben wird.

5.1 Qualitative vs. quantitative Angaben

Informationen über immaterielle Vermögenswerte werden in den Geschäftsberichten überwiegend qualitativ kommuniziert. In weniger als einem Fünftel der Fälle, in denen immaterielle Vermögenswerte in Geschäftsberichten erwähnt werden (insgesamt 744 Nennungen), sind die Informationen auch durch konkrete Kennzahlen quantifiziert. Hauptgrund hierfür ist sicherlich die Schwierigkeit, überhaupt geeignete Kennzahlen für die Erfassung immaterieller Vermögenswerte zu finden. Unternehmensspezifisch ermittelte und nicht standardisierte Kennzahlen bergen zudem die Gefahr, dass eine positive Entwicklung der spezifischen Auswahl der Kennzahl zugeschrieben wird und damit ohne große Wirkung bleibt, während eine negative Entwicklung wesentlich stärker wahrgenommen wird.

5.2 Wo wird über immaterielle Vermögenswerte berichtet?

Als nächstes wurde in der Untersuchung ausgewertet, in welchen Teilen der Geschäftsberichte quantitative bzw. qualitative Informationen über immaterielle Vermögenswerte hauptsächlich kommuniziert werden. *Abb. 3* zeigt, dass sowohl im Hinblick auf qualitative als auch auf quantitative Informationen der Lagebericht eine bedeutende Stellung einnimmt. Dies ist unter anderem damit zu erklären, dass sowohl IAS, US-GAAP, als auch die Vorschriften des Handelsrechts (österreichisches HGB bzw. deutsches HGB) klare Anforderungen an den Lagebericht stellen (wie beispielsweise einen Forschungs- und Entwicklungsbericht zu erstellen). Die Bedeutung des Anhangs bei den quantitativen Informationen resultiert vor allem daraus, dass Patente und Lizenzen in einem Anlagenspiegel erläutert werden.

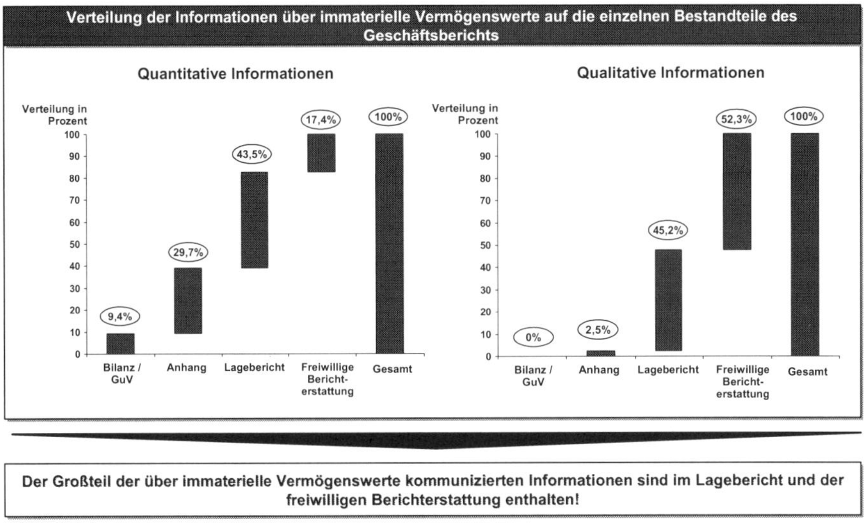

Abb. 3: Verteilung der Informationen über immaterielle Vermögenswerte auf die einzelnen Bestandteile des Geschäftsberichts

5.3 Kontextbezug der Informationen vs. isolierte Darstellung

Wie bereits ausgeführt, zeichnen sich immaterielle Vermögenswerte besonders durch deren Kontextabhängigkeit aus. Die Auswertung der Geschäftsberichte ergab, dass fast drei Viertel (72,4%) aller Informationen zu immateriellen Vermögenswerten isoliert genannt werden, d.h. nicht in einen Kontext eingebettet werden. Bei denjenigen Nennungen, bei denen eine Einbettung in einen Kontext erfolgte, wurde zwischen vier „Kontextkategorien“ unterschieden, wobei Kombinationen möglich waren (vgl. *Abb. 4*). Die Einteilung erfolgte danach, ob

- ein Zusammenhang mit konkreten Zielen hergestellt wurde („Um unsere Technologieführerschaft zu behaupten, investierten wir vermehrt in Grundlagenforschungsprojekte.“);
- die Bedeutung der Investition für den finanziellen Unternehmenserfolg kommuniziert wurde („Von den aufgeführten Maßnahmen zur Verbesserung der Kooperation mit unseren Zulieferern versprechen wir uns langfristig sinkende Bezugspreise, eine verbesserte Qualität und damit einen höheren Unternehmenserfolg.“);
- eine Verknüpfung mit konkreten Maßnahmen erfolgt („Mit unseren Ausbildungsaktivitäten eng verbunden ist ein erhöhter Qualifikations- und Verantwortungsgrad jedes Teammitgliedes. Hierarchische Zwischenebenen wurden abgebaut.“);
- ein Zusammenhang zur Unternehmensstrategie hergestellt wurde („Unser Erfolg als Dienstleister hängt davon ab, die besten Lösungen für unsere Kunden anbieten zu können. Eine hohe Kundenzufriedenheit ist für uns der zentrale Maßstab, dass wir diese Aufgabe gut erfüllen.“).

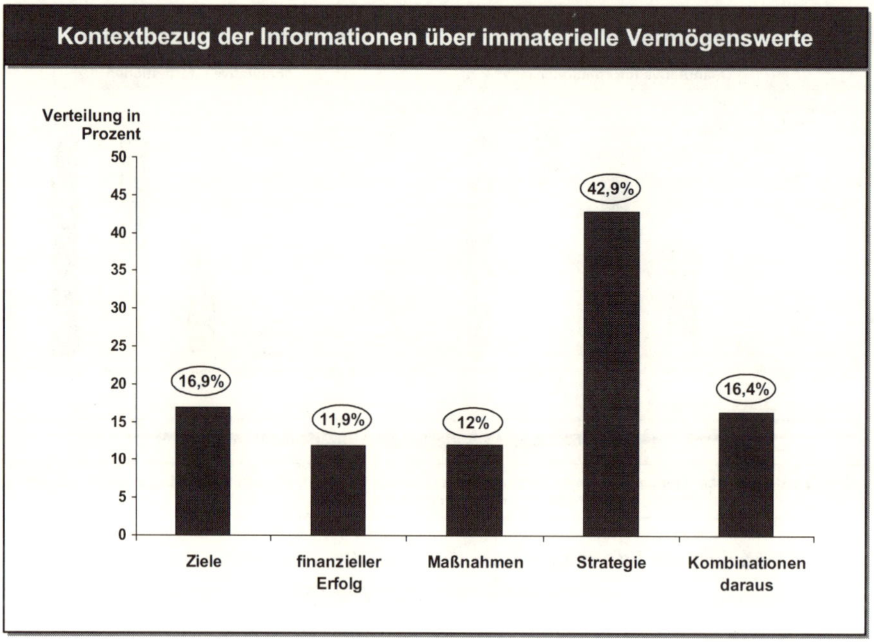

Abb. 4: Verteilung der deskriptiven Nennungen auf die einzelnen Kategorien

5.4 Welche Arten immaterieller Vermögenswerte werden überwiegend genannt?

Betrachtet man nun die einzelnen Kategorien, so ist festzustellen, dass von insgesamt 608 qualitativen Nennungen 238 Nennungen (39,2 %) in den Bereich Humankapital, 219 (36 %) in den Bereich des Strukturkapitals und 151 (24,8 %) in den des Beziehungskapitals fallen (vgl. *Abb. 5*).

Im Bereich Humankapital liegt der Schwerpunkt der kommunizierten Informationen bei der Ausbildung (11,5%), gefolgt von der Erfahrung (8,3%) und Mitarbeiterzufriedenheit (8%).

Im Bereich des Strukturkapitals dominieren Informationen zu „Network- und Informationssystemen“, gefolgt von Forschung und Entwicklung (13,7% bzw. 9,4%). An letzter Stelle, nach Management, Organisation und Unternehmenskultur, finden sich Patente und Lizenzen (2,6%).

Beim Beziehungskapital liegt der Schwerpunkt bei der Kundenzufriedenheit (10,6%). Nachgeordnet finden sich qualitative Informationen über die Marke (6,7 %), gefolgt von Reputation (4,2%) und Lieferantenbeziehungen (3,1%).

Abb. 6 zeigt die Rangfolge einzelner immaterieller Vermögenswerte hinsichtlich qualitativer Informationen bei den untersuchten Geschäftsberichten.

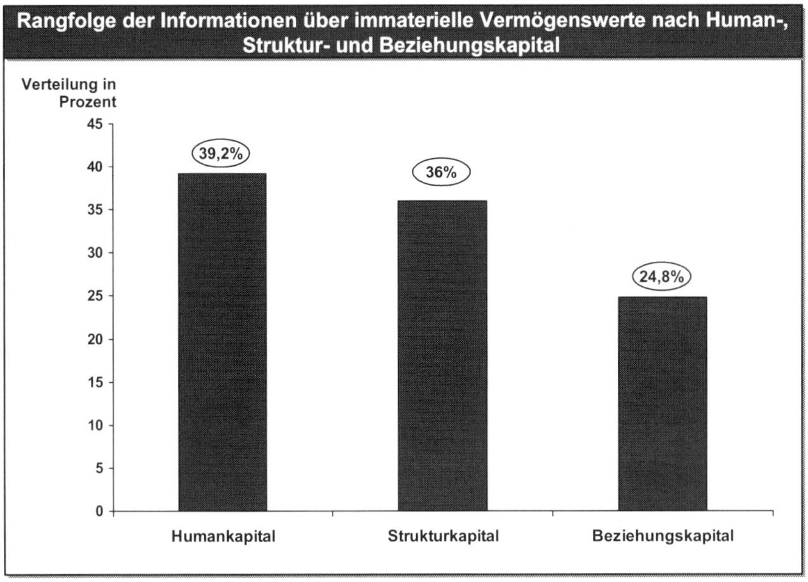

Abb. 5: Rangfolge qualitativer Informationen über immaterielle Vermögenswerte in den Kategorien Human-, Struktur- und Beziehungskapital

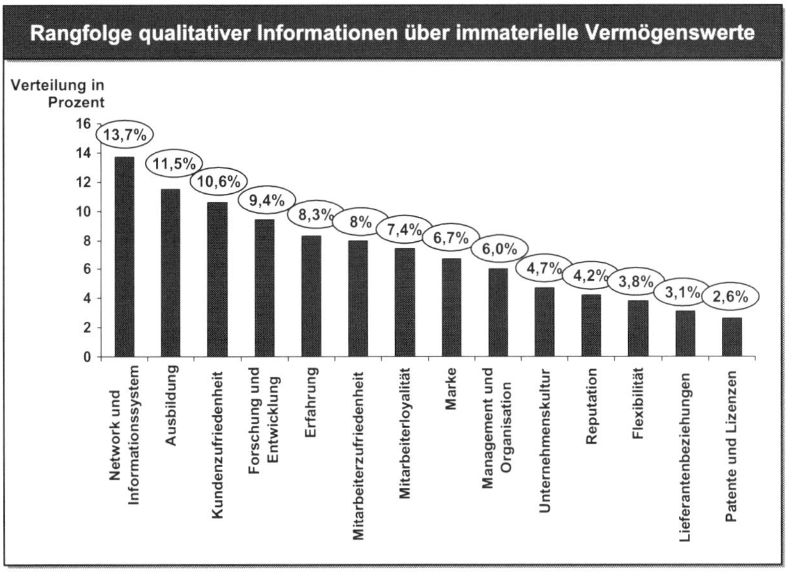

Abb. 6: Rangfolge qualitativer Informationen über immaterielle Vermögenswerte

6. Ausgewählte Beispiele zum Reporting immaterieller Vermögenswerte

Im Folgenden werden einige augenfällige Beispiele der Kommunikation immaterieller Vermögenswerte in Geschäftsberichten aufgeführt. Das dabei verwendete Analyseraster stellt auf die Systematisierung immaterieller Vermögenswerte in Humankapital, Strukturkapital und Beziehungskapital ab.

6.1 Humankapital

Die *Deutsche Bank* weist eine Vielzahl von Kennzahlen zum Bereich Humankapital in ihrem Geschäftsbericht aus. Einerseits werden im Bereich der Mitarbeiter prozentuell die Anzahl der Mitarbeiter mit Hochschulabschluss, Hochschulreife und sonstigen Schulabschlüssen angegeben, andererseits wird die Altersstruktur der Mitarbeiter aufgeschlüsselt sowie Leistungskennzahlen wie Mitarbeiter-Commitment-Index und Fehlzeitenquote kommuniziert. Besondere Beachtung wird laut Geschäftsbericht der Reputation des Unternehmens als attraktivem Arbeitgeber zugemessen. In diesem Zusammenhang wird über eine 2001 durchgeführte globale Studie über das Image der *Deutschen Bank* als Arbeitgeber mit 1200 Studenten an 50 angesehenen Universitäten berichtet. Ziel war hierbei nach Angaben des Unternehmens die Steigerung der Attraktivität als Arbeitgeber und gleichzeitig die Identifizierung und Etablierung von modernen Anreizsystemen in den Bereichen Work-Life Balance, Karrierechancen etc.

Bei der *OMV* wurden die bestehenden Personalsysteme 2001 konzernweit durch das Human Capital Management ergänzt. Als Performancemaßstab soll der Human Capital Index (HCI) die Unternehmenskultur messbar und somit auch steuerbar machen. Zu dessen Erhebung wurde eine ausführliche Befragung mittels Fragebogen zu den Themen Gesundheit/Arbeitsfreude, Kommunikation/Information und Mitarbeiter/Organisation in 12 Ländern und in 34 Standorten durchgeführt. Der Human Capital Index wird für 2001 mit 63 beziffert und durch eine Graphik anschaulich auf die oben genannten Teilgebiete aufgeschlüsselt (vgl. *Abb. 7*).

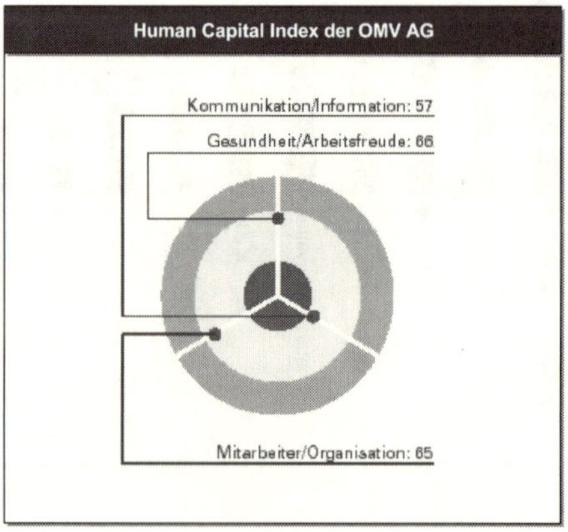

Abb. 7: Human Capital Index OMV (Quelle: Geschäftsbericht OMV, 2001)

Die *Deutsche Post* gibt an, besonders auf die Kreativität ihrer Mitarbeiter zu setzen. Für das Geschäftsjahr 2001 wurde der berechenbare Jahresnutzen aus der Umsetzung von Mitarbeiterideen im Bereich der Deutschen Post AG auf über 34 Mio. € beziffert (gegenüber 12,4 Mio. € im Vorjahr). Als weiterer Erfolgsfaktor wird die Beteiligung am Vorschlagswesen angesehen, welche 2001 laut Geschäftsbericht um 40% auf 85.000 Vorschläge stieg.

Im Lagebericht der *BBAG* wird nach Ländern aufgeschlüsselt darüber berichtet, wie viele Teilnehmer wie viele Seminare besucht haben. Daneben werden auch die Themen der Seminare erwähnt. Somit können genaue Informationen über das absolvierte Ausbildungsprogramm der Mitarbeiter aus dem Geschäftsbericht entnommen werden.

6.2 Strukturkapital

Altana, tätig in der Chemie- und Pharmabranche, kommuniziert in ihrem Geschäftsbericht, dass ihr Erfolg im Wesentlichen auf innovativen Produkten aus der eigenen Forschung und Entwicklung basiert. Als weitere entscheidende Einflussfaktoren wird die Kooperation mit anderen Biotechnologie-Unternehmen gesehen und es wird der Fokus auf die Kreativität der eigenen Mitarbeiter gelegt.

Die *Deutsche Bank* betont in ihrem Geschäftsbericht die Unternehmenskultur und deren Bedeutung für die erfolgreiche Eingliederung von akquirierten Gesellschaften. Nach Angaben des Unternehmens wurde in den letzten fünf Jahren eine eigene Identität geschaffen, in deren Werten sich alle Mitarbeiter wiederfinden. Diese Unternehmenskultur wird als Garant zur Erreichung der Unternehmensziele gesehen und als Basis, Mehrwert für Stakeholder und insbesondere Shareholder zu schaffen.

6.3 Beziehungskapital

Die *Lufthansa* unterstreicht in ihrem Geschäftsbericht den Zusammenhang zwischen der Kundenzufriedenheit einerseits mit neuer Technologie und kundenorientierten Maßnahmen auf der anderen Seite. Ziel ist es, den Kunden eine hohe Dienstleistungsqualität zu angemessenen Preisen zu bieten. Auf kundenorientierte Maßnahmen, wie Ausbau internetbasierter Vertriebskanäle, Kundenbindungsprogramme sowie auf kurz vor der Realisierung stehende Zusatzleistungen wie High Speed-Internet an Bord, wird im Geschäftsbericht ausführlich eingegangen. Laut der Angaben im Geschäftsbericht von *Lufthansa* wird die Kundenzufriedenheit regelmäßig durch ein Qualitätsmonitoring ermittelt, um einerseits die Akzeptanz der Produkte zu messen und anderseits die Möglichkeit zu nutzen, auf weitere Kundenwünsche eingehen zu können.

Die *Commerzbank* erwähnt in ihrem Geschäftsbericht, dass im Falle eines Dienstleisters, der Erfolg in erster Linie daran gemessen wird, wie gut Aufgaben erfüllt werden. Als Maßstab wird die Kundenzufriedenheit genannt, jedoch wird eine exakte Kennzahl zur Kundenzufriedenheit auf den weiteren Seiten des Geschäftsberichts nicht nach außen kommuniziert. Im Gegensatz dazu berichtet die *Deutsche Bank* einen operativen Kundenzufriedenheits- und Kundenbindungsindex im Bereich Personal Banking. Die *Deutsche Post* gibt Daten zur Kundenzufriedenheit bezogen auf Filialen (unterteilt nach subjektiver Zufriedenheit (Ipsos-Zufriedenheitsstudie) und objektiver Zufriedenheit (EMNID-Testkunden-Mess-System)) an.

Adidas-Salomon erstellt laut ihren Angaben im Geschäftsbericht Benchmarking-Vergleiche zwischen den einzelnen Geschäftsbereichen und anderen relevanten Unternehmen der Branche anhand einer Balanced Scorecard. Als Indikatoren werden wichtige finanzielle Kennzahlen sowie interne und externe nicht-finanzielle Indikatoren kommuniziert. Da *Adidas-Salomon* den Absatz seiner Produkte abhängig von den Präferenzen der Verbraucher und vom Markenimage sieht, werden laut Geschäftsbericht 2001 laufend Marktforschungen und Trendstudien in Bezug auf Design, Qualität, Image und Preisgestaltung durchgeführt.

7. Die Wissensbilanz als ein möglicher Lösungsansatz zur externen Kommunikation immaterieller Vermögenswerte

Die Wissensbilanz (Intellectual Capital Statement) kann als ein möglicher Ansatz gesehen werden, die immateriellen Vermögenswerte eines Unternehmens systematischer als bisher in der Praxis üblich abzubilden. So wie eine Bilanz dazu dient, vorwiegend materielle Werte einer Organisation darzustellen, verfolgt die Wissensbilanz das Ziel, das immaterielle Vermögen und Kapital eines Unternehmens in verschiedenen Kategorien zu dokumentieren. Dabei kann die Wissensbilanz sowohl als internes Steuerungsinstrument als auch als Ergänzung der externen Rechnungslegung verwendet werden.

Eine Möglichkeit, eine Wissensbilanz aufzubauen, beschreiben *Mouritsen* et al. Danach umfasst eine Wissensbilanz drei zentrale Bausteine (vgl. *Mouritsen* et al., 2002, S. 14ff.):

- Wissensbericht (Knowledge Narratives),
- Wissensziele (Management Challenges)
- Reporting.

Innerhalb des Wissensberichts wird beschrieben, wie die Organisation ihre Existenz in der Zukunft sicherstellt, ihre Produkte und Dienstleistungen den Anforderungen des Kunden gerecht werden. Er beinhaltet die Strategie des Unternehmens und deren Auswirkung auf die Stakeholder, den speziellen Nutzen, den der Kunde von dem Produkt oder der Dienstleistung des Unternehmens hat, und die Grundbedingungen der Produktion, im Speziellen welche Wissensressourcen zur Erstellung des Produkts oder der Dienstleistung benötigt werden.

Die Wissensziele beschreiben die Herausforderungen insbesondere an das Wissensmanagement, um den Anspruch an die Leistung des Unternehmens, die im Wissensbericht wiedergegeben wird, erfüllen zu können. In weiterer Folge werden diese Wissensziele auch in konkrete Aktionen übersetzt und durch Indikatoren gemessen.

Das Reporting wird unterteilt in die jährliche externe Wissensbilanz und in detailliertere interne Wissensbilanzen. Die externe Wissensbilanz ergibt sich durch die Verknüpfung von Wissensbericht und Wissenszielen, unter Verwendung eines geeigneten Layouts und aussagekräftiger Grafiken und Illustrationen. Sie soll den externen Stakeholdern durch ihren Aufbau und Stil die Identität der Organisation vermitteln. Die detaillierten internen Wissensbilanzen, die auch unterjährig erstellt werden können, dienen der internen Kommunikation und Steuerung.

Die *ARC Seibersdorf research GmbH* hat diesem Ansatz folgend im Jahre 1998 begonnen, eine eigene Wissensbilanz zu entwickeln, die das wertschaffende Zusammenwirken von Strukturkapital (Laboreinrichtungen), Beziehungskapital (Verbindungen zu anderen Forschungsinstitutionen in aller Welt) und Humankapital (hoch qualifizierte Mitarbeiter) nach innen und nach außen kommunizieren soll und seit dem Jahre 1999 regelmäßig jedes Jahr erstellt wird (vgl. *Abb. 8*). Die *ARC* stellt das größte Forschungszentrum Österreichs dar, das mit Universitäten, Fachhochschulen und anderen Forschungseinrichtungen in der ganzen Welt kooperiert. Sie wurde vom österreichischen Staat mit dem Ziel ins Leben gerufen, ein Forschungszentrum zu etablieren, das die Wettbewerbsfähigkeit Österreichs auf internationaler Ebene sichert. Die *ARC* ist zwischen der universitären Grundlagenforschung und der angewandten Forschung positioniert und setzt ihre Schwerpunkte in den Bereichen Gesundheit, Informationstechnologien, Intelligente Infrastrukturen, Weltraumanwendungen, Medizintechnik, Systemforschung Technik-Wirtschaft-Umwelt, Umwelt- und Lebenswissenschaften, Werkstoffe und Produktionstechnik. In diesen Bereichen werden sowohl Aufträge von der öffentlichen Hand als auch von der Privatwirtschaft vergeben. Dabei konnte für das Geschäftsjahr 2000 mit 504 Mitarbeitern ein Umsatz von 67,5 Mio. € erzielt werden.

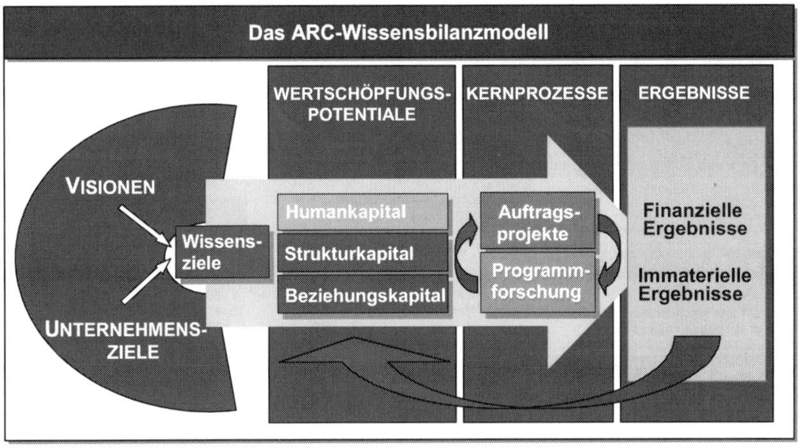

Abb. 8: Die Wissensbilanz der ARC Seibersdorf research GmbH (Quelle: Wissensbilanz ARC Seibersdorf, 2001, S. 18)

Das System der Indikatoren baut auf drei Basiselementen auf: Wertschöpfungspotentiale, Kernprozesse und Ergebnisse. Diesen Basiselementen werden die langfristigen Unternehmensziele und Visionen vorgelagert, welche sich in den so genannten Wissenszielen manifestieren. Die Festlegung der Indikatoren erfolgt nach diesen Wissenszielen, die sowohl als Aspekte der langfristigen Unternehmensstrategie, als auch als operative Ziele gesehen werden können. Die Gliederung der Indikatoren erfolgt nach den Bereichen Human-, Struktur-, Beziehungskapital (als das Wertschöpfungspotential), Programmforschung und Auftragsprojekte (als die Kernprozesse) und den erzielten Ergebnissen. Es gilt, zu beachten, dass nicht jeder Teilaspekt in jedem Wissensziel auffindbar ist. Den Abschluss der Darstellung bildet ein Gesamtüberblick über die wichtigsten Kennzahlen und Indikatoren für das gesamte Unternehmen.

Abschließend kann festgehalten werden, dass die Wissensbilanz versucht, die für die *ARC* besonders wichtige Ressource Wissen in Form von verschiedenen Kategorien und Kennzahlen darzustellen. Sie dient dabei der *ARC* sowohl als internes Steuerungsinstrument als auch als externes Kommunikationsmittel, um die Leistungen für die externen Anspruchsgruppen, wie Kunden und Gesellschaft transparent zu machen, die in einer traditionellen Bilanz nicht berücksichtigt werden können. Sie hilft damit, die bereits in der Einleitung dieses Beitrages beschriebenen Defizite des traditionellen Rechnungswesens und die damit verbunden Informationslücken zu reduzieren.

8. Konkrete Gestaltungsempfehlungen für eine Wissensbilanz

Eine Wissensbilanz verfolgt, wie bereits weiter oben angesprochen, das Ziel, internen und externen Adressaten des Reportings eine komprimierte Darstellung über den Bestand an immateriellen Vermögenswerten zu liefern. Dabei werden anhand von nicht-finanziellen Indikatoren (Messgrößen) immaterielle Vermögenswerte quantifiziert und für das Management sowie Investoren greifbar gemacht. Die sinnvolle Darstellung des Bestands an immateriellen Vermögenswerten erfordert die Erfassung von Veränderungen im Zeitablauf (siehe auch *Abb. 9*).

Indikatoren-Übersicht in einem Intellectual Capital Statement

	2001	Veränd. Gg. VJ.	2002	Veränd. Gg. VJ.
Kundenbeziehungen				
Anzahl Kunden	1.000	2,0	1.000	2,0
Umsatz der 10 größten Kunden	15.000	15,4	15.000	15,4
Auftragsbestand in Monaten	1,5	25,0	1,5	25,0
Humankapital				
Zahl der Mitarbeiter in F&E	5	25,0	4	0,0
Ausbildungsausgaben	10	11,1	9	12,5
Fluktuationsrate	3	-25,0	4	33,3
Führungsstruktur 1. und 2. Ebene				
30 bis 55 Jahre	1	-50,0	2	0,0
über 55 Jahre	6	20,0	5	66,7
Innovationskapital				
Wert der 5 wichtigsten Marken	1.000	233,3	300	-25,0
Anzahl Lizenzverträge	5	66,7	3	-25,0
Prozessstruktur				
Wertschöpfung je Mitarbeiter (Tsd. EUR)	5	25,0	4	-33,3
Investitionen in Informationstechnologie	15	15,4	13	0,0

(Quelle: Maul/Menninger: Das „Intellectual Property Statement" – eine notwendige Ergänzung des Jahresabschlusses. In: Der Betrieb, Heft 11 vom 17. März 2000, 53. Jahrgang, S. 529 – 533)

Abb. 9: Indikatoren-Übersicht einer Wissensbilanz (in Anlehnung an Maul/Menninger, 2000, S. 532)

Um eine Wissensbilanz zu einem wirkungsvollen Instrument der Unternehmenssteuerung und Kapitalmarktkommunikation zu machen, sind eine Reihe weiterer Gestaltungsempfehlungen zu beachten:

- Darstellung von Synergien im Kontext der Unternehmensstrategie: Im Wertschaffungsprozess ist vor allem die enge Verschränkung zwischen den einzelnen Indikatoren und der Unternehmensstrategie von Interesse. Immaterielle Vermögenswerte sind für sich isoliert betrachtet in vielen Fällen wertlos. Erst im Verbund ergibt sich eine einmalige Kombination von materiellen und immateriellen Ressourcen, die nicht länger durch den

Markt replizierbar ist, und damit dem Unternehmen Wettbewerbsvorteile erschließt. Ressourcenkombinationen sind Ausdruck der Strategie eines Unternehmens. Eine eingehende Darstellung dieser Wechselwirkungen erfolgt auch im Zuge der Erstellung von Ursache-Wirkungsbeziehungen im Rahmen der Balanced Scorecard (vgl. *Kaplan/Norton*, 1996) und erscheint im Falle von Wissensbilanzen als unverzichtbarer Bestandteil. Informationen über einzelne immaterielle Vermögenswerte sind daher in einen eindeutigen Zusammenhang zur Vision, Strategie und daraus abgeleiteten Teilzielen zu bringen, wodurch insbesondere für Investoren ein „Business-Modell" erkennbar werden soll. In der Praxis erfolgt dies häufig durch verbale Erläuterungen, aussagekräftige Grafiken und Illustrationen (vgl. *Mouritsen* et. al., 2002).

- „Strukturierte Individualität" in der Darstellung: Strategische Unternehmensführung verfolgt das Ziel, einzigartige Strategien zu formulieren und damit Wettbewerbsvorteile gegenüber Konkurrenten aufzubauen. Um diesem Umstand Rechnung zu tragen, müssen Wissensbilanzen flexibel gestaltet werden, um auch die individuelle Abbildung des Wertschaffungsprozesses zu ermöglichen. Starre Schemata in Form standardisierter Aufstellungen von Indikatoren erweisen sich dabei in vielen Fällen als nicht ausreichend. Aus dieser Anforderung heraus ergibt sich jedoch ein Spannungsverhältnis mit der häufig geäußerten Forderung, möglichst vergleichbare Darstellungsformen zu liefern, die dem Anlegerpublikum eine systematisierte Gegenüberstellung verschiedener Anlagealternativen am Kapitalmarkt ermöglichen. Hier ist in der praktischen Ausgestaltung der Mittelweg einer „strukturierten Individualität" zu finden. Weder ein unstrukturiertes Story-Telling allein, noch die ausschließliche Darstellung in Form einer standardisierten Sammlung von Indikatoren nach dem Vorbild der traditionellen Bilanz sind hierbei im Sinne einer „Extremlösung" sinnvoll.
- Die Notwendigkeit einer „Gebrauchsanweisung": Anders als im traditionellen Rechnungswesen existiert (noch) kein rechtlich vorgeschriebenes und damit institutionalisiertes Regelwerk zur Aufstellung und auch Verwendung von Wissensbilanzen. Unternehmen befinden sich hierbei (noch) in einem relativ konventionsfreien Raum, der jedoch auch Zurückhaltung seitens des Kapitalmarktes hinsichtlich der Akzeptanz mit sich bringt. Um sicherzustellen, dass Investoren eine Wissensbilanz in der intendiert richtigen Weise lesen und verstehen, muss deren interne Logik dem Adressaten entsprechend erklärt werden (vgl. *Mouritsen* et al., 2001). Hinzu kommt die Notwendigkeit, Erläuterungen zur Berechnung der einzelnen Indikatoren vorzunehmen, um Fehlinterpretationen und damit Fehleinschätzungen seitens der Investoren auszuschließen.

9. Fazit

Der Beitrag verdeutlicht, warum das traditionelle Rechnungswesen nicht in der Lage ist, den Wertschöpfungsprozess in modernen, zunehmend wissensbasierten Unternehmen adäquat abzubilden. Eine bloße Erweiterung der traditionellen Accounting-Modelle um immaterielle Vermögenswerte greift sowohl im Hinblick auf die interne Steuerung als auch auf das externe Reporting zu kurz. Der empirische Teil des vorliegenden Beitrages hat sich dabei auf die externe Dimension konzentriert.

Die Auswertung und Analyse der Geschäftsberichte der ATX 20 und DAX 30 Unternehmen des Geschäftsjahres 2001 hat folgende Schwachstellen der externen Reportingpraxis aufgezeigt:

- Die meisten Unternehmen stellen in ihrem Geschäftsbericht bereits Informationen über immaterielle Vermögenswerte zur Verfügung, es überwiegen dabei jedoch qualitative Beschreibungen. Das lässt die Schlussfolgerung zu, dass geeignete Kennzahlen zur Abbildung immaterieller Vermögenswerte noch in vielen Fällen fehlen.
- Nur relativ selten werden Berichte über immaterielle Vermögenswerte in einen Kontext zur Strategie des Unternehmens gestellt bzw. mit konkreten verbundenen Zielen und Maßnahmen kommuniziert. Dies reduziert den Aussagegehalt für die Empfänger entscheidend.
- Der Großteil der Unternehmen konzentriert sich nur auf die Darstellung einiger, ausgewählter Informationen über immaterielle Vermögenswerte, die als besonders wichtig erachtet werden. Die Bewertung und Vergleichbarkeit wird hierdurch wesentlich erschwert.

Einen ersten viel versprechenden Ansatz zur Reduktion dieser Schwachstellen stellt das Instrument der Wissensbilanz dar, wie es am Beispiel der *ARC Seibersdorf research GmbH* in diesem Beitrag erläutert wurde. Die Wissensbilanz ermöglicht eine systematische und vergleichbare Aufbereitung immaterieller Vermögenswerte und stellt diese gleichzeitig in einen strategischen Gesamtkontext. Bei geeigneter Implementierung erhöht sie damit wesentlich den Informationswert für externe Empfänger.

10. Literatur

Canibano, L., *García-Ayuso, M.* and *Sánchez, P.*, Accounting for Intangibles: A Literature Review, in: Journal of Accounting Literature, Vol. 19 (2000), No. 2, S. 102-130.

Chandler, A. D., Scale and Scope: The Dynamics of Industrial Capitalism, Cambridge 1990.

Epstein, B. J. and *Mirza, A. A.*, IAS 2001 - Interpretation and Application, New York 2001.

Kaplan, R. S. and *Norton, D. P.*, The Balanced Scorecard: Translating Strategy into Action, Boston 1996.

Kaplan, R. S. and *Norton, D. P.*, Transforming the Balanced Scorecard from Performance Measurement to Strategic Management: Part I, in: Acconting Horizons, Vol. 15 (2001), No. 1, S. 87-104.

Maul, K.-H. und *Menninger, J.*, Das "Intellectual property Statement" – eine notwendige Ergänzung des Jahresabschlusses?, in: Der Betrieb, 53. Jg. (2000), H. 11, S. 529-533.

Mouritsen, J.,Larsen, H. T.,Bukh, P. H. and *Johansen, M. R.*, Reading an Intellectual Capital Statement – Describing and Prescribing Knowledge Management Strategies, in: Journal of Intellectual Capital, Vol. 2 (2001), No. 4, 359-383.

Mouritsen, J., Bukh, P. N., Larsen, H. T. and *Johansen, M. R.*, Developing and managing knowledge through intellectual capital statements, in: Journal of Intellectual Capital, Vol. 3 (2002), No. 1, S. 10-29.

Pellens, B., Immaterielle Werte in der kapitalmarktorientierten Rechnungslegung, http://www.urwip.uni-linz.ac.at/veranstaltungen/ws02/vortrag_pellens.pdf, 24.10.2002.

Ruthner, R. und *Speckbacher, G.*, Wertorientierte Performancemaße und Anreizsysteme – Konzeptionelle Grundlagen und Einsatz in börsennotierten österreichischen Unternehmen, Wien 2002.

Speckbacher, G., Investitionssteuerung und -kontrolle, in: *Küpper, H.-U.* und *Wagenhofer, A.* (Hrsg.): Handwörterbuch Unternehmensrechnung und Controlling, 4. Aufl., Stuttgart 2002, Sp. 867-876.

Sveiby, K. E., The New Organizational Wealth, San Francisco 1997.

Implementing Strategic Planning with the *Skandia* Navigator

Marek Rydén [1] *and Ann-Charlotte Bredahl* [2]

---

[1] *Marek Rydén*, Skandia Insurance Ltd., Stockholm, Schweden

[2] *Ann-Charlotte Bredahl*, Skandia Insurance Ltd., Stockholm, Schweden

The gaps between a company's market value and the accounting reality has grown over the last decades showing a fundamental discrepancy between the story told on corporate balance sheets and the real one played out daily by organizations themselves. Skandia has tried to capture and understand this difference by developing the Skandia Navigator. The Skandia Navigator is a tool for implementing, follow up and communicating the strategy of the company. The article gives a background to how the Skandia Navigator works and a more in-depth part where requisites for a successful implementation are discussed and experiences from actual implementations are covered. This in-depth discussion is generic and does not only apply to Skandia even if the discussion primarily is based on internal experience. Skandia Insurance Ltd was founded in 1855 and was listed at the Stockholm stock exchange in 1863. The company has transformed itself from a traditional insurance company to one of the world's leading companies in financial services. Skandia is active in 25 countries around the globe and had annual sales of 138 billion SEK (15.3 billion Euro) in 2001.

1. Introduction

"To cope with the new business realities, we need to improve our navigational instruments. Those we have are too primitive and constrained by backward-looking notions. They deal with forms and frames of reference, but tell us too little about relations and futurizing dynamic factors. One of the weak links is the narrow time aspect. Conventional accounting can be downright misleading. Like an old magnetic compass on board a modern supertanker. Since the hull is made of iron, there's no way of knowing if the needle is pointing at the true directions. It may just point in the direction the ship is travelling. With such an instrument, it would be impossible to establish one's position, let alone set a course for a destination" [Accounting for Minds – An Inspirational Guide to Intellectual Capital by *Gottfried Grafström* and *Leif Edvinsson*, Larsherberts Offset (*Skandia* material)].

These words describe the reason why *Skandia* started with the concept of the *Skandia* Navigator. The management of *Skandia* found it necessary to find a way to maneuver the company in the transition from being a traditional old-fashioned insurance company to an innovative world-leader of modern financial services. The *Skandia* Navigator was the result of this endeavor.

The development of the *Skandia* Navigator coincided with the development of the generic balanced scorecard concept. However, the *Skandia* Navigator is in many ways similar or identical to a balanced scorecard and therefore this article is valid for all balanced scorecard practitioners.

The article will start by giving short background to the *Skandia* Navigator and how it is supposed to be implemented. This will be followed by a more in-depth part where requisites for a successful implementation are discussed and experiences from actual implementations are covered. This in-depth discussion is generic and does not only apply to *Skandia* even if the discussion primarily is based on internal experience.

There have always been occasional and temporary gaps between market perception and accounting reality, but somehow these gaps seem to have increased during the past decade. Nowadays there is a fundamental discrepancy between the story told on corporate balance sheets and the real one played out daily by organizations themselves. This fact does not necessary tell us that we should get rid of balance sheets and income statements, but rather that we need to widen our focus in order to capture a true and fair view of the company and to be prepared to meet the future. The question is: How do we do that? In the beginning of the 1990's, *Skandia* developed a concept to formulate, follow-up and communicate *Skandia*'s long-term strategy in a balanced way – it was called the *Skandia* Navigator.

2. The *Skandia* Navigator

The *Skandia* Navigator is *Skandia*'s tool for implementing its vision and strategy all the way through the organization. It is a management tool, a follow-up tool and a communication tool. A management tool in order to manage the business, a follow-up tool in order to follow-up on the strategy and see if we are working accordingly and a communication tool in that sense that it creates a common language for the whole of *Skandia* that helps us communicate where we want to be in the future and how we are going to get there.

The *Skandia* Navigator is a collection of critical measurements that all comprise a holistic view of performance and goal achievement. Divided into five focus areas or perspectives it captures different areas of interest. Each area visualizes the value creation process.

The *Skandia* Navigator facilitates a holistic understanding of the organization and its value creation. We need a structure to help us learn from each other and make sure that our daily efforts do contribute to *Skandia*'s overall strategic objectives. It is possible to know which indicators are truly strategic and thus make a difference on the bottom line. The targets defined for each indicator make it possible to apply an operational approach to the overall goals of the organization.

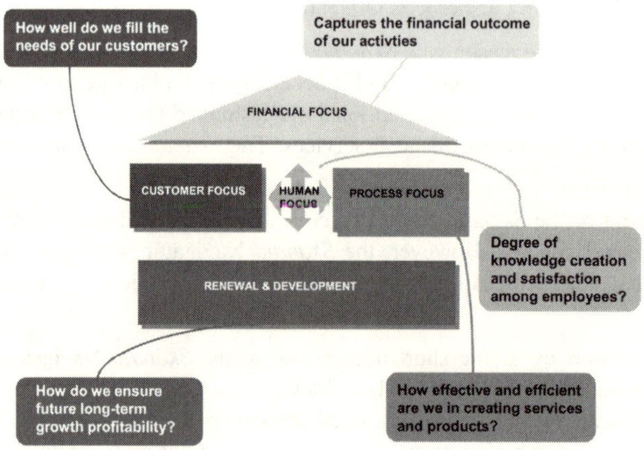

Fig. 1: Structure of the Skandia-Navigator

Customer focus

Customer focus gives an indication on how well the organization fills the needs of its customers via services and products. For example, how much of our sales derive from new customers when compared to old ones or how loyal are our customers? It represents a view that goes from the outside looking in. You can also see it from another perspective; how pleased are we with our customers? This can be relevant when it comes to other customers than end customers, for example distributors.

Human focus

Human focus is the heart of the organization and is essential in a value creating organization. The process of knowledge creation is visualized in this focus area. For example, do we have the right competence to fulfill our tasks and develop the company? It is also essential that the employees are pleased with their work situation; pleased employees lead to pleased customers, which creates a stable foundation for better results. Acting on results from employee satisfaction surveys is essential for a healthy organization.

Process focus

Process focus captures the actual process of creating the services and products the customers' desire. It covers questions like how do we handle our customer support, how efficient are our sales and marketing departments? This focus area is also connected to the internal processes. Are we working efficiently? It is important to emphasize the importance of transferring the employee's knowledge to structural capital [Structural capital: Customer capital and organizational capital, i.e. everything apart from the employees themselves, for example the brand, customer databases & computers etc. (Words of Value by *Skandia* and *Ericsson*)].

Renewal and development

Renewal and development aims at reassuring the organizations long-term renewal and in part its sustainability. What steps and actions are we taking now to ensure long-term growth profitability? What is required to attain and develop the knowledge needed to perceive and satisfy our customer's needs? How do we maintain efficiency and productivity needed in our processes that create actual value for our customers? Which core competencies do we need to develop in order to follow our strategy and fulfil our vision?

Financial focus

Financial focus captures the financial outcome of our company's activities in the other focus areas. Some like to see it as a receipt. This focus area would typically contain information that reflects the demands of profitability and growth from *Skandia*'s shareholders.

Environment

The environment where we operate (both business and social) affects *Skandia*'s total strategy and therefore all focus areas in the *Skandia* Navigator. This makes it essential to analyze and understand what effects this interaction will have for us. How *Skandia* is managed might also affect the environment, which makes it important for *Skandia* how we act.

2.1 The Process Model

The implementation starts with the process model. A prerequisite to the implementation is an already defined company vision. The aim of the model is to translate the vision and strategy into measurable indicators and activities.

The model is divided into several steps to ensure and simplify the translation process (see *fig. 2*).

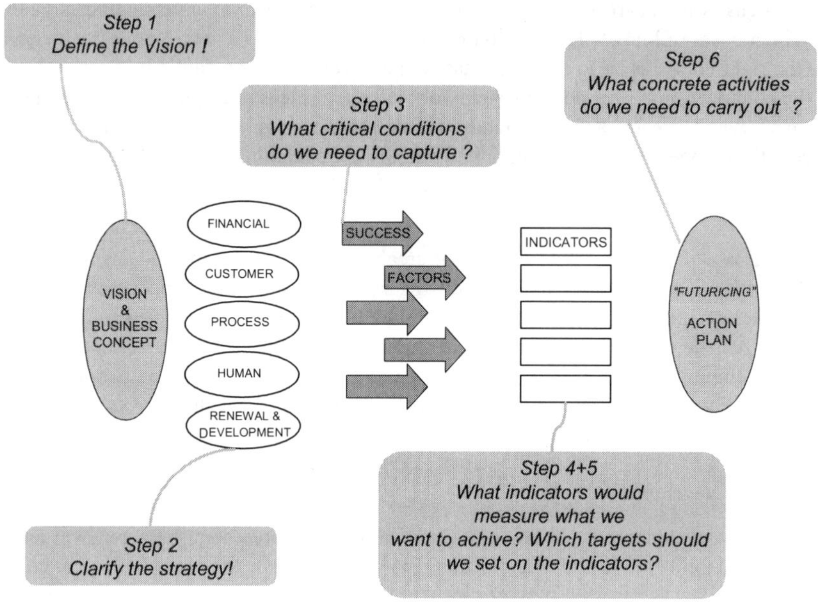

Fig. 2: The process.model

Before starting the strategic work with the process model it is important to define the company culture so it may support and enable the strategic work. Our values should permeate everything we do, from strategic work to our daily work tasks. For example, *Skandia* has worked a lot with its values, which has resulted in five common values for the whole company. The *Skandia* values are passion, commitment, contribution, courage and creativity.

Step 1 - Vision

In a future situation when the organization has reached its vision - what will it be like? That statement is captured in the vision. The vision is a picture of a desired future state of the company that does not need to be concrete.

Step 2 – Strategic objectives

Strategic objectives describe how the vision will be achieved; what strategy will the organization adopt in order to support vision fulfillment? These statements can be short and "catchy" or richer descriptions of the future state, and it is positive if they are not easily achievable goals (stretch goals).

Step 3 - Success factors

A success factor captures critical conditions that are crucial to the organization's ability to reach the strategic objectives in order to reach the vision. There are probably a large amount of factors that would qualify as success factors, but finding the really crucial ones and keeping the number to a minimum is beneficial. There are success factors in every focus area. The creation and documentation of the success factors is an important step in the process, therefore *Skandia* try to make them concrete and pragmatic. This analysis should identify both internal and external risks and opportunities and might include market analysis, competitor analysis & SWOT-analysis (Strength/Weakness/Opportunities/Threats).

Step 4 - Indicators

The fourth step is to create indicators that indicate if the organization reaches or works according to the defined success factors. An indicator is a measurement. Each success factor should deliver one or several indicators, which also secures that we have indicators in all focus areas. There is a risk of becoming too ambitious with measurements that the organization cannot collect, which results in a loss of momentum in the implementation of the *Skandia* Navigator. Indicators have to be both measurable and comprehensive. The maximum number of indicators is recommended not to exceed four indicators per focus area in a company Navigator.

Indicators should be divided into leading and lagging indicators. Leading indicators are indicators that affect the outcome of the lagging indicators such as Call Center Competence (leading) and Customer Satisfaction Index (lagging).

Skandia has a few mandatory indicators that we follow-up on corporate level. Aside from this each subsidiary/ business unit define their own indicators connected to their local business. The reason is that the subsidiaries, business units and the group company operate in such different business environments or conditions that it is more useful to create indicators that fit their own specific situation.

Step 5 - Visualizing the indicators

When you have created an indicator it is important that you set a target for it as well as a target date. This is crucial for the follow up work to create sustainability. In order for indicators to be useful they must be acted upon and used as instruments to follow performance. The act of measuring will make it more understandable to the entire organization how the long-term goals are to be fulfilled. It will be easier to act on the right signals.

Step 6 - Action plans

Indicators are used as diagnostic instruments, visualizing a situation at a given point in time. This information may very well be valuable in itself but the real leverage comes when acting upon the indicators, using action plans. They consist of documented activities that act upon stimulating indicators in a desired direction. If we secure that we will reach the set targets for the indicators we also secure that our organization is directed towards the success factors and thus the strategic objectives and the vision.

Step 7 – Cause and effect

An important step in the creation or the revision of a process model is to identify existing relations that can help us understand cause and effect relationships. Such a relation is the relation between the leading and lagging indicators. Other relations that can be identified in the process model are the relations in the different focus areas from strategic objective to the indicators. These relations indicate if we follow our strategy or not. Another important relation is the one between the financial focus and the other focus areas since we need to see bottom line effects.

2.2 Communicating the Strategy

To make the *Skandia* Navigator concept work in full effect it is crucial to implement the concept through the whole organization.

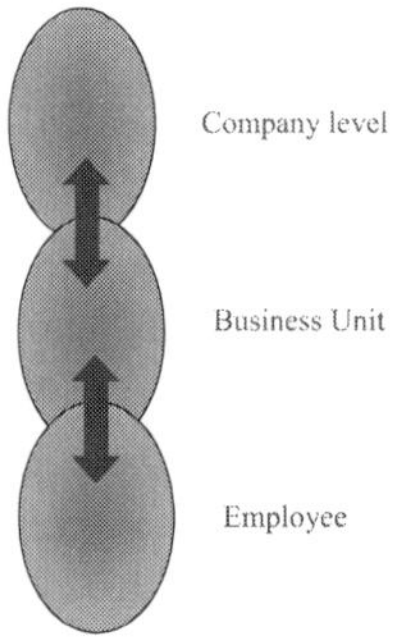

Fig. 3: Levels of Implementation

Skandia starts with the overall vision and strategy in order to create a common platform. In the process it is important to link the vision and strategy at all levels to each other, so that all "sub navigators" will be aligned with the overall company vision and strategy.

Using different levels will involve a greater number of employees in the strategic planning and help visualise for each employee his or her contribution to the company strategy. Navigators on an individual level should also be a basis for appraisals and linked to bonus plans.

Communicating a strategy through developing Navigators at all levels within the organization can be truly powerful but it is important not to go to fast doing this. If the organization is not mature enough it can create badwill among the employees instead of being a powerful tool. In some organizations this will not work at all and then one should not implement it further down; company level will be enough if the strategy is communicated in other ways.

2.3 Planning Process

In order to manage and communicate a company strategy it is necessary to have a planning process in place that supports knowledge sharing. Traditionally the planning process is equivalent with the budget process. In its worst apparition this basically means that the department heads get to ask for money based on how much they spent last year. No connection is made between planning and strategy and thus a very inefficient organization is created that does not support any kind of strategic work.

These organizations get more and more rare, primarily because these companies cannot survive in a modern competitive business environment, but there are still examples or variations of this to be found.

A good planning process must involve a link between the strategic work – the company's vision – and the actual numbers and actions planned for next year. If the process is structured well enough it will be possible to communicate the strategy to all departments at their internal planning, thus ensuring that the bottom-up plans support the top-down plans. In most *Skandia* companies local controlling functions work closely with the HR department in both creating and implementing local strategy, which has shown to be very successful since both competence development and the foundation for recruiting is closely linked to budget and set strategy.

At *Skandia* the implementation work is performed locally giving the local CEO plenty room for manoeuvre, which is important. Nevertheless the local companies need to keep their planning within the corporate course of action. The *Skandia* corporate planning process starts in September each year and is seen as the foundation of the *Skandia* business since it creates regularity in strategy work as well as focus on long-term planning. The basis is the naturally the *Skandia* Navigator concept and the Process Model, enabling a balanced view of our business.

The on-going process starts when the corporate guidelines are distributed from *Skandia* Group management to the CEOs and CFOs in the subsidiaries. These guidelines comprise *Skandia* Group strategy going forward (long-term), as well as focus for the year to come (short-term). During October and November corporate controlling department has meetings at the local offices, discussing their Navigator Process Model and related issues. The intention is to have a working level meeting at the local office, lasting one or two days, with representative's corporate controlling department. During December, each subsidiary will have to present its future strategy to *Skandia* Group Management. Once the plan is confirmed, it is the task of each subsidiary to implement the strategy organization-wide as mentioned above.

In order to make sure that each subsidiary implements the outlined strategy, the corporate controlling department performs continuous Navigator follow-ups throughout the year. Both through book closing meetings each quarter and through business review meetings once or twice a year – depending on business maturity. Following all meetings, reports are written and sent to *Skandia* management as a source for decision-making.

2.4 Staff Involvement

There is a need to involve people outside of the boardroom in the strategic planning. There are enormous and obvious advantages to the company if it manages to involve each employee in the strategic work. If all employees understand their role in the company and how it is aligned with the company's objectives there will be:

- Less mistakes and errors will be done because of misunderstandings,

- Decisions can be made faster more accurate on a lower level in the organization thus saving time and money,
- If everybody understands and is involved in the strategy the amount and quality input for improvements will increase,
- If more employees "feel" for the company, productivity will improve.

2.5 Bonus Schemes

A further way to strengthen the understanding of the company strategy is to align personal bonus schemes with the strategic planning. If the individual bonus is dependent on goals that are the result of the strategic planning there will be an automatic link from the overall company vision to the tasks of an individual employee thus securing that the strategy is communicated through the company to all levels.

At one *Skandia* company the individual bonus is dependent both on the financial result of the company and on a number of personal goals that are set to correspond to the strategic objectives. The financial result determines the size of the potential bonus and the personal goals determine the degree of fulfillment.

The personal goals are set through an assessment of the employee where weight is given to personal strengths and weaknesses, the function, development possibilities and to the team. The goals should be set to strive for personal development aligned to the company strategy. Team goals are encouraged since these lead a department towards the same goal.

The *Skandia* Navigator supports personal bonus schemes. Ideally the bonus for one employee can be set as a combination of key ratios or activities from all focus areas of the Navigator and as a combination of personal goals and department goals. However, through experience we have realized that it might be difficult to create an individual bonus plan built on key ratios or activities from all focus areas in the individual Navigator. Consequently we recommend splitting the bonus plan into three parts; company level, departmental level and individual level. The company level needs to be balanced building on the different focus areas while the departmental and individual level is naturally connected to the individual employee's area of working.

Many organizations still pay compensation (salary) without any variable part. This is a rigid way of looking at compensation and does not utilize the compensation as a way to motivate the staff and to communicate the strategy. At the same time many companies pay bonuses that have no connection to personal achievements. This is thought to motivate the staff but in reality it is difficult for an employee to see how he can earn his bonus when there is no direct connection to his tasks.

The risk when setting up personal goals as a part of the bonus is that these are not aligned with the company and that they are not followed up upon. It is very important that personal goals are discussed over the year. A result of setting goals with no follow up can be that an employee prioritizes an activity that will give him a bonus over an urgent activity that has emerged after the bonus goals were set.

It is also important to make the employees perceive the bonus as a bonus and not a guaranteed part of the salary. Fixed or expected compensations should be paid out as a salary, not a bonus.

2.6 Interactive or Written Communication

Some companies make a large mistake when they try to implement a balanced scorecard or when communicating a strategy; they focus on the tools that support the process instead of focusing on the concept. However, this does not mean that tools are not a good idea; on the contrary they can be vital tools when communicating a strategy. As long as the tools are not the means to build the strategy and the employees are the center of attention a number of interactive, written or graphic tools should be used in the implementation. Some examples are:

- Written articles in company newsletters or magazines that explain the strategic planning and how it works is an easy way of communication.
- An Intranet is an excellent way to publish information about your strategy. You can present the company vision, strategy, business plans and other projects on your Intranet. There is also the chance to follow up on the business here by publishing key ratios, sales figures and perhaps the balanced scorecard here. At some *Skandia* companies all employees can find extensive information about the company vision and strategic objectives, the annual strategic planning process and all key ratios on the Intranet. Some departments have their department goals on the Intranet so they can follow up on their personal bonus goals on a regular and easily accessible basis.
- There are a number of special tools on the market that are supposed to manage the strategy for the company. Many of these tools are good but it is very important to realize that an expensive tool will not solve the company's problems unless the employees can understand them and act on them.
- Posters that show the company vision or posters that regularly show different key ratios is a good and easy way to make information accessible to everyone without even looking for it actively.
- Screen savers can also be a source of automatic communication of company statements.

One risk with this approach is when the company does not fulfill what they have communicated. If you don't walk the talk, mistrust to the concept will quickly emerge.

2.7 Live Communication

A way of communication that is absolutely necessary is spoken communication. Staff or department meetings are one way to do this, but more efficient is to have regular follow up meetings with all employees. This is especially important if you work with personal bonus schemes.

Since it is not possible to involve all staff in all strategic meetings or discussions, it is the responsibility of each participant to communicate strategic information to his staff and to get feedback from the staff. This may sound obvious but there are many companies that fail with this.

2.8 Requisites for Success

Different ways to communicate strategic issues to the employees does not mean that you automatically will succeed in improving your company. *Skandia* has been working with the development and implementation of the *Skandia* Navigator since 1987. During that time

Skandia has collected a vast amount of experience. Based on these experiences the following discussion covers some essential issues that need to be considered when implementing a balanced scorecard.

2.9 Company Culture

Perhaps the most important aspect of a successful implementation of a strategic planning process is that the company culture must support the process. Mentioned above are the needs for employee involvement and ways to communicate. However, to inspire the staff to work actively towards a goal and take part in the strategic planning, there needs to be an honest wish from the management to involve all staff.

If a company wants its employees to be able to make correct decisions themselves in order to speed up the decision process and to work proactively, it has to give its employees responsibility. Mistakes need to be allowed – to learn from – and trust must replace micro management.

When changing a company culture in this direction there will always be some resistance and there will be people that cannot cope with this culture. In *Skandia* the experience is that the people that will not accept the new company culture will leave the company by themselves. It may sound dramatic but this is not a bad thing - people who do not feel comfortable with a culture that does support efficient strategic planning may not belong in that culture. Some will learn and perhaps change their ways but those who do not accept the culture can do themselves and their company a favor – change jobs.

Another vital thing is that management "walk their talk". If management supports the idea of a new company culture they must act accordingly. It is also the task of the management to make sure that middle and lower management adapt to the culture and that they apply it in all departments. With the help of professional staff satisfaction surveys it is often fairly easy to see if there are exceptions to the general culture. However, these deviations must be acted on and changed.

3. CEO Buy-in

The success of an implementation is completely dependent on the buy-in from the CEO. As previously mentioned, the *Skandia* Navigator is a management tool, a follow-up tool and a communication tool. Although a controlling department can use a balanced scorecard for their internal follow-up even without the support from the CEO, this is not what a balanced scorecard is about. The purpose is to use the balanced scorecard company wide, not only by a single department.

If the CEO is not convinced and understands the balanced scorecard concept, the implementation will not succeed throughout the organization and it will not survive in the long term. Even a single new key employee can destroy a successful implementation if the gains are not presented well by spreading badwill in the organization. One goal should be that an implementation should survive a shift of a CEO. The balanced scorecard concept should be a part of the company culture – of the way the company is managed. If it is, new employees will be automatically involved and presented to the gains. And if the CEO has accepted the concept, personnel transitions this will be even more trivial.

The initiative to implement a balanced scorecard often comes from someone else than the CEO, typically the local controlling department or a central unit like a mother company. Thus, one of the first things the implementer needs to think about is to get the buy-in from the local CEO. In *Skandia* we have made a number of both positive and negative experiences that all stress the importance of the buy-in:

Adapt the implementation to the local environment

The local CEO needs to be responsible for the implementation and shape it in his style, i.e. adapt it to his way of management. If he is not allowed this it will be difficult to get his support. It is also important not to forget the local culture. Businesses are run differently all over the world and a balanced scorecard sometimes needs to be introduced gently to get acceptance by the local management. The maturity of the company or business unit also needs to be considered. A very mature company has different management and communication needs than a small start-up organization. It will be easier and quicker to introduce the concept in a small organization that makes it a natural part of the company culture from the start. However, if you overdo the implementation in a small organization by overloading them with tools or analyses that they are not ready for the buy-in of the concept can be quickly lost.

The attitude and level knowledge

If the CEO does not have the right attitude it will be difficult to implement, but it is equally important that the person that is responsible for implementing has the right attitude. Otherwise it will be hard to gain buy-in. We have seen a number of examples of this in *Skandia* and it shows the need of having a person that is responsible for the balanced scorecard that is not only skillful, he must also be able to transfer the knowledge, explain the concept and receive understanding and buy-in for the balanced scorecard.

A high knowledge level of what the balanced scorecard concept is needs to be secured both with the CEO and others. If not, the risk is that the expectations of the role of a balanced scorecard do not coincide with the purpose. Too often the balanced scorecard is presented not as a concept but as a high-tech software tool that will solve all problems in the organization automatically. If the CEO is presented with this (from inside or outside of the organization) the buy-in can perhaps be gained very fast, but it will be lost even faster. Everyone needs to be aware of the purpose and the efforts it will take to succeed.

Follow up on decisions and do not promise things that cannot be fulfilled

It can be tempting to "smooth" the introduction of a balanced scorecard with promises of fast gains. This is an efficient way of losing buy-in since too high expectations will be created. When introducing any new concept it is important to be realistic both about the results and the efforts it will take to get there.

In the implementation it is not only important to refrain from promising too much but it is also important to follow up on what has been decided or said. One of the main purposes behind a balanced scorecard is follow up and by showing negligence to do this the risk of failure because of a lack of acceptance will increase. An example from *Skandia* is a company that introduced individual balanced scorecard for all employees. However, the individual scorecards were never followed up on or reviewed by anyone. This lead to a low acceptance because the efforts of maintaining an individual balanced scorecard were never appreciated.

4. Strategic Process Focus

A strategic process can be extremely time consuming and thus expensive if it is not coordinated. There is little to be won from a process that involves too many people and too much of their time. Instead the goal should be that the strategic planning is a natural part of the working day. If people know the company strategy they know how to act. There has to be some sort of structured planning process as well and it has to be coordinated and involve the right people. The end result of a planning cycle should be a revised business plan, which is understood and accepted by the whole company.

If the strategic planning follows a known structure year after year and if results are presented according to a well-known format, the strategic planning process will become more and more efficient over time. At *Skandia* we use the *Skandia* Navigator to drive the process and to present the results. It is important though, that the process is the focus and not the output. A balanced scorecard is like a dashboard of a car; you can build it to look fancy but that will not change the fact that it only presents the output of a mechanical process and the performance of the motor. In a business the balanced scorecard can also be made to look fancy but without well functioning underlying processes the results will never improve.

5. Communication Strategy

A company needs to work with its communication strategy in order to secure that all people are reached by all relevant information. At the same time it is important to avoid information overload. Too many e-mail, too many articles in too many newsletters and too much information on the Intranet will lead to a lack of focus and confusion.

Direct communication is not the only way of communication. At one *Skandia* office the strategic planning process is started every year with departmental blue-sky meetings where all staff is given the chance to influence by giving suggestions for how to drive the company forward. The meetings encourage new ideas and everyone is given a chance to give his opinion. Without the input from all levels of the hierarchy the information and the basis for decisions that the management receives will not be complete.

6. Experiences and Summary

As *Skandia* has worked with the concept of the *Skandia* Navigator for some years now we have collected our experiences from different parts of the organization. The most important aspect is that the work with implementing a balanced scorecard really is worthwhile. It can greatly improve your business' chances of success and help to manage the business through difficult market situations. The experience also shows that if you want a successful implementation you have to go all the way. Many companies try to implement this by just getting a balanced scorecard and type key ratios into it, but as discussed above, this does not work.

The work will not change your business over a day. It is hard work that needs a lot of patience and especially the management has to be dedicated – they have to "walk their talk". It is also a good idea to have an active spokesperson in the company and to take outside professional help when performing planning meetings (it is much more complex than many people think). You also need to start in the right end. If your company does not have a vision and strategic objectives, these have to be created before beginning with anything else.Try to involve as many people as you can in the process. A good idea is to take the strategic planning down to departmental level. In theory each department can have its own vision and strategic objectives and its own planning process (that is aligned with the company of course). We have tried this in a number of instances at *Skandia* and it has always come out very good. You will also encounter resistance on the way. Never put these "disbelievers" outside the process though, since that will only make them more convinced that they are right. Instead you should engage them in the process and make them influence the way it is structured. This will improve the process as well, since something in it obviously did not work perfectly. Another crucial part is to have managerial back up, acceptance and understanding. Without this you will have a very hard time to manage a strategic process. It will take some time, resources and money so without the support from management it will not succeed. An implementation never stops at a certain point in time. Eventually it becomes a natural part of the work tasks in the company. You will not be talking about "doing strategic planning", you will be doing it automatically and incorporating it into the company culture.

Futurizing Culture and Sustainable Earnings Capabilities

Mart Kivikas [1]

---

[1] *Mart Kivikas*, WEKE GmbH, Germany

It has become more and more evident that future value creation is more likely to happen outside our classical balance sheets as we know them today. In this case study, made in a mid-sized German company, we made a prototype of how a company can manage and better be prepared for future value creation by using its intellectual capital (IC). The first step was to define its existing success factors by asking management to write a booklet on why they had succeeded so far. The second step was then to test the validity of these factors with a method called IC-Rating™. With the result of the rating, a basis of new value creation areas was presented with key performance indicators that support the implementation of this IC-strategy. The success of this strategy can then be approximated with the IC-Multiplier, which visualizes how efficiently management is nurturing and developing its intellectual capital.

1. Introduction

A company culture and its core values are essential for sustained competitiveness and for future earnings capabilities. The purpose of this case study was to answer the question raised from the management at a midsized company in Germany of what actions could be taken today in order to secure a continuous positive evolution of the successful company culture. This presented case should be seen as a prototype of how to secure one of the drivers for sustainability and the future of the company.

In this case, it was evident that the company culture had played and continues to play a major positive role for the bottom line performance. However, in a situation where everything seems to be in line with expectations and no immediate threats can be seen, the management decided to be precautious and to find out where future value creation for the company could take place.

Major additional future earnings potentials are likely to be outside a classical balance sheet and more in the area of Intellectual Capital (IC), or the hidden values that are not to be seen in existing management controlling systems. Therefore a specialist company in this area, *Wissenskapital Edvinsson & Kivikas Entwicklungsunternehmen GmbH* (WEKE), was contacted and an overview of this cooperation is presented in this article.

In order to find out if the company culture was fit for the future and what management could improve in order raise its IC, it was decided to make an IC-Rating™ conducted in June 2003. The result and the recommendations in this paper are based on this rating.

IC-Rating™ is one of the most used methods today to prototype and to measure the IC in an organisation. It was introduced 1997 in Sweden under the supervision of *Leif Edvinsson* and since then, more than 200 ratings have been made world-wide under the realm of *Intellectual Capital Sweden AB*. The main purpose of the rating in this case was to visualise the company culture and to analyse the key areas of structural and human capital. This study was a prototyping study and limitied to internal perspecitves on the IC. With this as a base, some tentative recommendations for the future were made.

2. Theoretical Background to the Method Used

Today a majority of a company's investments are not to be seen on the classical financial balance sheet. Instead, they are what we call intangible assets. Investments in human resources are not at all in any balance sheet. If at all mentioned, then only as a note and as a liability for future pensions. Our work with IC tries to make these investments more transparent as a completion to traditional bookkeeping. At the same time we can show management and investors the importance of a high IC rating in order to stay competitive in the future.

The tools we use today to analyse and decide the profitability of a company are based on tools that go back to renaissance Italy. These tools are appropriate for a production-based society, not a knowledge society. It supports managers who try to make their existing asset sweat, but not managers who care about value creation and a companies IC.

WEKE defines IC to be a combination of human capital – the brains, insight, skills and potential of those in an organisation – and organisational structural capital – things like the capital wrapped up in processes, company culture, patents, know-how, brands, IT systems and its relational capital. A company's IC Rating increases as it improves its ability to transform knowledge and intangible assets into wealth creating resources. This involves the multiplying of human capital with structural capital.

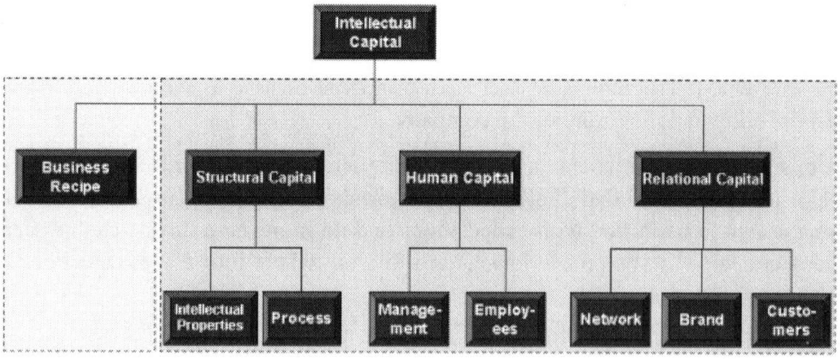

Fig 1: IC-Rating™ Platform

As a framework, the following IC-Rating™ platform is used (see *figures 1* and *2*). The platform divides IC into three major areas – human capital, structural capital and relational capital.

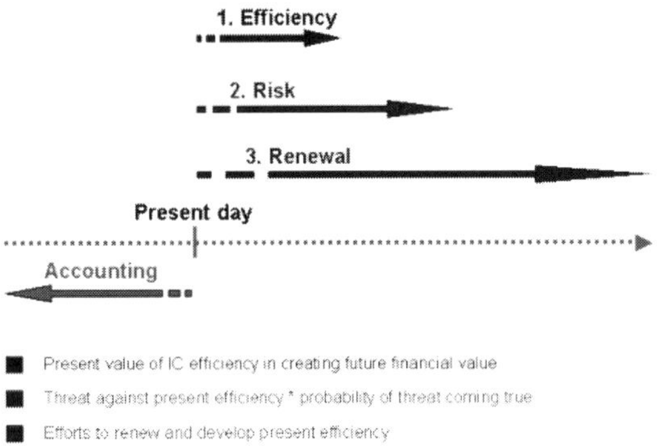

Fig 2: IC-Rating™ and its three focus areas

2.1 Human Capital

Human capital refers to know-how, capabilities, skills and expertise of the human members of the company, its management and employees. Some of the key functions tied to human capital management are drawn from the traditional practices of human resource and knowledge management. The tacit knowledge of the employees can be transformed into explicit knowledge, which leads to the conversion of human capital into structural capital which thereby becomes the property of the company (see also 2.4 "The IC-Multiplier"). The human capital is therefore of vital importance because it is the source of innovation and strategic renewal.

2.2 Structural Capital

Structural capital includes the capabilities developed inside the company to make value of its human capital and the networks outside it. These capabilities may include patents, business processes, internal communication tools and, maybe the most important one, the company culture. Structural capital can best be described as all that remains of tangible and intangible assets when the employees go home for the day.

A critical factor for the future will be the ability to find and use external structures and to integrate them into the companies internal processes. This can be done through what we call "insourcing" of competencies. Using the current or a stream as a metaphore, the management should try to stand in the middle of it and pick out whatever passes by that could be beneficial for the company. This not only means to catch up with new technological trends, but also to find out what affects people and the society. Social trends are becoming as important as any other trend today.

2.3 Relational Capital

Of the three categories, relational capital is the easiest one to measure through market share, customer retention, defection rates and per-customer profitability. Relational capital also includes connections outside the organization such as customer loyalty, goodwill, supplier relations and other stakeholders. The recent developments around Customer Relation Management (CRM) belong to this area.

2.4 The IC-Multiplier

The three components of IC need to effectively interact with each other for the organization to create value. The IC-Multiplier, a quantitative concept developed by *Edvinsson* (see www.wissenskapital.info), concentrates on the correlation between structural and human capital. The structural capital components are more manageable and can therefore be affected by management in a short period of time compared to the other two components.

In order for a company to be able to fully take advantage of its human capital, the necessary organisational structure needs to be provided. If the right structure is available, then this leverages the capacity and opportunities of the human capital. Because of this leverage effect, a high portion of structural capital becomes of vital importance. One way of increasing this leverage is by converting the human capital into structural capital (see *fig. 3*).

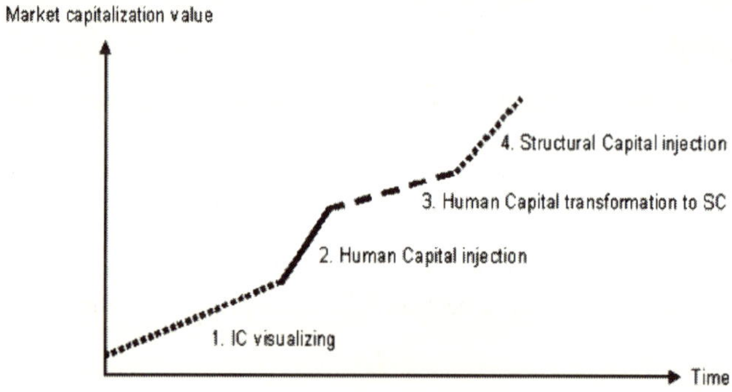

Fig. 3: Market capitalisations over time

The first step - IC visualizing - is to map and evaluate the existing intangible assets of the organisation. The second step - human capital injection - handles the increasing of the human capital within the organisation by attracting talent, the retention of it and for further recruitments. The third step - human capital transformation to structural capital - focuses on the packaging of knowledge so that it can be shared rapidly and globally. This also contains a shift of leadership focus from human capital to structural capital in order to visualize and stimulate this continuous flow. This transformation acts as a multiplier, The IC Multiplier, with sustained earnings potential for the organisation.

The IC Multiplier provides an indication on how well the human capital of the organization uses the structural capital to leverage its potential. The ratio is calculated by looking at the relationship between structural capital and human capital, i.e.

IC Multiplier = structural capital divided with human capital

This ratio shows that structural capital has to be larger than human capital, otherwise the company has not a multiplier but the opposite, which in turn could lead to an erosion of human capital. A small degree of structural capital also means that a company is less enduring and that the risk increases, since the employees can leave the company whenever they wish. This process, in which human capital leverage with structural capital, thus creating the value of IC, can be measured by using value added per employee.

The fourth step is the structural capital injection. By acquiring structural capital externally, the organisation gets a turbo effect on the already established IC Multiplier since it now combines different types of structural capital and uses it in the value creation. Furthermore this can often be rented through various forms of networking, alliances or contracting and thereby not burdening the traditional balance sheet.

3. The Analyse Based on the IC-Rating™

The rating of the human and structural capital was in this prototyping case limited to 15 internal interviews. The 15 employees interviewed were picked out from the rated company's management in cooperation with the project manager from WEKE. Different departments and ages groups were taken in account as well as years of service, organisational level and gender.

After analysing the results of the interviews that based on this sample, the overall result of the IC-Rating™ was close to best in class. Moreover, the rated company got the second highest grade (out of ten) in operational efficiency and a very high grade on company renewal. The risk factor was considered modest to low. The analysis and comments below will concentrate on areas which could be improved.

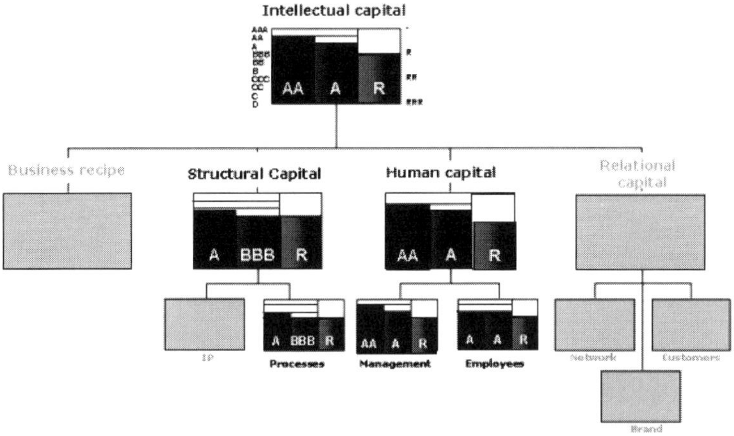

Fig. 4: The Result of the IC-Rating

After this overall presentation was made, the rated company received a more detailed analysis with comments on each of the areas. The focus of the comments were on the more critical issues and therefore the management could see in which areas action was needed and in which areas they would not have to invest more time. Some comments made of the interviewed were also added not to lose any relevant information that may have been overseen in the questions.

In this context we would like to get more into detail with what we call „Value added" and the „IC-Multiplier". As already mentioned above, the IC-Muliplier highlights how good the human capital is in using the structural capital to create value added. A proxy for this could be the value added per head. The value added should be high enough to pay for salaries, dividends, tax and future oriented research & development. How much of the value added is created by the human capital versus structural capital? As an early indication of this multiplier effect the salary expressed as percentage of the total value adding can be used. If the human capital component is larger, it highlights a high dependancy on talent insourcing for the future, i.e a risk of less sustainable earnings capability.

Efficiency	Risk	Renewal & Development
AAA	-	AAA
AA	R	AA
A	RR	A
BBB	RRR	BBB
BB		BB
B		B
CCC		CCC
CC		CC
C		C
D		D

The *efficiency* of the intellectual capital as well as the *efforts to renew and develop* are rated in a tenthgraded scale ranging from "AAA" down to "D". The *risk* of a decreasing efficiency of the intellectual capital is rated in four levels, ranging from negligible risk "-" to a very high risk "RRR".

EFFICIENCY		RENEWAL & DEVELOPMENT		RISK	
AAA	Extremely high efficiency	AAA	Extremely good r&d efforts	-	Negligible risk of decreasing efficiency
AA	Very high efficiency	AA	Very good r&d efforts		
A	High efficiency	A	Good r&d efforts	R	Moderate risk of decreasing efficiency
BBB	Relatively high efficiency	BBB	Relatively good r&d efforts		
BB	Average efficiency	BB	Average r&d efforts		
B	Relatively low efficiency	B	Relatively weak r&d efforts	RR	High risk of decreasing efficiency
CCC	Low efficiency	CCC	Weak r&d efforts		
CC	Very low efficiency	CC	Very weak r&d efforts	RRR	Very high risk of decreasing efficiency
C	Extremely low efficiency	C	Extremely weak r&d efforts		
D	Lack of efficiency	D	Lack of r&d efforts		

Fig. 5: The Scales of a Rating

The rough calculation for a company in this business segment (see *fig. 6*) indicates usually a favour of human capital instead of organisationally owned structural capital. This could be an early warning signal to develop more structural capital as discussed before.

Revenues in Euro	200.000.000
Value Added (only the direct costs of sales are deducted) in Euro	100.000.000
Value Added in %	50%
Cost of Personnel in Euro	40.000.000
Personnel Cost related to Value Added in %	40%
Number of Employees	700
Value Added per Employee (Euro)	142.900

Comments:

- Very rough estimate, to be used for a learning conversation only
- Apprentices not included in personnel
- Research and Development not included

Fig. 6: Calculation of Value Added per Employee

4. General Learnings and Future Perspectives

In this section an extract of the learnings made and the future value growth areas based on the IC-Rating™ is presented. Generally we recommend the continuation on the company's culture and the evolution of it. This would also reinforce a financially healthy bottom line in the future.

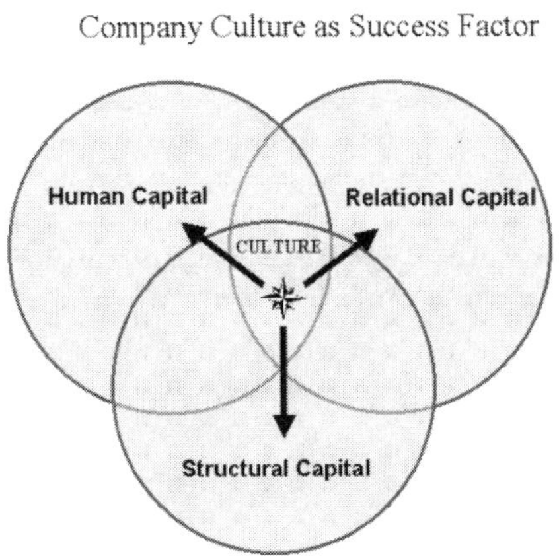

Fig. 7: The importance of company culture within the IC framework

Figure 7 shows what we found to be the excipient or glue, which makes this company successful today. In the center you have the driving force behind the positive evolution so far - the company culture. The future challenge will be to find a balance between the forces driving its value creating activities around its culture and to use this energy to create innovation and competitive advantages in the future.

The arrows can be seen as an indicator of how well management is able to communicate into the three areas of IC. As we will point out later, it is inefficient to put much emphasize into one area thereby neglecting some others. There must be a balance, excelling in one area does not mean the same in other areas.

As the rating clearly points out, the human capital is good but some improvements could be made in structural capital. This does not mean that the management should open up the doors to any kind of external cooperation or the funding of obscure social events. What this says is that the management should take this as a strategic issue and decide what it wants to do to strengthen its IC.

In the following subtitles, we show that it is a matter of management fine-tuning to an already successful approach. A good working organisation like this would not be helped with an imposition of new standards or regulations. Too much regulation could take away the innovating spirit of the company.

4.1 Raise the IC-Multiplier

As is being indicated, the structural capital is slightly higher than the human capital. In other words, the human capital is close to the structrual capital, i.e the multiplier effect is only around 1,5 and the aspiration should be to have at least 2 or more. How do we as leaders and managers supply such a leveraing culture that is also sustainable for the future?

Looking at *figure 3* above, we would assume that the company has reached the second level of human capital injection, but there is still something to be done before it can reach the third level, which should be the target for the near future.

To reach this, new Key Performance Indicators (KPI) could be developed that support investments in a higher structural capital and therefore a higher value added leverage per employee as well as reduced risk. A first start would be to verify the measure of the value added per employee and the relation between structural and human capital as was shown under chapter 3 above and make it explicit in the internal communication reports. This gets interesting the more down to the operational units one gets. A very quick analyse could find out best-in-class units and compare them with others. What are they doing differently? Have they always been so successful etc.? Can the assistance of a peer or a peer review system be developed for the bench learning between those units?

This basic benchmarking could be supplemented with an IC-Rating™ on other subsidiaries to quickly make it transparent for other companies within the group to discover what they can learn from one another.

Through this, the company does not have to search for best practices outside the group, but could start to use internal know how and experience to improve performance. This is much more motivating to the people involved and supports the collaborative atmosphere that already exists between the companies within the group.

4.2 Strategy to Strengthen the External Networks

Most of the value creation in the future will take place outside the company's boundaries. Therefore, the management could support and demand this more often in order to have some control of it and make it into a company (structural) asset.

One way to secure this is to intensify the cooperation with universities and other research institutions. Management could continue its efforts to find top students and to promote research, both internally and externally, in its area.

An obvious trend today in Europe is the emerging of regional clusters, which try to specialize in specific scientific areas in order to attract companies, investors and future taxpayers. The company may want to take an active part in some of these developments. Which clusters or regions would provide technology and know how for future needs? Politicians have become very flexible to tax subsidies when scientific clusters are mentioned. Such a cluster could be an excellent research and recruiting ground.

This could also be supported by finding internal "specialists" who then could be connected to communities from where they can get new impulses and ideas for innovations in their day-to-day operations. Through this approach, the company can make sure these innovations will stay within it own realms.

An arena like a Future Center (see below) could reinforce these efforts of connecting with the outside world in a more comprehensive way for the insourcing of intelligence, talents and new business opportunities.

4.3 Secure Existing Knowledge

In order to protect the company's knowledge when senior management retire, a senior network company for insourcing on part time bases retired executives and knowledge experts could be formed. This company can then sell its services as consultants to younger executives within the company and will therefore be promoted with a higher pension. Such schemes are often promoted from the state and usually have no down side risk to the former employer.

4.4 Insource Intelligence in Order to Keep Operationally Fit

The efforts in some areas to use external knowledge sources should be extended into as many functions as possible. The insourcing strategy also gives management a possibility to stand as a gatekeeper of evolution between the internal organisation and the outside world. If this possibility is made a strategical task, one does not have to follow the evolution, but can actually take an active part in creating and leading it.

4.5 Future Center

The Future Center, a concept developed by *Edvinsson*, during his work with *Skandia*, is an arena for experimental learnings about the future as an organisational lab for prototyping structural capital (see www.skandiafuturecenter.com). It is usually a place and space with a mind-stimulating atmosphere and should support time for creativity and cross-collaborative learnings between different individuals within or outside the company, as well as different generations and cultural groups. The visitors will have the impression of getting away from their ordinary working context and are invited as individuals. Organisational boundaries will be blurred.

The objective is to find a common language, to define a specific problem and to make an organisational rhythm more flexible and quicker. Politicians, academics and external knowledgeable persons can be included as insourcing, because they may play an important role for the future development of the company in concordance with society.

When the issues discussed are integrated in the internal processes of the company, the output of the future center should be to define guidelines for future actions and to give orientation in which direction the company should develop.

5. Key Message

In this case study we have made a prototype of how a company can manage and better be prepared for future value creation. With the IC-Rating™, a basis of new key performance indicators was presented that supports cultural issues for sustainable earnings development. The success of this strategy can then be approximated with the IC-Multiplier, which visualizes how efficiently management is nurturing and developing its intellectual capital. The next step will be to include external parties in the rating and to commence an external benchmarking based on the vocabulary and KPI's of the IC platform (see *fig. 1*).

A management system based on the rating could be seen as the second generation of measurement tools, which have understood the importance of innovation for value creation - the first one being the Balance Scorecard. At *WEKE GmbH* we are about to develop a third generation of mapping for the German speaking market – the corporate longitude.

Intangibles als Werttreiber

Klaus Möller [1]

---

[1] Dr. *Klaus Möller*, Lehrstuhl für Allgemeine Betriebswirtschaftslehre und Controlling, Universität Stuttgart

Die Rentabilität eines Unternehmens wird zunehmend von Intangibles geprägt. Im Zuge dieser Entwicklung verdrängen immaterielle Vermögenswerte zumindest teilweise die klassischen materiellen Vermögenspositionen hinsichtlich ihres Beitrags zum Unternehmenserfolg und zur Sicherung einer nachhaltigen Unternehmensexistenz. Aus diesem Grund müssen bei der Ermittlung der Werttreiber auch Intangibles einbezogen werden. Durch die Verbreitung wertorientierter Steuerungskonzepte gewinnt eine solche Integration große Bedeutung. Vor diesem Hintergrund entwirft der Beitrag ein integratives Konzept zur Berücksichtigung von materiellen und immateriellen Werttreibern bei wertorientierten Investitionsentscheidungen. Abschließend wird der Einsatz des Konzepts bei einem international agierenden Maschinenbau-Unternehmen dargestellt.

1. Einleitung

Immaterielle Werte bzw. Intangibles gehören inzwischen zur betriebswirtschaftlichen Realität. Die intensive Beschäftigung mit „kritischen Erfolgsfaktoren", „Kernkompetenzen" und „Werttreibern" hat Wissenschaft und Praxis dafür sensibilisiert, dass monetäre und nicht-monetäre Werte, qualitative Faktoren und finanzieller Erfolg eng verknüpft sind. In der New Economy wurden die Erkenntnisse genutzt, um Geschäftsmodelle und Vermögensbestandteile in unreflektierter Weise zu bewerten. Die Diskussion ist damit jedoch keineswegs zum Erliegen gekommen, sondern findet – nun eher von wissenschaftlicher Seite getrieben – Einzug in die Praxis. Die zielgerichtete Nutzung der Intangibles soll der Unternehmenssteuerung eine wirkungsvollere Erschließung von Wertschöpfungspotenzialen ermöglichen. Auch wenn es inzwischen erste Beispiele für auf Intangibles basierende umfassende Steuerungskonzepte gibt, so wird in der Unternehmenspraxis der Transformationsprozess nur langsam vollzogen. Intangibles werden zunächst ergänzend zu den bisherigen finanziellen Steuerungsgrößen genutzt. Ein Grund hierfür liegt in der Bewertungsproblematik: „Wie steuere ich etwas nicht Fassbares?" mündet sofort in die Frage: „Wie messe ich etwas nicht Fassbares?" Die komplexe Thematik der adäquaten Messung soll hier nur am Rande diskutiert werden. Im Mittelpunkt steht der Entwurf eines handhabbaren Konzeptes, mit dem Intangibles in die wertorientierte Unternehmenssteuerung integriert werden. Damit soll eine Beurteilung alternativer Investitionen bzw. Projekte auf Basis einer kombinierten wertorientierten und intangiblen Steuerungsgröße ermöglicht werden. Die Anwendung des Konzeptes wird abschließend am Beispiel eines internationalen Maschinenbaukonzerns dargestellt.

2. Anforderungen an Intangibles im Kontext der wertorientierten Unternehmensführung

Das Paradigma der wertorientierten Unternehmensführung verlangt eine konsequente Ausrichtung des gesamten Unternehmens auf die Strategie, damit im Sinne des Ziels „Steigerung des Unternehmenswertes" gehandelt wird (vgl. *Stewart,* 1990; *Günther,* 1997, S. 7ff. und S. 336ff.). Ausgangspunkt ist die Sichtweise, dass eine langfristige Steigerung des Unternehmenswertes die Voraussetzung dafür ist, die Überlebensfähigkeit zu sichern. Zahlreiche Veröffentlichungen und empirische Studien haben die Praxisrelevanz und große Verbreitung des Ansatzes dokumentiert. Bei der Integration von Intangibles sind – neben dem eigentlichen Problem ihrer Messung und Bewertung – insbesondere zwei Anforderungen zu berücksichtigen, nämlich die Berücksichtigung von Unsicherheit und der Komplex der Ursache-Wirkungs-Beziehungen.

2.1 Berücksichtigung von Unsicherheit

In den meisten Fällen geht man in der wertorientierten Planung davon aus, dass Vorhaben auch die gewünschten Ziele erreichen. Ergänzend betrachtet man häufig noch Szenarien maximaler Ausprägung der angenommenen Variablen (best/worst case). Die Realität, geprägt durch Komplexität und Dynamik, wird damit nur unzureichend abgebildet, da Modelle zwangsläufig mit einer Komplexitätsreduktion arbeiten. Der Faktor Unsicherheit darf jedoch keinesfalls unberücksichtigt bleiben. Insbesondere betrifft dies die Intangibles, da ihre Prognosegenauigkeit bzw. Bewertung zusätzlichen Unsicherheiten unterliegt. Neben der Möglichkeit, dass man ein festes Ergebnis in der Zukunft kennt, kann auch Unsicherheit bezüglich der zukünftig zu erwartenden Umweltzustände herrschen. Kann man die Unsicherheit mit Wahrscheinlichkeiten (entweder typische Häufigkeitsverteilungen oder subjektiv geprägte Erfahrung) ausstatten, so spricht man von Risiko. Ein besonders geeignetes Verfahren, um Zukunftsprognosen zu strukturieren, stellt der Ansatz der Realoptionen dar (vgl. bspw. *Hommel/Pritsch*, 1999). Aufschlussreich ist dabei der Korridor, der die verschiedenen Optionsverzweigungen nach oben und unten begrenzt: sofort lassen sich best und worst case ablesen. In Fällen, bei denen Wahrscheinlichkeiten nur schwer festzulegen sind, bietet sich die Szenario-Technik als Diskussionsgrundlage an. Zusätzlich kann man über Sensitivitätsanalysen aufdecken, wie sich das Ergebnis bei Variation eines einzelnen Risikoparameters verändert. Beide Ansätze stellen allerdings hohe Anforderungen an Problemaufbereitung und Lösungsaufwand. In der Praxis finden sie daher nur bei besonders wertintensiven oder risikoreichen Entscheidungen und damit nur in Einzelfällen Anwendung.

2.2 Ursache-Wirkungs-Beziehungen

Werttreiber sind wichtig, um einzelne Bereiche und Aspekte innerhalb eines Performance Measurement Systems darzustellen. Aber erst Verknüpfungen der Werttreiber – besonders zwischen nicht-finanziellen und finanziellen – liefern die für eine erfolgreiche Umsetzung der Strategie notwendigen Beziehungen. So wirken sich zum Beispiel Mitarbeiterzufriedenheit oder technologisches Know-how nicht unmittelbar auf finanzielle Ergebnisse aus. Ihr Wert im Unternehmen ist stark von Kontextfaktoren, wie der Unternehmensstrategie oder organisationalen Gegebenheiten, abhängig (vgl. *Ittner/Larcker*, 1998). Für eine gezielte Unternehmenssteuerung ist daher eine Analyse und Darstellung wichtig, über welche Kombinationsprozesse immaterieller Werte im Rahmen einer geeigneten Strategie letztendlich in finanzielle Ergebnisse transformiert werden sollen. Vorteilhaft sind hier quantifizierbare Ursache-Wirkungs-Beziehungen, das heißt das Wissen um die mengenmäßige Beeinflussung und nicht nur um die grundsätzliche Richtung (vgl. *Wall*, 2001, S. 72). Ursache-Wirkungs-Ketten können aufgrund ihrer begrenzten Abbildung der Realität nur ein Modell darstellen. Der Anspruch, alle Beziehungen vollständig zu erfassen, lässt sich praktisch nicht umsetzen. Nichtsdestotrotz können auf nicht ganz vollständigen Informationen basierende Hypothesen im Management-Alltag sehr hilfreich sein, um die relevanten Zusammenhänge zu verstehen.

2.3 Fokussierung des Untersuchungsziels

Im klassischen Sinn bedeutet Wirtschaften den Umgang mit knappen Ressourcen. Dazu gehört insbesondere die Entscheidung zum Einsatz dieser Ressourcen, die üblicherweise als Verteilungsproblem auftritt. Die Zugrundelegung eines wertorientierten Steuerungskonzeptes bedingt in diesem Zusammenhang sowohl eine einheitliche Auffassung zum obersten Unternehmensziel (Wertschaffung) als auch ein einheitliches Verständnis hinsichtlich der Art der Wertschaffung. In der Regel konkurrieren verschiedene (potentiell wertschaffende) Projekte um die knappen Unternehmensressourcen. Die Unternehmensführung (unterstützt durch das Controlling) muss hier eine „Rangliste" der Projekte nach deren Wertbeitrag bilden, um daraus die realisierbaren auszuwählen. Diese Entscheidung muss allen beantragenden Teileinheiten entsprechend kommuniziert werden. Entsprechend transparent und nachvollziehbar sollte der Evaluationsprozess sein. Auch für eine spätere Überprüfung der Erfolgsrelevanz empfiehlt sich eine systematische Entscheidungsfindung.

Die zentrale Fragestellung des vorliegenden Beitrags ist daher, wie bei der Ressourcenallokation eine aussagefähige Vergleichsbasis zwischen verschiedenen Projekten bei der Bewertung hergestellt werden kann. Als wesentliche Prämisse wurde dabei die besondere Relevanz der Intangibles zugrunde gelegt, die in die Bewertungssystematik aufgenommen werden sollen. Im Folgenden wird daher zuerst die konzeptionelle Basis der Bewertung von Intangibles dargestellt.

3. Erfassung und Bewertung von Intangibles

Wachstum und Rentabilität eines Unternehmens werden zunehmend von Intangibles anstatt von materiellen Vermögensgegenständen geprägt (vgl. *Lev,* 2001). Entsprechend sind diese nicht „fassbaren" Werte Gegenstand einer intensiven wissenschaftlichen Diskussion. Sie nähert sich von zwei Seiten dem Thema: Einerseits soll über den Einbezug der Intangibles in die Unternehmenssteuerung ein verbessertes Verständnis und daraus resultierend eine effektivere Führung der betrieblichen Wertschöpfungsprozesse erreicht werden. Andererseits sollen durch die Berücksichtigung von Intangibles im Berichtswesen neue Wertkategorien eingeführt und verankert werden, die zu einer Gestaltung der Rechnungslegung herangezogen werden können. Formal verändert sich damit die Berichterstattung (Bilanz, Gewinn- und Verlustrechnung) nach Inhalt und/oder Aufbau. Der hier vorgestellte Ansatz ist in die erste Kategorie einzuordnen, da er der strategischen Unternehmenssteuerung dient. Um eine systematische Einbindung der Intangibles zu gewährleisten, wird im Folgenden eine Definition und Kategorisierung vorgenommen sowie Besonderheiten bei der Erfassung und Bewertung herausgearbeitet.

3.1 Kategorisierung von Intangibles

In der Literatur existiert eine Vielzahl von Definitionsansätzen zu Intangibles, immateriellen Werten bzw. Intellectual Capital (vgl. u.a. *Edvinsson/Malone* 1997; *Teece*, 2000; *Lev* 2001; *Daum* 2002). Weitgehende Einigkeit herrscht hinsichtlich einer negativen Abgrenzung immaterieller Vermögenswerte gegenüber anderen Werten. So unterschieden sich immaterielle Werte von den materiellen Werten durch ihre fehlende körperliche Substanz. Finanzielle Werte, wie Geldforderungen, zeichnen sich zwar ähnlich wie immaterielle Werte durch die fehlende körperliche Substanz aus, gleichwohl lassen sie sich von den immateriellen Werten durch das Kriterium "monetär" abgrenzen (vgl. *Arbeitskreis "Immaterielle Werte im Rechnungswesen" der Schmalenbach-Gesellschaft*, 2001). Immaterielle Werte können somit als nicht monetäre Werte ohne körperliche Substanz umschrieben werden.

Hinsichtlich der Kategorisierung gibt es ebenfalls zahlreiche Vorschläge, die sich in Bezug auf die Anzahl der Kategorien und deren Verhältnis zueinander zum Teil deutlich unterscheiden. Ein inzwischen verbreiteter Ansatz (vgl. *Arbeitskreis "Immaterielle Werte im Rechnungswesen" der Schmalenbach-Gesellschaft*, 2001 und 2003) beschreibt sieben verschiedene Kategorien immaterieller Werte (vgl. *Abb. 1*).

Abb. 1: Kategorisierung immaterieller Werte

Diese sieben Kategorien beschreiben immaterielle Werte umfassend. Es kann durchaus der Fall auftreten, dass ein Wert mehreren Kategorien zugeordnet werden kann, da die Kategorien nicht frei von Überschneidungen sind. So ist beispielsweise ein effizientes Just-in-Time-Beschaffungssystem sowohl Bestandteil des Supplier Capital als auch des Process Capital.

3.2 Bewertungsstrategien für Intangibles

Intangibles sind meist sehr heterogen. Eine Bewertung ist folglich nur sehr schwer zu bewerkstelligen. Während der letzten 10 bis 15 Jahren wurde eine große Anzahl von Konzepten zur Messung und Bewertung von Intangibles entwickelt. Diese können in die folgenden vier Kategorien unterteilt werden (vgl. *Sveiby*, 2003):

- **Direct Intellectual Capital Methods (DIC):** Intangible-Komponenten werden identifiziert und einzeln monetär bewertet, um in Summe den Wert des Intellektuellen Kapitals zu erhalten.
- **Market Capitalization Methods (MCM):** Die Differenz zwischen Marktkapitalisierung und Buchwert eines Unternehmens wird als Wert des intellektuellen Kapitals interpretiert.
- **Return on Assets Methods (ROA):** Der Return on Assets des materiellen Vermögens eines Unternehmens wird mit dessen Branchendurchschnitt verglichen. Die Differenz wird mit dem durchschnittlichen materiellen Vermögen multipliziert, um den Ertrag zu erhalten, der auf Intangibles zurückzuführen ist. Durch Division mit den durchschnittli-

chen Kapitalkosten des Unternehmens erhält man eine Schätzung des Wertes der Intangible Assets.

- **Scorecard Methods (SC):** Die verschiedenen Intangible-Komponenten werden identifiziert und klassifiziert und anschließend mit Hilfe mehrerer Indikatoren gemessen und dargestellt. SC-Verfahren sind den DIC-Verfahren ähnlich, jedoch unterscheiden sie sich dadurch, dass keine Abschätzung eines monetären Wertes der Intangibles vorgenommen wird.

Die verschiedenen Bewertungsverfahren weisen unterschiedliche Schwerpunkte und bevorzugte Einsatzbereiche auf. Die monetär orientierten Verfahren (wie das ROA- oder das MCM-Konzept) sind besonders für Mergers & Acquisitions und für Aktienmarktbewertungen geeignet. Die Scorecard-Verfahren zeichnen sich dadurch aus, dass sie ein umfassenderes Bild über den Zustand des Unternehmens entwerfen können als rein finanzielle Größen, und dass sie verhältnismäßig leicht auf jeder Unternehmensebene anwendbar sind (vgl. *Pike, Roos,* 2003). Der hier verfolgte Ansatz legt eine auf Investitionen fokussierte Scorecard zugrunde, die auf Werttreibern basiert.

3.3 Investitionen in Intangibles

Im Rahmen einer wertorientierten Unternehmenssteuerung gilt es, den Wert des materiellen und immateriellen Vermögens zu steigern. Daher ist es für ein zielorientiertes Management von großem Interesse, inwiefern sich bestimmte Investitionen auf die Assets eines Unternehmens auswirken. Im Folgenden wird daher der Versuch unternommen, aus der Perspektive der Investitionen auf mögliche Intangible Assets zu blicken (vgl. *Gupta./Roos*, S. 299). Es wird unterstellt, dass finanzielle Ressourcen, die in einem Projekt Verwendung finden, in der Regel auch zu einem entsprechenden Nutzen beziehungsweise Wert führen. Dabei stellt sich die Frage, wie sich die Intangible Assets pro investierte Währungseinheit verändern. Die einfachste Näherung stellt ein linearer Zusammenhang dar. Realitätsnäher sind Beschreibungen durch Funktionen, die aus anderem Kontext bereits bekannt sind: Lernkurven oder Input-Output-Funktionen aus der Produktionstheorie in den verschiedenen Ausprägungen (vgl. *Abb. 2*).

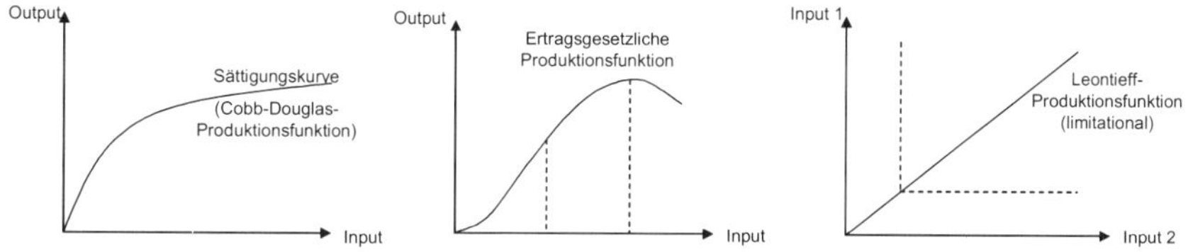

Abb. 2: Zusammenhänge zwischen Input und Output

Von enormer Bedeutung werden in diesem Kontext die Identifikation von Ursache-Wirkungsbeziehungen zwischen Inputfaktoren und Intangibles. „Unsere bisherigen finanziellen Systeme leisteten wertvolle Arbeit, wenn sich die Wettbewerbsstrategien auf die Akquisition und Steuerung materieller Vermögenswerte beziehen. Doch im Zeitalter der Wissensökonomie kann ein dauerhafter Mehrwert nur durch die Weiterentwicklung immaterieller Vermögenswerte geschaffen werden" (*Kaplan/Norton*, 2001, S. 59) – so begründen *Kaplan* und *Norton* die Notwendigkeit, immaterielle Werttreiber und die darauf aufbauen-den Ursache-Wirkungs-Beziehungen zu berücksichtigen. Die Balanced Scorecard (BSC) stellt eine Möglichkeit zur Verknüpfung zwischen materiellen und immateriellen Werten dar. Sie soll daher als konzeptioneller Rahmen zur Erfassung und Bewertung der immateriellen Vermögenswerte herangezogen werden.

4. Fallstudie: Projektbewertung unter Berücksichtigung von Intangibles

4.1 Einführung des Fallstudienpartners

Die *Fixing AG* (Name anonymisiert) ist ein deutsches Maschinenbauunternehmen mit 1,2 Milliarden Euro Umsatz und rund 10.000 Mitarbeitern, die auf dem Gebiet der Automatisierungstechnik sowie damit verbundenen Technologien und Dienstleistungen tätig ist. Ein bedeutender Teil des Umsatzes wird von mehr als 50 Landesgesellschaften im Ausland erzielt. Die *Fixing AG* wird im Rahmen eines wertorientierten Steuerungskonzeptes mit der Spitzenkennzahl EVA® (EVA® is a registered trademark of *Stern Stewart & Co.*) geführt. Seit über fünf Jahren ist eine Balanced Scorecard, die ständigen Verfeinerungen unterworfen ist, konzernweit im Einsatz. Die entwickelte Bewertungs- und Steuerungssystematik für Intangibles wurde eng an die bereits bestehenden Konzepte des Wertmanagement und der BSC angebunden, um nicht eine „stand-alone"-Lösung zu schaffen. Für die *Fixing AG* stand nicht eine Gesamtbewertung der Intangible Assets im Vordergrund, sondern eine ganzheitliche Vergleichsmöglichkeit für die Ressourcenallokation bei der Auswahl von internen Projektvorschlägen. Damit soll eine umfassende und nachvollziehbare Beurteilung und Bewertung alternativer Ressourcenverwendungen (Projekte) sowohl ex ante als auch ex post ermöglicht werden.

4.2 Integratives Bewertungskonzept

Ausgangspunkt für das Bewertungskonzept war die Frage nach den Auswirkungen qualitativer Faktoren (Intangible Value Drivers) auf die zukünftigen EVAs eines Unternehmens. Ein methodisch immanentes Problem quantitativer Kennzahlensysteme wie dem EVA oder der DCF-Methode nach *Rappaport* besteht darin, dass die Auswirkungen qualitativer Faktoren bei der Betrachtung häufig vernachlässigt werden. Die Gründe liegen neben dem Problem der Erfassung dieser qualitativen Faktoren und der Frage nach einer Quantifizierung dieser nur sehr schwer messbaren Größen auch darin, dass die Auswirkungen aufgrund komplizierter Ursache-Wirkungsbeziehungen nicht direkt erkennbar sind. Das hier entwickelte integrative Bewertungskonzept für Intangibles und EVA versucht die Vielzahl der qualitativen Werttreiber durch eine intensive Analyse der Ursache-Wirkungsbeziehungen der Geschäftsprozesse zu erfassen und die Ausprägungen der einzelnen Werttreiber auf einer qualitativen Skala in ein quantitatives Punktesystem zu übertragen und damit eine Verbindung zum EVA als monetäre Vergleichsbasis herzustellen. Das integrative Konzept unterstützt wesentlich eine ex-ante-Perspektive im Sinne einer Früherkennung. Verschiedene Szenarien innerhalb des Konzepts führen zu unterschiedlichen Wirkungen auf die Werttreiber. Zudem wird vereinfachend unterstellt, dass man über das Investitionsvolumen auf Durchschlagskraft und Stellenwert schließen kann. Um die Umsetzung des Konzeptes bei der *Fixing* AG sicherzustellen wurde das Konzept im Rahmen eines IT-Tools realisiert (vgl. *Möller/Walker*, 2003, S. 495).

Auswahl der Werttreiber

Für die Werttreiberliste wurden die vier vorhandenen Balanced Scorecard-Perspektiven Kunde/Markt, Prozesse, Mitarbeiter/Lernen/Wissen/Information und Finanzen der Top-Level-BSC beibehalten. Damit wurde gleichfalls die Verknüpfung der in der BSC definierten strategischen Ziele mit den Ansätzen der wertorientierten Steuerung hergestellt. Um die nötige Transparenz zu schaffen, wurden die vier Perspektiven durch weitere Untergruppen ergänzt. Zusätzlich wurde eine Unterteilung in beeinflussbare und nicht-beeinflussbare Werttreiber eingeführt. Nicht-beeinflussbare Werttreiber (z.B. Konjunkturdaten, gesetzliche Rahmenbedingungen) sind als gegeben anzusehen, während man beeinflussbare Werttreiber über Managementmaßnahmen im Unternehmen steuern kann. Neben dem Wissen um die Beeinflussbarkeit oder Nicht-Beeinflussbarkeit eines Werttreibers, ist es für eine effiziente Planung und Steuerung unabdingbar, die Werttreiber nach dem Werteinfluss (Wertwirkung) zu gewichten und dadurch eine Werttreiberhierarchie aufzubauen. Eine Schwierigkeit hierbei bestand darin, einen Kompromiss bezüglich der Anzahl der Werttreiber zu finden. Eine große Anzahl von Werttreibern hilft zwar, Situationen und Prozesse exakter abzubilden, führt aber auch zu einer erhöhten Komplexität und Unübersichtlichkeit. Letztendlich wurde eine Liste von Werttreibern erstellt. Diese Liste bildete den Ausgangspunkt für das Bewertungskonzept. Einen Ausschnitt für die BSC-Perspektive „Mitarbeiter" zeigt *Abb. 3*.

Mitarbeiter		Soft Facts		Wissensmanagement		Unternehmenskultur	
Arbeitsmarktsituation	N	Motivation	B	Informationsversorgung	B	Werte/Regeln	B
Qualifikationsniveau	B	Zufriedenheit	B	Wissensaufbau	B	Kommunikation	B
Umgang Know-how-Träger	B	Lernbereitschaft	B	Wissensverfügbarkeit	B	Zusammenarbeit	B
Sicherung Nachfolge	B	Kundenorientierung	B	Wissensdistribution	B	Kompetenzen	B
Image Arbeitgeber	B	Flexibilität	B	Wissenserhalt	B	Entscheidungsprozesse	B
...		...		...		...	

Abb. 3: Ausschnitt der intangiblen Werttreiber in der BSC-Perspektive Mitarbeiter (N = nicht beeinflussbarer Werttreiber; B = beeinflussbarer Werttreiber)

Struktur des integrativen Bewertungskonzepts

Bei der Bedienung des IT-Tools muss der Anwender Einstellungen an den Werttreibern auf drei Ebenen vornehmen. Es sind dies die Gewichtung der Werttreiber, die aktuelle Werttreiberausprägung sowie die Auswirkungen auf die Werttreiber durch die jeweilige Maßnahme. Diese Einstellungen werden über Skalen von „null" bis „zehn" eingegeben. Die Festlegung der Skalenbreite stellt einen Kompromiss dar, zwischen dem Wunsch, Einflüsse möglichst exakt einzugeben und der Subjektivität einer realitätsnahen Abbildung. Darüber hinaus kann für jeden Werttreiber eine geeignete mathematische Funktion als Näherung für die Input-Output-Relation ausgewählt werden.

Zunächst müssen die Werttreiber gewichtet werden, wobei die nicht beeinflussbaren nur eine geringe Bedeutung zugewiesen bekommen, da das strategische Management für diese Werttreiber keine Steuerhebel besitzt. Für verschiedene Länder und damit auch Regionen können einem Werttreiber unterschiedliche Gewichtungen zugewiesen werden. Der nächste

Schritt besteht in der Erfassung des gegenwärtigen Niveaus der Werttreiber. Nur so kann eine Zunahme/Verbesserung vernünftig erfasst und auch bewertet werden. Betrachtet man nochmals *Abb. 2* mit ausgewählten typischen Produktionsfunktionen, so erkennt man (abgesehen vom Ausnahmefall *Leontieff*-Produktionsfunktion, der eine lineare Steigung zugrunde liegt), dass es eine große Rolle spielt, von welchem Ausgangspunkt Zuwächse eintreten.

Die letzten Einstellungen schließlich sind die Auswirkungen auf die Werttreiber durch die durchgeführten bzw. geplanten Projekte. Neben den herkömmlichen finanziellen Parametern (Investitionen, Kosten und Erlöse), die direkt in den EVA eingehen, werden auf Projektebene vor allem die erwarteten Auswirkungen auf die einzelnen Werttreiber betrachtet. Entsprechend wird der Grad der Einflussnahme auf den jeweiligen Werttreiber festgelegt.

Sind diese Einstellungen erfolgt, können einzelne Projekte erfasst werden. Die IT-Applikation erlaubt die Eingabe von länderbezogenen Projekten, außerdem können Multigesellschafts-Projekte verarbeitet werden. Multigesellschafts-Projekte können für Gemeinschaftsvorhaben, aber auch zur Erfassung von Spillover-Effekten genutzt werden. Finanzielle Größen aus einer klassischen Investitionsrechnung (5-Jahres-Betrachtung) sind ebenfalls integriert, zumal der Projektetat auch auf die „Durchschlagskraft" eines Projektes schließen lässt. Die Verbindung zwischen Werttreiber-Ausgangslage, Zuwachs und finanziellen Größen stellt dann die Werttreiber-Veränderungs-Matrix her.

Aus den Ausprägungen der unterschiedlichen intangiblen Werttreiber über den Planungszeitraum und ihrer Gewichtungen errechnet sich dann eine Gesamtausprägung. Die Veränderung über den Planungszeitraum zeigt die erwartete Entwicklung der Werttreiber und ihre Beeinflussung durch die Projekte. Impliziert man, dass durch die Beeinflussung von intangiblen Werttreibern in ähnlichem Maße auch der EVA beeinflusst wird, und damit eine positive Entwicklung der Intangibles in Zukunft höhere EVAs erwarten lassen, so ergibt sich zwangsläufig auch ein Zusammenhang zwischen EVA und Intangible Value. Im vorliegenden Fall wurde ein linearer Zusammenhang zugrunde gelegt: ein Anstieg der Intangibles um 8% impliziert einen entsprechenden Anstieg des EVA in Höhe von 8%. Der integrierte EVA (tangibel und intangibel) steigt für das entsprechende Jahr von 100% des EVA auf 108%. Die sich ergebende Differenz zeigt demnach das Potenzial, das sich aus den Investitionen und Aufwendungen in intangible Faktoren für das Unternehmen ergeben kann. Zur Bewertung von Strategiealternativen und Konzepten wird die Summe der integrativen EVAs über den Planungszeitraum betrachtet und für die unterschiedlichen Alternativen verglichen.

Sind die gewünschten Projekte erfasst, können die Ergebnisse auf mehreren Ebenen verglichen werden, um einen maximierten kombiniert materiellen und immateriellen EVA zu erreichen. Die Projektdaten können auf Landes- und Regionenebene aggregiert werden. Eine Rangfolge der einzelnen Projekte – unter der zusätzlichen Berücksichtigung von „kann"- und „muss"-Projekten – ergibt sich aus einer automatischen Sortierung, wobei das Budget vom ranghöchsten Projekt abwärts vergeben wird: es ergibt sich eine gedankliche Begrenzungslinie, die nicht durchzuführende Projekte abtrennt. Neben dieser statisch-komparativen Betrachtung besteht noch die Möglichkeit, eine Sensitivitätsanalyse durchzuführen: einzelne Parameter (Beginn, finanzielle Größen, Wirkungen) der Projekte lassen sich ceteris paribus variieren. Aktiviert und deaktiviert man dazu noch einzelne Projekte, so kann man im Rahmen einer Simulation optimieren.

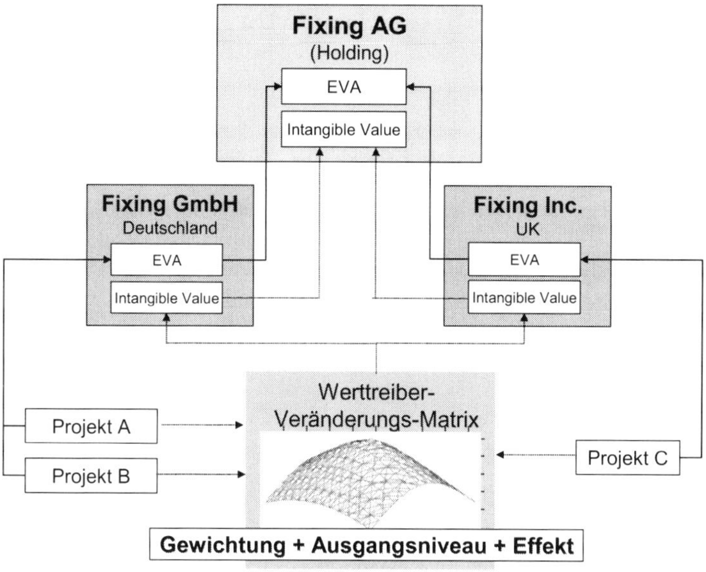

Abb. 4: Hierarchische Konsolidierung des integrativen Konzeptes

Das Projekt „Fortbildung der Fachberater“ führt beispielsweise in der Perspektive Mitarbeiter zu einer erhöhten Mitarbeiterqualifikation und dadurch zu einer verbesserten Nutzung von Mitarbeiterpotentialen. Besser ausgebildete Fachberater können kundenorientierter arbeiten, was dann wiederum in der Perspektive Kunde und Markt zu einem verbesserten Servicegrad und einer verbesserten Kundenbetreuung und damit verbunden zu einer erhöhten Kundenzufriedenheit führt. Dies bietet die Voraussetzung dazu, Marktanteil und/oder Umsatz zu erhöhen. Vergleicht man Input (Investitionen/Kosten) und Output (Auswirkungen auf monetäre und insbesondere auch nicht-monetäre Werte), so erhält man eine Form von Produktivität, die mit anderen Projekten verglichen werden kann.

5. Fazit

Vorgestellt wurde ein Konzept und seine praktische Umsetzung, das den Wertbeitrag intangibler Werttreiber aufzeigt. Dies erfolgt weniger im Rahmen einer exakten finanzmathematischen Modellierung, sondern vielmehr als Orientierungs-, Planungs- und Steuerungshilfe für das Management. Ziel ist eine effektive und effiziente Ressourcenallokation zwischen verschiedenen Projekten in Einklang mit den strategischen Zielen des Unternehmens unter der Annahme knapper Ressourcen. Entscheidungsprozesse können damit systematisiert, dokumentiert und transparent durchgeführt werden.

Es bleibt festzuhalten, dass der kombiniert materielle und immaterielle EVA lediglich eine Tendenz für zukünftig zu erwartende Werte darstellt und nicht der Ermittlung von verbindlichen Zielvorgaben dienen kann und soll. Um Transparenz und Handhabbarkeit zu gewährleisten, wurde die qualitative Einschätzung der Werttreiber auf einer vergleichsweise gering abgestuften Skala abgebildet. Gleiches gilt für Projekteinflüsse und die Gewichtung der Werttreiber. Weiter stützt sich die Schätzung der Werttreiberentwicklung auf eine statische

Lernkurve, deren Ausprägung erst weitere empirische Untersuchungen erfordern würde, um eine exakte Prognose zu liefern.

Die Subjektivität solcher Bewertungen ist offensichtlich. Bei dem Versuch, die nicht in finanziellen Größen erfassten Wirkungen von Projekten und Vorhaben messbar zu machen, bildet das integrative Konzept die Realität dennoch weitaus besser ab, als konventionelle Verfahren zur Ressourcenallokation. Durch die erhöhte Transparenz und Systematisierung lassen sich relativ einfach Vergleiche zwischen einzelnen Vorhaben anstellen. Annahmen und Vereinfachungen führen auch bei diesem Konzept zu Unschärfen. Allerdings sind diese Unschärfen im Vergleich zu einer meist sehr subjektiv geführten Diskussion weitaus geringer.

6. Literatur

Arbeitskreis "Immaterielle Werte im Rechnungswesen" der Schmalenbach-Gesellschaft für Betriebswirtschaft e. V., Kategorisierung und bilanzielle Erfassung immaterieller Werte, in: Der Betrieb, 54. Jg. (2001), H. 19, S. 989-995.

Arbeitskreis "Immaterielle Werte im Rechnungswesen" der Schmalenbach-Gesellschaft für Betriebswirtschaft e.V., Freiwillige externe Berichterstattung über immaterielle Werte, in: Der Betrieb, 56. Jg. (2003), H. 23, S. 1233-1237.

Daum, J., Intangible Assets, Bonn 2002.

Edvinsson, L. and *Malone, M. S.*, Intellectual Capital: Realizing Your Company´s True Value By Finding Its Hidden Brainpower, New York 1997.

Günther, T., Unternehmenswertorientiertes Controlling, München 1997.

Gupta, O. und *Roos, G.*, Mergers and Acquisitions Through an Intellectual Capital Perspective, in: Journal of Intellectual Capital, 2. Jg. (2001), H. 3, S. 297-309.

Hommel, U. und *Pritsch, G.*, Marktorientierte Investitionsbewertung mit dem Realoptionsansatz, in: Finanzmarkt und Portfolio Management, 13. Jg. (1999), H. 2, S. 121-144.

Horváth, P., Controlling, 8. Auflage, München 2001.

Ittner, C.D. and *Larcker, D.F.*, Are Nonfinancial Measures Leading Indicators of Financial Performance? An Analysis of Customer Satisfaction, in: Journal of Accounting Research, 36. Jg. (Sonderheft 1998).

Kaplan, R. S. und *Norton, D.P.*, Die Strategiefokussierte Organisation: Führen mit der Balanced Scorecard„ Stuttgart 2001.

Lev, B., Intangibles. Management and Reporting, Washington 2001.

Möller, K. und *Walker, U.*, Intangibles in der wertorientierten Planung, in: Controlling, 15. Jg. (2003), H. 9, S. 491-498.

Pike, S. und *Roos, G.*, Mathematics and Modern Business Management, Working Paper, presented at the Performance Management Association Symposium, July 27-29th, 2003.

Stewart, G.B., The Quest for Value – The EVA™ Management Guide, New York 1990.

Stewart, T. A., Der vierte Produktionsfaktor, München 1998.

Sveiby, K. E., Methods for Measuring Intangible Assets, http://www.sveiby.com/articles/IntangibleMethods.htm 2003.

Teece, D., Managing Intellectual Capital, New York 2000.

Wall, F., Ursache-Wirkungsbeziehungen als ein zentraler Bestandteil der Balanced Scorecard. Möglichkeiten und Grenzen ihrer Gewinnung, in: Controlling, 13. Jg. (2001), H. 2, S. 65-74.

Messen und Managen von Intangibles in unterschiedlichen organisationalen Kontexten

Martin Piber [1]

---

[1] Dr. *Martin Piber*, Institut für Organisation und Lernen, Abteilung für Controlling und Organisationskultur, Universität Innsbruck, Österreich

Thema dieser explorativen Feldstudie ist die Bedeutung von Intangibles und Intellektuellem Kapital für unterschiedliche Typen von Organisationen. Auf der Basis von 49 semistrukturierten Interviews in einem mittelständischen Produktionsunternehmen, einem Ingenieurbüro, zwei Krankenhäusern und einer Universität werden einerseits der Charakter des Intellektuellen Kapitals (IK) in diesen Organisationen beschrieben, und andererseits die dort verwendeten Strategien zur Schaffung von Transparenz in Bezug auf das Intellektuelle Kapital analysiert. Dies stellt die Grundlage für eine Diskussion über Controlling und Management von Intangibles und Intellektuellem Kapital dar.

1. Einleitung

In der so genannten Wissensgesellschaft können vergleichsweise geringe Wettbewerbsvorteile durch die traditionellen Produktionsfaktoren Arbeit, Kapital und Boden erzielt werden. Wesentlich wichtiger sind der Zugang und die Nutzung von individuellen Fähigkeiten, Kernkompetenzen und Wissen, das im Idealfall in die organisationalen Ressourcen und in ein Netzwerk von Partnern eingebunden ist. Im Rahmen dieser andauernden gesellschaftlichen Entwicklung stehen insbesondere wissensintensive Unternehmen im Wettbewerb um die vorhandenen Wissensressourcen. Hochqualifizierte Mitarbeiter suchen sich die Arbeitsbedingungen, unter denen sie ihr Wissen am besten verwerten und auch am besten dafür entlohnt werden. Arbeit ist kein austauschbares Gut mehr, sondern eine Ressource, die nicht nur erworben, sondern auch entwickelt werden muss. Damit entstehen einzigartige Ressourcen und Ressourcenbündel, die aufgrund ihrer multiplen Verankerung in der Organisation schwierig abzubilden und zu steuern sind: Intangibles. Dies hat freilich auch Konsequenzen für das Controlling: Es wird notwendig, den Bestand an solchen Ressourcen einzuschätzen, um Fehlentwicklungen rechtzeitig erkennen und einen Mehrwert für die Stakeholder generieren zu können.

Auch die Performance von Organisationen ist mehr und mehr vom Einsatz dieser intangiblen Ressourcen abhängig. Dazu gehören z. B. der Markenwert, die Qualität der Kundenbeziehungen, das Wissen und die gesammelten Erfahrungen der Mitarbeiter oder die Organisationskultur. Einerseits werden solche Ressourcen organisationsintern bewusst entwickelt und andererseits entstehen sie auch als ‚Nebenprodukt' organisationalen Handelns. Eine Organisationskultur entwickelt sich sogar ohne aktives Zutun. Großteils können diese Ressourcen nur sehr schwer aus ihrem Entstehungskontext herausgelöst und an einen anderen Ort transferiert bzw. veräußert werden. Deswegen steigt die Notwendigkeit, Partner zu finden, die die eigenen Assets perfekt ergänzen. Am ehesten transferierbar sind personifiziertes Wissen und Erfahrungen, für die am Arbeitsmarkt auch ein Aufschlag bezahlt wird. Für die meisten intangiblen Ressourcen ist die Bestimmung eines Marktwertes jedoch problematisch, was auch einer bilanziellen Aktivierung im Wege steht. Ebenso ist man weit davon entfernt, Wissen und Erfahrungen bilanztauglich aufzusummieren zu können. Von der gesetzlichen Unmöglichkeit bestimmte Ressourcen in der Bilanz zu aktivieren, wird auch der Begriff der ‚Intangible Assets' abgeleitet und stellt einen Gegenpol zu tangiblem Vermögen dar, das bilanziert wird.

Auch die Performance einer Organisation hängt sehr stark mit der Fähigkeit zusammen, Intangibles zu generieren und nutzbringend einzusetzen. Mit der Verwendung des Begriffes ‚Intellektuelles Kapital' wird auch implizit unterstellt, dass Intangibles – wie Kapital – in der Zukunft eine bestimmte Rendite abwerfen. Darüber herrscht in der Literatur Einigkeit. Umstritten ist jedoch der Grad der Steuerbarkeit von Intangibles. Allerdings bezieht sich ein Großteil des derzeit diskutierten und in der Praxis verwendeten Controllinginstrumentariums ohnehin vorwiegend auf tangibles Vermögen, das meistens in der Vergangenheit erworben oder selbst generiert wurde und finanziell leichter handhabbar ist. In Anbetracht der unterschiedlichsten Ausprägungen von Intellektuellem Kapital stellt dies nur einen bescheidenen Ausschnitt aus dem Dasein einer Organisation dar – quasi eine schwarz-weiß Fotografie auf der Basis von Geschäftsvorfällen in der Vergangenheit. Weitaus spannendere Ereignisse und Faktoren wie z. B. das Betriebsklima, die Fluktuation, persönliche Interessen, Gründe, warum Mitarbeiter in der Organisationen verbleiben oder sie verlassen, welche Schlüsselkompetenzen vorhanden sind oder Erzählungen von Kunden bleiben weitgehend ausgeblendet (vgl. dazu *Gabriel* et al., 2000). Aus diesem Grund wird die von *Johnson* und *Kaplan* (1987) schon in den 1980er Jahren geäußerte Kritik in Form einer rhetorischen Frage, ob das Rechnungswesen bzw. auch das Management Accounting ihre Relevanz verloren haben, nur allzu verständlich. Die Kritik ist auch heute nach wie vor hochaktuell und zeigt die Notwendigkeit, das vergangenheitsorientierte und auf Finanzzahlen fixierte Paradigma des Rechnungswesens zu überschreiten. Dies ist auch die Basis für die Beschäftigung mit Intangibles und Intellektuellem Kapital (IK). Insofern ist die nun schon seit mehr als 10 Jahren laufende Debatte dazu auch als Antwort auf *Johnson* und *Kaplan* zu deuten.

In diesem Beitrag wird anhand einer explorativen Feldstudie gezeigt, welche Bedeutung diese „vibrant, value-creating and critical mass that is intellectual capital" (*Roslender* und *Fincham*, 2001, S. 392) für unterschiedliche Typen von Organisationen hat. Auf der Basis einer Serie von 49 semistrukturierten Interviews in einem mittelständischen Produktionsunternehmen, einem Ingenieurbüro, zwei Krankenhäusern und einer Universität werden einerseits der Charakter des IK in diesen Organisationen beschrieben und andererseits die dort verwendeten Strategien zur Schaffung von Transparenz in Bezug auf das IK analysiert. Dies stellt die Grundlage für eine Diskussion über Controlling und Management von Intangibles und IK dar.

2. Intellektuelles Kapital und Intangible Assets

Die Wurzeln der IK-Debatte reichen bis in die Mitte der 1980er Jahre zurück, als einige nordeuropäische Dienstleistungsunternehmen begannen, über eine Ausweitung der Rechnungslegung nachzudenken (vgl. *Bukh* et al., 2001, S. 87ff.). Zur gleichen Zeit beschäftigten sich *Itami* (1987) und *Teece* (1986) mit der wirtschaftlichen Nutzung von Technologie und anderen intangiblen Vermögenswerten. Seitdem beherrschen *Sveiby* (1997), *Stewart* (1997), *Lev* (2001) sowie *Edvinsson* und *Malone* (1997) die Debatte zum IK und Intangible Assets. Am Anfang war das Thema eher von Praktikern besetzt, bevor die akademische Welt in der letzten Dekade dessen Bedeutung entdeckte.

Für das Controlling ist die Konzeptualisierung des IK sehr nützlich, weil dadurch strategische Ressourcen lange vor deren – gewinnbringenden – Verwertung thematisiert werden. Deswegen kann schon Alarm geschlagen werden, wenn ein Mangel bestimmter Ressourcen manifest wird und nicht erst, wenn dieser Mangel sich im Finanzergebnis niederschlägt. Damit spielen Messtechniken für IK im Rahmen unternehmerischer Frühwarnsysteme eine gewichtige Rolle. Zusätzlich wird die Wertschöpfung in den indirekten Bereichen der Wertschöpfungskette transparenter. In der Konsequenz können IK-Messsysteme finanzielle Tools wie Analysen zum Shareholder Value oder den Economic Value Added perfekt ergänzen.

Der augenscheinlichste empirische Beweis für die Existenz von IK und Intangibles ist das in vielen Fällen massive Überschießen der Marktwerte börsennotierter Unternehmen über die Buchwerte. Der Buchwert lässt sich leicht aus der Bilanz ablesen und der Marktwert wird bei gelisteten Unternehmen über die Börse bestimmt und bei nicht gelisteten punktuell (z. B. beim Verkauf) ermittelt. In den theoretischen Modellen wird die Lücke dazwischen unter anderem mit nicht aktivierbaren Investitionen in F&E, die Corporate Identity, Marken, organisationale Kompetenzen, Kundenkapital, Wissen und Humankapital erklärt.

IK umfasst alle unternehmerischen Ressourcen, die finanziell nicht oder nur schwer bewertbar sind und sich zumindest mit hoher Wahrscheinlichkeit positiv auf den zukünftigen Erfolg der Organisation auswirken. Obwohl viele Ansätze IK auch in einzelne Teilbereiche zerlegen, bleiben die meisten Beschreibungen doch eher vage. Dies legt nahe, dass einige Komponenten des IK nicht beschrieben werden können, oder dass diese nur im Kontext der spezifischen Situation erklärbar werden. Ergo scheint eine erschöpfende Aufzählung der IK-Kategorien unmöglich zu sein. *Eustace* (2001, S. 15) teilt Intangibles in Güter und Kompetenzen und *Ulrich* (1998) hebt neben der Kompetenz auch das Commitment der Mitarbeiter hervor. Diese Beispiele zeigen, dass IK mehr ist als die Summe seiner Teile – also auch mehr als die Summe aus verschiedenen Komponenten. Dennoch beziehen sich weite Bereiche der Literatur auf mehr oder weniger begründete Taxonomien von intellektuellem Kapital: Im ‚Intangible Assets Monitor' unterteilt *Sveiby* (1997) Intangibles in drei Hauptkategorien: interne Struktur, externe Struktur und (menschliche) Kompetenzen. *Edvinsson* und *Brünig* (2000, S. 28–32) sprechen von Humankapital, Strukturkapital und Kundenkapital. Auch die Europäische Union hat ein breit angelegtes Forschungsprojekt zum IK unterstützt. Ziel war die Entwicklung, Evaluierung und Verbreitung von Richtlinien zum Messen, zum Management und zum Reporting von IK. Das Projekt lief unter dem Titel '*MERITUM*' (Measuring Intangibles to understand and improve innovation management) von 1998 bis 2002 (vgl. *Meritum*, 2002). Das Ergebnis des *Meritum*-Projektes ist unter anderem eine Klassifizierung von IK in Humankapital, Strukturkapital und Beziehungskapital.

Eine Definition für Intangible Assets findet sich bereits in den IAS 38: "An identifiable non-monetary asset without physical substance held for use in the production or supply of goods or services, for rental to others, or for administrative purposes." (*Brennan/Connel*, 2000, S. 208). Dabei wird intangibles Vermögen als Gegenstück und Ergänzung zu tangiblem Vermögen gesehen. Deshalb wird der Begriff der ‚Intangible Assets' vorwiegend im externen Rechnungswesen im Kontext einer zu reflektierenden Aktivierungsfähigkeit dieser Assets verwendet. Der Begriff des „Intellektuellen Kapitals" hat seine Wurzeln in der Human-Resources-Debatte und stellt eine Erweiterung des Humankapitals dar (vgl. *Vickery*, 2000, S. 72). Da damit auch die Summe verschiedener Intangibles gemeint ist, werden die beiden Begriffe in diesem Beitrag synonym verwendet. Beide beziehen sich auf nicht-physische Ressourcen, die mit zukünftigen Erfolgen/Profiten in Zusammenhang stehen. Restriktiver werden Intangible Assets verstanden, da mit dem Asset-Begriff der oben erwähnte Rechnungswesen-Kontext hergestellt wird (vgl. *Meritum*, 2002, S. 61).

Auch in der Praxis hat sich die *Meritum*-Gliederung für IK in Humankapital, Strukturkapital und Beziehungskapital als praktikabel erwiesen. Deswegen wird für diese explorative Studie die *Meritum*-Taxonomie als Kontroll-Heuristik verwendet. Humankapital stellt das Wissen, die Erfahrungen und die Fähigkeiten der Mitarbeiter dar. Strukturkapital verkörpert das Wissen, das z. B. in Form von Prozessen, Routinen, Systemen, Datenbanken oder Kultur in der Organisation verbleibt, wenn am Abend alle Mitarbeiter physisch die Firma verlassen. Schließlich umfasst Beziehungskapital alle Ressourcen, die mit externen Beziehungen der Organisation in Zusammenhang stehen. Dazu gehören beispielsweise Kunden, Lieferanten, Partner oder Nachbarn (vgl. *Mertium*, 2002, S. 63).

Langfristig tragen diese Ressourcen wesentlich zum finanziellen Erfolg und der Weiterentwicklung der Organisation bei. Aus diesem Grund gibt es verschiedene Ansätze, um IK zu quantifizieren und zu steuern (vgl. dazu *Stoi*, 2003). Wie oben beschrieben ist das Residuum des Marktwertes über den Buchwert ein erster Anhaltspunkt für den finanziellen Wert des IK. Auf ähnliche Art und Weise versucht *Tobins* ‚q' (vgl. *Tobin*, 1969) die Diskrepanz zwischen tangiblen und intangiblen Gütern zu messen. Allerdings lässt sich die Differenz zwischen Markt- und Buchwert nicht auf bestimmte Intangibles zurückführen und wird auch stark von der Börsenpsychologie beeinflusst. Der ‚calculated intangible value' bezieht sich auf den Übergewinn im Vergleich zum Durchschnittswert des Gewinns von Unternehmen in der gleichen Branche. Das IK wird nach dieser Methode als Barwert der Übergewinne nach Steuern kalkuliert (vgl. z.B. *Stewart*, 1997). Die Methode von *Edvinsson* und *Brünig* (2000) summiert Investitionen in das IK auf. Dazu gehören z. B. die Festigung von Kundenbeziehungen, der Kundenservice, die Entwicklung von Marken, die Weiterbildung von Mitarbeitern und Investitionen in Patente. Die Summe dieser Investitionen wäre der theoretisch höchstmögliche Wert des IK. Allerdings wird argumentiert, dass das geschaffene Potenzial auch ausgenützt werden muss. Deswegen wird ein sogenannter Effizienzkoeffizient E ($0 < E < 1$) berechnet. Beispielsweise zeigt sich etwa die Effizienz der Investitionen in Kundenbeziehungen unter anderem auch durch den Anteil der Stammkunden. Der Effizienzkoeffizient E wird schließlich als Durchschnitt (nötigenfalls auch gewichtet) verschiedener Performancekennzahlen und -indizes kalkuliert und zeigt den Ausbeutegrad für das IK. Dieser Ausbeutegrad wird dann mit der Investitionssumme multipliziert, wodurch sich eine tangible Zahl für die Bewertung der Intangibles ergibt. Freilich ist die Aussagefähigkeit des Ergebnisses eingeschränkt. Erstens werden nur finanzielle Investitionen in das IK erfasst. Unberücksichtigt bleiben z. B. Erfahrungen, die Mitarbeiter learning by doing machen, weil dafür per se nichts investiert werden muss. Zweitens ist es sowohl problematisch die auf diese Art und Weise errechneten finanziellen Werte der Intangibles untereinander, als auch mit tangiblen Assets zu vergleichen. Schlüsse können deshalb nur aus der Veränderung des IK im Zeitablauf gezogen werden.

Auch die Balanced Scorecard stellt ein Tool zum Management von Intangibles dar (vgl. *Kaplan* und *Norton*, 1996). Dabei wird die Organisation mit einem System an Performanceindikatoren überzogen, die vor allem Einblick in die nicht-finanziellen Erfolgsperspektiven geben sollen. Durch das Nachdenken über Ursache-Wirkungs-Zusammenhänge wird außerdem ein besseres Verständnis des Kontextes und der erfolgsbestimmenden Rahmenbedingungen erreicht. Dies führt auch zu einer besseren Kommunikationsbasis über Abteilungsgrenzen hinweg. Damit können die ‚Reporting'-, die ‚Contexting'- und die ‚Cross-communicating'-Funktion der BSC unterschieden werden (*Piber*, 2001, S. 890ff.). Konstitutives Element der Scorecard sind Kennzahlen wie z. B. die Anzahl der erfüllten Schlüsselaufgaben eines Bereiches, der Umsatz mit Key Accounts, der Anteil an Kleinstaufträgen, die Weiterbildungsstunden pro Jahr, Fehler pro 1000 Einheiten, verlorene Kunden oder auch Kunden- und Mitarbeiterzufriedenheitsindizes. Dabei geht es nicht um die Berechnung einer Spitzenkennzahl, sondern um ein besseres Verständnis des Wertschöpfungsprozesses und seiner Determinanten. Jedenfalls wird mit der Scorecard erreicht, dass die Aufmerksamkeit des Managements auch auf nicht-finanzielle aber sehr wohl zukunftsträchtige Indikatoren gelenkt wird. Werden diese Zahlen kontextualisiert, entstehen ein Lernprozess und eine sehr gute Basis für eine breite Diskussion über IK im Management. Dadurch werden Intangibles zwar nicht tangibel aber transparenter und zugänglicher für Management-Interventionen.

3. Methodik

Da die Beschreibung unterschiedlicher Ausprägungen intellektuellen Kapitals explorativen Charakter hat, liegt dem Beitrag ein qualitatives Forschungsdesign zu Grunde. Damit ist es möglich, verschiedene Dimensionen des IK sowie deren Messung und Steuerung unabhängig von bestehenden theoretischen Konzeptionen transparent zu machen. Dazu sind offene Fragen ohne vorstrukturierte Gliederungen am besten geeignet (vgl. z. B. *Denzin/Lincoln*, 1994, S 1ff.). Die Untersuchung zielt im Kern auf das Verstehen von unterschiedlichen Kategorien von IK in verschiedenen organisationalen Settings. Wesentlich ist dabei auch der spezifische Kontext jeder Organisation und jedes Interviewpartners. Nicht nur deshalb war zu erwarten, dass jeder Beteiligte seine eigene Wirklichkeit konstruiert und sich auch mehrere Bilder der Organisation und Sichtweisen von IK ergeben würden. Dazu kommt, dass die Forscher vielfach Beobachtungen der Interviewpartner beobachten und diesen anschließend Sinn zuschreiben (vgl. *von Glasersfeld*, 1995).

In Summe wurden 49 semistrukturierte Interviews ohne zeitliche Vorgaben in 5 Organisationen geführt, transkribiert und in 5 Fallstudien zusammengefasst. Die Interviews wurden zwischen Juni 2001 und Juni 2003 in Österreich und Italien geführt. Bei der Auswahl der Organisationen wurde einerseits Wert auf die Wissens- und IK-Intensität und andererseits auf Diversität in Hinsicht auf Größe, Eigentümerschaft (privat und öffentlich) und Branche (Dienstleistung und Produktion) geachtet. Als Heuristik für die IK-Intensität wird dazu auf den Begriff der Expertenorganisation von *Grossmann* (1997), unter den beispielsweise Universitäten und Krankenhäuser subsumiert werden können, und den Begriff des wissensintensiven Unternehmens von *Starbuck* (1997) zurückgegriffen. Im Sample von 5 Organisationen (Namen vom Autor verändert) sind damit eine Universität, ein größeres und ein kleineres Krankenhaus, ein Ingenieurbüro und ein mittelständisches Produktionsunternehmen vertreten:

- Universität (öffentlich-rechtliche Universität mit 2500 MitarbeiterInnen)
- *Sanicare* (öffentliches Krankenhaus, 1500 MitarbeiterInnen)
- *T-Health* (öffentliches Krankenhaus, 480 MitarbeiterInnen)
- *Berger & Co* (privates Ingenieurbüro, 100 MitarbeiterInnen)
- *Superflex* (privates, mittelständisches Produktionsunternehmen, 80 MitarbeiterInnen)

Aufgrund der Notwendigkeit, die Expertise verschiedener Berufsgruppen zu kombinieren, sowie der langen und intensiven Ausbildungszeiten für medizinische Berufe repräsentiert ein Krankenhaus den Archetypus einer Expertenorganisation. Deshalb kann dort auch eine breite Palette von IK vermutet werden. *T-Health* ist ein kleineres Bezirkskrankenhaus mit einzelnen Spezialisten, die über die Landesgrenzen hinaus eine hohe Reputation genießen. *Sanicare* ist ca. dreimal so groß und hat als Gesamtinstitution einen guten Ruf, sodass sich auch viele Patienten aus entfernteren Regionen dort behandeln lassen. *Berger & Co* ist ein Ingenieurbüro, das in den letzten Jahren stark gewachsen ist, und in fünf Hauptgeschäftsfeldern wie z. B. die Projektierung von Eisenbahntrassen und dem Tunnelbau tätig ist. Das kleine Produktionsunternehmen *Superflex* ist ein innovativer Nischenanbieter im Markt für Sanitärinstallationen und hat sich mit anwendungsfreundlichen Produkten und einer flexiblen Fertigung einen Namen gemacht. Beispielsweise werden Komplettbäder inklusive Wänden, Installationen und Einrichtung erzeugt, die schlüsselfertig auf die Baustelle geliefert werden und vom Kunden nur mehr an Strom, Wasser und Abwasser angeschlossen werden müssen. Alle Organisationen werden von Fachleuten im Vergleich zu den jeweiligen Referenzgruppen als erfolgreich bezeichnet.

Die Interviewpartner wurden aus allen hierarchischen und funktionalen Bereichen der Organisationen ausgewählt. Weil bestimmte Funktionsbereiche eine besondere Affinität zum IK haben, wurden der/die Geschäftsführer/In (bzw. der/die DirektorIn), der/die ControllerIn, der/die EDV-Verantwortliche, ein/e VertreterIn der HR-Abteilung und der/die Verantwortliche für Organisationsentwicklung oder Qualitätsmanagement in jeder Organisation interviewt. Dazu kommen organisationstypische Spezialistinnen und Experten wie z. B. ÄrztInnen, Schwestern und Pfleger in den Krankenhäusern, ProfessorInnen, AssistentInnen und Lehrbeauftragte in der Universität, AutoCAD-SpezialistInnen und Ingenieure bei *Berger & Co* sowie Maschinenbetreuer und Vertriebsspezialisten bei *Superflex*. Das wesentliche Forschungsinteresse war die Frage, wie IK von den Interviewten konstruiert wird und wie Transparenz in verschiedenen IK-Dimensionen erreicht werden kann oder zumindest versucht wird, zu erreichen. Ohne Vergleiche a priori ausschließen zu wollen, werden alle Organisationen als Einzelfälle verstanden (vgl. *Stake*, 1994, S. 41f.). Sie sind mit verschiedenen Anspruchsgruppen konfrontiert und in unterschiedliche soziologische, wirtschaftliche, und politische Kontexte eingebunden.

Im ersten Teil des Interviews wurde der Begriff des IK bzw. der Intangible Assets nicht verwendet, um den Interviewten nicht schon möglicherweise zu Beginn des Interviews das Gefühl zu geben, als Unwissende wahrgenommen zu werden. Die Interviewten sollten die Möglichkeit haben, ohne theoretische oder praktische Einschränkungen einen breiten Begriff des IK zu entfalten und in Bezug zur breiten Palette verschiedener bekannter aber auch vielleicht noch unbekannter Dimensionen des IK zu stellen. Im zweiten Teil des Interviews wurde der Begriff des IK dann eingeführt und das bis dahin geführte Interview diesbezüglich gespiegelt. Die Fragen wurden so formuliert, dass alle Komponenten der *Meritum*-Taxonomie gleichsam angesprochen werden konnten.

4. Beschreibung des Intellektuellen Kapitals in den Fallstudien

Im folgenden Abschnitt werden Auszüge aus den Interviews präsentiert. Sie beziehen sich auf

- unterschiedliche Kategorien von IK (geordnet in Bezug auf die Meritum-Taxonomie),
- die in den Interviews charakterisierten Erfolgsfaktoren der Organisationen,
- mögliche Ansatzpunkte für Verbesserungen – mit und ohne Einsatz zusätzlicher finanzieller Ressourcen,
- die Art und Weise, wie Wissen und Erfahrung in der Organisation aufgebaut wird,
- Möglichkeiten verschiedene Aspekte der täglichen Arbeit bzw. deren Erfolg niederzuschreiben und zu messen und
- die wahrgenommene Wichtigkeit von IK in den untersuchten Organisationen.

	Universität	Sanicare	T-Health	Berger & Co	Superflex
Human-kapital	• hohe Relevanz • Fachwissen der Mitarbeiter	• hohe Relevanz • fachliche Kompetenz • gesammelte Erfahrung • Managementkompetenz	• hohe Relevanz • fachliche Qualifikation des Personals • medizinischer und emotionaler Teil der Dienstleistung	• hohe Relevanz • Ausbildungsstand und Projekterfahrung der Mitarbeiter	• mittlere Relevanz • Vertrauen der Mitarbeiter in die Firma • Erfahrung mit Spezialmaschinen
Beziehungs-kapital	• geringe Relevanz • Zufriedenheit der Studierenden • Kooperation mit der Praxis • Einbindung in die Scientific Community	• mittlere Relevanz • Netzwerk mit Lokalpolitikern, praktischen Ärzten und anderen Gesundheitsorganisationen • Patienten als Co-Produzenten ihrer Gesundheit	• mittlere Relevanz • Ruf der Ärzte • Zeit, die mit den Patienten verbracht wird • Beziehungen zu pensionierten Ärzten, um den Verlust von Erfahrungen zu vermeiden	• hohe Relevanz • Kundenzufriedenheit • Image bei potentiellen Bewerbern • gute Referenzen	• hohe Relevanz • Kundenzufriedenheit • Flexibilität bei Spezialaufträgen • Qualität der Kontakte zu Installateuren und zum Außendienst
Struktur-kapital	• geringe Relevanz – Effizienz der Struktur wird allg. bemängelt • keine Strukturen, um die Kommunikation und einen Wissensaustausch zwischen Instituten und Fakultäten zu organisieren • ineffiziente Strukturen für die Forschung	• hohe Relevanz • Qualitätsmanagementsystem auf der Basis einer ISO-Zertifizierung • IT-Struktur • Evidence Based Medicine	• geringe Relevanz • patientenorientierte Organisation • System der Integration neuer Mitarbeiter • Qualitätszirkel	• hohe Relevanz • Archiv mit Projektdokumentationen • Pausen als Mögl., zum ungezwungenen Wissensaustausch' • gute administr. Unterstützung • Expertenliste - ermöglicht hohe Spezialisierung • Qualitätsmanagement- u. Dokumentenver-waltung	• geringe Relevanz • Arbeitsklima • Strukturen zur abteilungsübergreifenden Kommunikation
Charakterisierung von Faktoren, die positiv mit dem ‚Erfolg' der Organisation verbunden werden	• Kommunikation • Herausforderung, der Verbindung von Lehre und Forschung • Verbinden von Freiheiten für die Forschung und organisationalen Zwängen • nicht zu viel Vorwissen der Mitarbeiter (!) (Möglichkeit zur Sozialisierung) • Bezug der Universität zur Region	• 'Qualitätskultur' • effizientes Prozessmanagement • gute Organisation der Verantwortungsbereiche • EDV-Organisation	• Ruf der Ärzte • Verminderung der Angst der Patienten • Zufriedenheit der Mitarbeiter und Patienten • Teamwork auf allen Ebenen • gutes Klima • beste medizinische Versorgung	• Preisdurchsetzungsmacht im Verkauf • Macht auf dem Personalbeschaffungsmarkt • Kompetenz und Erfahrung der Mitarbeiter • effiziente Projektorganisation – vermeidet Pönalezahlungen • Arbeitsklima • Teamorientierung • Neuentwicklung von Kernkompetenzen in vertretbarer Zeit	• Ausbildung von Schlüsselkräften • Bekanntheitsgrad bei verschiedenen Anspruchsgruppen • Flexibilität in der Produktion • Innovationsgrad und -geschwindigkeit • Qualität der Kundenbetreuung • Teamarbeit • Arbeitsklima

	Universität	Sanicare	T-Health	Berger & Co	Superflex
Wie müssten finanzielle Ressourcen eingesetzt werden, um die Zielerreichung des Krankenhauses/der Universität/ des Unternehmens zu verbessern?	• zusätzliches Personal, um administrative Tätigkeiten der Forscher reduzieren zu können • längerlaufende Verträge für Assistenten ermöglichen, damit langfristige Projekte bearbeitet werden können • breiteres Spektrum in der Lehre • weniger Studenten in den Kursen • bessere Bezahlung • Akzente in der Weiterbildung setzen	• Ausweitung der Pflegekapazitäten • bessere Weiterbildungsangebote für alle Mitarbeiter • finanzielle Anreize für die MitarbeiterInnen, um Wartezeiten für Patienten zu verringern • ausreichende Bezahlung	• Verwendung neuester Technologien • Investition in Hard- u. Software • den ÄrztInnen und Schwestern ermöglichen, mehr Zeit mit den Patienten zu verbringen • Anstellung von ÄrztInnen mit ausgewiesenen Spezialgebieten • Engagement von Spezialtrainern für das Personal • Bezahlung ist ausreichend • „Mit Geld kann man fast überhaupt nichts tun!"	• finanziell aufwendige Expansion im Ausland • Organisation von Fach-Exkursionen • Verstärkung der Weiterbildung Schaffung von Anreizen für die Umsetzung von Ideen • Bezahlung ist ausreichend	• Weiterbildung • knappe Personalsituation würde mehr Mitarbeiter erfordern • Bezahlung ist ausreichend • Einführung eines Vorschlagswesens
Wie kann die Zielerreichung des Krankenhauses/der Universität/des Unternehmens ohne finanzielle Ressourcen verbessert werden?	• Kommunikation • Koordination von Forschern, die am selben oder an verwandten Themengebieten arbeiten • Problem der ‚privaten Bibliotheken'	• „denken kostet nichts" • sorgfältigere Planung • Outsourcing von Verwaltungsaufgaben	• Qualitätszirkel • mehr Augenmerk auf die Personalbeschaffung richten • Reduktion des ‚organisational slack'	• Organisation des Archivs: Verbesserung der Suchfunktion • Kommunikation • Verbesserung des Datenbankzugriffes	• Schaffung von Vertrauen in die Führung • Verbesserung der Kommunikation zwischen den Abteilungen
Auf welche Art und Weise wird Wissen und Erfahrung erworben?	• learning by doing • Spezialliteratur • Weiterbildung • in Krisensituationen entwickelt sich automatisch ein intensiver Lernprozess	• Ausbildung • learning by doing • Fachliteratur • Lernen von KollegInnen	• Kongresse,auch ‚dumme' Fragen stellen	• Erfahrung hin zu • Weiterbildung • Arbeit in Projekten	• Erfahrung spielt die größte Rolle • Wissensaustausch mit KollegInnen
Wie bewerten die Interviewten ihre eigene Arbeit?	• Anzahl der Publikationen • Reputation in der Scientific Community • Qualität der Serviceleistungen für Studierende • Rückmeldungen von den Studierenden	• Erledigung der Arbeit in der vorgegebenen/ geplanten Zeit • Nachfrageintensität der eigenen Beratertätigkeit • Bewertung durch die eigenen MitarbeiterInnen und Kollegen • Bild des Krankenhauses in der Presse	• Erfüllung aller Zeitvorgaben (auch selbst gesetzte) • umso weniger Beschwerden, umso besser • allgemeines Gefühl der Zufriedenheit, das sich in Gesprächen mit KollegInnen äußert • letztendlich nur gefühlsmäßig	• Kundenzufriedenheit • Abwicklung der Projekte im geplanten Zeitrahmen • finanzieller Erfolg • Selbstverwirklichung • Rückmeldungen von Kundenseite	• Ausschussquoten • Kundenzufriedenheit • Arbeitsklima • Geschwindigkeit, in der wir uns auf eine Spezialanfrage eines Kunden einstellen können • Ausmaß der erreichten Preiserhöhung
Welcher Teil der täglichen Arbeit, des Wissens und der Erfahrung kann niedergeschrieben werden?	• ca. 50 % • Publikationen halten nur ein Teil der persönlichen Erfahrungen fest • fachl. Ausbildung; der größte Teil des didaktischen Wissens • nicht: Angelegenheiten der Personalführung	• ca. 50-70 % • Standardprozesse • persönlicher Arbeitsstil	• ca. 50-80 % • es kann sehr viel niedergeschrieben werden, aber es ist ein Problem der Ressourcen • umso komplexer der Fall, umso schwieriger ist es, die wesentlichen Fakten festzuhalten	• weniger als 60 % • wird im Rahmen von Projektdokumentationen gemacht • Archiv	• ca. 60 % • sehr viel! • kommt im Rahmen von Arbeitsplatzbeschreibungen vor

	Universität	Sanicare	T-Health	Berger & Co	Superflex
Welcher Teil der täglichen Arbeit, des Wissens und der Erfahrung kann gemessen werden?	• ca. 20 % • z. B. Anzahl der Publikationen, Prüfungen, Studienabschlüsse, Anzahl der Beratungsgespräche, Anrufe, Sitzungen, Kommissionen • problematisch: „Man kann auch nicht sagen, in einem Aufsatz mit 20 Seiten steckt doppelt so viel, wie in einem 10-seitigen – das stimmt einfach nicht.“	• 0 - 100 % • schwierig zu quantifizieren • sogar die Qualität der persönlichen Beziehungen kann indirekt quantifiziert werden, wenn man z. B. zählt, wie oft man mit Kollegen auch außerhalb der Dienstzeit eine Pizza essen geht	• fast nicht möglich • sehr schwierig • Quantifizierungen würden zu falschen Schlüssen führen	• ca. 10 % • Nennungen in der Projektdokumentation • Anzahl der Referenzprojekte • verbrauchte Zeit ist eine relevante Messgröße • Erfahrung korreliert mit den projektierten Tunnelkilometern, der Anzahl der geplanten Brücken oder den verbauten Tonnen Stahl • Punktesystem im Rahmen von Ausschreibungen	• ca. 10 % • z. B. Ausschussquote • Außendienst: Anzahl der akquirierten Aufträge • Innendienst: wie viele Fehler werden gemacht • Geschäftsführung: Betriebsergebnis
Wichtigkeit des IK für die Organisation	• sehr wichtig • IK wird permanent generiert	• sehr wichtig • Ausbildung • Gruppenarbeit • Fokus auf die Bedürfnisse der Patienten richten • die im Laufe der Zeit gesammelte Erfahrung	• sehr wichtig • Know-how in der Medizin • Arbeiten in Teams • Kommunikation über alle Berufsgruppen hinweg	• Die Nutzung und Bildung von IK ist selbstverständlich • In manchen Fällen wird Wissen gehortet und bewusst nicht weitergegeben	• wichtig, aber nicht zentral • Kundenzufriedenheit • Motivation • Struktur der Produktion

Abb. 1: Auszüge aus dem empirischen Material

5. Diskussion der Fallstudien

Grundsätzlich konnten in den Fallstudien alle Kategorien von IK aus der *Meritum*-Taxonomie gefunden werden. Dabei wurde dem Humankapital die größte Bedeutung zugedacht. Die Ausnahme ist *Superflex*, weil dort auch viele angelernte Mitarbeiter tätig sind. Deshalb spielt Humankapital dort vorwiegend im Zusammenhang mit persönlichen Erfahrungen eine Rolle. Beziehungskapital hat für die Privatunternehmen *Berger & Co* und *Superflex* große und für die beiden Krankenhäuser mittlere Relevanz. In der Universität werden zwar die guten Kontakte mit der Scientific Community erwähnt, aber vor allem die Verbesserungsfähigkeit der Beziehungen zu den Studenten betont. Drei Gesprächspartner sprachen sogar von einer Vernachlässigung des Beziehungskapitals. Im Rahmen des Strukturkapitals unterscheiden sich die Fälle am stärksten: Für *Sanicare* und *Berger & Co* haben Systeme, standardisierte Arbeitsprozesse und strukturierte Datenbanken eine große Bedeutung, während in den anderen Fällen der Fokus auf Human- und Beziehungskapital gerichtet ist. Eine Darstellung der in den Interviews besonders betonten Komponenten des IK ist in *Abb. 2* zusammengefasst.

In der Universität steht das Humankapital klar im Vordergrund. Die meisten Interviewten sahen das Spezialwissen des Forschungs- und Lehrpersonals als Schlüsselressource. Beziehungskapital stand in keinem Gespräch im Mittelpunkt. Da für die MitarbeiterInnen von Instituten (9 von 15 InterviewpartnerInnenn) auch das Institut den entscheidenden Bezugs- und Identifikationspunkt darstellt, sind die meisten Strukturen auf Instituts- und in geringerem Ausmaß auf Fakultätsebene angesiedelt – nicht aber auf Ebene der Gesamtuniversität. Es passiert sogar, dass verschiedene ForscherInnen an ähnlichen Projekten arbeiten – ohne die KollegInnen, die möglicherweise sogar im gleichen Gebäude arbeiten, zu kennen. Deswegen werden diesbezügliche Initiativen in mehreren Interviews ausdrücklich begrüßt. Die Bedeutung guter Beziehungen zu den Studenten wurde vor allem von Führungskräften angesprochen, aber nur in sehr geringem Ausmaß vom Lehrpersonal. Generell hat die Organisation eine sehr starke Binnenorientierung. Im Vordergrund stehen die Institutsstrukturen, für die die meisten organisationalen Kapazitäten reserviert sind. Übergreifende Strukturen auf Fakultäts- und Universitätsebene werden weitgehend vernachlässigt. Es besteht nur geringe Motivation seitens des wissenschaftlichen Personals, sich diesbezüglich zu engagieren.

Universität	Sanicare	T-Health	Berger & Co	Superflex
• hochspezialisiertes Fachwissen • strukturelle Freiheit der Forschung: Reflexion ausdifferenzierter Forschungsfragen abseits von Machtstrukturen in der Gesellschaft • Qualität in Forschung und Lehre	• perfekte Organisation • hervorragendes Qualitätssicherungssystem • standardisierte Arbeitsprozesse • Evidence Based Medicine • guter Ruf des Krankenhauses	• soziale Kompetenz der Mitarbeiter • Eingehen auf die sozialen Bedürfnisse der Patienten • ausgezeichnete Ärzte • gute Kooperation zwischen Verwaltung, ÄrztInnen und PflegerInnen • Know how für Qualitätszirkel • Wissensmanagement	• hochqualifizierte Mitarbeiter • guter Ruf in der Branche (ergo auch bei der Personalbeschaffung) • Organisation des Archivs • Kundenzufriedenheit • gute Projektorganisation • Vertrauen der Mitarbeiter	• organisatorische Flexibilität • Erfahrung mit dem Set-up von Spezialmaschinen • Kontakt zwischen Entwicklung und Vertrieb • Wertschätzung der Arbeiter in der Produktion

Abb. 2: Betonte Komponenten des intellektuellen Kapitals

Als generelle Herausforderung wird die Erhaltung von ausreichend kritischem Potenzial an der Universität bezeichnet. In der Sicht des Direktors einer Fachbibliothek hat die Universität auch den Zweck, einen Meinungspluralismus zu erhalten und insbesondere Minderheitsmeinungen in frühen Stadien zu schützen. Ein Professor der Philosophie sagte: „Die Universität lebt von ihrem intellektuellen Kapital." Allerdings besteht auch die Sorge, dass ein zu starker Bezug auf vergangenheitsorientiertes Wissen reaktionär oder sogar innovationsfeindlich wirken könnte. Auch die Tradition der Selbstverwaltung wird nicht immer als Vorteil gesehen: Ein Professor der Rechtswissenschaften erklärte, dass Managementaktivitäten für die Fakultät und die Universität (die im Wesentlichen von Professoren wahrgenommen werden) für ihn eine Zeitverschwendung darstellen. Auch andere Interviewpartner stellen dem Management der Universität kein gutes Zeugnis aus. Dies stimmt auch mit der geringen Relevanz des Strukturkapitals überein. Fazit: Die Universität ist eine Organisation, die im Interesse der Freiheit der Forschung nicht nur in Bezug auf IK ‚undermanaged' ist.

In den beiden Krankenhäusern wird das Konzept des IK als nützlich für das Bewusstmachen und Fördern der individuellen und organisationalen Kompetenzen eingeschätzt. Das Beziehungskapital fokussiert bei *Sanicare* primär auf die regionale Ebene, während bei *T-Health* die behandelten Patienten die größere Rolle spielen. Große Unterschiede zeigen sich beim Managementverständnis: Bei *Sanicare* wird stärker auf die Strukturierung und Systembindung des IK Wert gelegt als bei *T-Health*. Die Verwendung von ISO- und anderen (Labor-)Standards sowie von ‚Evidence Based Medicine' stellen nützliche Systeme der organisationalen Strukturierung individuellen Wissens dar. Dabei werden weitgehend vorgefertigte Strukturen übernommen, die sich bereits in anderen Organisationen bewährt haben. Spezifisch für *Sanicare* ist auch die regionale Perspektive auf Gesundheit im Sinne einer Harmonisierung intra- und extramuraler Ressourcen. Das heißt, dass eine verstärkte Zusammenarbeit mit anderen medizinischen Einrichtungen in der Region bezüglich der Budgetierung, der Großgeräteausstattung, einer gemeinsamen IT-Plattform und des Personals angestrebt wird. Dadurch lässt sich auch die Ausrichtung des Beziehungskapitals auf andere Gesundheitsdienstleister erklären.

Bei *T-Health* spielt eindeutig das Humankapital die größte Rolle. Der ärztliche Direktor erklärte z. B., dass die Behandlung eines alten und eines jungen Patienten mit derselben Diagnose völlig unterschiedlich ausfallen kann. Deswegen ist es vor allem bei der Therapie wichtig, jeweils den ganzen Fall zu betrachten. Dabei sind einerseits die persönliche Erfahrung des medizinischen Personals und andererseits die Beziehung zu den Patienten von höchster Bedeutung. Da die Patienten die medizinische Komponente der Behandlung aufgrund der mangelnden Fachkompetenz nicht beurteilen können, wird die emotionale Komponente umso stärker bewertet. Außerdem merken die Patienten sehr wohl, ob sie nach ‚Schema F' behandelt werden. Deshalb ist das Eingehen auf spezielle Bedürfnisse auch wichtig für das Image des Krankenhauses. Die persönliche Note der Behandlung wurde von den Interviewten mit Begriffen wie ‚persönlicher Bezug', ‚Einfühlungsvermögen' oder ‚Geist des Hauses' beschrieben. Deshalb ist bei *T-Health* das Beziehungskapital vorwiegend auf die Patienten ausgerichtet. Da eine intensive Auseinandersetzung mit den Patienten zeit- und kostenintensiv ist, erfordert dies ein effizientes Prozessmanagement. Ferner wird Wert auf gute Kontakte zu den niedergelassenen Ärzten der Region und zu den Landeskrankenhäusern gelegt. Zur Qualitätssicherung finden regelmäßige Qualitätszirkel statt, die auf Initiative jedes Mitarbeiters gestartet werden können.

Für die Firma *Berger & Co* ist IK ein Ansatz, der von allen Befragten als hochrelevant für das Unternehmen gesehen wird: „Das Problem ist, dass wir nichts produzieren – wir verkaufen unser Wissen." (Aussage des Geschäftsführers von *Berger & Co*). Dieses Wissen ist vorwiegend im Humankapital verankert. Die Besetzung von freien Stellen über den Arbeitsmarkt ist für das Ingenieurbüro schwierig, weil die notwendige Erfahrung erst innerbetrieblich erworben wird: „Das Wissen, das man hier braucht, kann man nicht in 14 Tagen erwerben, sondern das dauert 4 Jahre. Dies ist unser eigentliches Potenzial." (Aussage eines Projektleiters von *Berger & Co*). Da es am Arbeitsmarkt kaum Personal mit der erforderlichen Kompetenz gibt, ist es für die Firma essenziell, die Mitarbeiter im Unternehmen zu halten. Eine andere wichtige IK-Komponente ist das Gefühl für zukünftige Märkte. Das heißt, neue Technologien ein bis zwei Jahre vor den Hauptkonkurrenten zu entdecken. In diesem Zeitraum ist es auch noch möglich, Experten am Arbeitsmarkt dafür zu finden. Immer wieder betont werden auch die Kundenbedürfnisse, das Einhalten des Zeitrahmens für übernommene Projekte und die Entwicklung neuer Kernkompetenzen. Das Ingenieurbüro hat eine flache Hierarchie, obwohl der Geschäftsführer die meisten Anteile an der Gesellschaft hält.

Der Fokus auf das Humankapital zeigt, dass die Ingenieure eine machtvolle Position innehaben, da sie kurzfristig nicht ersetzbar sind. Würden sie das Unternehmen verlassen, ginge einerseits wertvolles Wissen verloren, und bestünde andererseits auch die Gefahr, dass Kunden mitgenommen werden. Aus diesen Gründen wurde bei *Berger & Co* ein umfangreiches Archiv für Projektergebnisse installiert, um auf das in den Projekten generierte Know-how unabhängig von einzelnen Mitarbeitern zugreifen zu können. Allerdings hat sich gezeigt, dass der Zugriff auf dieses Archiv für Mitarbeiter, die am jeweiligen Projekt nicht beteiligt waren, mit Hürden verbunden ist. Vier Interviewpartner regten eine Verbesserung des EDV-Zugangs zum Archiv an. Dennoch wird im Archiv ein großes Potenzial gesehen, sobald relevante Informationen auch leicht aufgefunden werden können. Ein weiterer Aktivposten in der IK-Bilanz ist das Vertrauen der Mitarbeiter in die Geschäftsführung und den Eigentümer. Da *Berger & Co* ein relativ kleines Unternehmen ist, kann der Haupt-Eigentümer noch persönliche Beziehungen zu den Mitarbeitern aufrechterhalten. Daraus resultiert ein großes Vertrauen in die eigene Zukunft in der Firma. So fühlen sich die Mitarbeiter auch verantwortlich für die Entwicklung des ganzen Unternehmens.

Im Fall *Superflex* ist die Flexibilität des Unternehmens der Schlüssel zum Erfolg. Deswegen finden sich die meisten Intangibles im Bereich des Strukturkapitals. Spezialkomponenten nach Sonderspezifikation des Kunden können innerhalb von 24 Stunden gefertigt werden. Dieser Wettbewerbsvorteil kann auf die hohe Motivation der Mitarbeiter in der Produktion, eine effiziente Prozess-Organisation, eine gute Kommunikationsbasis der beteiligten Abteilungen und die Transparenz des Produktionsprozesses für die Kunden zurückgeführt werden. Die Mitarbeiter in der Produktion sind motiviert, weil sie das Gefühl haben, wichtig für die Firma zu sein: „Nicht der finanzielle Aspekt zählt, sondern die Leute sehen und merken, dass die [die Geschäftsführung] bemüht sind, uns besser zu stellen. Dann haben sie einfach das Gefühl, sie gehören zur Firma dazu." (Aussage des Produktionsleiters von *Superflex* über seine Mitarbeiter). Ebenfalls positiv wirkt sich das gute Verhältnis der Mitarbeiter untereinander aus. Einige Interviewpartner beschrieben die offiziellen Kaffeepausen als wesentlichen Faktor für das Betriebsklima und als Basis für die gute Kommunikation.

Auf der anderen Seite beklagten sich 4 Interviewpartner über schlechte Erfahrungen mit der alten Geschäftsführung. Insbesondere wurden völlig unverständliche Entlassungen qualifizierter Mitarbeiter und die mangelnde Führungskompetenz angesprochen. Obwohl sich die alte Geschäftsführung mittlerweile in den Aufsichtsrat zurückgezogen hat, existiert noch immer ein bitterer Nachgeschmack, der die volle Identifizierung bestimmter Teile der Belegschaft mit dem Unternehmen verhindert. Dies wird dadurch bestätigt, dass die Mehrheit der Befragten eine flache aber dennoch starke Hierarchie verspürt. Resümierend kann gesagt werden, dass *Superflex* die einzige Organisation im Sample ist, für die Humankapital ‚nur' bedingt relevant ist. Obwohl das Know-how der Experten in der Organisation geschätzt wird, sind die wesentlichen IK-Komponenten die exzellenten Beziehungen zu den Kunden und die ausgefeilte Produktionsstruktur, die von hochmotivierten Mitarbeitern getragen wird.

6. Messen und andere Arten zur Schaffung von Transparenz

Beim Versuch Intangibles transparent bzw. sogar messbar zu machen, lassen sich in den Fallstudien unterschiedliche Strategien erkennen. *Sanicare* verwendet bereits seit längerer Zeit ein umfangreiches Kennzahlensystem und das Controlling ist gut ausgebaut. Die Adressaten der Informationen sind vorwiegend die Gesundheitsbehörden und das Management. Ärzte und PflegerInnen werden damit nur in Ausnahmefällen konfrontiert. Inhaltlich geht es dabei vor allem um Leistungsziffern, Kosteninformationen und Auslastungsgrade. Ein Performance Measurement System, das auch Qualitätskennzahlen wie z. B. die Patientenzufriedenheit, die durchschnittliche Aufenthaltsdauer, die Wartezeiten in verschiedenen Stationen des Patientenpfades oder die Anzahl von Behandlungsfehlern enthält, ergänzt die Informationsbasis. Ein effizienter Einsatz der Ressourcen, Qualitätsmanagement und ein funktionierendes Controlling spielt für das Management des Krankenhauses eine zentrale Rolle. Dies soll in Zukunft weiter ausgebaut werden. Es geht, wie es ein Manager ausdrückte, um das Ausleuchten der Schattenzonen der Organisation zugunsten eines humanen Krankenhauses. Insgesamt kann festgehalten werden, dass eine quantitative Effizienz- und Qualitätsmessung bei *Sanicare* allgemein akzeptiert wird.

An einem ähnlichen Kennzahlensystem arbeitet auch die Universität. Allerdings steht dort vorerst noch das Reporting nach außen im Vordergrund. Deswegen sieht das Kennzahlensystem der Universität eher wie ein Leistungskatalog aus. Obwohl mit der Hilfe von Lehrveranstaltungsevaluierungen und Punktesystemen für Publikationen auch die Qualität von Lehre und Forschung beurteilt werden soll, wurde dies von drei Professoren skeptisch kommentiert. Auf die Frage, wie IK in der Universität gemessen wird, wurde in einem Interview auch auf die spezielle Erfahrung von Universitäten mit der Evaluierung von Humankapital verwiesen: In Prüfungen und Klausuren wird schon seit Jahrhunderten das Wissen bzw. die Kompetenz der Studierenden überprüft und auf einer Notenskala gemessen. Allerdings wurde das vorherrschende Prüfungssystem mit schriftlichen und mündlichen Prüfungen über bestimmte Stoffgebiete nur als „das geringste, aber gerade noch vertretbare Übel" (Aussage eines Lehrbeauftragten der Universität) bezeichnet.

Im Reporting System des Ingenieurbüros werden im Rahmen des Controllings ausschließlich finanzielle Informationen verarbeitet. Allerdings wird bei der Interpretation der Daten auch der Kontext mitberücksichtigt. Entscheidungen werden erst nach einer ausführlichen Diskussion in der Geschäftsleitung getroffen. Der Geschäftsführer hat die Wichtigkeit nichtfinanzieller Kennzahlen im Gefühl: „Ich fühle, dass für unseren Erfolg sehr wichtige Faktoren gemessen werden sollten, aber ich kann sie nicht benennen." Anders gelagert ist die Situation bei *Superflex*: Dort werden auch qualitative Informationssysteme verwendet. In einer Lieferantendatenbank werden alle Kontakte mit den Lieferanten aufgezeichnet. Die Datenbank enthält auch qualitative Felder, die von den Einkäufern gepflegt werden. Im Jahr 2001 wurde auch ein CRM-Software-Tool erfolgreich implementiert. Allerdings fließen die Daten aus diesen Tools nicht in das Reporting ein und bleiben daher als ExpertInneninformation bei den Ein- bzw. VerkäuferInnen. Eine organisationsweite kausale oder argumentative Verknüpfung mit finanziellen Kennzahlen kann deshalb nicht erfolgen. Diese Vernetzung verschiedner Informationen findet nur in einzelnen Köpfen auf verschiedenen Ebenen und ohne System statt. Trotzdem wird von den Ein- und Verkäufern das qualitative Informationssystem unisono als ebenso wichtig eingestuft wie quantitative Informationen über Umsatz und Geschäftserfolg.

Bei *T-Health* war in den Gesprächen Zahlen gegenüber ein tiefgreifendes Misstrauen zu spüren – bei Ärzten, Krankenschwestern und Pflegern sehr stark, beim Verwaltungspersonal etwas weniger stark ausgeprägt. Sogar der Controller des Spitals erklärte: „Mit Zahlen kann man alles bewiesen." Auch der Verwaltungsdirektor sieht nur einen sehr begrenzten Anwendungsbereich für Zahlen: „Eine Quantifizierung würde [oft] zu falschen Schlüssen führen." Vor allem wurde betont, dass bei der Interpretation von Zahlen ohnehin immer der Kontext mitbedacht werden muss. So hängen beispielsweise die durchschnittlichen Kosten für Medikamente pro Belagstag sehr stark vom Einzugsgebiet des Krankenhauses ab und ein Vergleich verschiedener Spitäler auf dieser Basis wäre nicht möglich. Ergo würde es vielfach keinen Sinn machen, Zahlen ohne Hintergrundwissen über die spezielle Situation aus ihrem Kontext herauszulösen. Diese Vorsicht bei Zahlen war durchwegs bei allen ExpertInnen in den Krankenhäusern und in der Universität zu spüren. *Abbildung 3* zeigt eine Zusammenfassung der wichtigsten Kennzahlen zum Messen von Intangibles, die in den untersuchten Organisationen verwendet werden.

Wenn nun die wichtigsten Kategorien des IK (*Abb. 2*) und die verwendeten Kennzahlen (*Abb. 3*) verglichen werden, zeigt sich, dass wichtige Kategorien Intellektuellen Kapitals mit metrischen Mitteln nicht oder nur unzureichend gemessen werden (bzw. messbar sind). Auch die ärztliche Leitung von *T-Health* ist sich bewusst, dass die soziale Kompetenz, die eine einfühlsame und verständnisvolle Behandlung erfordert, nur rudimentär mit der Zeit, die mit den Patienten verbracht wird, gemessen werden kann. Dennoch wurde die soziale Kompetenz in vielen Interviews beschrieben und auch qualitätsmäßig beurteilt. Das heißt, dass es auch andere Mess-Methoden geben muss. Ebenso kann der Einfluss der Organisationskultur nicht gemessen werden. Trotzdem kann auch IK, dass sehr tief in der Kultur der Organisation verwurzelt ist, transparent gemacht oder zumindest verstanden werden: „[I]f we want to understand a society, or some part of a society, we have to discover its repertoire of legitimate stories and find out how this evolved [...]" (*Czarniawska*, 1997, S. 16). Besser verständlich wird dies mit einer Aussage des Verwaltungsdirektors von *T-Health*: „Wir hatten Ordensschwestern in unserem Spital. Leider haben sie uns mittlerweile verlassen – was ich außerordentlich bedauere. Aber man spürt, dass ihr Geist und ihre Tradition noch durch die Mauern wehen. Ich sage nicht, dass sie nicht wirtschaftlich denken konnten – und ob! Sie hatten einfach einen guten Hausverstand. Auf gewisse Weise waren sie cleverer als andere. Dennoch gab es auch Meinungsverschiedenheiten. Sie hätten nie darüber nachgedacht, ob teure Medikamente verschrieben werden sollten oder nicht."

Universität	Sanicare	T-Health	Berger & Co	Superflex
• Studentenzufriedenheit (Fragebogen) • Forschungsoutput (Punktesystem) • Anzahl der Diplomarbeiten u. Dissertationen • Anzahl der Lehrveranstaltungsstunden • Anteil der Vorlesungsstunden am Gesamtlehrangebot	• Anzahl der Patienten, die aus anderen Regionen kommen • Anzahl der Patienten, die an andere Häuser überwiesen werden müssen • Patientenzufriedenheit (Fragebogen) • Anzahl der Operationen • Summe der Belagstage • Auslastungsgrad der Operationssäle • Wartezeiten der Patienten in internen Abteilungen	• Zeit, die mit den Patienten verbracht wird • Anzahl der Qualitätszirkel • positive Nennungen in den Medien • Bettenauslastungsgrad • Summe der Belagstage • erwirtschaftete LKF-Punkte pro Jahr, pro Abteilung • Kosten für Medikamente • Patientenzufriedenheit (Fragebogen)	• finanzielle Kennzahlen • Anzahl der Referenzprojekte • Messung der Erfahrung der Mitarbeiter: z. B. projektierte Tunnelkilometer • durchschnittliche Projektdauer • Projektfeedback (vorwiegend qualitativ)	• finanzielle Kennzahlen • Ausschussquoten je Maschine bzw. Schicht • % der Aufträge, die in 24 h produziert werden können • Anzahl der Fehler • prozentuelle Veränderung der Einkaufspreise

Abb. 3: Typische Kennzahlen in den untersuchten Organisationen

Damit zeigt sich, dass in den untersuchten Organisationen IK teilweise auf sehr subtile Art und Weise transparent gemacht wird. Dies passiert – weitgehend unbewusst – im Dialog, durch Erzählungen oder einfach durch ein intuitives Gespür. Ähnlich schwer ist es, für den Verkaufsleiter von *Superflex*, eine Kennzahl für die Kompetenz und Gewandtheit seiner Verkäufer zu finden. Umsätze und Deckungsbeiträge bieten zwar einen guten Ansatzpunkt. Ein Umsatzerfolg kann aber auch durch die hohe Dichte an potenziellen Kunden im Verkaufsgebiet oder einfach durch Glück begründet sein. Deshalb werden bei *Superflex* die Verkäufer bewusst in Mitarbeitergesprächen und unbewusst in alltäglichen Dialogen beurteilt, wodurch der Verkaufsleiter im Laufe der Zeit ein sehr gutes Gespür für deren Kompetenz bekommen hat. In bestimmten Fällen kann das IK nicht eindeutig beschrieben, aber intuitiv gefühlt werden, wie es ein Forscher an der Universität formulierte. Demnach kann festgehalten werden, dass Intangibles auf drei Arten transparent gemacht werden können: metrisch, dialogisch und intuitiv (vgl. dazu im Detail *Habersam/Piber*, 2003 und *Abb. 4*).

Dialogische und intuitive Transparenz kann bewusst durch kontinuierliche Kommunikation, Erzählungen, Sektches, Theaterstücke und Visualisierungen unterstützt werden (vgl. *Mouritsen* et al., 2001). *Roslender* und *Fincham* (2001, S. 393ff.) vertreten die These, dass es für IK eine Art ‚Self-Accounting' geben könnte. Dazu können z. B. beschreibende Erzählungen verschiedener Organisationsmitglieder gezählt werden. Auch bei der Erstellung eines Intellectual Capital Statement (vgl. *Daum*, 2003 und *Bukh* et al., 2001) wird Transparenz durch eine Kombination verschiedener Methoden erreicht: "Intellectual capital statements acquire relevance not because they are logical in a strict mathematical sense like financial key ratio analysis such as the DuPont pyramid. Rather they are relevant, because they can be made to support [...] a broad story about the identity of the firm." (*Bukh* et al., 2001, S. 98f.). Dennoch spielen auch Kennzahlen und Performance Measurement Systems eine wichtige Funktion beim Messen von Intangibles. Im Prinzip kann Unscharfes auch mit Kennzahlen nur unscharf abgebildet werden. Allerdings kann durch die Verwendung eines gut ausgewählten Samples unscharfer Kennzahlen, ein scharfes Gesamtbild entstehen – scharf aber nur im Hinblick auf die bewussten und unbewussten Prämissen, die bei der Auswahl des Kennzahlensamples implizit vorhanden sind. Resümierend kann festgestellt werden, dass mit rein metrischen Mitteln nur ein sehr kleiner Ausschnitt der „vibrant mass" des IK transparent wird.

Art der Transparenz	**Methode**	**Beispiel**
metrisch	über Kennzahlen	Messung der Kundenzufriedenheit über einen Index
dialogisch	im Gespräch	Der Verkaufsleiter beurteilt die Kompetenz eines Verkäufers in einem Mitarbeitergespräch.
intuitiv	über Gefühle oder Intuition	Gespür für den Einfluss der Unternehmenskultur auf die Qualität der Dienstleistung

Abb. 4: Verschiedene Arten der Transparenz von intellektuellem Kapital

In Summe kann in den untersuchten Fallstudien bei *T-Health* und *Berger & Co* ein Übergewicht von dialogischer und intuitiver Transparenz festgestellt werden, während die Steuerung bei *Sanicare* und der Universität vorwiegend mit metrischen Kennzahlen erfolgt. Bei *Superflex* stehen metrische, dialogische und intuitive Komponenten etwa gleichgewichtig nebeneinander. Generell gilt, dass dialogische und intuitive Beurteilungsprozesse auch unbewusst stattfinden und deswegen nie ausgeklammert werden können.

7. Controlling und Management des Intellektuellen Kapitals

Bezüglich des Managements von IK können in den untersuchten Organisationen ganz allgemein zwei grobe Richtungen unterschieden werden: Personalisierung und Kodifikation (vgl. *Hansen* et. al., 1999, S. 107ff, die von ‚codification' und ‚personalization' sprechen). Im Rahmen der Kodifikationsstrategie wird versucht, Kompetenzen, Erfahrungen, Beziehungen und Kontakte der Mitarbeiter in Systeme und Strukturen zu gießen, um diese auch allgemein zugänglich zu machen. Damit kann ein Verlust an IK beim Abgang von Mitarbeitern verhindert werden. Grundsätzlich wird diese Strategie von *Sanicare* und *Superflex* verfolgt. Im Fall von *Sanicare* zeigt sich die Systemgebundenheit wichtiger IK-Komponenten vor allem durch die Verwendung von Evidence Based Medicine und externer (Labor-)Standards und die strikte Definition wichtiger organisationaler Prozesse. *Superflex* versucht z. B. Erfahrungen der Mitarbeiter mit bestimmten Maschinen in die jeweiligen

Arbeitsplatzbeschreibungen zu integrieren. Darüber hinaus werden Informationen von Zulieferern und Kunden über eine Datenbank systematisch erfasst. Somit steht hinter der Kodifikationsstrategie der Versuch, Human- und Beziehungskapital in personenunabhängiges Strukturkapital umzuwandeln.

Im Rahmen der Personalisierungsstrategie wird den Mitarbeitern die verantwortungsvolle (und auch machtvolle) Rolle zugedacht, Wissen und Erfahrungen für die Organisation zu generieren und zu konservieren. So ist z. B. bei *T-Health* der Großteil des IK auf der individuellen Ebene des medizinischen Personals verankert. Dahinter steht die implizite Annahme, dass nur ein geringer Teil des Humankapitals systematisch erfasst werden kann: „Es gibt sicher Dinge, bei denen ich das Gefühl habe, dass das nicht jeder machen kann, weil er einfach nicht versteht, woran es fehlt. Solche Dinge kann man auch nicht weitergeben. Das ist so ein Fingerspitzengefühl, das ich nicht näher beschreiben kann." (Aussage eines Arztes von *T-Health*). Aufgrund des Organisationsverständnisses und des Einzelkämpfertums vor allem in der Forschung kann auch die Universität zu dieser Gruppe gezählt werden.

Berger & Co ist gerade dabei, die Projektdokumentation neu zu organisieren und allen MitarbeiterInnen zugänglich zu machen. Diese Tendenz zur Kodifikation schließt aber auch individuell verankertes IK nicht aus. Die Kodifikationsstrategie ist meistens mit größeren IT-Investitionen verbunden und kann nicht für dialogische und intuitive Transparenz sorgen. Die Personalisierungsstrategie erfordert umfangreichere Investitionen in Weiterbildung und Personalentwicklung. Sie hat den Nachteil, dass die Verfügbarkeit des personalisierten IK vom Verbleib der Mitarbeiter in der Organisation abhängt.

Zwei Fälle geben Anlass, über eine Konstruktion nachzudenken, die als 'negatives IK' bezeichnet werden kann. Im Fall von *Superflex* sind es die mangelhaften Führungsqualitäten der alten Geschäftsführung, die noch immer in den Köpfen vieler Mitarbeiter herumgeistern und bei einzelnen Mitarbeitern sogar verhindern, dass sie sich mit dem Unternehmen identifizieren können. Deshalb hat die Organisation auch einige Jahre nach dem Wechsel in der Geschäftsführung immer noch diesen historischen Ballast zu tragen, der sich einerseits motivations- und anderseits auch leistungshemmend auswirkt. Auch an der Universität findet sich derlei historischer Ballast: Es scheint bereits seit Generationen von Professoren nicht en vogue zu sein, sich für das Management der Universität bzw. der Fakultäten zu engagieren. Dies wurde in den Interviews mehrfach angesprochen und dürfte mit ein Grund des Managementproblems der Universität sein. Deshalb kann auch diese Tradition als negatives IK interpretiert werden.

Es zeigte sich, dass sich bezüglich der unterschiedlichen IK-Dimensionen auch unterschiedliche Controlling-Ansätze anbieten. Im Fall von Humankapital stehen persönliche Fähigkeiten und Wissen im Vordergrund. Ergo muss die Struktur in der Organisation so gestaltet sein, dass sich die ExpertInnen in der Organisation wohl fühlen und damit das Risiko eines Verlustes gering bleibt. Bei *Berger & Co* versucht der Geschäftsführer deswegen auch eine persönliche Beziehung und Vertrauen zu seinen Mitarbeitern aufzubauen. Bei *Superflex* hat die ideelle aber auch finanzielle Wertschätzung des Managements für Schlüsselpersonal zu stabilen Verhältnissen geführt. In Bezug auf das Strukturkapital zeigt sich der größte Spielraum für Managemententscheidungen. Bei *Berger & Co* sowie bei *Superflex* wurde bewusst eine Struktur zur Nutzung und Weiterentwicklung von IK geschaffen. Das Management von Beziehungskapital kann nicht leicht verallgemeinert werden, da es hier die geringsten Überschneidungen zwischen den untersuchten Organisationen gibt. Der Schlüssel zum Erfolg liegt aber sicher darin, sich geschickt und vor allem rechtzeitig in relevante Netzwerke einzuklinken. Beim Kundenkapital ist der direkte Einfluss des Managements gering. Vielmehr zeigt sich, dass das Vorhandensein von IK im Bereich des Human- und des Strukturkapitals langfristig indirekt auch zum Aufbau von Beziehungskapital führt, da qualifizierte Mitarbeiter und effiziente Strukturen z. B. auch Kunden zufrieden stellen.

Darüber hinaus muss das Management der Kommunikationsprozesse und –strukturen eigens erwähnt werden. Diese fördern einerseits den Wissenstransfer als Leveragefaktor für das Humankapital und andererseits das reibungslose Funktionieren der Strukturen und Prozesse. Diese Quintessenz kann aus allen fünf Fallstudien abgeleitet werden. Ansatzpunkt für das Management ist hierbei das Lokalisieren funktionierender und fehlender Kommunikationsbrücken in der Organisation. Die Fallstudien zeigten dabei positive und negative Aspekte: In drei Organisationen (*Superflex*, Universität und *Berger & Co*) wurden (Kaffee-)Pausen als wichtigste(!) Möglichkeit zum Wissensaustausch genannt. Als positiv haben sich vor allem abteilungsübergreifende Treffen und die Freiwilligkeit der Teilnahme erwiesen. Zwölf Interviewpartner erklärten auch, dass der Wissensaustausch umso besser funktioniert, umso besser das private Verhältnis der Mitarbeiter untereinander ist. Dies klingt trivial und ist nicht verwunderlich – weil aber Kaffeepausen auch für private Kontakte der Mitarbeiter förderlich sind, wird der oben genannte Zusammenhang dadurch plastisch veranschaulicht. In diesem Sinne wurden bei *Superflex* und *Berger & Co* Kaffeepausen, die bei weitem über das arbeitsrechtlich erforderliche Maß hinausgehen, vom Management institutionalisiert.

Auf der negativen Seite beklagten sich Interviewpartner aus allen Organisationen übereinstimmend über mangelnde Informationen von KollegInnen. Vor allem wurde moniert, dass Informationen weitgehend als Hol- und nicht als Bringschuld gesehen werden. Das Problem verstärkt sich, wenn einerseits die Abteilungsgrenze überschritten wird, und andererseits unterschiedliche Berufsgruppen involviert sind. Im Krankenhaus ist dies zwischen ÄrztInnen und PflegerInnen und bei *Berger & Co* zwischen dem Management und den TechnikerInnen der Fall. TechnikerInnen sind grundsätzlich daran interessiert, höchste Qualität zu bieten ohne auf die Kosten zu achten, während Manager auf einen fairen ‚value for money' bedacht sind. Da beim Austausch von Informationen und Wissen auch persönliche Interessen im Spiel sind, können sicher nicht alle Schwierigkeiten des Wissens- und Informationsaustausches mit dem Einrichten von Kommunikationskanälen gelöst werden. Bemerkenswert ist aber die Vehemenz, mit der in allen Organisationen Probleme bei der Entfaltung von IK auf fehlende Kommunikationsmöglichkeiten zurückgeführt werden.

8. Fazit

Die Konzeptualisierung von IK lässt sich in den Fallstudien sehr gut nachvollziehen. Humankapital ist in allen untersuchten Organisationen präsent und wird über personalwirtschaftliche Instrumente entwickelt. Strukturkapital spielt bei *Sanicare* und *Superflex* eine zentrale Rolle, wo mittels einer Kodifizierungsstrategie versucht wird, IK in stabilen Strukturen zu konservieren. Schließlich gibt es in allen Organisationen mit Ausnahme der Universität eine hohe Relevanz von Beziehungskapital – allerdings im Zusammenhang mit unterschiedlichen Partnern: die Studenten, die Scientific community und Forscher in Partnerinstitutionen in der Universität, die Patienten, andere (teilweise in einem Wettbewerbsverhältnis stehende) Gesundheitsinstitutionen und Ärzte für die Krankenhäuser sowie vorwiegend die Kunden für *Berger & Co* und *Superflex*.

Generell sind die Intangibles in allen Sparten hochgradig organisationsspezifisch, was eine metrische oder finanzielle Bewertung erschwert. Trotzdem kann über Kennzahlen- und Performance-Measurement-Systeme ein guter Einblick in den Zustand des IK gewonnen werden. Die größten Schwierigkeiten ergeben sich beim Human- und beim tief in der Unternehmenskultur verwurzelten Strukturkapital. Diese Bereiche können nur mit dialogischen und intuitiven Methoden transparent gemacht werden. Diesbezügliche Bewertungsprozesse finden auch unbewusst statt. Sie können aber auch bewusst durch Erzählungen, Geschichten, Sketches, Theaterstücke und Visualisierungen gefördert werden. Es macht keinen Sinn, diesen Bereich des IK mit einem Feuerwerk an Kennzahlen fassen zu wollen, da Zahlen zu eindeutig sind und keinen Spielraum für Ambiguitäten offen lassen. Vielmehr ist eine Ko-Existenz verschiedener Methoden angebracht: In einem Intellectual Capital Statement verleihen sich Performance-Measurement-Systeme, Erzählungen und Visualisierungen gegenseitig Sinn und schaffen damit den nötigen Freiraum, mehrdeutige und mehrdimensionale Informationen zu transportieren. Ebenso kann ein besseres Verständnis von Intangibles und Intellektuellem Kapital nicht nur den Aussagewert sondern auch die Relevanz finanzieller Kennzahlen im Controlling und Rechnungswesen aufwerten.

9. Literatur

Bukh, P. N., *Larsen, H. T.* und *Mouritsen, J.*, Constructing intellectual capital statements, in Scandinavian Journal of Management 17. Jg. (2001), H. 1, S. 87-108.

Czarniawska, B., Narrating the organization: Dramas of institutional identity, Chicago 1997.

Daum, J. H., Intellectual Capital Statements: Basis für ein Rechnungswesen- und Reportingmodell der Zukunft? In Controlling 15. Jg. (2003) H. 3/4, S. 143-153.

Denzin, N. K. und *Lincoln, Y. S.*, Introduction: Entering the Field of Qualitative Research, in *Denzin, N. K.* und *Lincoln, Y. S.* (Hrsg.) Handbook of Qualitative Research. London, New Delhi 1994, S. 1-17.

Edvinsson, L. und *Brünig, G.*, Aktivposten Wissenskapital: Unsichtbare Werte bilanzierbar machen, Wiesbaden 2000.

Edvinsson, L. und *Malone, M. S.*, Intellectual Capital: Realising Your Company's True Value by Finding its Hidden Brainpower, New York 1997.

Eustace, C., The Intangible Economy: Impact and Policy Issues. Report of the High Level Expert Group on the Intangible Economy, Luxembourg: European Communities, 2001.

Gabriel, Y., *Fineman, S.* und *Sims, D.*, Organizing and Organizations: an Introduction, London et al. 2000.

Glasersfeld, E. von, Radical Constructivism: A Way of Knowing and Learning, London 1995.

Habersam, M. und *Piber, M.*, Exploring Intellectual Capital in Hospitals, in European Accounting Review 12. Jg. (2003), H. 4.

Hansen, M. T., *Nohria, N.* und *Tierney, T.*, What's your strategy for managing knowledge, in Harvard Business Review 77. Jg. (1999), H. 2, S. 106-116.

Itami, H., Mobilizing Invisible Assets. Cambridge, MA 1987 (Übersetzung von Roehl), Originalversion 1980 auf japanisch veröffentlicht.

Johnson, H. T. und *Kaplan, R. S.*, Relevance lost: The rise and fall of management accounting. Boston 1987.

Kaplan, R. und *Norton, D.*, The Balanced Scorecard: Translating Strategy into Action, Boston 1996.

Lev, B., Intangibles: Measurement, Management and Reporting, Washington 2001.

Meritum, Guidelines for managing and reporting on intangibles, in: *Cañibano, L.*, *Sanchez, P.*, *Garcia-Ayuso, M.* und *Chaminade, C.* (Hrsg.), Fundaciòn Airtel Mòvil 2002.

Mouritsen, J., *Larsen, H.T.* und *Bukh, P.N.*, Intellectual Capital and the 'Capable Firm': Narrating, Visualising, and Numbering for Managing Knowledge, in AOS 26. Jg. (2001), H. 7/8, S. 735-762.

Piber, M., Controlling the Uncontrollable, in *Bescos, P.-L.* und *Cauvin, E.* (Hrsg.), Proceedings of the 1st Conference on Performance Measurement and Management Control, Nice, 2001, S. 884-899.

Roslender, R. und *Fincham*, R., Thinking Critically about Intellectual Capital, in Accounting, Auditing & Accountability Journal 14. Jg. (2001), H. 4, S. 383-398.

Stake, R. E., Case Studies, in *Denzin, N.K.* und *Lincoln, Y.S.* (Hrsg.) Handbook of Qualitative Research, London 1994.

Starbuck, W.H., Learning by Knowledge-Intensive Firms, in *Prusak, L.* (Hrsg.), Knowledge in Organisations, Boston et al., 1997, S. 147-175.

Stewart, T.A., Intellectual Capital, London 1997.

Stoi, R., Controlling von Intangibles, Controlling 15. Jg. (2003), H. 3/4, S. 175-183.

Sveiby, K.E., The new organizational wealth, San Francisco 1997.

Teece, D., Profiting from technological innovation - implications for integration, collaboration, licensing and public policy, in Research Policy 15, 1986, S. 285-305.

Tobin, J., A general equilibrium approach to monetary theory, in Journal of Money Credit and Banking (1969), H. 1, S. 15-29.

Ulrich, D., Intellectual Capital = Competence x Commitment, in Sloan Management Review 39. Jg. (1998) H. 2, S. 15-26.

Vickery, G., Accounting for Intangibles: Issues and Prospects, in *Buigues, P.*, *Jacquemin, A.* and *Marchipont, J.-F.* (Hrsg.) Competitiveness and the value of intangible assets. Northhampton, Mass. 2000, S. 72-99.

Autorenverzeichnis

Arbeitskreis „Immaterielle Werte im Rechnungswesen" der *Schmalenbach-Gesellschaft für Betriebswirtschaft e.V.*: Prof. Dr. *Walther Busse von Colbe*, Ruhr-Universität Bochum; Dr. *Rolf Uwe Fülbier*, Ruhr-Universität Bochum; Dr. *Torsten Ganske* (Arbeitskreisleiter), *Deutsche Telekom AG*; Prof. Dr. *Thomas Günther*, Technische Universität Dresden; Prof. Dr. *Axel Haller*, Johannes Kepler Universität Linz/Österreich; Dr. *Matthias Hendler*, *KPMG Deutsche Treuhand-Gesellschaft AG*; Prof. Dr. *Isabel von Keitz*, Fachhochschule Münster; *Guido Kerkhoff*, *Deutsche Telekom AG*; Dr. *Harald Köster*, *Henkel KGaA*; Prof. Dr. *Hans Peter Möller*, *RWTH* Aachen; Prof. Dr. *Bernhard Pellens* (Arbeitskreisleiter), Ruhr-Universität Bochum; Dr. *Achim Reinhold*, *ZDF*; WP/StB *Rüdiger Reinke*, *KPMG Deutsche Treuhand-Gesellschaft AG*; Dr. *Wolfgang Ritter*, *SAP AG*, und *German Weinberger*, *Deutsche Post AG*.

Prof. Dr. *Jürgen Blum* ist Rechtsanwalt und Geschäftsführendes Vorstandsmitglied des *Zentrums für Wissenschaftsmanagement Speyer e.V.* Er war lange Zeit administrativer Vorstand des *Deutschen Zentrums für Luft- und Raumfahrt e.V.* (DLR) in Köln. Unter seiner Verantwortung sind für das DLR die Wissensbilanzen 2000 und 2001 entstanden, die die ersten in der Bundesrepublik Deutschland waren.

Robert Borrmann ist Unternehmensberater der *Horváth & Partner GmbH* Stuttgart und unterstützt seit März 2001 den *Hermann von Helmholtz-Gemeinschaft Deutscher Forschungszentren e.V.* beim Aufbau eines wissenschaftsadäquaten Controllingsystems.

Ann-Charlotte Bredahl is Strategic Process Coordinator at *Skandia Insurance Ltd.*, Stockholm, Schweden.

Prof. Dr. *Hans-Georg Bruns* ist Liaison Board Member (Germany) im *International Accounting Standards Board*, London, UK.

Jürgen H. Daum ist Management-Berater, Buchautor und Experte für Unternehmenssteuerung. Als Senior Business Consultant bei der *SAP AG Walldorf* berät er CFOs in den Bereichen Finance&Business Architecture, Unternehmenssteuerung, Performance Management und Informationssysteme (www.juergendaum.de).

Ralph Dietrich ist Doktorand am Institut für Revisions-, Treuhand- und Rechnungswesen an der Johannes Kepler Universität Linz in Österreich.

Prof. Dr. *Leif Edvinsson*, is the world's first holder of Professorship of Intellectual Capital, at the University of Lund, Sweden, the first Director of Intellectual Capital at *Skandia*, Sweden, and was appointed Brain of the Year 1998.

Lisa Fernström is a consultant with *Intellectual Capital Services Ltd.*, London, UK.

Ronald Fiedler ist Project Manager bei *Stern Stewart & Co.* in München.

Dr. *Stefan Güldenberg* ist Universitätsassistent am Institut für Unternehmensführung, Abteilung Unternehmensführung, Controlling und Beratung der Wirtschaftsuniversität Wien, Österreich.

Prof. Dr. *Edeltraud Günther*, ist Inhaberin des Lehrstuhls für Betriebliche Umweltökonomie an der Technischen Universität Dresden.

Prof. Dr. *Thomas Günther* ist Inhaber des Lehrstuhls für Betriebliches Rechnungswesen/Controlling an der Technischen Universität Dresden.

Dina Gray is a PhD student at the Centre for Business Performance at *Cranfield School of Management*, UK.

Prof. Dr. *Axel Haller*, Institut für Revisions-, Treuhand- und Rechnungswesen, Johannes Kepler Universität Linz, Österreich.

Prof. Dr. *Roland Helm* ist Inhaber des Unilever-Stiftungslehrstuhls für Allgemeine Betriebswirtschaftslehre, insbesondere Absatzwirtschaft, Marketing und Handel an der Friedrich-Schiller-Universität Jena.

Prof. Dr. *Reinhard Heyd*, Fachhochschule Nürtingen, Universität Ulm.

Sandra Kirchner-Khairy, Lehrstuhl für Betriebliches Rechnungswesen/Controlling, Technische Universität Dresden.

Mart Kivikas is CEO and co-founder of *WEKE GmbH*, Germany.

Dr. *Martin Kloyer* ist Habilitand am Lehrstuhl für Allgemeine Betriebswirtschaftslehre, Produktion und Industriebetriebslehre an der Friedrich-Schiller-Universität Jena.

Dr. *Catharina Kriegbaum-Kling*, MBA, ist Mitarbeiterin der *TRUMPF GmbH + Co. KG* in Ditzingen.

Prof. Dr. *Baruch Lev* is Professor of Accounting and Finance at *Stern School of Business*, New York University, USA.

Dr. *Andreas Markom* ist Mitarbeiter der *Österreichischen Akademie für Arbeitsmedizin* in Klosterneuburg und Koordinator des EU Forschungsprojektes *Plexus*.

Prof. *Bernard Marr* is Research Fellow at the Centre for Business Performance at *Cranfield School of Management*, UK, Visiting Professor at the University of Basilicata, Italy, and Managing Director of *BWMC Ltd.*

Dr. *Rudolf C. Meiler* ist Direktor im Personalbereich der *ThyssenKrupp Stahl AG* in Duisburg.

Dr. *Klaus Möller* ist Habilitand am Lehrstuhl für Allgemeine Betriebswirtschaftslehre und Controlling an der Universität Stuttgart.

Prof. Dr. *Andy Neely* is Associate Director of AIM, the *Advanced Institute for Management*, Chairman of the Centre for Business Performance at *Cranfield School of Management*, and Managing Director, *The Performance Practice*.

Andreas Nilsson ist Assistent am Institut für Informatik an der Universität Göteborg und Projektmitarbeiter im EU Forschungsprojekt *Plexus*.

Martin Nohava ist Mitarbeiter der *Österreichischen Zertifizierungs- und Begutachtungs GmbH* in Wien und Projektmitarbeiter im EU Forschungsprojekt *Plexus*.

Dr. *Martin Piber* ist wissenschaftlicher Mitarbeiter am Institut für Organisation und Lernen, Abteilung für Controlling und Organisationskultur der Universität Innsbruck, Österreich.

Dr. *Stephen Pike* is Director for R&D at *Intellectual Capital Services, Ltd.*, London, UK.

Prof. *Göran Roos* is Visiting Professor at Cranfield School of Management, Visiting Professor at Helsinki School of Economics, Intellectual Capital Adjunct at Mt. Eliza Business School and founder and CEO of *Intellectual Capital Sevices, Ltd.*, London, co-founder and chairman of *AssetEconomics, Inc.*, New York, USA.

Raoul Ruthner ist Universitätsassistent am Institut für Unternehmensführung, Abteilung Unternehmensführung, Controlling und Beratung der Wirtschaftsuniversität Wien, Österreich.

Marek Rydén, is Senior Business Controller at *Skandia Insurance Ltd.*, Stockholm, Schweden.

Bernhard Schmiedinger ist wissenschaftlicher Mitarbeiter bei der *Profactor Produktionsforschungs GmbH* in Steyr und Projektmitarbeiter im EU Forschungsprojekt *Plexus*.

Dr. *Susanne Schunder* ist Mitarbeiterin der *Österreichischen Akademie für Arbeitsmedizin* in Klosterneuburg und Projektmitarbeiterin im EU Forschungsprojekt *Plexus*.

Prof. Dr. *Hans-Gerd Servatius* ist Vorstand der *Horváth AG* in Düsseldorf. Daneben lehrt er Technologie- und Innovationsmanagement an der Universität Stuttgart.

Prof. Dr. *Gerhard Speckbacher*, ist Ordinarius für Unternehmensführung an der Wirtschaftsuniversität Wien und leitet das Institut für Unternehmensführung, Abteilung Unternehmensführung, Controlling und Beratung.

Prof. Dr. *Roman Stoi* lehrt Controlling, Unternehmensführung und Rechnungswesen an der Berufsakademie Stuttgart.

Curt Stoll ist Geschäftsführer der *KACU systemic GmbH* in Auenwald und Projektmitarbeiter im EU Forschungsprojekt *Plexus*.

Dr. *Ulrich Stolz* ist Partner der Geschäftsführung bei der *KACU systemic GmbH* in Auenwald und Projektmitarbeiter im EU Forschungsprojekt *Plexus*.

Dr. Rainer Strack ist Geschäftsführer und Direktor bei *The Boston Consulting Group* in Düsseldorf.

Michael G. Thuy, CPA, CIA ist Senior Financial Analyst bei der *Philip Morris International Management S.A.*, Lausanne, Schweiz.

Klaus Valentin ist wissenschaftlicher Mitarbeiter bei der *Profactor Produktionsforschungs GmbH* in Steyr und wissenschaftlicher Koordinator des EU Forschungsprojektes *Plexus*.

Ulrich Villis ist Manager bei *The Boston Consulting Group* in München.

Markus Zeimes, WP, StB, ist Manager der *KPMG Deutsche Treuhand-Gesellschaft AG* und derzeit als Technical Director beim *Deutschen Rechnungslegungs Standards Committee e.V.*, Berlin, tätig.

Annikka Zurwehme, Lehrstuhl für Betriebliches Rechnungswesen/Controlling, Technische Universität Dresden.